AF699035

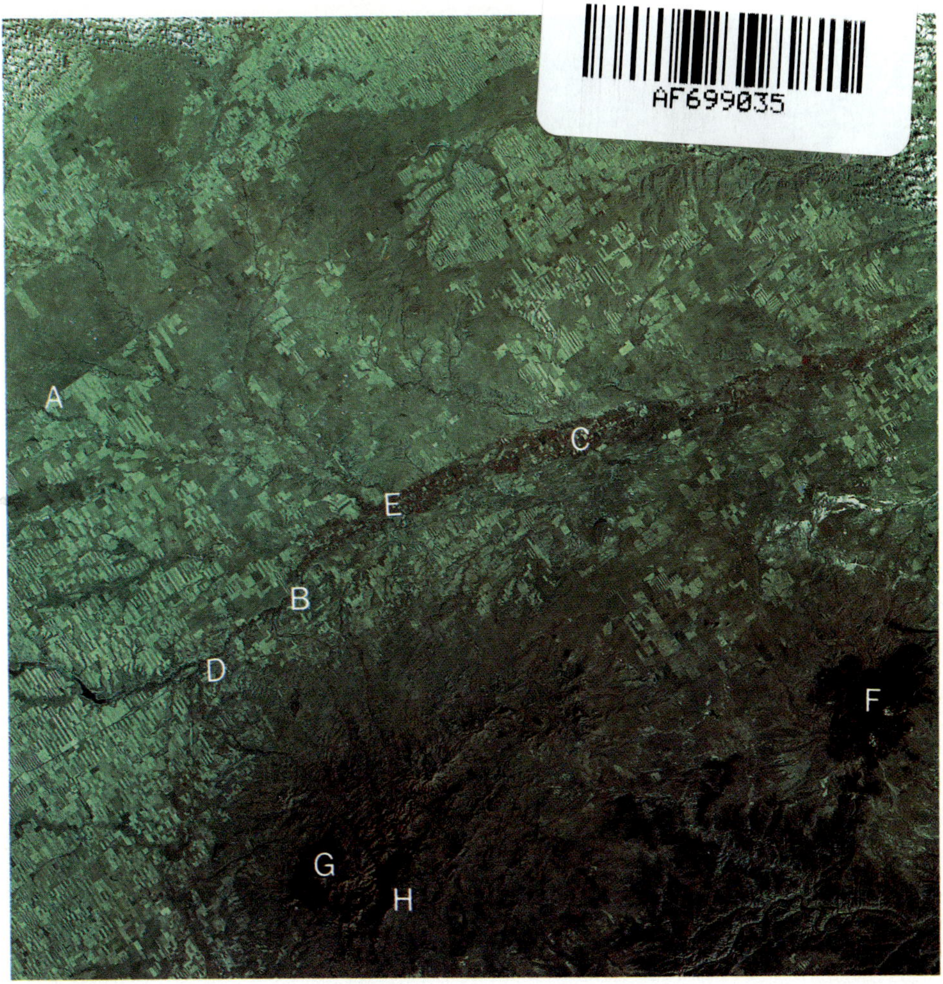

Color-IR photo frame number 1633 taken by the Large Format Camera on NASA Shuttle Mission STS-41G, October, 1984. Photo scale is approximately 1:1,358,778.

This scene includes portions of the region of north central Montana known as the "High Line" and adjoining portions of Alberta and Saskatchewan, Canada. The U.S.-Canada border (A) appears on the upper left as a diagonal line caused by changes in land use patterns along the border of the two nations. The Milk River (B), a major northern Montana drainage, runs in an east-west direction through the approximate center of the scene. Note the occurrence of irrigated fields along portions of the river's flood plain (C). The two Montana High Line communities of Havre and Chinook appear at D and E respectively.

The mountain ranges in the lower portions of the scene are the Little Rockies (F) and the Bear Paw Mountains (G). Dr. William T. Pecora, former director of the U.S. Geological Survey and Undersecretary of the Interior in whose honor this symposium was held, spent a considerable time during the early years of his career in the Bear Paw Mountains while studying for his Ph.D. at Harvard University. The prominent ridge (H) in the southern portion of these mountains bears his name.

WITHDRAWN

THE TENTH WILLIAM T. PECORA MEMORIAL REMOTE SENSING SYMPOSIUM PRESENTED UNDER THE SPONSORSHIP OF

IN COOPERATION WITH

The Pecora Symposium was established in 1975 to foster the exchange of scientific and applications findings on the use of remotely sensed data for resource management programs. The symposium series honors the memory of William T. Pecora, former Director of the U.S. Geological Survey and Undersecretary, Department of the Interior. Dr. Pecora played a major role in the development and establishment of satellite remote sensing systems.

The focus of Pecora 10 is on applications of remote sensing in forest and range resource management. The purpose of this symposium is to expose participants in all aspects of forest and range management to opportunities to use this technology.

SYMPOSIUM COMMITTEE:

William M. Ciesla, Symposium Co-Chairperson
Richard S. Driscoll, Symposium Co-Chairperson
Dr. Robert Haas, Program Chairperson
Richard J. Myhre, Poster Session Chairperson
Dr. James Smith, Tutorials Chairperson
Raymond A. Byrnes, Logistics Chairperson
Lawrence R. Pettinger, Special Activities Chairperson
Dr. Richard Mroczynski, Exhibits Chairperson
Craig A. Sommer, Local Arrangements Chairperson
Patricia M. Ciesla, Spouse Events Chairperson

SPONSORED BY:

Society of American Foresters

Society for Range Management

American Society for Photogrammetry and Remote Sensing

IN COOPERATION WITH:

United States Geological Survey

National Aeronautics and Space Administration

National Oceanic and Atmospheric Administration

United States Department of Agriculture

Pecora 10

REMOTE SENSING IN
FOREST AND RANGE RESOURCE MANAGEMENT

PROCEEDINGS

August 20, 21, 22, 1985
Student Center
Colorado State University
Fort Collins, Colorado

© 1985 by the American Society for Photogrammetry and Remote Sensing. All rights reserved. Reproductions of this volume or any part thereof (except short quotations for use in the preparation of reviews and technical and scientific papers) may be made only after obtaining the specific approval of the publisher. The publisher is not responsible for any opinions or statements made in the technical proceedings.

SD
387
.R4
W55
1985

ISBN 0-937294-67-5

Published by the
AMERICAN SOCIETY FOR PHOTOGRAMMETRY AND REMOTE SENSING
210 Little Falls Street
Falls Church, Virginia 22046

PREFACE

The William T. Pecora Symposium Series was established in 1975 to foster the exchange of scientific data and application of remote sensing in resource management. The symposium series honors the memory of William T. Pecora, former director of the U.S. Geological Survey and Under-secretary of the Department of the Interior. Dr. Pecora played a major role in the development and establishment of satellite remote sensing systems.

The theme of Pecora 10 was application of remote sensing in forest and range resource management and was held at the Student Center, Colorado State University in Fort Collins, Colorado. This symposium marked the tenth anniversary of the Pecora Symposium Series and was the first of the series to be held in a location other than the EROS Data Center in Sioux Falls, South Dakota.

These Proceedings contain the texts of the plenary and technical papers presented at the symposium. Also included are brief summaries of poster papers and forum discussion sessions. These papers reflect the current state-of-the-art. The Pecora Symposium Committee hopes that these Proceedings will make a significant contribution toward the application of remote sensing in all aspects of forest and range resource management.

WILLIAM M. CIESLA
Program Cochairperson

RICHARD S. DRISCOLL
Program Cochairperson

THE WILLIAM T. PECORA AWARD

THE WILLIAM T. PECORA AWARD, presented annually in recognition of outstanding contributions of individuals or groups toward the understanding of the Earth by means of remote sensing, is sponsored jointly by the National Aeronautics and Space Administration and the Department of the Interior. The award was established in 1974 to honor the memory of Dr. William T. Pecora, former Director of the U.S. Geological Survey, and later, Undersecretary, Department of the Interior. Dr. Pecora was a motivating force behind the establishment of Earth resource sensing from space. He was a Government leader with broad vision and deep appreciation for the use of satellite programs in continually inventorying and managing our national resources.

PREVIOUS RECIPIENTS

1974 William A. Fischer, U.S. Geological Survey, Department of the Interior

1975 William Nordberg, National Aeronautics and Space Administration, and Carlos Brockmann, Director of the LANDSAT-Bolivia Project

1976 Awarded jointly: Environmental Research Institute of Michigan and Laboratory for Applications of Remote Sensing of Purdue University

1977 Robert N. Colwell, School of Forestry, University of California, Berkeley and Michel T. Halbouty, Consulting Geologist and Petroleum Engineer, The Halbouty Center

1978 David S. Johnson, National Environmental Satellite Service,Department of Commerce

1979 John M. DeNoyer, U.S. Geological Survey, Department of the Interior and Virginia T. Norwood, Senior Scientist, Hughes Aircraft Company

1980 Verner E. Suomi, Professor of Meteorology, University of Wisconsin

1981 Leonard Jaffe, National Aeronautics and Space Administration and James R. Anderson (posthumously), U.S. Geological Survey, Department of the Interior

1982 Alexander F.H. Goetz, National Aeronautics and Space Administration and Lawrence C. Rowan, U.S. Geological Survey, Department of the Interior

1983 Floyd F. Sabins, Jr., Senior Research Associate, Chevron Oil Field Research Company

1984 Archibald B. Park, GLOBEX, Inc.

1985 Charles Elachi, Jet Propulsion Laboratory

WILLIAM T. PECORA, the Man We Honor

Dr. William T. Pecora was an internationally known geologist whose professional career as a scientist and government leader spanned nearly four decades. He was the eighth director of the U.S. Geological Survey from 1965 to 1971 and Undersecretary of the Interior from 1971 until his untimely death on July 19, 1972.

Born February 1, 1913, in Belleville, New Jersey, Dr. Pecora attended public schools in Newark, New Jersey. He graduated with honors in geology from Princeton University in 1933, and received his doctorate in geology from Harvard University in 1940. He also received honorary degrees from Franklin and Marshall College in 1969 and the Colorado School of Mines in 1970.

In addition to his academic and professional achievements, Dr. Pecora was an accomplished athlete. He was the U.S. Intercollegiate Fencing Champion in 1933 and was a member of the 1936 U.S. Olympic Team.

Dr. Pecora joined the Geological Survey in 1939 and soon became a widely acclaimed expert in mineralogy, petrology and geochemistry. His work emphasized the use and development of scientific principles as guides in exploration for mineral, fuel and water resources. He authored more than 50 scientific publications and received international recognition for his work on rare minerals and the geology of volcanic regions. He conducted research throughout the United States and in many parts of Latin America. As a result of his work in Brazil, nine new minerals were identified for the first time.

In 1965, Dr. Pecora was elected to the National Academy of Sciences for his achievements in original research and as a Fellow in the American Academy of Arts and Sciences. In 1968 he received the Department of Interior's highest honor, the Distinguished Service Award. The following year he received the Rockefeller Public Service Award for career servants.

Under Dr. Pecora's leadership the Geological Survey moved into several new areas. The Survey assumed leadership of the Interior Department's Earth Resources Observation Systems (EROS) Program, involving the use of orbiting satellites to gather Earth resources data for analysis. He also established the Survey's National Center for Earthquake Research at Menlo Park, California.

As Undersecretary of the Interior, he coordinated and guided the preparation of the environmental impact statement on the trans-Alaskan oil pipeline proposal. He also helped Interior Secretary Rogers Morton formulate the Department's position on the numerous and complex problems relating to development of the nation's mineral, water and energy resources.

A decisive and frequently colorful administrator, Dr. Pecora proved unusually effective at advancing programs and policies in which he strongly believed. Long-time associates credited him with a blend of precision, wit and personal warmth that made him a memorable figure in the Federal bureaucracy. He was recognized for his adeptness in explaining complex and puzzling scientific phenomena in layman's terms.

The William T. Pecora Memorial Remote Sensing Symposium Series was initiated in 1975 in recognition of Dr. Pecora's commitment and support of this technology. Past symposia have emphasized remote sensing applications in cartography, petroleum and mineral exploration, wildlife management, and hydrology. The 1985 symposium on applications of remote sensing in forest and range resource management marks the tenth anniversary of this series.

1985
William T. Pecora Award Citation
Dr. Charles Elachi

In recognition of his outstanding contributions as a scientist, engineer, and manager in developing and demonstrating synthetic aperture imaging radar techniques for Earth sciences and practical applications.

Dr. Charles Elachi has been a leader in satellite-borne synthetic aperture imaging radar remote sensing for studying the solid Earth and oceans. The remarkable capabilities of this technique have been demonstrated by three space systems flown under NASA sponsorship: (1) the Seasat satellite, launched in 1978; (2) SIR-A, carried into orbit on the Shuttle in November 1981; and (3) SIR-B, on the Shuttle flight in November 1984. These three systems were designed and developed at the Jet Propulsion Laboratory under Dr. Elachi's guidance and expert leadership.

The Seasat radar was designed primarily to observe ocean features, but also proved capable of producing highly useful images of land. The Seasat mission was the first flight in space of an imaging radar and its success assured a sound basis for further development and application.

The SIR-A and SIR-B missions have been progressively more impressive. The SIR-A instruments delineated land surface features in great detail and also penetrated the extremely dry sand of the Eastern Sahara to a depth of several meters to reveal alluvial deposits and drainage patterns which had not been detected by conventional imaging. These findings opened new areas of investigation. The SIR-B mission provided images at a variety of incidence angles and illumination geometrics, enabling the construction of stereographic images from the data.

Demonstration of the capability of imaging radar involves considerably more than the flight apparatus, for by its nature the technique produces huge quantities of raw data. Dr. Elachi led the effort to update earlier optical processors and is playing a key supportive role in developing advanced digital techniques for SIR-B data.

The satellite missions have received wide public acclaim. But during the decade prior to these highly visible contributions, Dr. Elachi conducted pioneering work in theoretical analyses and airborne systems to show the potential of imaging radar for oceanography and meteorology on Earth and for studying Venus. The breadth of these contributions, represented in over one hundred publications in a variety of technical fields, arises from his unique skill in interpreting radar images and his intimate knowledge of Earth science. Beyond these notable technical abilities, Dr. Elachi has unusual skill in communicating with his co-workers, engineers and scientists from other disciplines, and with management.

In recognition of these accomplishments, the National Aeronautics and Space Administration and the Department of the Interior take great pleasure in presenting the William T. Pecora Award to Dr. Charles Elachi.

Donald Paul Hodel
Secretary of the Interior

James M. Beggs
Administrator, National Aeronautics and Space Administration

SYMPOSIUM CALENDAR

MONDAY, AUGUST 19, 1985

8:00 a.m.-9:30 a.m. SYMPOSIUM REGISTRATION
9:00 a.m.-5:00 p.m. PRE-SYMPOSIUM TUTORIALS
- Introduction to Remote Sensing
- Introduction to Geographic Information Systems
- Vegetation Damage Mapping
- Geobotanical Remote Sensing
- Advanced Remote Sensing

1:00 p.m.-6:00 p.m. SYMPOSIUM REGISTRATION

TUESDAY, AUGUST 20, 1985

7:30 a.m. SYMPOSIUM REGISTRATION

9:00 a.m. OPENING AND WELCOME

9:15 a.m. PLENARY SESSION
- Evolution of Remote Sensing in Range Management, Speculations on its Future

10:00 a.m. COFFEE AND ROLLS
10:30 a.m.
- Status of Remote Sensing in Forestry
- New Horizons in Remote Sensing for Forest and Range Resource Management

11:45 a.m. LUNCHEON — Guest Speaker

1:30 p.m. CONCURRENT TECHNICAL SESSIONS
- Integrated Resource Inventory
- Land Use/Land Cover Mapping

3:00 p.m. COFFEE BREAK

3:30 p.m. CONCURRENT TECHNICAL SESSIONS
- Geographic Information Systems Applications
- Change Detection and Monitoring

6:00 p.m.-7:00 p.m. PECORA 10 MIXER

WEDNESDAY, AUGUST 21, 1985

8:00 a.m. SYMPOSIUM REGISTRATION
8:30 a.m. CONCURRENT TECHNICAL SESSIONS
- Vegetation Damage Assessment
- Mapping Soil and Water Resources

10:00 a.m. COFFEE AND ROLLS

10:30 a.m. CONCURRENT TECHNICAL SESSIONS
- Fire Fuels Mapping
- Mapping and Monitoring Wildlife Habitat

12:00 noon LUNCH

1:30 p.m. SPECIAL POSTER SESSION (Rooms 228 & 230)

3:00 p.m. COFFEE BREAK

3:30 p.m. SPECIAL FORUM SESSIONS
- Interfacing Remote Sensing with Spatial Data Systems
- Integrating Remote Sensing into Operational Systems
- Remote Sensing for Detection and Monitoring of Atmospheric Deposition Damage
- Research and Development Priorities in Remote Sensing for Forest and Range Resource Management

6:00 p.m. PRE-BANQUET SOCIAL

7:00 p.m. AWARDS BANQUET
- Dinner
- Keynote Speech
- William T. Pecora Award Presentation

THURSDAY, AUGUST 22, 1985

8:00 a.m. SYMPOSIUM REGISTRATION

8:30 a.m. TECHNICAL SESSION
- Global Resource Assessment

10:00 a.m. COFFEE AND ROLLS

10:30 a.m. TECHNICAL SESSION
- New and Emerging Technology—Data Acquisition

12:00 noon LUNCH

1:30 p.m. TECHNICAL SESSION
- New and Emerging Technology—Data Processing

3:00 p.m. COFFEE BREAK

3:30 p.m. SUMMARY SESSION
- Forum Summaries
- Symposium Overview
- Closing Remarks
- Adjourn

FRIDAY, AUGUST 23, 1985

8:30 a.m. POST-SYMPOSIUM FIELD TRIP
- Bus Trip to Poudre River/Pingree Park
- Lunch
- Area Interest Stops
- Dinner at Pingree Park

TABLE OF CONTENTS

Front Matter

Plenary Session

Integrated Resource Inventory

Land Use/Land Cover Mapping

Geographic Information Systems Applications

Change Detection and Monitoring

Vegetation Damage Assessment

Mapping Soil and Water Resources

Fire Fuels Mapping

Mapping and Monitoring Wildlife Habitat

Special Poster Session

Forum Sessions

Global Resource Assessment

New and Emerging Technology—Data Acquisition

New and Emerging Technology—Data Processing

PLENARY SESSION

Moderator: L. Pettinger

EVOLUTION OF REMOTE SENSING IN RANGE MANAGEMENT, SPECULATIONS ON ITS FUTURE

Charles E. Poulton, Consultant
Natural Resources Assessment and Management
121 Muir Avenue
Santa Clara, CA 95051 (408)985-7704

ABSTRACT

In view of the diverse audience, a common ground is established through definitions and position statements. The kinds of resource data needed by range managers and the applications of the information is briefly reviewed. The evolution of rangeland resources methodology is briefly traced from its beginning in 1910 through the present with emphasis on use of remote sensing rather than specific measurement methodologies. A number of case studies, mostly from developing nations, that illustrate information needs and applications are briefly summarized. Ground verification, legend development and data gathering methods, particularly the effective combination of low level aerial observation with ground measurement, are discussed as components of Ecological Resource Analysis. Problems encountered and the potential of remote sensing as a tool in ecological assessment and monitoring of rangeland resources is evaluated from the author's experience. Some suggestions are made about requirements and opportunities for its expanded and successful use in future years.

INTRODUCTION

Definitions, A Common Ground

Since this is a diverse professional audience, I wish to start with a few definitions and position statements so we may be assured of proceeding from a point of common understanding. **First,** I define remote sensing to include all photographic, multispectral, vidicon and microwave sensor systems with their analytical technologies regardless of platform used.

Second, some people equate rangelands and the involvement of the range profession only with domestic livestock production. This is a misconception. Rangelands and the involvement of the practitioners of the art and science of range management encompasses all those lands that provide food and cover for herbivorous animals of all kinds.

Third, inventory and monitoring are concerned with **Ecological Resource Assessment** which has specific as well as general applicability. The job is not approached from a restricted, single-discipline point of view. This assessment looks initially at the specific **rangeland ecosystems**, at the vegetation-landform-soil systems, and proceeds to map, identify and characterize each integrated system of organisms taken with their abiotic living space. The rangeland ecosystem is a restricted and particular piece of the landscape which has unique vegetation and soils characteristics -- a concept especially appropriate to the multidiscipline use of remote sensing in ecological resource assessment.

Fourth, rangeland ecosystems encompass vegetation-soil systems from the very sparcely or transiently vegetated deserts, through the grasslands and steppes, wetlands and marshes, tundra, tall shrublands, shrub and tree savannas, to the open woodlands and those forests that allow enough light to reach the ground to produce subordinant shrub and/or ground layers of vegetation with value to herbivores.

Fifth, Survey or inventory deals with the mapping of rangeland ecosystems and the characterization, measurement or indexing of their intrinsic features or qualities. **Monitoring** is concerned with the detection or measurement of extrinsic factors and with the changes brought about in the ecosystem by these extraneous influences.

Sixth, rangeland resources development and management should be the leading beneficiary of remote sensing technolgoy among the natural resources disciplines. Since many of these ecosystems are more open, vegetationally, more intrinsic features affect image characteristics. Thus it is highly probable that more such features will be detectable directly from remotely sensed data. Additionally, many kinds of needed information are correlated with image features so that they may be inferentially identified and qualitatively or quantitatively indexed by association, even when the parameter is not directly registered in the image.

Seventh, the knowledge derived from inventory is used for Policy and Objective Setting, for Resource Alloction, and for Development, Improvement and Management Planning. Monitoring information is used to adjust or improve these decisions and especially to fine-tune management programs. **The goal of rangeland management** is to maximize total production and/or benefits and to assure the perpetual quality and viability of the resource.

Kinds of Information Range Managers Need

While making some comparisons of range survey maps and data from the second, fourth, sixth and seventh decades of this century I concluded that there is no agreement among range professionals or U.S. Government Agencies with responsibility for management or technical assistance on rangelands as to what needs to be mapped, written up by way of legend descriptions or documented in quantitative data from rangeland ecosystems. Since this is not the forum for resolving such matters, I shall briefly summarize the kinds of maps and information I know a serious rangeland resource manager needs. Perhaps then, you can think with me on how remote sensing technology can contribute to that need.

The starting point is an Ecological Resource Assessment that maps, identifies, and characterizes the specific vegetation-soil systems that occur on any given landscape. These may be mapped and described at various levels of refinement, depending on the intended use of the data. At any given hierarchical level, mapping may be done of pure types or complexes of types, depending on survey goals and complexity of individual landscapes. The descriptive legend should characterize each kind of mapping unit, providing information on species composition and production, normally associated landform conditions and soils or soil conditions. Tabulations include areas and productivity by types of mapping units within each contiguous management or jurisdictional area. Because productivity and other characteristics of the vegetational component may change over time, it has been traditional to repeat these kinds of basic inventories each one or two decades.

From the survey data various determinations are made that relate to management. One considers present site production in relation to potential. Another concerns ecological stability or range and soil condition. Production data, condition, use patterns, season of use and many other factors are interpreted to derive an animal or herd support capacity for the land. I prefer to put this determination in the context of an estimated initial stocking rate. This becomes the managers target in manipulating animal numbers, herd structure, timing and frequency of grazing use. The often mentioned and misunderstood variable, Carrying (or Grazing) Capacity, is a derived variable and a function of management. It is not measurable but can be arrived at through a series of stocking rate approximations as tested under progressively improving management strategies.

Because it is vital in the design and fine-tuning of management strategies to know the nature and successional direction of vegetation change on a shorter time scale than a decade or two, observations heavy on ground measurement technology are made on an annual, biennial or triennial basis. This activity is referred to as the indexing or measurement of Range Condition and Trend -- the detection of plant succession.

As an early-warning of undesirable successional change and animal condition, the range technician, rancher or wildlife biologist determines the level of use or the residue of useable fodder. Under intensive management, observations or measurements of this parameter may be made twice per year, never less than once near the end of the grazing season or period. A rancher seriously interested in optimum animal gain and continued high range productivity may observe the amount and pattern of forage residue on a monthly or weekly basis. This activity has traditionally been ground-work intensive. Additional thematic maps or overlays may be made of range condition and use patterns. All this inforamtion is used to fine-tune the management strategy for the specified area to which the data and information applies.

Because rangeland areas are subject to changes in land use -- conversion to wilderness, agriculture, urban, industrial, mining, and transportation and utility corridors -- it is essential to map the soil components of the ecosystem and existing land uses. The combined ecological interpretation of the vegetaiton and soils features of the ecosystem is used to guide land-use change and range improvement and development activities.

Stereo examination of aerial photography together with evaluation of the ecosystem maps, productivity, utilization patterns and condition is the basis for locating fences, trails and other animal control facilities designed to increase the efficiency and uniformity of forage use.

Finally, water for domestic livestock, wild animals and birds is a key factor in animal control and management manipulation. Thus an important aspect of rangeland resources inventory is the location of springs, watershed evaluation, potential reservoir siting, and the assessment of ground water or well drilling, especially in arid environments where water is at a premium. Particularly in developing nations, a ground water hydrologist is frequently added to the range analysis team.

REMOTE SENSING IN RANGE MANAGEMENT

Historical Development

How and When Range Surveys Began. Range surveys had their operational beginning in the United States between 1905 and 1910 when James T. Jardine, newly appointed Chief of the Office of Grazing Studies in the Forest Service, embarked on a program to encourage District Forest Rangers to inventory and map the rangelands on their districts (Chapline, 1985). His program had varying levels of success depending on the workload and interest of individual Rangers. Where completed at all, Grazing Atlas maps were colored to designate 10 cover types using general appearance and dominant growth form of the plants as criteria for mapping unit-classes. Horizontal or vertical color striping indicated types with a poor stand of forage plants and overgrazing, respectively.

By the summer of 1910, Jardine had become disenchanted with progress and he and W. R. Chapline designed a procedure to be carried out by special survey teams rather than individual Rangers. This procedure became known as the **Reconnaissance Method of Range Survey.** The first range surveys with this method was conducted entirely by ground methods on the Coconino National Forest, Arizona, in 1911. Where available, topographic maps were used for delineating type lines. Either the same or the next year, additional crews started work on the Manti National Forest in Utah and on the Wallowa National Forest in Oregon.

Interestingly, I have been able to locate one copy of an original Ranger map prepared in 1910 for part of the Caribou National Forest in Southeastern Idaho and almost all of the original field maps with writeup sheets from the Manti survey in 1912 - '14.

The 1930's to World War II. The Reconnaissance Method remained unchanged until 1937 when the number of first-order cover types was changed from 10 to 18, to better accommodate arid vegetations in the Great Basin and Southwest. This took place concurrent with the initiation of the interagency Western Range Survey.

With creation of the Agricultural Adjustment Administration early in the New Deal era, 1:20,000 aerial photography was extensively used for administering crop compliance; and by the beginning of World War II, 90 percent of the Agricultural land in the nation had been photographed. Unfortunately rangeland was included only to block out flight lines peripheral to crop land. This did however, make some photos available especially to Soil Conservation Service Range Conservationists. They quickly took the opportunity and pioneered the idea of using uncontrolled mosaics at scales as large as 1:15,840 for range surveys and the preparation of Ranch Management Plans.

In the last half of the 30's, research began to focus on improving the methods and the data gathering system was changed from a plotless to a 100-sq-foot circular plot area. On occasion, in unsurveyed country, the entire map was generated by running compass lines across the relief and sketching in drainages, cultural features and type lines as they were crossed in traverse. "Formline maps" were sometimes created in advance of field work by hand sketching partial topographic lines on a planimetric base to aid the vegetation typing process. These were used as late as 1938 on the Dixie National Forest in Utah.

In 1935 and '36, black-and-white aerial photography became sparingly available at the common scale of 1:20,000 (Moyer, 1950). Unfortunately in those days rangelands were valued at $2.50 to $3.00 per acre and expenditures for these new pictures was rarely justified. Adding to the negative attitude and based on studies begun in 1939, Reid, et al., (1942) stated that aerial photography taken especially for range survey could not be justified. This decision was based solely on statistical considerations with no awareness of the potential of photo interpretation to save field time and costs. The methodology remained an on-the-ground technique with no attempts or agency encouragement seriously to get into photo interpretation. Stereoscopes were available to crews using the new photography around the end of the decade, but they were mainly used to help find section corners, better perceive drainage patterns and to settle arguments about location of watershed divides. Pretyping by stereo examination and image interpretation was not a standard practice.

By the early 40's, aerial photography became sufficiently appealing to range people that they began to use small-scale Army Mapping Service photography where larger scale photographs were not available, accepting without challenge whatever print quality came out of the photo laboratory.

World War II to the Space Age. In 1949 Henriques wrote of the advantages of controlled mosaics for range resource mapping and of the substantial use of aerial photos in regular range management within the Bureau of Land Management, U.S. Department of Interior. Their use started immediately after World War II when the Bureau began a major push to inventory all their lands and meet their Congressional mandate to "establish the carrying capacity of the land for adjudication purposes." Similarly, the Bureau of Indian Affairs, USDI, has used aerial photography from about the same period to make combined range and soil surveys on Indian lands.

It was not until the late 1950's that aerial photography, other than the AMS photos, became generally available for the western range, and first time coverage for range purposes was still being acquired at more suitable scales as late as 1970. Black and white was still the standard although some range research applications were being made of color, black-and-white infrared and color infrared for range applications in the mid- to late 1960's (Carneggie, et al., 1967).

During this postWar period, the California Soil-Vegetation Survey Program of the Pacific Southwest Forest and Range Experiment Station probably did more than any other program to foster photo interpretation as an element of mapping of both vegetation and soil features on range and forest lands. But, it was not until 1967 and '68 that anyone in range developed stereograms, aids and keys for photo interpretation of vegetation types and soil conditions on rangelands or made a serious attempt to set down guidelines for quality control on black-and-white and color aerial photography for range applications (Carneggie, et al., 1967; Culver and Poulton, 1968; Poulton and Isley, 1970; and Poulton, Faulkner and Martin, 1971).

Blasting into the Space Age. In 1966 the first application of multispectral scanner technology to range resource analysis was made by Poulton, Colwell and Carneggie using the then classified 18 channel system developed at the Willow Run Laboratories, Univeristy of Michigan.

In December 1966, Poulton and colleagues from the University of California and the University of Arizona tested the usefulness of Gemini IV photography for range vegetation analysis in Arizona (Carneggie, et al., 1967). This lead to continuing projects from the Apollo 7 experiments through Skylab under Poulton's leadership at Oregon State University and Carneggie's leadership at the University of California, Berkeley. Driscoll and Reppert of the Rocky Mountain Forest and Range Experiment Station became involved in applications research quite early in this period.

Somewhat concurrently, advanced work, initially with photography and later with Landsat, was going on in Australia under the leadership of Ray Perry and his associates. The thrust of all these projects was to develop the new satellite and highflight technology for rangeland resources application, but the range agencies and the profession was slow to pick up on the advances until the mid-1970's and later.

Immediately with the availability of the first Landsat imagery in 1972, a new era was born in the remote sensing of rangeland resources, especially in the developing countries of the world. From this beginning Landsat imagery has proved to be an operational system, contrary to certain bureaucratic contentions; and the geologists, mineral industries and developing nations of the world have been the primary beneficiaries.

The majority of range resources applications of Landsat in the United States have been in a researchy, try it and see context which, incidentally, is still the prime motivator (Bureau of Land Management, Forest Service and Soil Conservation Service, 1985, Personal Communication). On the other hand, all of the applications of Landsat in the developing world have addressed practical resource development, management, land use and resource alloction problems. In these nations the need has been matched by information derived at scales of 1:200,000 and smaller. Another significant factor favoring Landsat in these countries is the frequent difficulty of obtaining good quality aerial photography without being variously obstructed by the local military establishment.

EXAMPLES OF REMOTE SENSING APPLICATIONS

It should be helpful in the deliberations of this conference to briefly review a cross-section of problem-solving applications of remote sensing in range management. In doing this, if I fail to mention your favorite project, please accept my apology. I have made a conscious effort to select projects not well known from the North American remote sensing literature.

Detailed Surveys with Aerial Photography

Oregon State Rangelands Survey. By the mid 1960's the State of Oregon still owned approximately 607,000 acres of rangeland scattered through approximately 13 million acres in southeastern and central Oregon. Only 36 blocks were larger than 640 acres, the largest being 14,459 acres and the smallest 0.18 acres. So little was known about these lands, that the Division of State Lands and the Oregon Legislature were in a dilemma regarding further disposal vs. the management and income-generation potential of these lands.

Obviously this was an ideal candidate for a photo interpretation approach which the Range Management Program at Oregon State University contracted to do. A feasibility study showed that conventional survey would require approximately 4 man-years and $1. per acre or more for professional resource analyst time and travel. A procedure was developed to make the inventory with a ground check on about 20 percent of the parcels and primary reliance on photo interpretation. The main project was begun in the summer of 1967 and completed early in the summer of 1968. The statistical report was completed in the summer of 1969 at a total cost of only $0.14 per acre including data summarization and report writing.

I suspect this is the most intensive and comprehensive reliance on photo interpretation of rangeland features from aerial photography prior to 1969, perhaps to date. Interpretation was intentionally pushed to the limit yet average accuracy of all factors interpreted was 72 percent. Many ranged in the mid-80 to low 90 percentile with a few at 100 percent based on actual ground checks. We started the project with the idea that interpreters would be carefully trained and do their own field checking. We soon learned, however, that we had two interpreters who's capability and accuracy was so superior that we finished the survey with these two people doing all the interpretation (with occasional field-check trips) and organized 2-person field- check crews who communicated regularly with the interpreters to sharpen collective abilities.

National and Regional Surveys in Developing Countries

Rangeland Productivity Analysis, Iran. Patterned somewhat after the continuing forest inventory system in the United States, Earth Satellite Corporation, under the leadership of Phil Langley, made an assessment of the current and potential range feed production in Iran during 1973-'74. A mosaic of the newly available Landsat imagery was laid for the entire country at 1:1,000,000 scale. The country was stratified into desert, dry steppe and mountainous regions. A sample of Landsat scenes was then drawn for the two vegetated strata (true deserts excluded). Each of these scenes was mapped at a scale of 1:500,000 with a combined landforms-vegetation legend. Next a subsample proportional to area of full-frame stereo models was drawn from existing 1:50,000 aerial photography within each Landsat scene. These were more intensively mapped with the same landform-vegetation legend breaking vegetation down to dominant growth form. A sample strip was then selected across each stereo model, flown with color infrared, 70 mm photography at a scale of 1:3,000 and mapped according to the same hierarchical legend with still more refined attention paid to the vegetation component. Ground examination points were selected to represent the major vegetation and landform conditions across this strip and examined by a helicopter crew. Vegetation composition, soils and landform data where taken and forage production was determined by a 10-plot, double-sampling cluster in which one random plot was both estimated and clipped. The work was timed for near peak vegetation production. While on the ground, experienced range scientists estimated probable percentage of recovery above present production based on vegetation composition, soil conditions and an assumed 10 year recovery period. These data were then expanded back by proportional probability sampling theory to provide areas by dominant vegetation types and current and potential production by types for the nation as a whole and for political subdivisions. This is the first example of such a nation-wide survey of rangelands. It would not have been possible in a reasonable time period and cost without the Landsat component.

Range and Agricultural Potential, Oman. In 1981-'82 the Ministry of Agriculture and Fisheries in Oman became interested in the expansion potential of their country for range livestock and agricultural production. The need for information was met by interpreting 1:250,000 false color composite Landsat scenes mapping soil forming material, landforms and indications of the potential for additional shallow groundwater development.

Following this mapping the country was extensively traversed and overflown to check the mapping and to provide a ground verified basis for characterizing all SFM delineations according to a 4-class potential category for both range (R0 to R3) and Agriculture (A0 to A3). In each case the "0" category was barren sand or steep rocky lands with little soil mantle or vegetation cover. The "3" category was good to excellent potential, "2" was fair or marginal for increased technological development and "1" was submarginal for agriculture and suited only for traditional nomadic grazing.

On this basis, nine areas were delineated as having some potential for additional agricultural development and two areas were identified for possible application of range improvement and grazing management intensification. In the interim period the Ministry has followed recommendations and completed an Australian type land systems survey of the best rangeland area and has completed a reconnaissance soil survey of most of the agricultural areas.

Land Use and Ground Water Studies, Tanzania. Expanding population and the pressure for increased food crop production has created some problems in the Masai Steppes of Tanzania. The Government desired to identify priority areas for possible expansion of agriculture and the areas that should remain in range use. The problem was addressed similarly to the Oman work except that a Land Systems/Land Units approach was taken to mapping. Interparetations were backed by over 3,000 miles of ground travel and aerial reconnaissance of areas inaccessible from the ground. A ground water exploration guide map was prepared showing favorable locations for more detailed groundwater studies. A land use capability overlay showed candidate areas for more detailed survey soil surveys and possible agricultural development and those areas that should remain in range use. As might be suspected these two land uses were frequently in direct competition for the same areas.

Multistage Soil Surveys, Sudan and Morocco. In these countries FAO has used 1:1,000,000 Landsat mosaics to map soil associations and prioritize areas for more detailed surveys at a scale of 1:250,000, a 2-stage survey with the same imagery. The initial reconnaissance survey required 12 man months and cost $0.04 per sq. km. plus field travel. It was estimated that the same work would have required one and one-half years and a crew of four to complete with aerial photography.

Multistage Forest and Range Survey, Brazil. The project used 1:1M and 1:250K B&W Landsat images with supporting aerial photography to map dense and open forest, gallery forests, evergreen and deciduous (using two seasons of imagery), open woodland (cerrado) in 3 classes, grasslands, floodplain vegetation, and deforestation for pasture development. Mapping units ranged from 75 to 10,000 ha. They also compared the new analysis with earlier radar imagery interpretation to map 11 land-use classes along the Transamazonia Highway.

Global Monitoring of Rangeland Resources

FAO-UNEP Senegal Project. The United Nations Environment Program (UNEP) is devoting about $2,000,000 per year to the development of a Global Environmental Monitoring System (GEMS). In a joint project under operational leadership of the Food and Agricultural Organization (FAO), UNEP opted to give initial attention to rangeland monitoring. They recently completed a pilot project in Senegal, West Africa.

The project used 1:200,000 Landsat FCC's to map vegetation/land form/soil systems. Sample strips of aerial photography were flown across transition zones. Within these, ground study sites were selected and photographed at the largest feasible photo scale in 70 mm format, 5 to 10 stereo models long.

In addition a Systematic Reconnaissance Flight (SRF) program was used in which carefully defined belt transects were observed from low-flying aircraft to supplement ground data and provide count statistics on such things as waterholes, animal and human population densities, and vegetation characteristics. The results of SRF observations and counts were summarized by 1 sq. km. grid cells over the project area. All these data were then correlated to Landsat image characteristics so that extrapolation of ground and SRF information could be done by Landsat interpretation to the greatest extent possible.

In cooperation with the Senegal project, Tucker and colleagues at Goddard Space Flight Center used NOAA-7/AVHRR data from 7 dates in 1981 to generate a biomass production map of the project area in 8 production classes from ¼ 200 to 1,400 + kg/ha. Using the Normalized Difference method with AVHRR Ch. 1 (0.55 - 0.68 microns) and Ch. 2 (0.725 - 1.10 microns), Tucker obtained correlations of 0.81 for green biomass and 0.75 for dry biomass, in my judgement sufficient for practical management applications.

Significant outgrowth of the project has been establishment of national Ecological Monitoring Units to carry out parallel rangeland resources monitoring in other African countries. Ten countries were moving in this direction in 1982.

FAO and UNEP are currently laying the groundwork for an international consultancy to develop the next phase of global rangeland resources monitoring program within the United Nations organizations.

Desert Locust Control.

The desert locust, *Schistocera gregaria*, can potentially ravage 30 million sq. km. of range and agricultural land in 60 countries. Combating this pest requires monitoring precipitation conditions optimal for egg development, green vegetation for food and shelter during the nymphal stage and suitable temperatures for development and flight. A uniquely successful program for monitoring these conditions has been cooperatively developed by FAO, USAID and NASA. It involves obtaining the climatic and soil temperature data from analysis of geostationary satellite imagery received daily at FAO in Rome, Italy, and cross checked with rare and scattered ground records. Vegetation biomass development was indexed from Landsat and NOAA/AVHRRX satellite data.

A methodology has been developed for automatically extracting location and amount of vegetation biomass from Landsat data and then calculating a Potential Breeding Activity Factor for specific areas. These projections then become the basis for control programs. Through this project, the desert locust control program has become eminently more cost effective and a serious outbreak has not occurred since the inception of the FAO monitoring program about 25 years ago.

GROUND VERIFICATION AND DATA GATHERING

Systematic Reconnaissance Flight

I perceive the single greatest problem in remote sensing applications to be the general deficiency of ground verification. I wish to share with you a technique developed in East Africa and which, in my opinion, warrants general adoption as a part of practically all large area remote sensing applications programs. A technique which, in fact, can directly solve some of the most critical problems in monitoring rangeland resources condition and use and increase many fold the quality and usefulness of satellite image interpretation whether done visually or by computer-aided technology.

The Method. The Government of Kenya, with the assistance of UNDP/FAO, and Canadian Aid developed, through the Kenya Wildlife Management Project, a highly refined and effective method of aerial observation called Systematic Reconnaissance Flight (SRF). Space does not permit a complete description of the method. Briefly it involves flying of transects with sophisticated navigation instrumentation and radar control on flight altitude to maintain a near constant observational swath width low oblique to the plane's position. Aerial oblique photography supplements direct observation and data recording as needed. Where data are being acquired for management application, the strips typically match either a 10, 5, 2.5 or 1 km grid pattern. Data are fed into a computerized data management system that summarizes and presents information in square grid units appropriate to the flight pattern. Sample statistics are expanded to total area estimates for any parameter observed (Gwynne and Croze, 1975). Locational control is referenced to the UTM grid so that the data can be positively related to specific image areas on either aerial photography or satellite imagery.

Examples of Application. Two examples of application would be enlightening. In 1976 a team from the Kenya Wildlife Project flew a project area of 13,975 sq. km in Zaire at a cost of $45.10 per 1,000 sq. km in a 10 km grid and a cost of $101.00 per 1,000 sq.km in a 5 km grid. Wildlife population data were obtained along with large amounts of supporting ecological information with 38.8 flying hours plus compilation and reporting time. The Government of Zaire had previously plannned to obtain similar information in a three-year project and at a much higher personnel and dollar cost. Previous to the SRF survey, the white rhino population was thought to be about 25 animals and declining. The survey confirmed, among many other facts, a population of 400 with positive indications of a healthy and expanding herd (Savidge, Woodford and Croze, 1976)

In an East African project the number and distribution of Masai livestock was estimated and mapped in relation to the growth of range vegetation as indexed both from Landsat interpretations and from SRF observations. This SRF survey confirmed the accuracy of the Landsat interpretation and gave the first comprehensive picture of livestock numbers and distribution ever achieved for a major part of Masailand in Kenya (Gwynne and Croze, 1975).

Multispectral Sensing of Biomass

Serengeti Range Productivity Survey. Airborne radiometers were used in the red and near infrared spectra to calculate ratios which were then correlated with identical ground measurements, and these in turn correlated with yield measurement data. A map of both green and dry season forage production was made of the whole Serengeti National Park in Tanzania as an input to wildlife management (Gwynne and Croze, 1980). Ground measurements gave correlations over 0.90 and aircraft measurements 0.86 between the band ratioed spectral data and biomass (McNaughton. 1974. Personal Communication).

Time Sequence Photography in Change Detection

Catastrophic Change. Such changes are often associated with settlement, serious drought and natural disasters. During the 1971-'73 Sahelian drought, De Wilpelaere and Toutain (1976a, 1976b) documented degradation around a typical village of northern Upper Volta between 1955 and 1974-'75. The area normally received 400 mm ppt. It was normal for five years preceeding the 1955 photography. 1974 was a normal year but the preceeding three years had been only 25 percent of normal. From measurements on aerial photography for the three years, the authors determined that barren sands around the village covered 22 ha. in 1955, 89 ha. in 1974 and expanded to 250 ha. in 1975 while the area of herbaceous steppe vegetation in 1974 was 55 percent less than 1955 and 61 percent less in 1975.

In a similar situation in northern Kenya, Conant (1981) used the standard deviation of spectral reflectance values in the four bands of Landsat imagery and a 5/7 ratio to detect degradation around refugee settlements. He compared data for 1973 presettlement with 1978 post-settlement and found increases in standard deviations some of which were significant at the 0.001 level. This corresponded with degradation patterns confirmed by ground examination and discussions with displaced persons.

Gradual Vegetation Change. Along with the evolution of ground methods, many remote sensing approaches for documenting change have also been tried, shelved or discarded and occasionally picked up again. Earliest attempts involved repetitive photographing of landscapes from permanently marked camera points. Later low oblique or vertical photo plots, taken from the ground, were added as positive documentation of how the species change in position, vigor or composition.

Occasionally the quadrats were planimetrically mapped by various methods but the practice was most time consuming unless limited to freehand sketching of the location and size of major species. All earlier efforts were limited to single-frame photographs although stereo photographs of quadrats have been periodically reported in the literature. More recently the idea has been revived as a detailed photo transect procedure for range condition and trend documentation. Both the full transect and each of a number of small plots are photographed in stereo from eye-level. When working under a forest canopy the procedure also provides for photographic documentation of tree crown canopy closure. Certain ground data are recorded along with the photographs (Hall, 1976). This method seems quite time intensive, especially if one attempts to convert the stereo models into hard data by comparative interpretation or photogrammetric measurement. The method does provide excellent documentation that can be checked by supervisors to maintain performance control on staff work. In view of the thousands of ground plots recorded over five to 10 or more years without statistically detecting a range condition difference attributable to animal use and management, photographic methods begin to be an attractive alternative even if nothing is done but compare the photos and make subjective statements based on an understanding of principles of plant ecology. It would be enlightning to make a time and cost comparison of this method with an approach based on 1:400 scale aerial photographs.

Very Large-Scale Aerial Photography in Range Management

Considerable work has been reported on the use of stereo aerial photography taken at scales from 1:400 to 1:3,000 for documenting vegetational change (Tueller, et al., 1972; Warren and Dunford, 1983). These methods are particularly suited to tall grass, shrub steppe, and savanna ranges. They permit photogrammetric measurement of many features; and with permanent marking or referencing of transect ends, may be accurately rephotographed.

The most important source of variation to control in vegetation measurement is among-location variance. If one concentrated on careful stratification of ecosystems preparatory to data gathering and then used rapid, possibly even less accurate methods or fewer measurements per location, the larger number of locations observed would detect changes where now they missed. More effective use of remote sensing and photogrammetry is the best way to move in this direction.

WORKING RESOURCE ECOLOGY AND REMOTE SENSING TOGETHER

Prerequisites to Image Interpretation

Whether one is interpreting aerial photography or satellite imagery visually or computer-aided, the greatest single determinant of quality of end product is the analyst's ability as a resource ecologist and how well he knows what to expect on any given landscape.

In intensively managed rangeland areas, the greatest need of the user is to have the specific vegetation-soil systems accurately mapped and thoroughly characterized. One of the best examples in the United States of the work necessary to guide this kind of resource assessment and monitoring is found in the plant association and condition guides prepared by the U.S. Forest Service, Region 6, under the leadership of F. C. Hall and L. A. Volland. Their ecological approach is fundamentally sound. An Ecolgical Resource Analysis, regardless of the discipline providing leadership or the primary reason for which it is done, can and does serve multiple user interests if carried out under sound principles of ecological classification. This does not mean that such an ecological assessment meets all the informational requirements for every resource discipline. Supplemental data will usually have to be gathered by the individual disciplines. The Ecological Resource Assessment establishes a common stratification base which all can use at substantial savings in cost of sampling and data gathering.

Legends and Their Application

For almost every remote sensing application an hierarchical legend system that progresses from the most general to the specific is easiest and most effective to use -- and most cost effective to develop. In developing countries it is often adequate to work to a second or third level in such legend systems. Here highly detailed ecological knowledge is often not required because at these levels most vegetational discriminations are made on plant lifeform and structural criteria. Occasionally mapping in developing countries is done to a fourth level where the discriminating criteria are broad floristic features that almost any natural scientist can easily learn to handle. The legend level most commonly used in traditional range surveys would fall between what I would consider the fourth and fifth levels. The present method used on most National Forests for range surveys approximates a fourth level but it requires, in the primary mapping units a judgemental input that cannot be related to image characteristics -- therefore not amenable to automatic computer recognition. At fifth and sixth levels, high capability of the analyst is required as a plant sociologist and student of vegetation-soil relationships. In many cases, these latter levels of discrimination will require drawing on associated evidence, more refined input from human experience and a higher percentage of ground verifications.

In the developed or industrial world such as North America and Europe the need is at least for a level determined on floristic criteria and in most cases, in my opinion, on plant sociological or vegetation-soil relationship criteria. The actual range analysis work currently being done by some of the Federal agencies does not suggest agreement with this position. They are settling for much less. Many of the surveys currently being done do not qualify as an Ecological Resource Assessment.

PROBLEMS AND OPPORTUNITIES, A LOOK TO THE FUTURE

Current Status of Use and Application

Aerial photography has been used by range practitioners for forty five years yet its full potential to provide resource information at very low cost has rarely been realized. Other photographic methods for documenting range condition have come and gone. The permanent camera point and photo transect have persisted as useful techniques, but large-scale aerial photography has not caught on.

In spite of many successful Landsat applications, especially in developing countries, and the potentials suggested by innovative research, there have been disappointments with products from the system.

With a recent upsurge in the availability of individual desktop capability to do one's own computer aided Landsat interpretation and with the higher resolution of Landsat Thematic Mapper and with the French SPOT system ready in the wings at 20- and 10-meter resolution, we are on the verge of an exciting new era in remote sensing of earth resources. Are we ready fully to capitalize on this opportunity?

Following are some of the questions that have bothered me: Why are not range practitioners in the United States making more operational use of Landsat and large-scale aerial photography? Why are so few developing skill as photo and image interpreters? Why are not innovative uses of the systems growing more rapidly among those who make the decisions and do the work at field level? What are the major factors contributing to this situation?

Causes of Problems in Remote Sensing Acceptance

In the words of one of our outstanding news analysts and commentators on the west coast, "This is the way it seems to me!"

1) We, the universities and responsible employers have not provided the required training, encouragement, equipment and support to the rank and file of potential users;
2) too frequently interpretations have not been adequately verified by ground checks; thus,
3) justified or overreaction of users who discover interpretation errors has created adverse user reactions and severely damaged credibility and acceptance of the technology;
4) lack of adequate ecological expertise on an interpretation staff, not knowing what to expect, severely limits information extraction;
5) simple oversell or overstatement of system capability, overexpectation from the magic of Landsat and the computer; or using the wrong system or analytical procedures to derive information is most damaging;
6) responding to a misunderstood need or unrealistic expectation of the user translates into lost credibility;
7) failure to achieve an appropriate mix of preinterpretation ground work, visual interpretive skills, computer- or machine-aided interpretation and ground verification in generating products for the manager and decision maker is not fair to the system being used;
8) not recognizing system limitations in terms of the user's needs and thus attempting unrealistic discriminations causes criticism;
9) inadequate project follow through with the information user relegates assessment results to the shelf and file;
10) indecision, uncertainty and lack of commitment and consistent direction of the United States' role in earth resources remote sensing R&D, promotion, education and marketing has abdicated leadership;
11) an effective marketing strategy for sophisticated technology is needed that works **from** users needs **back** to what the technology can do to meet real needs and solve problems. Working from what technology has to the perceived user needs will continue to limit future growth of applications.

The Future of Remote Sensing in Rangeland Resources Management

What's in my crystal ball? The future of this technology in service to range practitioners can be very bright or a substantial disappointment. It can help them do a better job in less time and at less total cost. It can contribute much more than it has to better management, greater productivity and a more secure and stable resource base. It can help provide smooth and rational transitions to other land uses where apropriate. The determinant of which it is to be lies in how well we as a total group of educators and trainers, R&D experts and innovative, visionary users retrench to deal with all the above kinds of problems and shortcomings in our recent history.

Whatever your position is, whether you agree with what I have said or not, let us resolve not to leave the outcome to chance, starting today! Make this a great day! Have a super 10th Pecora Memorial Symposium!

REFERENCES

Chapline, W. R. 1984. Personal Communication.

Carneggie, David M., D. G. Wilcox and R. B. Hacker. 1971. The use of large scale aerial photographs in the evaluation of Western Australian

rangeland. Dept. Agr. West. Aust., So. Perth, Tech. Bul. 10, 37 pp.
Carneggie, David M., C. E. Poulton and E. H. Roberts. 1967. The evaluation of rangeland resources by means of multispectral imagery. Ann. Progr. Rept. for National Aeronautics and Space Administration, 76 pp. Forestry Remote Sensing Lab., Univ. of Calif, Berkeley.

Carneggie, David M., and Jack N. Reppert. 1969. Large scale 70-mm aerial color photography. Photogram. Engr., 35(3):249-257.

Clouston, J. G. 1950. The use of aerial photography in range inventory work on the National Forests. Photogram. Engr., 16(3):329-331.

Culver, Roger N., and Charles E. Poulton. 1968. Application of ecology and remote sensing in the analysis of range watersheds. Progr. Rept., to USDI, BLM., Work Plans 1 nd 3, 91 pp., illus., Range Management Program, Oreg. Agr. Expt. Sta., Corvallis.

Conant, Francis Paine. 1981. Refuge settlements and vegetation change: A multistage Landsat data analysis of a semi-arid region in Kenya. Paper presented at the First Thematic Conference: "Remote Sensing of Arid and Semi-Arid Lands," International Symposium on Remote Sensing of the Environment, Cairo, Egypt. 3 - 9 November 1981.

De Wispelaere, G. et Toutain. 1976a. Un example de degradation du couvert vegetal sur one dune continentale fixee dans le Sahel Voltaique entre 1955 et 1975. Paris. Rev. Photo-Interpretation. Ed. Technip.

________________ 1976b. Estimation de l'evolution du couvert vegetal en 20 ans consecutivement a la secheresse dans le Sahel Voltaique. Paris. Rev. Photo-Interpretation. Ed. Technip.

FAO/USAID. 1980. Remote sensing techniques and methodologies for monitoring ecological conditions for desert locust population development. Final Technical Report. Rome, Food and Agriculture Organization of the United Nations. (September 1980) 66 pp.

Gwynne, M. D. and H. Croze. 1975. East African Habitat Monitoring Practice: A review of methods and application. Paper presented to Symposium on the Evaluation and Mapping of Rangeland in Tropical Africa. Bamako, Mali, 3-8 March 1975. 47 pp., illus. (mimeo). UNDP/FAO Kenya Habitat Utilization Project, Nairobi, Kenya.

Gwynne, M. D. and H. Croze. 1980. East African Habitat Monitoring Practice: A Review of Methods and Application. In UNEP. 1980. Selected works on Ecological Monitoring of Arid Areas. GEMS PAC Information Series No. 1. (April 1980). Nairobi, United Nations Environment Program. pp. 95-135.

Hall, Frederick C. 1976. Range trend sampling by photographs. USDA, Forest Service. Pacific Northwest Region, R-6 Regional Guide 2-1, 50 pp. (July, 1976).

Henriques, Douglas E. 1949. Practical applications of photogrammetry in land classification as used by the Bureau of Land Management. Photogram. Engr., 15(4):540-545.

Meyer, M. et al. 1973. A 35-mm aerial photography system for forest and range resource analysis. Minn. For. Resources Note No. 240. 4pp. Univ. of Minn., Minneapolis, Minn.

Milligan, Kevin. 1982. Assessing Human, Anmimal and Environmental Conditions by Low Altitude Aerial Survey: The ILCA Experience in the Sahel. Paper presented to the Sahel Resource Inventory Mapping and Coordination Workshop, Bamako, Mali. 20-24 May 1982. Kaduna, Nigeria, ILCA, PMB 2248. 16 pp., illus.

Moyer, Ralph H. 1950. Some uses of aerial photography in connection with the production and marketing programs of the U.S. Dept. of Agric. Photogram. Engr., 16(3):305-307.

Poulton, Charles E., David P. Faulkner and Niels L. Martin. 1971. A Procedural Manual for Resource Analysis: Application of Ecology and Remote sensing in the Analysis of Range Watersheds. Final Project Report, Vol. 1, 97 pp; Vol. 2, 72 pp. Oregon Agricultural Experiment Station, Corvallis. (In fulfilment of contracts No. 14-11-001-3610 and -4246 (n), USDI, Bureau of Land Management).

Poulton, Charles E. and Arleigh G. Isley. 1970. An analysis of State-owned Rangeland Resources for Multiple-Use Management in Southeastern Oregon. Oregon Agricultural Experiment Station, Corvallis. Report to the Oregon State Land Board, 125 pp., tabs., illus.

Reid, Elbert H., G. D. Pickford and Talmage N. Nelson. 1942. An appraisal of range survey methods from the standpoint of effective range management. USDA Forest Serv., Pacific Northwest For. & Range Expt. Sta. Range Res. Rept. PNW 2, 66 pp.

Reppert, Jack N. and Richard S. Driscoll. 1970. 70-mm Aerial Photography--A remote sensing tool for wild land research and management. **In** USDA, Forest Service. Range and Wildlife Habitat Evaluation, a Research Symposium, pp. 190-193. USDA, Forest Service, Misc. Publ. No. 1147, 220 pp.

Savidge, J. M., M. H. Woodford, and H. Croze. 1976. Report on mission to Zaire (KEN/71/526 - ZAI/70/001), 3-10 April 1976. UNDP/FAO, Nairobi, Kenya Wildlife Management Project Report W/K1593. 34 pp., illus.

Tucker, C. J., J. R. G. Townshend and T. E. Goff. 1983. Continental Land Cover Classification Using Meteorological Satellite Data. NASA Technical Memorandum 85060. 20 pp. Greenbelt, MD, 20771. NASA Goddard Space Flight Center.

Tucker, C. J., C VanPraet, E Boerwinkel and A. Gaston. 1983. Satellite Remote Sensing of Total Dry Matter Production in the Senegalese Sahel. Remote Sensing of Environment, 13:461-474.

Tueller, P. T., G. Lorain, K. Kipping and C. Wilkie. 1972. Methods for measuring vegetation changes on Nevada Rangelands. Univ. Nevada Agric. Expt. Sta. T-16, 55 pp. Reno.

UNEP. 1980. Selected Works on Ecological Monitoring of Arid Areas. The Global Environment Monitoring System. GEMS PAC, Information Series No. 1. 135pp. Nairobi, United Nations Environment Program.

Warren, Peter L., and Christopher Dunford. 1983. Vegetation Sampling with Large-scale Aerial Photography. University of Arizona Remote Sensing News Letter 83-2 pp. 1-6. (Fall-Winter 1983)

REMOTE SENSING: ITS STATE-OF-THE-ART IN FORESTRY

Robert C. Heller*
21 Eastwood Drive
Orinda, CA 94563

ABSTRACT

Foresters were using remote sensing long before the term "remote sensing" even was coined. Before 1945, most surveys of resources were done by visual observations. After 1945, photographic surveying by aircraft came into wide use. Since then remote sensing technology has progressed from aerial photography to thermal scanners to imaging radar to satellite sensing to geographic information systems. Today, photointerpretation of aerial photographs remains the principal source of information for forest management, protection, and assessment. Thermal scanning is used chiefly to detect and map large forest fires. Imaging radar is most useful where cloud or other objects obscure and prevent the use of aerial photography or satellite sensing. Satellite sensing and thematic mapper techniques have shown major improvements in quality. The inclusion of remote sensing into geographic information systems is expected to be the wave of the future.

INTRODUCTION

Even before someone originated the term "remote sensing," forest managers were already using the technique in a variety of land management applications. From the very beginning, they relied on the visual observations to survey natural resources and determine the appropriate management. Since then, we have progressed from that earliest form of remote sensing to aerial photography to thermal scanners to imaging radar to satellite sensing to geographic information systems (Aldrich 1979, Heller and Ulliman 1983).

Before 1945, most forest observations were done by visual inspection from mountain tops, fire towers, or aircraft. These observations were mainly for fire detection and location. In 1921, seaplanes were flown over western Canada to detect insect damage to forest stands. In 1927, pilots in army biplanes observed and photographed bark-bettle killed trees in California (Eaton 1942). In the late 1930's, Pitcairn autogiros were used to locate Dutch elm diseased-trees in New Jersey, and in 1939, to locate southern-pine beetle killed trees in Virginia. After 1945, foresters experienced a great surge in the use of black-and-white aerial photographs--partly because many military personnel had received photo intelligence training

---

*Formerly Project Leader--Remote Sensing Research, Pacific Southwest Forest and Range Experiment Station, Forest Service, U.S. Department of Agriculture, Berkeley, California; and Research Professor Emeritus, College of Forestry, Wildlife, and Range Sciences, University of Idaho, Moscow.

during World War II. Photographs became readily available in many parts of the country for use by Forest Survey foresters and others to locate and sample timber stands. Books on applying aerial photography to forestry stimulated foresters to use air photos stereoscopically for measuring stands (Avery 1962, Spurr 1948).

Although remote sensing has benefited heavily by new technology, that technology has--in my judgment--outpaced our ability to adapt to the revolutionary changes in the past 40 years. A wide gap exists in the transfer of remote sensing technology to the user--due in part to resistance to change by large organizations.

This paper summarizes the state-of-the-art developments in six broad remote sensing systems, and describes their application to forest management, inventory, and protection.

REMOTE SENSING SYSTEMS

Visual observations

The eyes and the mind of a trained forester make an excellent sensing device. Even today, in the mountainous West, yearly pest damage surveys are made by visual observation. Sketch mapping surveys from slow flying, high wing aircraft at 1,000 to 5,000 feet above terrain permit outbreaks to be plotted on 1/4- to 1/2-inch planimetric maps. Such methods have been used since 1947 and provide valuable historical data for entomologists and pathologists (Wear and Buckhorn 1950). Sketch mapping is subject to some plotting errors and oversight of small or lightly infested areas. While the technique requires observers with strong stomachs, it is probably one of the fastest and most economical methods to track ephemeral changes in forest conditions.

Aerial photography

Aerial photography continues to be a primary source of forest resource information for foresters. It is widely used throughout the world for forest management and protection purposes.

Forest management and inventory. Foresters use photographs for timber harvest planning, silvicultural projects, transportation planning, timber volume estimates, field orientation, and many other purposes.

The USDA Forest Service is using mainly color photographs from Kodak aerocolor negative film 2445* taken at scales of 1:12,000 to 1:24,000. Color infrared (CIR) photographs taken in the course of the National High Altitude Photograph program have not been widely used, possibly because of the small scale (1:58,000) and the seasonal timing. What is optimal timing for photogrammetric mapping is often poor for vegetation analysis.

---

*Trade names and commercial enterprises or products are mentioned solely for information. No endorsement by the U.S. Department of Agriculture is implied.

Some special resource photography is taken at scales of 1:8,000 to 1:12,000 with Kodak Aerochrome infrared film, type 2443, processed to a transparency (except for the Southern Region which processes this film to a negative and produces CIR prints for field use). Other special projects have used the Itek-IRIS II advanced optical bar camera (24-inch focal length on 5- by 38-inch format) or the HR 732 Large Format Reconnaissance Camera (24-inch focal length on 9- by 18-inch format). Kodak High Definition Aerochrome Infrared film, type SO-131 is commonly used from National Aeronautics and Space Administration (NASA) U-2 aircraft and produces scales from 1:30,000 to 1:120,000.

The Forest Service has started a new photo interpretation training program to improve the skills of workers using aerial photographs. Because photo interpretation is not required at all forestry schools, the skill level is below standard. Often photos are used only as a map. In Oregon and Washington, the USDI Bureau of Land Management uses natural color prints (scale 1:12,000) for their natural resource inventory and mapping.

Foresters in other countries also continue to use aerial photographs. For example, color prints made from CIR film processed to a negative (scale 1:15,840) shows promise for forest inventory in the Canadian Boreal Provinces (Zsilinsky and Keimer 1984). However, mostly black and white aerial photographs are taken at scales of 1:10,000 to 1:50,000.

In western Europe, foresters work with smaller forest properties than those found in the United States and Canada. Scales of photography vary from 1:10,000 to 1:50,000. Photography is used mostly for forest planning, mapping, and inventory. A permanent inventory and monitoring system for all European forests with aerial photography to furnish much of the data for this ambitious program was recently proposed (Hildebrandt 1983). Satellite data would contribute a first level of information to the inventory. A recently published manual on applications of aerial imagery in the temperate world summarizes state-of-the-art use from a European viewpoint (Stellingwerf, et al. 1983).

As of 1980, most of the forested areas of the People's Republic of China were covered by black-and-white photographs at the 1:50,000 scale. Forest boundaries are transferred and shown on their 1:50,000-scale planimetric maps in green, as is done on U.S Geological Survey quadrangle maps. Small-format (35- and 70-mm) photography has a long history of developmental work in the United States (Aldrich et al. 1959, Klein 1970). Rapid recycling, fast shutter cameras when used with color films at large scales produce extremely sharp, detailed photos useful in sampling surveys of tree damage and tree identification. One of the greatest proponents of 35-mm color aerial photography for forest applications has been Merle Meyer (Meyer and Grumstrup 1978, Meyers et al. 1981). He recently demonstrated the usefulness of his techniques for evaluating a large aspen forest in Wisconsin for aspen regeneration potential and stocking levels (Weih et al. 1985). But adoption of the technique to forestry management operations in the United States has been limited.

Use of 70-mm photography has been more widespread elsewhere. Canadian inventory specialists use 70-mm color transparencies with a radar altimeter to sample inaccessible areas. Australian foresters have found 70-mm color transparencies useful for evaluating prescribed burns of Eucalyptus forests and for identifying important tropical rainforest species (Myers and Benson 1981). Foresters in West Germany and Switzerland also use sample 70-mm color transparencies for forest inventory purposes.

Forest protection. Outbreaks of pest damages from insects and disease are devastating in effect and are difficult to budget and plan for because of their ephemeral nature. Because of the rapidness of response and cost efficiencies, remote sensing techniques are widely used to detect and evaluate the seriousness of forest damages.

The Forest Service's forest pest management program is a consistent user of standard and special photo products. In practice, once the boundaries of an insect or disease outbreak are determined, it conducts sample surveys in blocks or strips. Either normal color or color infrared transparency films are exposed at moderately large scales of 1:4,000 to 1:8,000. Surveys undertaken in 1983 include estimates of spruce decline in the New England States, mountain pine beetle surveys in the Rocky Mountains, incidence of prairie dog towns in Nebraska and South Dakota, and fir mortality and root rot disease in Oregon (Myhre 1984).

Another program of the Forest Service--its nationwide Forestry Applications Program--has undertaken many cooperative surveys with the National Aeronautics and Space Administration. These surveys used either the Itek-IRIS II panoramic camera or the HR 732 Large-Format Reconnaissance Camera (24-inch focal length on a 9- by 18-inch format). Usually Kodak High Definition Ektachrome Infrared film (SO-131) is exposed in the panoramic camera from one of NASA's U-2 or ER-2 aircraft, which produces scales of about 1:30,000 at the nadir. During a period of severe drought in 1977-1978 in California, the Sierra Nevada was completely photographed in CIR to detect and plot the ponderosa pine trees killed by the western pine beetle. The transparencies were interpreted on Richards MIM-4 light tables with Bausch and Lomb Zoom 240 heads. It is unlikely that any other method could capture the epidemic conditions as efficiently as this one. Similar panoramic photographs were taken in 1980 over 12 million acres (4.8 million ha) in the Front Range of Colorado. A three-stage sampling design was used to estimate both the numbers of trees killed and volume lost. It resulted in a sampling error of about 15 percent (Dillman et al. 1981b).

In 2 days, 70,400 square miles (119,200 sq.km) in Delaware, Maryland, New Jersey, and Pennsylvania were photographed by an ER-2 airplane using the panoramic camera and CIR film to cover areas of severe and moderate gypsy moth (Lymantria *dispar* L.) defoliation. Areas of defoliation were interpreted monoscopically on light tables and transferred to USGS 7-1/2-minute topographic maps. Except in northern higher elevation areas where defoliation was delayed, all cooperators were satisfied that this kind of photography was a viable alternative to aerial sketchmapping. Total cost per square mile was $1.58 (Ciesla et al. 1984).

Until recently, European countries have not been concerned with remote sensing applications for insect and disease problems. The more intensive forest management practices allowed protection specialists to be aware of problems and institute controls very rapidly on the smaller ownerships. However, in the last few years the concern over industrial pollution killing forests across large areas in Europe has grown, and remote sensing techniques appear to offer a solution to evaluate the seriousness of acid rain.

One-third of all forests in the Federal Republic of Germany have been affected by acid rain (Hildebrandt and Kadro 1984). In the southern German States, one half of the forests are affected to some degree. Old conifers are damaged more than young conifers or deciduous trees. These estimates were derived by interpretation of CIR aerial at scales of 1:5000 to 1:6000, photos mainly taken in 8-km sample strips by 24-inch (304-mm) focal length cameras. Six private contractors took 8400 photos during 1 week in July 1983. Interpretation of six to nine sample points (of a possible 25) were made on alternate photos. In all, 16,000 samples were taken for all of Germany. Specifications for four levels of damage were described for the interpreters, based on the damage level in the 20 conifers nearest to the sampling point; or in the case of young conifer stands or deciduous stands, based on a defined area of the stand. From the time of photography to evaluation and analysis of results, the survey took about 8 months and included a field inventory which sampled the interpreted photo points. Perhaps a similar survey in the United States and Canada could estimate the seriousness of acid rain in the northeast.

Orthophotomaps covering much of the United States are now available. West German foresters began using them in forest planning and management activities in the late 1960's. Inexpensive prints at a scale of 1:24,000 can be ordered from the U.S. Geological Survey and are valuable aids for delineating stand boundaries and making accurate area measurements. Consulting foresters might benefit most from the use of these orthophotomaps.

Thermal scanners

Thermal scanning data have been used by foresters mainly for fire detection and mapping of large fires. The Pacific Northwest Region of the Forest Service has called upon the NASA high-altitude, real-time, fire-detection and mapping system flown from the U-2/ER-2 platform from Ames Research Center, Moffett Field, California. In addition, the Boise Interagency Fire Center, in Idaho, which cooperates with both State and Federal agencies nationwide, coordinates the use of the scanning systems developed in the early 1960's.

The Center dispatches scanner aircraft and fire fighters to large project fires wherever needed.

The Daedalus Thermal Infrared Mapping System (TIMS) has been used by the Southern Region of the Forest Service to map subsurface gravel deposits for road construction on the Kisatchie National Forest in Louisiana. A drilling rig was used as a ground truth vehicle to confirm the anomalies on the scanner data. Flights were made at 10,000 feet (3.050 m) from a Learjet stationed at NASA's National Space Technology Laboratory, Slidell, Mississippi. The presence of gravel deposits covered by soil showed up as being warmer (lighter in tone on photo prints).

Imaging radar

Side-looking airborne radar (SLAR) imagery is most useful in tropical regions where acquisition of aerial and space imagery is difficult or impossible due to adverse weather conditions. Most often the presence or absence of forest is the only vegetation parameter consistently identifiable on SLAR images (Henderson 1979).

The eruption of Mount St. Helens on May 18, 1980, affected a large portion of the Gifford Pinchot National Forest in Washington. For nearly 1 month after the eruption, the area was covered by dust and clouds, preventing aerial photography. With the emergency assistance of the U.S. Air Force, high resolution radar (HRR) imagery was made available 3 days after the eruption. Interpretations made immediately thereafter defined crater characteristics, lake levels, new lakes and the extent of timber damage. Image tone and texture of the damaged trees were the most important criteria in making the interpretations. The amount and quality of any information derived from visual interpretation of radar imagery depend on the knowledge and experience of the analyst in interpreting image tone, texture, and shadow (Dillman et al. 1981a). Under similar disaster conditions, SLAR imagery may be an advantageous remote sensor to use.

Satellite sensing

Some 13 years have lapsed since the first earth satellites were flown. And much has been written about the effect of information recorded by the LANDSAT satellite for forest inventory and management. And yet, 13 years later, little operational use has been made of the data by government agencies, State forestry departments, or private industry. A few innovative programs are under way that may influence future use in forestry, especially those programs integrating geographic information systems with remotely sensed data.

In the early seventies, every investigator and university had a color combiner using 70-mm film chips. But today I know of no one analyzing imagery on color combiners. Of the film products, we all found that the best resolution was available on the 1:1,000,000 false color composite (FCC) transparencies and that the 1:250,000 FCC prints were helpful for orientation on USGS maps. The latter were also useful to establish visible control points for deskewing digital data to match map points and existing data bases.

Thematic Mapper (TM) images have almost the same resolution qualities as high altitude aerial photography and should serve as an excellent planning map because of the large area it covers (13,225 sq. mi.).

While working with Chinese foresters in 1980, I found that the FCC prints (1:250,000 scale) were useful in identifying tree plantations and conifer stands in northeastern China. In countries, with little or no computer expertise, the FCC's probably will continue to be used as the primary satellite product.

Computer-aided processing of digital data is well summarized in the Manual of Remote Sensing, which contains most of the up-to-date techniques and forestry applications (Hoffer 1983). The data products available now from the EROS Data Center at Sioux Falls, South Dakota are much improved over the products received in the early 1970's. Considerable preprocessing enhancements and geometric corrections are done routinely.

For large, mini, and microcomputers, a variety of computer software systems such as LARSYS, ORSER, VICAR-IBIS, EDITOR, ERIPS, IDIMS, and APPLEPIPS are available to users to produce forest maps and tabular data for their properties. All systems use much the same spectral pattern recognition techniques for classifying the digital data, such as minimum distance to means, parallelepiped, Bayes decision rule, maximum likelihood classifiers. Many enhancement computer programs are also available for preprocessing the digital data, such as ratioing, principal component analysis, atmospheric corrections, contrast stretches, and a variety of vegetation indexes. Finally, the forester-analyst must decide whether to use the classifiers in a supervised (training stage) mode or unsupervised mode (natural spectral clustering) or a combination of both. None of these are trivial operations, and in the past many classified data output products did not satisfy the needs of forest managment.

Unless forest managers have trained personnel and appropriate in-house computer equipment, they may have to rely on contractors for the desired output products. Some examples of use of contractors are these:

1. The Okanogan National Forest in Washington has contracted for Landsat imagery analysis to map present and future productivity on 36 7-1/2-minute quadrangles.

2. The Flathead National Forest, Montana Department of Forestry, and the EROS Data Center are using Landsat data to support their forest land management planning effort.

3. The San Juan National Forest in Colorado is using Landsat data to help prepare its lead management plan.

4. The Sierra National Forest in California contracted with the Geography Remote Sensing Unit of the University of California, Santa Barbara, to classify and stratify its

forest land for harvest planning and inventory. Forest types were consistently mapped and area estimates were useful for timber management planning. Elevation, aspect and slope were independent variables which were integrated with Landsat spectral values and were found to improve classification accuracies of forest type (Strahler et al. 1983).

5. Low resolution reflectance data (1.1 km at nadir) from the Advanced Very High Resolution Radiometer (AVHRR) were used to compute a vegetation index for Bureau of Land Management rangelands in northern Arizona (Miller and Moore 1983). The amount of greenness and date of senescence of annual grasses provided input to the Bureau's national wildfire management program.

Geographic Information Systems

The most successful applications of remote sensing data are those which incorporate remote sensing into a geographically referenced land base. The opportunities for updating, monitoring, and interacting with existing ground data provide the forest manager with a flexible and responsive system for making intelligent decisions.

All of the satellite sensing applications I have described used such a base map on which the classified forest data were overlain. Several computer programs are available that can "rubber sheet" the skewed raster scan pixels (picture elements) to latitude-longitude, Universal Transverse Mercator, or most any map projection. Again, this is not a trivial operation, but is one that enhances the usefulness of the classified forest data to a known location on the ground.

An outstanding example of an operational geographic information system that is interactive with forest managers is the Forest Resource Information System (FRIS) developed by Barker (1983). The system is being used by Champion International in Florida. The company has in-place remote work stations that provide tabular, cartographic, and digital imagery support for management. The Thematic Mapper is the main data source, and users see a great improvement in vegetative analysis with TM data over the four-channel Multispectral Scanner data. Barbe* believes that rather than developing new and more advanced sensors, more effort should be spent on Landsat's continuity in format, timeliness of delivery, and availability.

Another geographic-referenced computer system developed for foresters is LANDPAK (Langley 1983). This system incorporates forest type polygons derived from photo interpretation of air photos with any map projection. It also allows the forest manager to query the system for determining erosion-prone areas, timber harvesting opportunities, and identification of high-volume timber stands. A final output product is a multicolor coded map

*G. Robinson Barker, FRIS Manager, Champion International, Jacksonville, Florida. (Personal communication, January 4, 1985).

drawn by a computer-driven plotter to any desired scale. The system has been used successfully on several forests in northern California.

CONCLUSIONS

From this brief account of state-of-the-art developments in forest remote sensing systems, these conclusions can be drawn:

1. Visual surveys from aircraft will continue as a cost-effective technique for forest damage assessments.

2. Aerial photography and photo interpretation will be a primary data source for forest inventory, management, and damage assessments. Aerial photographs also serve as a needed "ground truth" reference for computer-aided analysis--both in the training mode and for checking accuracies of the classified polygons. More photo interpretation training of foresters is needed in colleges and workshops. The Remote Sensing Working Group of the Society of American Foresters is recommending that photo interpretation be a required course at forestry schools.

3. The primary use of thermal scanners is for detection and mapping of large forest fires. The Boise Interagency Fire Center acts as dispatcher of firefighters materials and thermal scanning aircraft on a nationwide basis.

4. Imaging radar is most useful where cloud or other obscuration prevents the use of aerial photography or satellite sensing.

5. Earth satellite sensing by Landsat MSS and Thematic Mapper data have shown great improvements in quality. The slow distribution of TM data and increased expense of all products make these data less accessible for experimenters. The possible commercialization of the space program is producing much uncertainty as to the long range availability of data. Low resolution data from the AVHRR appears to be useful for broad scale mapping applications.

   Computer-aided analysis programs for forest applications are available, but the information needs must be balanced with cost and expertise within an organization. The adoption of satellite information into forest planning and management operations has been slow, but appears to be gaining support in government and industry.

6. The inclusion of remote sensing data into a geographic information system is the wave of the future, in my opinion. With all data, both ground and remotely sensed, on the same geographic-referenced base, the forest manager can gain access to layers of data for up-to-date management decisions.

ACKNOWLEDGMENTS

I thank the following people who generously responded to my requests and provided information that greatly helped me in preparing this paper: Ray Allison, William H. Clerke, Jack Levitan, and F. P. Weber, all of the Forest Service, U.S. Department of Agriculture; W. W. Glenn, Bureau of Land Management, U.S. Department of Interior; and G. Robinson Barker, Champion International.

REFERENCES

Aldrich, R. C. 1979. Remote sensing of wildland resources: a state-of-the-art review. Gen. Tech. Rept. RM-71. USDA For. Serv. Rocky Mt. For. and Range Exp. Stn., Fort Collins, Colo. 56 pp.

Aldrich, R. C., R. C. Heller, and W. F. Bailey. 1959. Large-scale 70-mm color photography techniques and equipment and their application to a forest sampling problem. Photogramm. Eng., Vol. 25 (5), pp. 747-754.

Avery, T. E. 1962. Interpretation of aerial photographs. Burgess Publishing Co., Minneapolis, Minn.

Barker, G. R. 1983. Forest resource information system--FRIS. Final project report. St. Regis Paper Co., Jacksonville, Florida.

Ciesla, W. M., R. E. Acciavatti, J. G. D. Ward, R. A. Allison, and F. P. Weber. 1984. Demonstration of panoramic aerial photograph for mapping hardwood defoliation over a multistate area of the northeastern United States. Rept. 84-3. USDA For. Serv. For. Pest Mgt. Staff, Methods Application Group, Ft. Collins, Colo. 21 pp.

Dillman, R. D., R. E. Hinkle, and R. L. Kerr. 1981a. High altitude radar assessment of the damage caused by the volcanic eruption of Mount St. Helens. Rept. 1623A. Lockheed Eng. and Mgt. Serv. Co., Inc., Houston, Tex. 16 pp.

Dillman, R. D., S. S. Shen, B. B. Eav, and W. B. White. 1981b. Operational test of panoramic aerial photography for estimating annual mortality of ponderosa pine caused by the mountain pine beetle. Rept. 16377. Lockheed Eng. and Mgt. Serv. Co., Inc., Houston, Tex. 20 pp.

Eaton, C. B. 1942. The adaptation of aerial methods to the forest loss survey. USDA For. Serv., Pacific Southwest For. and Range Exp. Stn., Berkeley, Calif.

Heller, R. C., and J. J. Ulliman. 1983. Forest resource assessments. In: Manual of Remote Sensing. Amer. Soc. Photogramm., Falls Church, Va, pp. 2229-2324.

Hoffer, R. M. 1983. Computer-aided analysis techniques. In: Manual of Remote Sensing. Amer. Soc. Photogramm., Falls Church, Va, pp. 2289-2310.

Henderson, F. M. 1979. Land-use analysis of radar imagery. Photogramm. Engr. and Remote Sensing, Vol. 45, no. 3, pp. 295-307.

Hildebrandt, G. 1983. Considerations on a permanent inventory and monitoring system for European forests. In: Proceedings: EARSOL/EAS Symposium on Remote Sensing Applications for Environmental Studies, April 1983, Brussels, Belgium, pp. 13-18.

Hildebrandt, G., and A. Kadro. 1984. Aspects of countrywide inventory and monitoring of actual forest damages in Germany. In: Proceedings, XVth Congress of Int. Soc. Photogrammetry and Remote Sensing, Commission VII, June 1984, Rio de Janeiro, Brazil, pp. 246-266.

Klein, W. H. 1970. Mini-aerial photography. J. Forestry, Vol. 68, no. 8, pp. 475-478,.

Langley, P. G. 1983. Integrating a geographic information system into the inventory program. In: Proceedings, Renewable Resource Inventories for Monitoring Changes and Trends. College of Forestry, Oregon State University, Corvallis. pp. 279-282.

Meyer, M. P., N. Miller, and R. Hackett. 1981. Application of 35-mm aerial photography to forest land use change in the Upper Peninsula of Michigan. IAFHE RSLO Res. Rept. 81-2. College of Forestry, Univ. Minnesota, St. Paul. 62 pp.

Miller, W.A., and D. G. Moore. 1983. Time-series vegetation monitoring with NOAA satellite data. In: Proceedings, Renewable Resource Inventories for Monitoring Changes and Trends. College of Forestry, Oregon State University, Corvallis. pp. 86-89.

Myers, B. J., and M. L. Benson. 1981. Rainforest species on large-scale color photos. Photogramm. Eng. and Remote Sensing, Vol. 47, no. 4, pp. 505-513.

Myhre, R. J. 1984. The Forest pest management nationwide aerial photography program: first-year accomplishments. Rept. 84-6. USDA For. Serv. For. Pest Management Staff, Methods Applications Group, Fort Collins, Colo. 11 pp.

Spurr, S. H. 1948. Aerial photographs in forestry. New York, Ronald Press.

Stellingwerf, D.A., G.S. Smith, and J. M. Reeijn. 1983. Applications of aerial photographs and other remote sensing imagery in forestry. ITC Publ. 3. Int. Institute for Aerial Survey and Earth Science, Ensched, The Netherlands. 50 pp.

Strahler, A. K., T. L. Logan, J. Franklin, H. Bowlin, and J. Levitan. 1983. Automated forest classification and inventory in the Eldorado National Forest. USDA For. Serv. Pacific Southwest Region, Timber Mgt. Staff, San Francisco, Calif.

Weih, R. C., G. D. Nord, and M. P. Meyer. 1985. Application of 35-mm aerial photography to evaluate post-harvest aspen regeneration. J. Forestry (in press).

Wear, J. F., and W. J. Buckhorn. 1950. Organization and conduct of forest insect aerial surveys. USDA For. Serv. Pacific Northwest For. and Range Exp. Stn., Portland, Oreg. 40 pp.

Zsilinsky, V., I. Ross, and D. Klimes. 1984. The capabilities of colour infrared film as a negative. In: Proceedings, XVth Congress of Int. Soc. Photogrammetry and Remote Sensing, Commission VII, June 1984, Rio de Janeiro, Brazil. 19 pp.

NEW HORIZONS IN REMOTE SENSING FOR FOREST AND RANGE RESOURCE MANAGEMENT*

D. T. Lauer
U.S. Geological Survey
EROS Data Center
Sioux Falls, South Dakota 57198

ABSTRACT

Forest and range resource scientists were among the first to recognize the potential of aircraft and satellite remote sensing for management of timber, forage, water, and wildlife resources. Today, data from a variety of sensor systems are being put to practical use for inventorying, monitoring, and assessing forest and range resources. In the future, improved sensor systems providing new kinds of data will be available. Likewise, new types of data handling and processing systems can be anticipated.

Among the new or anticipated aircraft and satellite systems and/or data are National High-Altitude Photography II, U.S. Geological Survey-acquired Side-Looking Airborne Radar, the Landsat thematic mapper, the National Oceanic and Atmospheric Administration Advanced Very High Resolution Radiometer, the French Systeme Probatoire d'Observation de la Terre (SPOT) satellite, the European Space Agency Earth Resources Satellite, the National Aeronautics and Space Administration Large Format Camera and Shuttle Imaging Radar (SIR-A, -B, and -C), and a variety of other systems in existence or planned by the Soviets, Japanese, Canadians, Chinese, Brazilians, Indonesians, and others.

Application examples are presented that illustrate uses of 1-kilometer-resolution AVHRR data, 80-meter Landsat multispectral scanner data, 30-meter Landsat thematic mapper data, and 10-meter SPOT-simulator data. These examples address fire fuel monitoring, land cover mapping, rangeland assessment, and soils landscape mapping.

INTRODUCTION

Dr. William T. Pecora was a Government leader with an uncanny ability to look to the future. Although his exposure to satellite remote sensing was limited to data acquired from the early meteorologic satellites and from the hand-held cameras used by the Gemini and Apollo astronauts, he had a deep appreciation for the potential uses of satellite remote sensing systems for inventorying, monitoring, and assessing natural resources.

---

* Publication authorized by the Director, U.S. Geological Survey, on August 8, 1985.

Since Dr. Pecora's death in 1972, shortly before the launch of the first Landsat system, extraordinary advances have been made in orbiting platforms, sensors, and data processing capabilities. Among these advancements were (1) the Skylab experiments in 1973 and 1974, in which tens of thousands of frames of imagery and non-image data were acquired by a multispectral photographic camera, an Earth terrain camera, an infrared spectrometer, a multispectral scanner, a microwave radiometer/scatterometer, an altimeter, and an L-band radiometer; (2) Landsats 1-5 in which well over a million scenes of data have been acquired, processed, archived, and made available for public use. The multispectral scanner (MSS) has been the principal sensor aboard the Landsat series, but data also have been acquired by two different configurations of a return beam vidicon (RBV) system on Landsats 1-3 and most recently by the thematic mapper (TM) system on Landsats 4 and 5; (3) the National Oceanic and Atmospheric Administration (NOAA) satellite series (formally TIROS) from which excellent-quality Advanced Very High Resolution Radiometer (AVHRR) data have been made available in the visible, near-infrared (IR) and thermal-IR spectral regions; (4) the Heat Capacity Mapping Mission, launched in 1978, from which scanning/imaging radiometer data were collected in both the visible/near-IR and thermal-IR regions; (5) the Seasat satellite system, also launched in 1978, which acquired synthetic-aperture radar data transmitted and received with horizontal polarization at an L-band frequency; (6) Shuttle Imaging Radar acquired by the Space Shuttle in 1981 and 1984, which provided synthetic-aperture radar data using an L-band frequency and high off-nadir viewing angles; and (7) a variety of photographic and electronic systems placed into Earth orbit by the Union of Soviet Socialist Republics, the European Space Agency, the Federal Republic of Germany, India, and the People's Republic of China.

This paper will look at new horizons in remote sensing for forest and range management. First, an examination will be made of new and proposed remote sensing systems and the types of data to be provided from these systems. Second, a few brief application examples are presented that illustrate how forest and range managers and scientists currently use remotely sensed data to help carry out their land management and resource assessment responsibilities. These examples address fire fuels monitoring, land cover mapping, rangeland assessment, and soils landscape mapping. Lastly, the implications of new and proposed remote sensor systems and their applications are discussed, particularly in the context of computerized data bases and information systems.

NEW AND PROPOSED REMOTE SENSING SYSTEMS

Aircraft Systems

In addition to the continued acquisition and use of conventional aerial photographs (black-and-white panchromatic black-and-white IR, true color, and false-color IR), data from a variety of new remote sensing systems have recently become available to support forest and range applications. Several other systems have been proposed and are anticipated in the near future. These new and proposed systems, most of which were reported in greater detail by Doyle (1983) and Watkins et al. (1984), are discussed below.

National High-Altitude Photography (NHAP) II -- Phase I of the NHAP program has been completed over a 6-year period under the sponsorship of a consortium of Government agencies. Both black-and-white panchromatic and false-color IR aerial photographs were acquired in a systematic manner over the entire lower 48 States. These data were acquired by jet aircraft flying at an altitude of approximately 40,000 feet above mean terrain. Data were acquired along north-south flight path lines, centered over 7½-minute U.S. Geological Survey quadrangle maps under non-growing season conditions (i.e., leaf-off). The black-and-white photographs were acquired at an approximate scale of 1:80,000 and the false-color IR at an approximate scale of 1:58,000. Since Phase I of NHAP has proved to be successful, Phase II has been initiated which will provide repeat coverage of the entire country under growing season (leaf-on) conditions. These data will be particularly useful for small-scale vegetation mapping and type stratification of large areas.

Side-Looking Airborne Radar -- In 1980, the U.S. Geological Survey initated a program to acquire SLAR data over large sections of the conterminous United States and Alaska (Southworth, 1984). The data are acquired using X-band frequencies, horizontal transmit and receive polarization, depression angles of 10° to 33°, and approximately 10-12 m spatial resolution. SLAR data have been acquired over approximately 1,000,000 km^2, principally over the Appalachian region and the Aleutian Islands. SLAR data have been used for resource assessment, mainly as a reconnaisance tool, in areas that are perennially covered by clouds.

Video Camera -- Interest in acquiring images over site-specific areas at very low cost has encouraged the development and use of video camera systems mounted in single-engine, fixed-wing aircraft. Image data recorded on magnetic tape can be acquired in black-and-white visible, black-and-white near-IR, true color, and false-color IR formats. These data are easily displayed using low-cost tape recorders and television monitors and lend themselves to visual interpretation of site-specific areas (e.g., wildlife habitat assessment).

Other Systems -- Optical mechanical scanners are being used in both low-and high-flying aircraft. These sensor systems provide high-resolution multispectral data throughout the visible, reflective-IR, and thermal-IR regions. Investigations are in progress on the development and use of multilinear array (MLA) sensor systems, which use thousands of charge-coupled detectors in one sensor device and can provide both high resolution and multispectral data. New capabilities for discriminating resource features are anticipated from MLA technology.

Spacecraft Systems

United States -- The most significant recent advancement made in United States satellite remote sensing systems is the TM, first launched on Landsat 4 in July 1982, and, more recently, on Landsat 5 in March 1984. The TM sensor provides data in seven spectral regions with approximately 30-m spatial resolution. Successful applications of TM data in numerous vegetation-related resource investigations have been reported.

The Landsat series will continue when the EOSAT Company, a joint venture of RCA and Hughes Aerospace, builds and launches Landsats 6 and 7. Landsat 6 will carry a TM sensor and will provide on-board emulation of MSS data. The multispectral-type data provided by the TM sensor will have approximately 30-m ground resolution and a capability will be included for acquiring panchromatic data at 15-m resolution.

The Large Format Camera (LFC) had its successful debut in October 1984 on the Space Shuttle. The camera has a 30.5-cm focal length and 23x46-cm frame format. The use of high-resolution, fine-grain film and forward motion compensation allowed for the acquisition of very high quality photographs of the Earth's surface. Black-and-white, color, true-color, and false-color IR film were spliced together on a single roll of film from which 2,400 frames of imagery were acquired at a nominal Space Shuttle orbital altitude of 300 km. The ground resolution of an LFC photograph is approximately 15 m. The next flight of the LFC on Space Shuttle could be in 1986 (equatorial orbit), and one is scheduled for late 1988 (polar orbit).

An MLA sensor is being designed by NASA to acquire image data in six spectral bands within the visible and near IR regions. It is hoped that this sensor can be tested on the Space Shuttle in 1988, and if tests are successful, the MLA could be placed on the multimission spacecraft in a free-flying, near-polar, sun-synchronous orbit in the 1990's.

France -- The Systeme Probatoire d'Observation de la Terre (SPOT) is a general-purpose remote sensing satellite developed by the French Centre National d'Etudes Spatiales (CNES) in cooperation with Sweden and Belgium. Current plans are for SPOT-1 to be launched in November 1985, but that launch date could slip possibly as much as 1 year. The SPOT imaging sensors consist of two high-resolution visible (HRV) linear arrays aligned to record data to the right and left of satellite ground track. Selective positioning of the sensor's pointable mirror will allow imaging of the same area at different angles on successive passes, thus providing stereoscopic images of the terrain. Multispectral data in the visible and near-IR regions will be provided as well as broad-band panchromatic data. Spatial resolution at image nadir will be approximately 20 m for the multispectral data and 10 m for the panchromatic data. The French Government has committed to launch SPOT-2 later this decade and is making plans for future SPOT systems in the 1990's.

Europe -- The European Space Agency (ESA) will launch the ESA Resource Satellite-1 (ERS-1) in 1988 or 1989. The system will be designed principally for ocean observation and will carry only one imaging device, a synthetic aperture radar (SAR) C-band system. Images produced by this system will be acquired at a maximum off-nadir angle of 23° and with approximately 30-m spatial resolution. Other ESA satellite systems are being planned for the 1990's and will include a SAR system and an optical imaging instrument with visible and near-IR, and short wavelength IR spectral bands with approximately 30-m spatial resolution and one panchromatic band with approximately 15-m spatial resolution.

Japan -- The Japanese Government plans to launch two Earth observation satellites by 1990. The first of these, the Marine Observation Satellite-1 (MOS-1), is scheduled for launch in late 1986. Three sensor systems are planned for MOS-1 and include a multispectral electronic self-scanning radiometer which will collect data in four spectral bands in the visible and near-IR regions with approximately 50-m spatial resolution. A mechanical scanner device called the visible thermal IR radiometer is included and will collect data in one visible band, with approximately 900-m spatial resolution and three thermal IR bands, with approximately 2.7-km spatial resolution. Also included in the MOS-1 system will be a microwave scanning radiometer with 20 to 30-km spatial resolution. A second Japanese remote sensing satellite system is the Earth Resources Satellite System-1 (ERS-1). It is still in the design phase and is expected to be launched in the early 1990's. A visible and near-IR radiometer will collect data in four visible and near-IR bands with approximately 25-m spatial resolution. An L-band SAR also will be included in this system and will provide data with 25-m spatial resolution.

Canada -- The Canadian Center for Remote Sensing has been given approval to proceed with development of a SAR to be launched in the early 1990's. The SAR sensor will operate in the C-band with a steerable look angle ranging between 20° and 45° with approximately 25-m spatial resolution. The Canadian Government also is considering including other sensors such as a scatterometer or a visible/IR radiometer.

Other Systems -- The German company Messerchmitt-Bolkow-Blolm (MBB), in cooperation with the West German Government and NASA, built a Modular Optoelectric Multispectral Scanner (MOMS). The MOMS is an MLA sensor that provides data in two bands (visible and near-IR) with approximately 20-m spatial resolution (at 300-km altitude). The MOMS was carried on the Space Shuttle and deployed on the free-flying Shuttle Pallet Satellite in June 1983 and again in November 1984. The Indian Space Research Organization plans to launch a series of advanced Earth-resource satellites beginning in the late 1980's. It expects to use MLA sensor devices capable of collecting data in at least four wavelength bands in the visible and near-IR regions with approximately 40-m spatial resolution. Brazil has announced its intentions to develop and launch a national remote sensing satellite early in the 1990's that would include a multispectral sensor with 50-m ground resolution capability. The Netherlands Agency for Aerospace Program, in cooperation with the Indonesian Government, is considering the launch of a tropical Earth resources satellite. The U.S. Geological Survey has completed a feasiblity study for an automated mapping satellite (MAPSAT), primarily to provide stereoscopic image data adequate for compilation of 1:50,000-scale maps with 20-m contour intervals (Colvocoresses, 1984). A variety of imaging devices is being planned for experimental use on board the Space Shuttle, including the Shuttle Imaging Spectrometer Experiment (SISEX). Image data would be collected by the SISEX simultaneously in 128 bands from 0.4- to 2.5-μm wavelength with an approximate ground resolution of 30 m.

APPLICATIONS OF SATELLITE DATA

For more than a decade satellite remotely sensed data from the Landsat and meteorological satellite systems have been available in both image and digital formats. These data have been put to operational use by resource scientists and land managers and planners for purposes of aiding in the inventory, mapping, and monitoring of the Earth's renewable resources. The compatibility of data from these diverse systems has been demonstrated, and the potential future applications of data having higher spatial resolution, increased spectral sensitivity, and/or stereoscopic characteristics from new and future satellite systems are anticipated to be substantial. The examples presented here illustrate current applications, and they provide an indication of what might be realized using data from new and advanced land resource satellite systems.

Data acquired from polar-orbiting meteorological satellites have been used for monitoring growth and accumulation of herbaceous vegetation in the southwestern region of the U.S. (Miller and Moore, 1983). This information is required by fire managers to estimate fuel loadings and for monitoring vegetation fuel conditions during the growth cycle of annual grasses. The visible and near-IR bands collected by the AVHRR on board the NOAA-6 and -7 satellites were processed to compute a normalized-difference (ND) numerical index, which is related to the greenness of vegetation. The AVHRR data used in this study were acquired during March and April of 1982, and provided a sequence of data that covered the spring growth cycle of the annual grasses from dormancy through senescence. These data were received within 1 week of the overpass and registered to a georeferenced data base. An ND index was computed for images acquired on five dates. When the data were analyzed to determine their usefulness in supporting wildfire management decisions, it was found that the data could be used to monitor the seasonal growth patterns of green herbaceous vegetation and to establish the date that senescence of the annual grasses began. Fire management personnel concluded that the data are valuable for wildfire Initial Attack Management System input.

Landsat MSS data have been operationally used in all parts of the world for regional mapping of surface features, in particular land cover. For more than 5 years, U.S. Geological Survey scientists have been working with other Federal and State resource management agencies in Alaska using Landsat and digital elevation data to map land cover (Shasby et al., 1985). Prior to this time, very little land cover information has been available for the vast regions of Alaska. The enactment of the Alaska National Interest Land Conservation Act, in December 1980, imposed mandates on all major land management agencies in Alaska to prepare comprehensive resource and management plans which assess wildlife habitat, oil and gas exploration and development, wild and scenic rivers, land disposal, timber production, and archaeological and cultural resources. Land cover information is a critical ingredient for making these plans. The methodology for making land cover classifications in Alaska employs digital Landsat spectral data and digital elevation data in a spatial data base approach. Since the spectral data used for land cover classification are incorporated into a digital data base, other types of information in the data base can be used to improve classification results. For example, in some cases Landsat winter scenes are registered to the data base and are used to separate land cover classes that might otherwise have similar spectral signatures on a spring or summer scene. Digital elevation data are used in a similar way to create new strata masks based on combinations of elevation, slope, and aspect. Thus, the accuracy of the Landsat classification results are improved by using other types of digital data in the classification process. The vegetation of Alaska and other terrain categories have been mapped at one pixel resolution (approximately 80 meters) over one-half the entire State of Alaska. The remainder of the State is expected to be completed in the next 5 to 8 years.

When Landsat classification results are incorporated into a digital data base, as noted in the work being done in Alaska, other types of information in the data base can be used to improve Landsat classification results. For example, digitized physiographic, soils, or elevation data can be used to make new strata masks, and cluster statistics can be derived within these masked areas. Miller and Shasby (1982) showed that, for vegetation classification in Arizona and fuel-type classification in Montana, the accuracy of results were improved from 54 to 73 percent and from 52 to 72 percent, respectively, when digital elevation data were used in the post-classification process.

Studies recently were conducted to evaluate the capability of Landsat TM data for providing land resource information within and adjacent to an urban environment using data taken over the Washington, D.C., area (Sadowski, et al., 1984). The results were derived from TM and coincident MSS data that were acquired by Landsat 4 on November 2,

1982. Digital analyses of the TM data as compared to the MSS data showed the potential for TM data to provide improved land resource information due to the addition of new spectral data channels. The short-wavelength IR bands (5 and/or 7) were especially useful because of (1) the large range of variability for digital values in five separate land resource classes (forest, herbaceous vegetation, bare soils in previous surfaces, and water), and (2) low correlations with other spectral bands for vegetated land cover and water. The short-wavelength IR bands may also play a role in increasing the dimensionality of data space that has been transformed into its principal components. Also, low correlations between the blue band (band 1) and other spectral bands suggest its utility for studying variations within areas of water as well as vegetated land areas.

The mean data values of five land resource classes occurring in the Washington, D.C., area were shown to be more distinctly separated in TM spectral bands 2, 3, and 4 than for the equivalent MSS bands 1, 2, and 4. In addition, the relative variances associated with the classes were more often lower for the TM bands than for the corresponding MSS bands. These observations indicate better multispectral discrimination of land resource classes by TM bands 2, 3, and 4. However, an artifact in the MSS data may have contributed substantially to the greater relative variances observed in the MSS spectral bands.

The large variability noted for the digital values in the short-wavelength IR bands (5 and 7) likely contributed to the results of a study of black-and-white images of the six reflective TM bands, in which interpreters indicated a preference for images of bands 5 and/or 7 over other bands for making distinctions in 8 out of 11 of the interpretation categories that were addressed. Results of interpreting six color composite images of the Washington, D.C., scene showed highest interpreter preferences for a standard false-color IR composite (bands 2, 3, and 4), or a color composite consisting of bands 3, 5, and 4 in 8 out of 11 categories. A natural-color composite image (bands 1, 2, and 3) showed high interpreter preferences for three interpretation categories involving distinctions within predominantly nonvegetated areas.

Landsat TM and MSS data acquired on September 24, 1982, were evaluated for purposes of making natural resource assessments within a native prairie area of South Dakota (Haas and Waltz, 1984). In this study, substantially more information was derived from TM than from MSS data. Greater resolution of TM data aided in locating roads, small stock ponds, and many other land features that can be used as landmarks. The improved spatial resolution of TM data also permitted more efficient visual interpretations of land use, better identification of resource types, and improved assessment of ecological status of natural vegetation. TM data also provided a new source of spectral information that was useful for natural resource assessment. Digital data analyses indicated that the effective dynamic range of TM band 5 data and the unique interaction of reflectance in TM bands 4 and 5 were important in digital classification of vegetation and land cover. New short-wavelength-IR spectral bands, TM bands 5 and 7, aided in distinguishing water resources, wetland vegetation resources, and other important terrain features. The added information was useful for both manual interpretation and digital data classification of vegetation resources and land features.

A cooperative investigation recently completed by the U.S. Geological Survey, the Soil Conservation Service, and the Bureau of Land Management has shown that digital elevation and Landsat data provide additional information to the soil surveyor that could reduce the field time necessary to conduct a soil survey (Horvath et al., 1984). Slope-interval maps were produced from digital elevation data and topographic maps for three 7.5-minute quadrangle areas in north-central Wyoming. These slope-interval maps were then overlain on orthophoto maps and used to produce photo-interpreted physiographic maps. The resulting physiographic maps were digitized into a data base and used with Landsat MSS data to produce tabular summaries that describe each map polygon in terms of physiographic unit, slope, aspect, elevation, area, and spectral values (brightness and greenness transformations). A good relationship was found between the physiographic units and soil mapping units defined during a conventional soil survey, and between the Landsat spectral categories and existing vegetation communities. Field evaluations confirmed the feasibility of using this approach for producing physiographic maps as aids for mapping soil and range boundaries.

SUMMARY AND CONCLUSION

Data collected with the AVHRR and the MSS have provided resource scientists and land managers and planners with significant opportunities for resource assessment and management analysis at the State, district, and planning unit levels. The increased spectral and spatial resolution of data collected by the TM and by France's proposed SPOT satellite offer a potential for more detailed mapping and assessment of land resources at the management unit level. Examples of the utility of the AVHRR data for regional applications include (1) monitoring vegetation phenology and growth, (2) mapping National Fire Danger Rating System fuel types, (3) monitoring fuel condition and loading, (4) assessing drought occurrence, (5) mapping recently burned areas, and (6) performing time series flood assessments. The MSS data have been used, in a complementary fashion, with the AVHRR data, for (1) developing generalized land cover maps, (2) mapping range capability units in support of a range inventory, (3) improving the efficiencies of conducting a soil survey, (4) monitoring Federal mineral trespass, (5) assessing vegetation changes, (6) providing a base for allocating field samples, and (7) planning various management activities including locating firewood collection areas, re-establishing wildlife habitat, and improving range conditions.

The increased spectral sensitivity and spatial resolution of data collected by the TM and the proposed SPOT system offer the potential for even more detailed mapping and assessment of land resources. Potential applications of TM data include: (1) mapping land cover at the plant community level; (2) assessing ecological status of rangeland vegetation; (3) locating and assessing stock water sources; (4) evaluating grazing distribution and use; and (5) determining desirability of wildlife habitat. Likewise, potential applications of the proposed SPOT system include: (1) providing stereoscopic images for soil mapping; (2) providing increased spatial detail for vegetation mapping; (3) assessing wildlife habitat; (4) locating and assessing stock water sources; and (5) determining area accessability. Additional research is needed to show how to use combinations of these various types of data by employing appropriate statistical linkages from one data type to another in a manner that insures optimum efficiency. Resource managers will begin to realize the full potential of these new data sources as both the data and research results become available and especially as these data are used in conjunction with other forms of digital data.

A rapid evolution in applications of aircraft and satellite remotely sensed data has occurred during the last decade and has significant implications to the disciplines of forestry and range management. Today, data from a variety of sensors are often amalgamated into comprehensive digital spatial data bases in which other types of map, map-like, and tabular data, either available in digital form or easily transferred to digital form, are also stored. (Several examples illustrating how remotely sensed data can be used as inputs to a spatial data base were presented in the previous section of this paper.) Advanced, computer-assisted spatial data processing and analysis techniques are then applied to these data bases for deriving complex, detailed resource information. A variety of spatial information processing techniques, including overlay analysis, neighborhood characterization, and distance measurements, can be applied to a digital, spatial data base. It is through this technique of combining various types of data in a spatial data base that remote sensing is having and will continue to have its greatest use.

In conclusion, it is important to note that improved data handling and processing systems must evolve to complement future sensor system development. As data from new systems provide increased spatial resolution and additional spectral bands, improvements must be made in the speed and efficiency of computer processing systems--as being demonstrated by the Massively Parallel Processor currently under development at NASA's Goddard Space Flight Center. However, forest and range resource managers are most often located at remote field offices and must work with a wide variety of spatially referenced data. They will seek stand-alone data processing and analysis systems that can handle both image (raster) and cartographic (vector) data. These systems will have to use off-the-shelf hardware components and transportable and expandable software that are relatively low cost.

REFERENCES

Colvocoresses, A. P., 1984, Report of the Committee for Acquisition and Processing of Space Data for Mapping Purposes, Working Group 4/3: International Society for Photogrammetry and Remote Sensing: Reston, Virginia, U.S. Geological Survey Open-File Report 84-314, 34 p.

Doyle, F. J., 1983, Status of satellite remote sensing programs, in United Nations Cartography Conference for Asia and the Pacific, Bangkok, Thailand, 1983, Proceedings: (in press).

Haas, R. H., and Waltz, F. A., 1984, Evaluation of thematic mapper data for natural resource assessment, in William T. Pecora Symposium, 9th, Sioux Falls, South Dakota, 1983, Proceedings: Sioux Falls, Augustana Research Institute, Augustana College, p. 122-133.

Horvath, E. H., Klingebiel, A. A., Moore, D. G., and Fosnight, E. A., 1984, Alternate data sources for soil surveys on rangeland: U.S. Geologic Survey Open-File Report 83-880, 16 p.

Miller, W. A., and Moore, D. G., 1983, Time-series vegetation monitoring for large geographic regions, in Conference on Renewable Resource Inventories for Monitoring Changes and Trends, Corvallis, Oregon, 1983, Proceedings: Corvallis, Oregon State University, College of Forestry, p. 86-89.

Miller, W. A., and Shasby, M. B., 1982, Refining Landsat classification results using digital terrain data: Journal of Applied Photographic Engineering, Springfield, Virginia, Society of Photographic Scientists and Engineers, v. 8, no. 1, p. 35-40.

Sadowski, F. G., Sturdevant, J. A., Anderson, W. H., Seevers, P. M., Feuquay, J. W., Balick, L. K., Waltz, F. A., and Lauer, D. T., 1984, Early results of investigations into Landsat 4 thematic mapper and multispectral scanner applications: in Landsat 4 Scientific Characterization Early Results Symposium, Greenbelt, Maryland, 1983, Proceedings: Greenbelt, NASA Goddard Space Flight Center, NASA Conference Publication 2355, v. IV, p. 281-297.

Shasby, M. B, Gaydos, Leonard, Fitzpatrick-Lins, Katherine, Carneggie, D. M., Lauer, D. T., Benjamin, Susan, and Ambrosia, Vincent, 1985, Interim program for land cover mapping in Alaska utilizing Landsat digital data, in William T. Pecora Symposium, 10th, Fort Collins, Colorado, 1985, Proceedings: will be published.

Southworth, C. S., 1984, The side-looking airborne radar program of the U.S. Geological Survey: Photogrammetric Engineering and Remote Sensing, Falls Church, Virginia, American Society of Photogrammetry, v. 4, no. 10, p. 1467-1470.

Watkins, A. H., Lauer, D. T., Bailey, G. B., Moore, D. G., and Rohde, D. G., 1984, A review of developments in space remote sensing for monitoring resources; oral presentation at 35th International Astronautical Congress at Lausanne, Switzerland (unpublished report).

INTEGRATED RESOURCE INVENTORY

Moderator: M. Goldblatt

THE POTENTIAL IMPACT OF THEMATIC MAPPER, SPOT AND MICROPROCESSOR TECHNOLOGY ON FOREST TYPE MAPPING UNDER LAKE STATES CONDITIONS

Dr. Thomas M. Lillesand
Paul F. Hopkins
Martin P. Buchheim
Ann L. Maclean

Environmental Remote Sensing Center (ERSC)
University of Wisconsin-Madison
1225 W. Dayton Street
Madison, Wisconsin 53706

ABSTRACT

The relatively coarse spatial, spectral, and radiometric resolution of the Landsat Multispectral Scanner (MSS) has severely limited the day-to-day application of MSS data to forest type mapping and condition assessment activities in the Lake States. However, forest managers continue to face restricted monetary and personnel resources while performing their management activities. Thus, it becomes important to assess the potential utility of Thematic Mapper (TM) and SPOT data in light of increasingly complex management information needs.

Based on the analysis of actual TM data on the one hand, and simulated SPOT data on the other, this paper provides an assessment of each form of data for type mapping under conditions typical of the Lake States. Substantial improvements over MSS are indicated for TM and SPOT data, using both manual and computer-assisted interpretation techniques. Likewise, microprocessing technology permits digital analysis of the data at the local level of management. However, "non-traditional" approaches to data analysis may more fully exploit the information value of the higher resolution data.

INTRODUCTION

Well over a decade of research with the data from the Landsat Multispectral Scanner (MSS) has dramatically improved the scientific community's capacity to numerically process multispectral satellite data, to "automate" the process of vegetation mapping, and to facilitate the storage and merger of these data with other resource data sets in computer-based geographic information systems (GISs). Regrettably, this research experience has also indicated that, under Lake States conditions, MSS data are adequate only for the most general level of forest resource assessment (Mead and Meyer, 1977; Hopkins, *et al.*, 1985). The design of the MSS has simply not been amenable to site specific data acquisition due to the complex mixes of types and sites found in such regions. Hence, there is a great deal of interest in assessing the utility of the high resolution data from the Landsat Thematic Mapper (TM) under such conditions, and, in anticipating the utility of the data from the planned SPOT satellite system in the same vein.

This paper discusses three case studies conducted in Wisconsin dealing with the analysis of TM and simulated SPOT data, respectively. The locations of these three study sites are shown in Figure 1. (TM data were used in Areas 1 and 2, and simulated SPOT data was used in Area 3.) The primary objective of these studies was to determine the accuracy and specificity to which type mapping could be performed with these data sources. The methods employed, and the results obtained, in each of these studies are presented in the next three sections of this discussion.

Concurrent with the above studies, our research group has been developing and testing a network of IBM AT micro-computers to support processing of TM and SPOT data. The latter section of this paper briefly describes this corollary activity.

By design, this paper is comparatively broad in scope and general in nature. Other papers have been prepared to convey additional detail about the various activities described herein (Buchheim, _et al_., 1984; Scarpace, _et al_., 1985). Furthermore, in that we are continuing our research on all of the fronts described in this discussion, this paper should be considered as merely a progress report and an overview of our efforts to date. As we point out, substantial additional research remains to be conducted in order to fully exploit the tremendous potential of the data sources and analysis procedures soon to become available to the forest manager.

TM CASE STUDY: AREA 1

Study Site/Image Data Analyzed. This study site is part of the northern unit of the Kettle Moraine State Forest in Fond du Lac and Sheboygan Counties in east-central Wisconsin (see Figure 1). This site contains a complex variety of cover types in addition to forested land--including water, wetlands, many agricultural types, and some developed areas. Both upland and lowland types of hardwood and softwood species are evident in the forest cover. Upland hardwood forest types predominate, represented by oak (_Quercus spp_.) shagbark hickory (_Carya ovata_), sugar maple (_Acer saccharum_), and bigtooth aspen (_Populus grandidentata_) with several other species (ash, basswood, beech, birch, etc.) in lesser amounts. The lowland hardwood areas consist mostly of ash (_Fraxinus pennsylvanica_) and red maple (_Acer rubrum_) along with willow (_Salix spp_.) and alder (_Alnus spp_.). The upland conifer types include plantations of red pine (_Pinus resinosa_), eastern white pine (_Pinus strobus_), and Norway spruce (_Picea abies_); while the lowland conifers consist mostly of tamarack (_Larix laricina_) with some cedar (_Thuja occidentalis_) and small amounts of spruce (_Picea mariana, principally_) and balsam fir (_Abies balsamea_).

Reference data for this site consisted of compartment and stand maps provided by the Wisconsin Department of

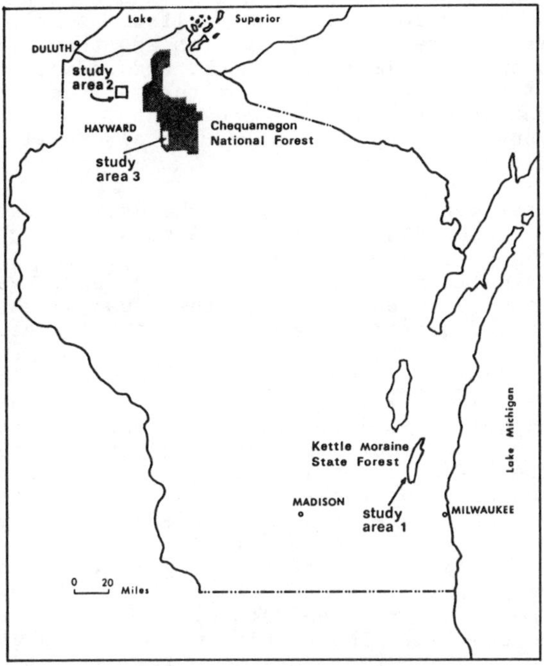

Figure 1. Location of three study areas in Wisconsin.

Figure 2. TM band 5 subscene covering study area 1.

Natural Resources (DNR). On these maps the upland hardwoods were typed into oak, northern hardwoods, central hardwoods, and aspen while lowland hardwoods were typed as swamp hardwoods or lowland brush (willow, alder). The upland conifers were typed by species and the lowland conifers as swamp conifer, tamarack, and cedar. Field visits verified the reliability of the reference maps but did show that oak species were prevalent in <u>all</u> upland hardwood types (hickory and maple were also well represented).

The image data were acquired by the Landsat 5 spacecraft on 18 July 1984 under NOAA's "special acquisition" program. The 512 x 360 pixel subscene extracted to cover the study site is of excellent quality with only a few very small clouds. Figure 2 shows band 5 of the subscene which emphasizes the vegetation variability of the area.

<u>Analysis Techniques Employed</u>. The primary technique employed was a supervised, maximum likelihood classification using all seven TM bands. A reference data set of 6003 pixels was derived by outlining 229 reference polygons on the subscene for 26 possible categories. Three reference subgroups were then created, each containing polygons from each category. Two reference subgroups were combined to form the training data and the third was reserved as an independent test sample. Three classifications were performed, each using a different combination of two subgroups as training data with the third subgroup reserved for testing. The results of the three classifications were combined to remove any bias resulting from the partition of the polygons into reference subgroups. In this way, all sample pixels were used for testing but were still independent of the training data.

Of the 26 categories used in the classifications, 8 were agricultural and 7 were other nonforest categories (water, developed land, marsh, cloud, and cloud shadow). Since the emphasis of the study was forest cover mapping, minimal attention was given to optimizing the classification results for these other categories. The only concern was providing distinction from forest.

Nine forest cover types (white pine, red pine, Norway spruce, lowland conifer, oak/northern hardwood, central hardwood, lowland hardwood, aspen, and tamarack/alder) were represented by the 11 forest categories. The oak and northern hardwood categories were combined after statistical analysis showed little class separability and field visits verified the prevalence of oak in the northern hardwood stands. Although only slightly more separable from oak/northern hardwoods, the central hardwoods and aspen were maintained as separate categories to evaluate how accurately they would be classified. The tamarack/alder category was sampled from areas of alder marsh which included scattered tamarack.

<u>Classification Results</u>. Accuracy statistics for the computer-assisted classification are reported in Tables 1 and 2. Table 1 provides information on detailed forest and

nonforest cover types while Table 2 aggregates these cover types into five forest categories and one nonforest category. In both tables, the tamarack/alder category contained only scattered tamarack, effectively making it a hardwood category. The forest/nonforest accuracy was 98%, the hardwood/softwood accuracy was 94%, and the average accuracy for the five aggregated forest classes was 90%. The average accuracy for the nine most detailed forest classes was 69%.

Discussion. The excellent accuracy for distinction of forest from nonforest was expected. However, excellent classification results are also evident for the more detailed forest classes. The nine most detailed forest classes even show promise, especially when the extreme similarity between the oak/northern hardwood, central hardwood, and aspen cover types is remembered. While the central hardwoods could not be distinguished, the results for classifying aspen seem promising even though aspen also accounted for most of the omission errors for the oak/northern hardwoods category. At any rate, when these three upland hardwoods classes are aggregated, the classification accuracy becomes excellent (99%). The separation of red and white pine is promising but the results for Norway spruce are disappointing. Most of these conifers occur in small plantations and the spruce stands are still quite young. Larger and more numerous stands of upland conifers may yield better results (as seems to be the case in the TM Case Study for Area 2). As with the upland hardwoods, when these three conifer classes are aggregated excellent classification accuracy (94%) results. The lowland conifers are also classified reasonably well with the only major confusion being with lowland hardwoods. This confusion is not surprising given the tendency of such areas to contain numerous pockets where lowland hardwoods predominate.

The utility of the thermal infrared data (band 6) was tested by performing identical classifications using only the other six channels. The comparative classification results were mixed by category with the six band results just slightly better overall. Curiously, the six band classification (no thermal data) yielded accuracy improvements of 5-9% in the lowland hardwoods, marsh, and cloud categories while the seven band results were better by 1-5% in the lowland conifer, tamarack/alder, central hardwoods, water, and developed categories. For this case study the thermal infrared data provided no additional value over the other six channels.

During the course of analyzing the training data, several other observations were made concerning the TM bands. First, band 5 often provided the most separation between forest classes and certainly competes with band 4 as the most important channel in this respect. Bands 1, 2, and 3 exhibited little contrast for the forest categories but were still important for distinguishing forest from nonforest. Consequently, it appears that at least one of these bands should remain to effect distinction of forest from nonforest. Finally, band 7 seemed to provide some distinguishing information both within forest and between forest and nonforest.

Several impressions were further developed related to the visual analysis of TM image data. First, and most obvious, the improved spatial resolution of TM (compared to MSS) allowed more effective location and delineation of sample polygon areas. Correspondence between stand maps and the imagery was readily obtained with the evidence of rural roads, rights-of-ways, etc. For location within the image, a standard false color composite (FCC) using bands 2, 3, and 4 provided a good compromise between cultural and vegetation detail. At the expense of cultural detail, an FCC using bands 3, 4, and 5 provided the most vegetation information. Finally, any color composite can be used to visually interpret Anderson level I cover types. With more suitable color composites, especially a band 3, 4, 5 FCC, detail in forest cover at level II was obvious and further detail was evident (especially between lowland and upland types).

TM CASE STUDY: AREA 2

Study Site/Image Data Analyzed. This site is a 15 km^2 area in Douglas and Bayfield counties east of Solon Springs in northwest Wisconsin (see Figure 1). Upland forest cover types are predominant with small areas of wetlands and lowland conifers also present. Sugar maple, northern red oak (*Quercus rubra*), yellow and white birch (*Betula papyrifera* and *B. alleghaniensis*), aspen (*Populus grandidentata* and *P. tremuloides*), jack pine (*Pinus banksiana*) and red pine compose the upland types. Lowland conifers present include tamarack, black spruce, and white cedar.

Imagery for the site was acquired on 21 June 1984 by Landsat 5. The data are of excellent quality with thin cloud cover over approximately 4% of the site. Figure 3 shows a band 7 image of the site, which emphasizes the spectral differences between hardwoods and conifers.

Analysis Techniques Employed. To date, only computer-assisted interpretations have been performed using a supervised maximum likelihood classification algorithm. The classification system developed by Anderson *et al.* was utilized for mapping the various forest cover types, and in addition, any areas of jack pine budworm (*Choristoneura pinus pinus*) defoliation were classified.

Classification Results. Of the seven bands available, only bands 2, 3 and 4 were analyzed in this preliminary study. Information classes included hardwoods, jack pine, red pine, lowland conifers (black spruce and tamarack), lakes, clearcuts, roads, gravel pits, clouds and jack pine budworm defoliation. Detail to Level III was obtained for the coniferous classes, with high classification accuracies for jack pine and red pine. Three levels of jack pine budworm defoliation were also separated: dead trees, severe defoliation (> 50% foliage removed), and moderate defoliation (20-50% foliage removed). It was not possible,

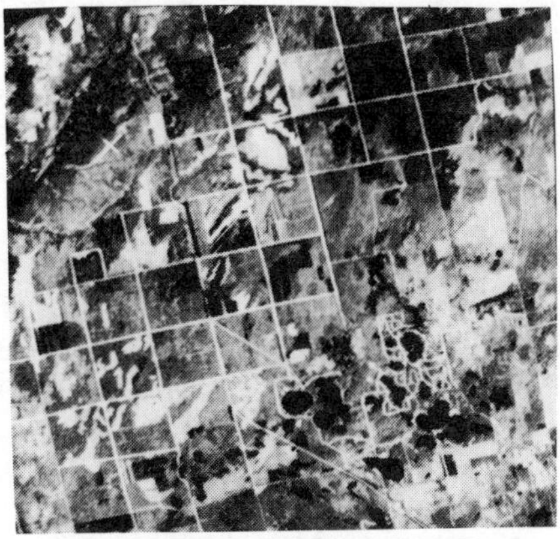

Figure 3. TM band 7 subscene covering study area 2.

Figure 4. Simulated SPOT panchromatic image covering study area 3.

using three bands, to distinguish the various hardwood species, or newly logged areas from unpaved roads. An accuracy assessment of the cover types is presented in Table 3. The overall accuracy was 93%; average cover class accuracy was 90%. The accuracy assessment was done by an evaluation of homogeneous test areas which were selected based on ground reference information and low altitude color infrared photographs acquired shortly after the Landsat overpass.

SPOT SIMULATION CASE STUDY: AREA 3

Study Site/Image Data Analyzed. This study area is a 5 km by 12 km site located in the Chequamegon National Forest in northwest Wisconsin (Sawyer County) (see Figure 1). This site contains both upland and lowland forest types, as well as nonforested wetland areas. The upland types are dominated by sugar maple - yellow birch, white pine - northern red oak - red maple, white birch - balsam fir, and aspen stands, as well as several red pine plantations. The lowland forests are dominated by black spruce and tamarack, with white cedar and balsam fir in scattered areas.

The data covering this site were acquired on 18 June 1983 as part of the 1983 U.S. SPOT Simulation Campaign. Using an airborne scanner, SPOT Image Corporation provided coverage simulating both the panchromatic (10m resolution) and multispectral (20m resolution) modes of data anticipated from the future SPOT satellite (SPOT Image Corp., 1983). The resulting data were provided in both digital and hardcopy formats. Simultaneously acquired 1:42,000 scale color infrared transparencies were also provided to aid in the interpretation of the SPOT data. All data were of generally excellent quality, with less than two percent cloud cover over the entire study site. Figure 4 is a representative sample of the panchromatic coverage of the test site.

Analysis Techniques Employed. In an effort to evaluate a wide range of applications of the SPOT imagery, both manual and computer-assisted image interpretation techniques were evaluated. The manual interpretations used the original photographic prints, supplied by SPOT Image Inc., of the panchromatic and multispectral data. (The multispectral photographic product was actually a composite derived from the panchromatic and multispectral data). In addition to type specific information, discrimination of eastern spruce budworm (*Choristoneura fumiferana*) defoliation was also attempted.

The computer-assisted interpretations involved the testing of a range of supervised and unsupervised classification algorithms available at ERSC. As with the visual interpretation, discrimination of spruce budworm defoliation levels was attempted.

Manual Interpretation Results. Detail to level II of the Anderson system was easily obtained with the manual interpretation of the panchromatic data. Excellent textural and tonal contrast aided in the separation of hardwoods and conifers. Additional level III differentiation between conifer species was obtained in some instances. It was not possible, however, to absolutely distinguish small agricultural areas from similar size recently logged areas.

The multispectral image product allowed species delineation (level III) over most of the scene. Distinctions were made between black spruce, tamarack, and red pine. In some areas it was also possible to delineate stands of white pine from surrounding hardwoods. Although the heterogeneous nature of the upland hardwood stands in the study area made precise differentiation of deciduous types difficult, the textural and contextual information apparent in the image product enabled some visual separation between the aspen type and the sugar maple-yellow birch type. Recently logged areas were easily distinguished from agricultural areas on the multispectral image. A breakdown of the various classes interpreted from the multispectral image is shown in Table 4.

Evaluation of the defoliation subarea was limited, due to the early summer date of imagery acquisition (18 June 1983), to assessment of the previous year's defoliation. Two levels of damage were visually quantified: "severe" (75 to 100 percent of the foliage removed with top kill) and "dead trees." It is postulated that a later date of acquisition would allow evaluation of the current year's defoliation into multiple levels. The ideal date of acquisition for such an analysis would be between mid-July and mid-August when peak browning of the clipped and webbed needles occurs (Maclean and Giese, 1984).

Computer-Assisted Classification Results. Discrimination of detail to level II of the Anderson system, and in some cases to level III, was attempted during the computer-assisted classification process. Initial attempts at classification revealed discrimination problems between certain cultural features, such as roads, plowed fields, quarries, and barren logged areas, and highly reflective wetland areas. Given the nature of the study, it was deemed appropriate to eliminate these nonforest related discrimination problems by partitioning (or stratifying) the scene into cultural and non-cultural features through interactive manual delineation of the major cultural features in the imagery. The remainder of the classification iterations were then carried out utilizing the manually partitioned data set. This enabled attention to be focused on discrimination of forest cover rather than land cover in general.

As mentioned earlier, a variety of image analysis and classification techniques are being used to evaluate the SPOT simulation imagery. These procedures have been applied to the original three bands of data, and to the original data combined with ratios derived from the original data. In every case, final classifications were produced using the maximum likelihood classifier available at ERSC.

Detailed and accurate classifications at levels II and III of the Anderson system were obtained with both supervised and unsupervised classification techniques. Results from the supervised classification of the original and ratio-enhanced data are presented in Table 5. Overall and class average accuracies, based upon assessment of homogeneous test areas, were 93% and 85%, respectively. Accuracies for the other combinations of analysis techniques and data types were nearly identical to those presented herein, although the unsupervised technique, by itself, was noticeably less accurate than the supervised technique. The unsupervised clustering algorithm employed, which delineated clusters based upon a user-defined variance maximum within a 3-by-3 moving window, was incapable of adequately characterizing the spectral variability of the data without using additional hybrid image analysis techniques (supplementing with supervised training data). Classification accuracies for this modified, hybrid unsupervised approach were similar to those presented in Table 5.

Examination of the results presented in Table 5 gives at least a preliminary indication of the successes and failures of the SPOT computer-assisted classifications. The upland deciduous and coniferous stands were very well separated, as were the dominant lowland types, black spruce and tamarack. A good test of upland conifer type separability was not possible because only red pine was sufficiently abundant to allow adequate sampling. Areas of confusion, in cases such as the conifer/wetland and mixed/wetland misclassifications, were simply due to spectral overlap between classes. In other cases, however, the problems were more a function of either the distributional properties of the types within the study area, as with the white cedar/balsam fir class, or the gradational nature of the classification system used, as with the deciduous-mixed-coniferous misclassifications.

A WORD ON MICROPROCESSOR TECHNOLOGY

It is clear that the improved resolution of TM and SPOT data will heighten the interest of the forest management community in satellite remote sensing--particularly with the prospective commercial operation of these systems. Of equal importance, however, is the prospective impact microprocessor technology will have on the ability to analyze these data at the local level of land management. Several image processing/GIS systems are already available commercially and the computational power of the available hardware is changing daily.

Our research group continues to be actively engaged in the development and testing of IBM AT based image processing systems (Scarpace, et al., 1985; Ahearn, et al., 1985). We believe these systems will have a pervasive impact on forest management, especially as their capabilities continue to improve and their relative cost continues to decline. Most forest management agencies are acquiring microprocessors for general use as a matter of course. Extension of the use of these systems into image processing and GIS support roles will be natural.

CONCLUSIONS/SUMMARY

The findings discussed in this paper are preliminary due to the limited nature of the techniques employed (in the case of TM data) and due to the nature of the data (in the case of simulated SPOT data). Even so, the results indicate the capabilities which might be expected from these new sources of image data. The respective utilities and complementarity of TM and SPOT data will undoubtedly vary depending upon the exact nature of the information required and the characteristics of the area under study, making extensive generalizations premature. However, preliminary evaluations of data over three different Wisconsin test sites suggest the following:

- <u>TM and SPOT data represent substantially improved sources of cover type information under Lake States conditions.</u> Whereas with MSS data even broad land cover classification can be difficult under these conditions, both sources of the higher resolution data enabled highly accurate classification of forest vs. nonforest, hardwood vs. conifer, and varying degrees of species differentiation (depending upon the site involved). Furthermore, both data sources showed potential for improved differentiation of damage levels due to stress agents such as the jack pine and spruce budworm.

- <u>Full exploitation of the information content inherent in TM and SPOT data will require the development of analysis procedures which go beyond the "traditional" procedures currently utilized to process MSS data.</u> The conventional computer-assisted techniques utilized herein were (in many cases) incapable of extracting all of the information visually apparent, and manually identifiable, in the data. Potential alternatives include hybrid manual/computer-assisted procedures wherein computer-generated, cover type-specific enhancements are applied to the imagery on a video display. Using some form of cursor the enhanced data can be simultaneously interpreted and digitized into a GIS-compatible format. Likewise, existing data base layers in a GIS can be used to supplement the classification process directly and/or to stratify the image data for subsequent processing. (This concept is similar to that employed to separate the cultural features from the forested areas in our SPOT simulation study). Textural and contextual, spatially-oriented classifiers will also take on added importance in data analysis.

While continued research and development is required to fully exploit TM and SPOT data, the capability to process these data in a microprocessor environment is already a reality. Furthermore, the comparative cost and efficiency of such systems continues to improve daily. When these factors are coupled with the prospective commercial provision of TM and SPOT data on a timely and cost effective basis it is clear that we are entering a new era of remote sensing and forestry.

ACKNOWLEDGEMENTS

This research was supported by the College of Agricultural and Life Sciences, School of Natural Resources, University of Wisconsin-Madison, and by the UW-Madison Institute for Environmental Studies. The reference data used in this research was supplied by the Wisconsin Department of Natural Resources and the U.S. Forest Service.

REFERENCES

Ahearn, S.C., R.W. Kiefer, and T.M. Lillesand. 1985. Tropical Deforestation and the Implications of Microcomputer-Based Image Processing Technology: Case Study Indonesia. Elsewhere in this volume.

Buchheim, M.P., A.L. Maclean, and T.M. Lillesand. 1984. Forest Cover Type Mapping and Spruce Budworm Defoliation Detection Using Simulated SPOT Imagery. SPOT Simulation Applications Handbook, pp. 259-266.

Hopkins, P.F., O. Long, and S. Menon. 1985. Mapping Forest Land in New York by Digital Analysis of Remote Sensing Data. SUNY College of Environmental Science and Forestry Remote Sensing Research Applications Report, Syracuse, NY.

Maclean, A.L. and R.L. Giese. 1984. A Methodology Using Low-Altitude Thermal Imagery Interfaced Digitally with Color Infrared Photography for Detecting, Mapping, and Evaluating Budworm Defoliation. Proceedings: ASP-ACSM 1984 Annual Convention, Vol. 2, pp. 580-588.

Mead, R.A., and M.P. Meyer. 1977. Landsat Digital Data Application to Forest Vegetation and Land Use Classification in Minnesota. Proceedings: Machine Processing of Remotely Sensed Data Symposium, pp. 270-279.

Scarpace, F.L., R.W. Kiefer, T.M. Lillesand, and P.R. Weiler. 1985. An IBM PC/AT Microcomputer-Based Image Processing Laboratory. Proceedings: Semiannual Meeting of the American Society for Photogrammetry and Remote Sensing. Indianapolis, Indiana (in press).

SPOT Image Corp. 1983. SPOT Simulation Campaign Auxiliary Information Package. Washington, D.C., pp. 90.

Table 1. Classification accuracies for TM Case Study 1.

VERIFIED	OBSERVED															
CLASS:	pw	pr	sn	lc	ta	oh	ch	as	lh	ag	wo	wm	dv	sh	cl	% Correct
pw	276	35	25	3	0	0	0	0	1	6	0	1	0	0	0	80
pr	45	256	16	0	0	0	0	0	0	12	0	3	3	0	0	76
sn	52	8	64	1	0	0	0	0	2	14	0	2	0	0	0	45
lc	1	0	0	145	0	0	0	0	45	4	0	6	0	0	0	72
ta	0	0	0	0	139	0	0	0	7	0	0	0	0	0	0	95
oh	0	0	0	0	0	831	47	239	2	4	0	0	0	0	0	74
ch	0	0	0	0	0	56	13	1	1	2	0	0	0	0	0	18
as	0	0	0	0	0	14	12	69	0	0	0	0	0	0	0	73
lh	0	0	0	3	17	2	1	0	370	12	0	13	0	0	0	88
ag	0	0	0	0	0	0	0	0	7	2236	0	18	89	0	0	95
wo	0	0	0	0	0	0	0	0	0	0	378	3	0	0	0	99
wm	0	0	0	11	0	0	0	0	8	33	0	83	0	0	0	62
dv	0	0	0	0	0	0	0	0	0	0	0	0	125	0	0	100
sh	0	1	0	0	0	0	0	0	0	0	0	0	0	60	0	98
cl	0	0	0	0	0	0	0	0	0	0	0	0	6	0	64	91

Overall Accuracy: 85 %   Average Class Accuracy: 78 %

Classes:

pw: white pine	oh: oak/N. hardwoods	wo: open water
pr: red pine	ch: central hardwoods	wm: marsh (emergent veg.)
sn: Norway spruce	as: aspen	dv: developed
lc: lowland conifers	lh: lowland hardwoods	sh: cloud shadow
ta: tamarack/alder	ag: agriculture	cl: cloud

Table 2. Aggregated class accuracies for TM Case Study 1.

VERIFIED CLASS:	uc	lc	ta	uh	lh	nf	% Correct
uc	777	4	0	0	3	41	94
lc	1	145	0	0	45	10	72
ta	0	0	139	0	7	0	95
uh	0	0	0	1282	3	6	99
lh	0	3	17	3	370	25	88
nf	1	11	0	0	15	3095	99

(OBSERVED classes across the top.)

Overall Accuracy: 97 %   Average Class Accuracy: 91 %

Classes:
- uc: upland conifers
- lc: lowland conifers
- ta: tamarack/alder
- uh: upland hardwoods
- lh: lowland hardwoods
- nf: nonforest

Summary of Further Class Aggregations:

Average Hardwood/Softwood Accuracy: 94 %

Average Forest/Nonforest Accuracy: 98 %

Table 3. Classification accuracies for TM Case Study 2.

VERIFIED CLASS:	dv	uh	wo	lc	cc	gp	dm	ds	dd	pr	pj	% Correct
dv	32	0	0	28	0	0	0	0	0	0	0	53
uh	0	173	0	0	11	0	0	3	0	0	0	92
wo	0	0	89	0	0	0	0	3	0	0	0	97
lc	0	0	0	93	0	0	0	0	0	6	3	91
cc	29	0	0	0	173	0	0	0	5	0	0	84
gp	0	0	0	0	0	34	0	0	0	0	0	100
dm	0	0	0	0	0	0	206	8	0	0	0	96
ds	0	0	0	0	0	0	0	192	3	0	0	98
dd	0	0	0	0	0	0	8	3	42	0	0	79
pr	0	0	0	0	0	0	0	0	0	231	3	99
pj	0	0	0	0	0	0	0	0	0	8	269	97

(OBSERVED classes across the top.)

Overall Accuracy: 93 %   Average Class Accuracy: 90 %

Classes:
- dv: roads/developed
- uh: upland hardwoods
- wo: open water
- lc: lowland conifers
- cc: clear cuts
- gp: gravel pits
- dm: moderate defoliation
- ds: severe defoliation
- dd: dead trees
- pr: red pine
- pj: jack pine

Table 4. Information classes for the manual interpretation of the SPOT multispectral film product. (Dashed lines indicate levels of detail not attainable.)

Level I	Level II	Level III
forest	deciduous	aspen sugar maple/yellow birch
	coniferous	red pine white pine balsam fir
	mixed	---------
water	lakes open bogs	(no further breakdown attempted)
wetland	forested	black spruce tamarack white cedar/balsam fir
	nonforested	bog vegetation (leather leaf/ericads) sedges/grasses alder thicket

Table 5. Computer-assisted classification accuracies for the SPOT Simulation Case Study.

VERIFIED	OBSERVED										
CLASS:	uh	uc	um	wo	wb	sb	tm	cw	wl	ws	wa	% Correct
uh	1185	0	11	0	0	0	0	0	0	0	0	99
uc	0	141	0	0	0	7	0	0	0	14	2	86
um	35	37	85	0	0	2	0	0	0	0	19	48
wo	0	0	0	411	0	0	0	0	0	0	0	100
wb	0	0	0	5	44	0	0	0	0	0	0	90
sb	0	1	8	0	0	388	0	3	0	7	0	95
tm	1	0	0	0	0	0	137	0	0	0	8	94
cw	0	0	0	0	0	20	0	17	0	0	0	46
wl	0	0	0	0	0	0	0	0	123	1	0	99
ws	0	0	0	0	0	3	0	0	0	203	0	98
wa	15	2	19	0	0	0	0	0	0	2	122	76

Overall Accuracy: 93 % Average Class Accuracy: 85 %

Classes:

uh: upland hardwoods
uc: upland conifers
um: upland mixed
wo: open water
wb: bog lakes
sb: black spruce
tm: tamarack
cw: white cedar *
wl: leather leaf wetland
ws: sedge/grass wetland
wa: alder thicket wetland

* -- constituted less than 1 % of the image

AN OPERATIONAL INTERAGENCY GIS: THE GLACIER NATIONAL PARK/FLATHEAD NATIONAL FOREST PROJECT

David B. Wherry, Judy A. Hart
Washington State University Computing Service Center
Pullman, WA 99164-1220

Carl H. Key
Glacier National Park
West Glacier, MT 59936

Stan A. Bain
Flathead National Forest
P.O. Box 147, Kalispell, MT 59901

ABSTRACT

Interagency agreement between Glacier National Park (GNP) and Flathead National Forest (FNF) has allowed these neighboring federal agencies in Montana to share the costs and benefits of a regional, Landsat-based, Geographic Information System (GIS). Benefits of the interagency approach have included cost sharing, increased Park/Forest dialog, and a database addressing much of the ecological region encompassing both agencies. Establishing in-house GIS expertise and maintaining in-agency project management were identified as vital success factors in creating an operational GNP/FNF GIS program. Park and Forest personnel were trained in GIS system use through the Digital Image Analysis Laboratory at Washington State University Computing Service Center (WSUCSC). Throughout the project, GIS processing has been performed with WSUCSC's VICAR/IBIS image processing system via telecommunications from GNP and FNF agency offices. A pilot GIS database was tested in GNP and in FNF's Hungry Horse Ranger District. When evaluated by Park and Forest managers, speed and accuracy of GIS processing and product generation met all acceptability criteria. In its operational phase, GIS databases are now being constructed for the four remaining FNF ranger districts and for GNP in its entirety.

PROJECT HISTORY

The Interagency Agreement to develop a Geographic Information System (GIS) grew out of a cooperative, Environmental Protection Agency (EPA) funded effort to create a Landsat classification of the Flathead Valley, northwestern Montana. That study, a portion of the Flathead River Basin Environmental Impact Study (Anon, 1983), was co-sponsored by the National Aeronautics and Space Administration, Ames Research Center (NASA/ARC). Through ARC, professionals from seven federal, state and local agencies received training in the digital analysis of Landsat data.

The Landsat classification, completed in 1982 and based entirely on multispectral scanner data, was only moderately successful given the study area's highly variable topographic relief and the diverse interests of participants. Flathead National Forest (FNF) and Glacier National Park (GNP) (Figure 1), lead participants in the project, did not find the classification particularly applicable to their management and research needs. Further, with the close of EPA funding, the State of Montana terminated support for digital image processing, and no means existed to access the classified dataset, given even limited usefulness.

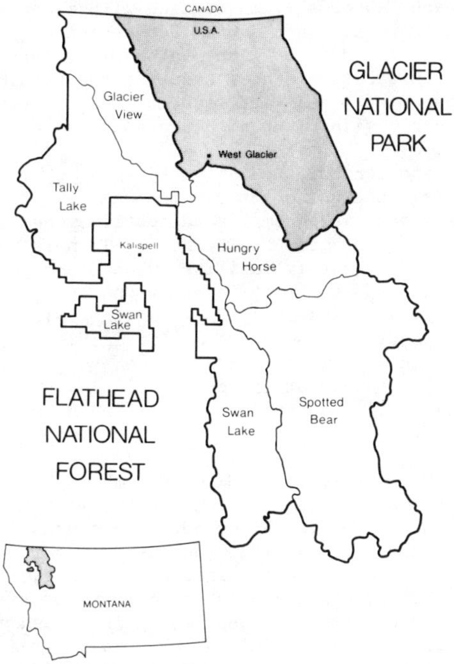

Figure 1: Glacier National Park and Flathead National Forest bridge the Rocky Mountains in northwestern Montana.

Recognizing the potential for applying Landsat data in the context of ancillary geographic themes, and with support from the FNF Forest Supervisor and the GNP Superintendent, Bain and Key developed the concept and conditions of an Interagency Agreement between Park and Forest (USDA - Forest Service and USDI - Park Service, 1982). The agreement established a mechanism to share costs, labor effort, and materials to continue the Landsat work already begun. With Bain and Key as appointed coordinators in their respective agencies, the project's scope broadened to develop a flexible, multi-layered GIS available to both agencies and capable of addressing diverse informational needs. Additional training, technical consultation, and access to appropriate computer facilities were secured in 1983 through the Digital Image Analysis Laboratory (DIAL) at Washington State University Computing Service Center (WSUCSC).

INTERAGENCY AGREEMENT VALUES

The values of an Interagency Agreement for GIS development are several. Costs are reduced by shared data gathering and processing. The GIS can be designed, conducted, and financially supported at the field level. Integrated with agency operations, results reflect salient management needs and a substantial expertise regarding local ecosystem components. Generated GIS data themes tend to be most pertinent and accurate.

The sharing of information and ideas is a natural by-product of the interagency approach, especially in dealing with a new technology, where a compatriot with whom to share one's successes and bewilderments is a valuable resource. In addition, the Interagency Agreement brings with it two sets of administrative biases causing the rethinking of many GIS decisions by the project team and leaving open a variety of options perhaps not attainable within the single agency concept.

In-house support and control of GIS work tends to maintain continuity through operational phases after development. This is often not possible when GIS work is accomplished outside the sponsoring agency through contractual arrangement. Trained field personnel, the creators of the system, are intimately familiar with GIS capabilities and are best able to both assess what information is needed and how to appropriately query the system. Work on the GIS does not have to terminate when contracting funds are not available as agency personnel can usually maintain some level of GIS use even in low-budget years. The local agency can proceed at its own pace, slowly building a tailored system, and efficiently altering plans, based on current needs or previous experience.

The Interagency Agreement supports databases that are regional in scope. Although management objectives differ across administrative boundaries, resources are held in common throughout any given region. The GIS database need not stop at political boundaries, but may extend to the limits of ecological significance. This affords managers a common basis of reference and also provides a basic framework for resource management and research involving several agencies. Interagency conflicts over management strategies are minimized by joint referral to common stratified databases, assembled and accessed with standardized procedures. Managers and researchers have access to resource data from adjacent areas facilitating assessment of potential outside influences and comparative analyses between areas.

INTERAGENCY PROJECT LIMITATIONS

Interagency project limitations are twofold. First, lands outside Park and Forest political boundaries but within the ecological region are, largely, excluded from the GIS database. In the case of GNP, Park lands are designated an International Biosphere Reserve with Waterton Lakes National Park, Canada. Inclusion of adjacent Canadian lands in the GIS would be beneficial, but except for Landsat data, park geographic reference data ends at the Canadian border. In the U.S. where GNP and FNF hold nearly two-thirds of the Flathead Valley region within their borders, attempts to reduce this limitation have been made. Neighboring land management agencies have been invited to enter into the Interagency Agreement, with the intent of reducing existing voids in the database for the common benefit. Provision is made in the current agreement to accommodate additional participants. The current GNP/FNF GIS strategy requires each agency to develop its own geographic data planes within its boundary, up to and including all border quadrangles. Mutually standardized procedures, formats and software are used by the agencies in the development tasks.

As a second limitation, operational budgets fluctuate yearly and asynchronously between agencies leading to unequal levels of effort and independent rates of completion. In GNP and FNF, however, these problems are tempered as both agencies have one full-time analyst devoted to GIS development. Salaries are built into the agency's base funding allowing continued progress, even at low budgetary levels.

Aside from the above, no limitations are imposed in the Interagency Agreement except that planned activities have mutual consent from both participants.

TECHNICAL SUPPORT

Processing and manipulation of the GIS database is accomplished through DIAL facilities at WSUCSC in Pullman, Washington. At the foundation of DIAL capabilities is the Video Image Communication and Retrieval/Image Based Information System (VICAR/IBIS), a batch oriented system resident on the Center's Amdahl 470V/8 mainframe computer. Complementing VICAR/IBIS is the International Imaging System 511, an interactive image processing tool hosted by a PDP 11/34 mini-computer. Both GNP and FNF utilize agency computer hardware (Figure 2) for GIS data layer digitization and for Remote Job Entry and Output via teleprocessing to WSUCSC's mainframe environment. Agency personnel also visit the DIAL facility on occasions when interactive capabilities are required.

Glacier National Park

- IBM PC micro computer
  - 512K RAM
  - Intel 8087 math coprocessor
  - 360K floppy disk drives (2)
  - 10MB hard disk drive
  - Hayes 1200B modem
  - AMDEK color monitor
  - Okidata dot matrix printer
- Numonics Digitizing Tablet
  - Model 2401-3648
  - 36x48 inch work area
  - RS-232 interface to IBM PC
- Numonics X-Y plotter
  - Model 5424
  - 23x34 inch plot size
  - RS-232 interface to IBM PC

Flathead National Forest

- Data General NV8000 mini computer
  - Kinex 4800 baud modem
  - Dasher 410 terminal
  - Model B600-60 line printer
- Teletype model 43 terminal
  - 300 baud modem, acoustic coupler
- Numonics Digitizing Tablet
  - Model 2400-2436
  - 24x36 inch work space
  - RS-232 interface to Data General

Figure 2: GNP and FNF computer hardware configurations. While purchased for general agency computing support, these multi-purpose computer configurations are used at Park and Forest offices for in-house GIS data plane development from agency resource maps and for two-way teleprocessing with WSUCSC's DIAL facility.

VICAR/IBIS is similar to many other GIS configurations, except that all database layers are stored, manipulated, and queried in raster (i.e., digital image) format. Landsat and digital terrain models are readily available in raster format. Other mapped themes are vector digitized to x,y coordinate representation and converted to image based maps (Jaffray, Hansen & Hart, 1984).

GIS DEVELOPMENT AND TESTING

Pilot GIS development and testing encompassed all of GNP and the Hungry Horse Ranger District (HHRD), one of five ranger districts in FNF. The pilot phase developed GIS capabilities consistent with operational agency requirements set forth by Park and Forest personnel. GIS utility was tested in the 480,000 acre HHRD where 13 thematic GIS layers were generated to complement Landsat stratified land cover and digital terrain data components. In GNP, only Landsat and digital terrain components were considered over the entire Park during the pilot testing. However, a six 1:24,000 quad area bisected by the Park and Forest border was developed and tested with 13 GNP thematic layers.

Standard Digital Products

Defense Mapping Agency (DMA) digital elevation data and Landsat Multi-Spectral Scanner (MSS) data were acquired on computer compatible tapes from the National Cartographic Information Center and EROS Data Center. Dataset formats were tailored with specialized logging programs for compatibility with VICAR/IBIS processing requirements (Hart & Wherry, 1984).

Digital Terrain Processing. Processed digital elevation data served as the planimetric datum for the GIS database. Four 1 by 1 degree DMA blocks were digitally mosaicked into a 2 by 2 degree configuration and transformed to a Universal Transverse Mercator (UTM) projection with 50 by 50 meter pixels.* Mosaic seams (inter-block gaps) were reduced with VICAR/IBIS noise reduction programs and slope magnitude and slope direction (aspect) planes were modelled from the elevation mosaic.

Landsat MSS Processing. Landsat acquisition 22354-17420 (03JUL84 - EDIPS) was selected for its data quality and near cloud free coverage of GNP and FNF's HHRD. A methodology similar to the multi-cluster blocks technique described by Hoffer (1979) was employed by Key and Bain for construction of the MSS classification statistics set. Landsat training areas of known land cover characteristics were processed by a clustering algorithm. Evaluation and manipulation of the resulting spectral statistics yielded 99 statistical classes which were used to drive the classification of the MSS imagery.

Landsat classified data were transformed to the GIS planimetric datum and then combined with elevation and aspect images. The resulting dataset contained 3,800 categories expressing the combined occurrence of one of three elevation zones, one of eight aspect divisions, and one of 99 spectral classes at each pixel location. Contingency tables were constructed to correlate each elevation-aspect-spectral category with the appropriate class on the GNP/FNF land cover legend (see Figure 3). Reassignment of each elevation-aspect-spectral category to one of 189 land cover classes completed the Landsat processing.

---

* DMA elevation models were acquired, mosaicked, and UTM transformed by NASA/ARC personnel as part of previous project work in the Flathead Valley region. All additional DMA data were retrofitted through DIAL.

WATER
Water Type

SNOW
Snow Type

LAND RESOURCES
Agricultural
Agricultural Type (irrigated, non-irrigated, pasture, plowed)
Agricultural Class within type
Barren Lands
Moisture Condition (dry, mesic, moist)
Elevation (lowland, subalpine to alpine)
Type of substrate
Natural Vegetation
Major Life Form (coniferous forest, shrub, herbaceous)
Moisture condition (dry, mesic, moist)
Elevation (lowland, mid-elevation, subalpine, alpine)
Canopy closure or density (low, medium, high)
Species composition or character of exposed substrate
Plant vigor (especially beetle killed lodgepole pine)

Figure 3: Major discriminating categories applied in the 189 class land cover legend. The categories are hierarchical, representing relative significance to the Landsat classification (after Butterfield and Key, 1985).

Planes Generated by Park and Forest Personnel

By terms set forth in the Interagency Agreement, GNP and FNF personnel independently developed map based planes for their respective agency's GIS databases. While the thematic character of several planes was common to both agencies, several distinct map themes were utilized by each agency in response to the unique mandates governing Park and Forest management goals (Table 1). In addition to Landsat land cover and digital terrain GIS planes, 14 planes were identified for development in GNP and 27 in FNF. Nearly all themes were recorded on standard 1:24,000 scale quads. Data were encoded at GNP and FNF offices utilizing x,y coordinate digitizer tablets (Figure 2). Edge matching was performed to assure continuity of the map surface from quad to quad. Thematic features were stored in a UTM (single zone extended) coding system spatially correlating the digitized map data to the GIS planimetric datum. In a final set of tasks, UTM values were converted to image coordinates which drove the transformation from vector (digitized) map characteristics to the raster map rendition. Once in raster format, image processing operations were applied to attach attributes to polygon, line, and point features thus giving the GIS layer its informational character (Wherry and Friedman, 1978).

As the last task of the pilot phase, testing was performed to demonstrate a selection of GIS capabilities. Testing included the quantification of linear features (distances) and aerial acreages from assorted planes and synthesis of new informational surfaces from a composite of database themes (Hart, Wherry, and Bain, 1985). GIS products included UTM projected maps and tabular printouts.

GIS Thematic Plane	Project Phase GNP	Project Phase FNF	Plane # GNP	Plane # FNF	Data Source	Data Organi-zation	Vector Data Format	Significance of Plane
Landsat Land Cover	P	P	G/F1		EDC CCT	Raster		189 land cover classes from MSS imagery
Digital Elevation Model	P	P	G/F2		NCIC CCT	Raster		Topographic elevation by 1 foot increments
Digital Aspect Model	P	P	*G/F3		G/F2	Raster		Topographic aspect by 5 degree increments
Digital Slope Model	P	P	*G/F4		G/F2	Raster		Topographic slope by 1 degree increments
Land Ownership	P	P	G5	F5	1:24,000	Vector	Polygon	Park/Forest lands and private lands
Drainage Basins	P	P	G6	F6	1:24,000	Vector	Polygon	Topographically defined watershed boundaries
Administrative Boundaries	O	P	G7	F7	1:24,000	Vector	Polygon	Areas with distinct management goals
Wild Scenic River Boundaries	P	P	G8	F8	1:24,000	Vector	Polygon	Recreational use area perimeter
Landtypes		P		F9	1:63,360	Vector	Polygon	Ecological limitations and opportunities
Timber Compartments		P		F10	1:24,000	Vector	Polygon	Management units for timber stand inventory
Timber Harvest History		P		F11	1:24,000	Vector	Polygon	Harvest activity type and date for all areas cut
7½' Quad Boundaries	P		G9		1:24,000	Vector	Polygon	Used to extract GIS information by quad(s)
Precipitation		P		F12	1:63,360	Vector	Polygon	Precipitation surface modelled from isohyets
Surface Hydrology	P	P	G10	F13	1:24,000	Vector	Line	1st-5th order streams (peren. or ephem.), major rivers and lakes
Roads	P	P	G11	F14	1:24,000	Vector	Line	Classifications incl.; function, maintenance, service, jurisdiction, and management
Trails	P	P	G12	F15	1:24,000	Vector	Line	Up to 5 different trail maintenance levels
Facilities	P		G13	F16	1:24,000	Vector	Point	Ranger stations, campgrounds, picnic areas & lookouts
Distance to Hydrology	P	P	*G14	*F17	G10 F13	Raster		Distance decay corridors (150 m. increments)
Distance to Roads	O	P	*G15	*F18	G11 F14	Raster		Distance decay corridors (150 m increments)
Distance to Trails	O	O	*G16	*F19	G12 F15	Raster		Distance decay corridors (150 m. increments)
Distance to Facilities	O		*G17	*F20	G13	Raster		Distance decay corridors (150 m. increments)
Fire Management Zones	O		G18		1:24,000	Vector	Polygon	Boundaries used in fire management
Fuel Groups		O		F21	1:24,000	Vector	Polygon	Vegetation and landtype characteristics
Forest Plan Units		O		F22	1:24,000	Vector	Polygon	Boundaries for administering the forest plan
Timber Stand Boundaries		O		F23	1:24,000	Vector	Polygon	Stand characteristics from aerial photography
3rd Order Subdrainage Basins		O		F24	1:24,000	Vector	Polygon	3rd order watershed boundaries
Fishery Reaches		O		F25	1:24,000	Vector	Line	Fishery habitat coded by stream reach
Geology		O		F26	1:63,360	Vector	Polygon	Generalized geologic influences
Oil & Gas Lease Units		O		F27	1:24,000	Vector	Polygon	Lands leased for oil & gas exploration
Fire Risk		O		*F28	G/F1-4	Raster		Landcover volatility & fire spread ratios
Elk Calving Areas		O		*F29	G/F1-4, F17	Raster		Ranked potential elk calving locations
Grazing Potential		O		*F30	G/F1,4	Raster		Potential wildlife grazing areas
Grizzly Bear Components		O		*F31	G/F1-3	Raster		Areas of specific habitat conditions

For Project Phase: P = Pilot; O = Operational. For Plane #: * = Planes synthesized from the GIS database.

Table 1: Digital map themes resident in the GNP/FNF GIS database.

GIS APPLICATIONS

With continued support through the DIAL facilities at WSU, GNP and FNF are growing into operational GIS capabilities. In the Park, Key is directing the development of thematic data planes encompassing all of GNP. In FNF, Bain is supplementing the HHRD pilot database with additional thematic layers and coordinating, from the U.S.F.S. Regional Office, the development of GIS capabilities for the four remaining ranger districts in the Forest.

Needs for an operational GIS approach to resource information analysis have been identified in the Park and Forest. Like most large wilderness areas of the western United States, a paucity of resource information existed in GNP prior to 1982, the onset of GIS development. The Park had no usable plant community map, for example (Colony, 1974). FNF had compiled a more in depth resource base to meet its multiple-use oriented mandate and administer its Forest Plan (USDA - Forest Service, 1983). For both agencies, established planes, such as topography, were compiled on paper. Their usefulness, however, was limited and inconvenient for applications requiring a synthesis of several reference planes, such as the overlay process for mapping wildlife habitat. Further, each time the data had to be analyzed in a somewhat unique manner, an expensive labor effort was required to requantify the database by the new criteria. As a consequence, many higher orders of information, such as forest fuel or grazing potential, which could have been compiled from existing references, were not developed.

To meet these needs, applications of the GIS have been designed to fall under three basic categories; inventory, modeling, and monitoring.

Inventory

The Interagency GIS affords an opportunity to fill voids in the resource inventory base. The pilot Landsat classification satisfies much of the need for a plant community map, and refinement in the near future should complete the process. The convenient quantification of area and length provides needed inventories for surface waters, roads, trails, drainage basins, timber stands, land ownership, and topographic strata. When digitally overlain, multiple data planes can be applied to inventory site-specific parameters, such as the distribution and abundance of huckleberry (*Vaccinium* spp.) types within a particular drainage.

Modeling

Modeling efforts involve manipulation of multiple GIS data planes providing output planes of more complex resource information. For example, modeling of distance corridor planes for roads, trails and streams, contribute a comparative means to address questions of man's influence on the environment and wildlife. Habitat component mapping, dependent on several contributive themes, like plant cover, distance from water, topography and fire history, is a significant current application for wildlife management and especially for threatened and endangered species. Wide ranging species, such as the grizzly bear, require regional databases to uniformly map habitat across agency boundaries (Butterfield and Key, 1985). In FNF, identity of potential elk calving locations is critical to management of that species (Langley, 1983). Fire behavior modeling, and the generation of fuel distribution data is of major interest as both agencies attempt to reestablish the natural role of fire while protecting property and human lives (Kessell, 1979; Barrett, 1983; and Key, 1983). Management toward minimizing bear/human confrontations (Keating, 1983) can benefit by modeling human-use patterns and bear habitat selection in the GIS context. Optimum strategies for keeping bears and humans apart, either temporally or spatially, may then be identified. From a research perspective, the GIS provides a medium for understanding macroscopic phenomena and ecosystemic processes. GNP is initiating a Biosphere Reserve study on the dynamics and genetics of insular populations using bighorn sheep as the subject species. GIS capabilities will be used to map distributions of seasonal habitats.

Several other modeling applications are being considered and will likely be addressed in the near future. Included in GNP are: mapping potential peregrine falcon nest sites; mapping woodland caribou habitat; modeling potential impacts from acid deposition; and modeling erosion impacts from natural fire. Identification of trail hazards, generation of a transportation plan, and precipitation/sedimentation runoff models are earmarked applications for FNF's HHRD.

Monitoring

The digital format of GIS data facilitates comparison between past and present conditions. Each geographic plane, whether derived through inventory or modeling, serves as a baseline by which changes can be detected and measured in the future. Monitoring further serves to test hypotheses and the validity of models over time. GIS maintenance, updating data planes as new information becomes available, is imperative for monitoring purposes. Applications in FNF will utilize temporally distinct Landsat scenes and will monitor timber disease and timber sales, for example. GNP will utilize monitoring to detect human influences on plant communities and wildlife distributions, thereby arriving at valid assessments of threats to Park resources. Monitoring will also provide understanding of relatively long-term natural processes such as plant succession after fire, and the progression of wide-spread forest pathogens.

SUMMARY

Development of Landsat based GIS techniques in GNP and FNF was facilitated by a cooperative interagency concept. Park and Forest personnel believe that without early EPA and NASA project involvements and the subsequent Interagency Agreement, no GIS effort would have been possible in either agency. Through contracts with WSUCSC's DIAL facility for GIS training, consulting, and computing support, both the Park and Forest were able to commence GIS development immediately with no capital expenditure for GIS equipment and no local GIS system installation and tune-up delays. In addition, the program has been developed by agency personnel establishing in-house GIS expertise and capabilities.

Now in an operational phase, GNP and FNF GIS implementation has passed the critical test of user acceptance and is becoming an integral part of strategies for achieving resource management and research goals. Incorporating inventory, modeling and monitoring techniques, the interagency GIS will provide a cost and time efficient method for addressing Park and Forest resource problems now and in the future.

REFERENCES CITED

Anon, 1983. "Flathead River Basin Environmental Impact Study" Flathead River Basin Steering Committee, Kalispell, MT, 184 pp.

Barrett, S. W., 1983. "Fire History of Glacier National Park: North Fork Flathead River Drainage" Supplement 22-C-2-INT-20, USFS Northern Forest Fire Laboratory, Missoula, MT, 68 pp.

Butterfield, B. R., and C. H. Key, 1985. "Mapping Grizzly Bear Habitat in Glacier National Park Using a Stratified Landsat Classification: A Pilot Study" Proceedings Grizzly Bear Habitat Symposium, Missoula, MT, (in press) Interagency Grizzly Bear Committee.

Colony, W. M., 1974. "Some Problems Arising in the Use of Maps to Maintain Basic Resource Inventory Data" USDI Park Service, Glacier National Park, West Glacier, MT, 9 pp.

Hart, J. A. and D. B. Wherry ed., 1984. VICAR/IBIS User Reference Manual, Volumes 1-4, Washington State University Computing Service Center, Pullman, WA, 876 pp.

Hart, J. A., D. B. Wherry and Stan Bain, 1985. "An Operational GIS for the Flathead National Forest," in Proceedings Seventh International Symposium on Computer-Assisted Cartography (AUTO-CARTO 7), Washington, DC. 10 pp.

Hoffer, R. M., 1979. "Computer-Aided Analysis of Remote Sensor Data - Magic, Mystery, or Myth?" in Proceedings of Remote Sensing for Natural Resources, University of Idaho, Moscow, ID, pp. 156-179.

Jaffray, M. J., R. S. Hansen, and J. A. Hart, 1984. "A Geographic Information System Application for Spokane County" in Proceedings 22nd Annual Conference of the Urban and Regional Information Systems Association, Seattle, WA, pp. 154-163.

Keating, K. A., 1983. "Trends in Grizzly Bear Numbers and Behavior (1910-1980), Relative to Sighting and Confrontation Rates in Glacier National Park, Montana" Glacier National Park, MT, 44 pp.

Kessel, S. R., 1979. Gradient Modeling, Springer-Verlag, New York, 432 pp.

Key, C. H., 1983. "Development of a Fire Information Map and Data Base for Glacier National Park" Supplement 22-C-INT-22, USFS, Northern Forest Fire Laboratory, Missoula, MT, 55 pp.

Langley, P. G., 1983. "Multiresource Inventory Methods Pilot Test, Final Report" USDA - Forest Service, Nationwide Forestry Applications Program, Houston, TX, Report No. NFAP 309.

USDA - Forest Service, USDI - Park Service, 1982. Interagency Agreement between Flathead National Forest, USDA - Forest Service and Glacier National Park, USDI - Park Service, USDA - Forest Service, FS-01-82-020, USDI - Park Service, CA-1430-82-18, 3 pp.

USDA - Forest Service, 1983. Proposed Forest Plan, Flathead National Forest, Kalispell, MT, p. II-28.

Wherry, David B., and S.Z. Friedman, 1978. "Cartographic Applications of an Image Based Information System," in Proceedings of the International Conference on Computer-Assisted Cartography - AUTO-CARTO III, published by the American Congress on Surveying and Mapping, pp. 148-168.

LEAF-OFF, REMOTELY-SENSED DATA AS A SOURCE OF FOREST RESOURCE INFORMATION

William D. Hudson
David P. Lusch
Center for Remote Sensing
Michigan State University
East Lansing, MI 48824-1111

ABSTRACT

Forest resource analysts have traditionally relied, almost exclusively, upon aerial photography and other remotely-sensed data acquired during the growing season. The ready availability of leaf-off, high-altitude, color infrared aerial photography (NHAP), as well as multi-temporal Landsat data, for most of the country make these two important additional sources of forest resource information. Leaf-off aerial photography is recommended as a supplement to "traditional," leaf-on photography wherever deciduous-coniferous mixtures occur. In addition to mid-growing season (summer) coverage, analysis of fall, winter, or spring Landsat data should also be considered as a valuable source of forest resource information.

SEASONAL CONSIDERATIONS

On the basis of varying user needs, aerial photographs may be divided into two broad groups: that acquired during the growing season (leaf-on) or during the dormant (leaf-off) season (Avery and Berlin, 1985). Leaf-off aerial photography is typically utilized for compiling topographic maps, identifying landforms, delineating soil boundaries, and a host of other tasks which require minimum obscuration of ground features by vegetation. In contrast, foresters, range managers, and others interested in analyzing the vegetation typically prefer leaf-on aerial photography. There are exceptions to this generalized division, however. For example, the Coast and Geodetic Survey routinely acquires leaf-on photography to permit accurate tree-height determinations in connection with their airport obstruction charting program (Swanson, 1964; Craunt, 1968). Another exception is photo interpretation of conifer-dominated woodlands (e.g. the boreal forest). Inventories of these forest ecosystems are frequently done with leaf-off photography (Aldred and Kippen, 1967; Aldred and Lowe, 1978; Nielson et al., 1979).

Whenever airphoto missions are planned, seasonal considerations should form a part of the specifications and generalized guidelines are available (Avery and Meyers, 1962; Sayn-Wittgenstein, 1967; Avery, 1970). Regarding tree species identification, Sayn-Wittgenstein (1961; 1978) has summarized the major phenological events with respect to the timing of airphoto acquisitions. Specialized photo-interpretation tasks, such as forest regeneration assessments (Kirby, 1980; Goba et al., 1982), should carefully formulate their temporal specifications (Colwell and Marcus, 1961).

Standard forestry photointerpretation procedures (e.g. Zsilinszky, 1966; Hudson, 1984) have stressed the use of leaf-on photography wherever deciduous trees are an important component of the vegetative assemblage. Leaf-on aerial photography is a logical choice whenever differences among hardwood species are required, although there may be exceptions (Newman and Shain, 1976).

Many inventories, especially in areas with a heterogeneous mixture of deciduous and coniferous forest types, may benefit from the use of multi-seasonal (leaf-on and leaf-off) photographic coverage (Hill and Evans, 1982). The recent availability of leaf-off, high-altitude, color infrared (CIR) aerial photography (National High-Altitude Photography Program, NHAP) over the entire continental U.S. may provide this additional source of forest resource information (Antill, 1982).

LEAF-OFF AERIAL PHOTOGRAPHY

Throughout the northern Lake States, a mosaic of deciduous forest types alternate with conifer-dominated forests -- the hemlock-white pine-northern hardwood association (Braun, 1950). Dominance by a single type varies and, complicated by disturbances, results in an often complex intermingling of various proportions of deciduous and coniferous species.

In a forest setting such as this, the unique capabilities of leaf-off CIR aerial photography provides an invaluable supplement to leaf-on airphotos. Of particular concern to the photo interpreter, are those instances where a deciduous overstory completely obscures the presence of a coniferous understory. For example, we have encountered stands which would be classified from leaf-on aerial photography as completely deciduous (aspen-birch or balsam poplar). Intrepretation of these areas on NHAP leaf-off photography, however, revealed a well-stocked understory of coniferous species (northern white-cedar, white spruce, and balsam fir). Subsequent field verification of one of these stands indicated that the coniferous species accounted for the majority of the basal area (100 sq. ft./acre) and volume, compared to the deciduous overstory (only 20 sq. ft./acre).

Although an experienced photo interpreter may have been able to infer the presence of these coniferous understories (based on site, overstory composition, and a knowledge of local environmental and successional relationships), the leaf-on airphotos provided no information by which to fully characterize these stands. This last point is particularly important. Even when the overstory does not completely obscure the understory, the intrepreter is frequently unable to accurately measure the understory. Wherever coniferous and deciduous species intermix, even if one doesn't "over-top" the other, we have found that the use of leaf-off photography enables the interpeter to better quantify the spatial arrangement of the stand. In instances where the coniferous species (e.g. white or red pine) are taller than the surrounding hardwood stand, the spatial extent of the conifers is highlighted on the leaf-off photography.

Although leaf-on aerial photography will continue as the "standard" for many forest photointerpetation tasks in areas where deciduous species are important, resource managers should not overlook the added information which may be derived from leaf-off photography. Especially now with the availability of NHAP leaf-off aerial photography for the entire U.S., we recommend its use as a supplement to "traditional" leaf-on photography.

LEAF-OFF LANDSAT DATA

The acquisition of multi-spectral, multi-temporal (including leaf-off) data from the Landsat series of satellites has provided a voluminous source of potential forest resource data. Although seasonal recommendations vary, to date the majority of forestry applications of Landsat data have relied on the analysis of scenes acquired during the growing season (e.g. Mead and Meyer, 1977; Bryant et al., 1980; Roller and Visser, 1980). Several Landsat applications conducted at the Center for Remote Sensing, Michigan State University will be used to illustrate the utility of leaf-off satellite data to provide forest resource information.

An evaluation of the accuracy of mapping small forestlands in southwestern Michigan from Landsat MSS imagery compared two acquisition dates and two image products (Karteris et al., 1981). For a winter (February), snow-covered scene, a black and white, positive transparency of band 5 was compared with a standard false-color composite produced by the EROS Data Center (EDC). For a second scene, acquired in the fall (September), a standard EDC false-color composite and a custom-made, diazo-enhanced color composite were compared. The diazo color composite was contrast-stretched to enhance the forested areas using the densitometric procedure outlined by Lusch (1981).

Separate forest/non-forest maps were compiled by visually interpreting each of the four Landsat images. Forest areas as small as one hectare (2.5 acres) were delineated. The overall mapping accuracies ranged from 74.0 to 98.5 percent and were higher for the winter scene than for the fall scene; the highest accuracy was achieved with the winter false-color composite. The diazo enhancement of the fall scene improved the mapping accuracy over the standard false-color composite. A spatial analysis of the error units showed that most of them were less than 4 hectares (10 acres) in size and that over 83 percent of all commission and omission errors were along forest/non-forest boundaries.

Franklin, et al. (1983) evaluated the utility of computer-enhanced Landsat imagery for mapping coniferous forest types in the northern Lower Peninsula of Michigan. They visually interpreted a false-color composite of a spring (April) scene which had undergone radiometric restoration, contrast enhancement, edge enhancement, and synthetic line generation. Prior to the actual image interpretation, the analysts were given intensive training which included the development of photo keys illustrating the appearance of the different coniferous forest types on Landsat false-color composites. Additionally, the interpreters systematically compared several examples of each forest type on high-altitude color infrared photography with their appearance on the Landsat color composite; forest inventory measurements (species, stocking, diameter, and height) for each of these training stands were available.

The visual interpretation procedures were tested over two sites to determine the feasibility of identifying four coniferous cover types (red pine, jack pine, pine mixtures, and swamp conifers). The mapping accuracies achieved for each test site are summarized in Tables 1 and 2. Overall classification accuracies were 84.8 and 72.7 percent, whereas the accuracies of interpreting the individual cover types ranged from a low of 32.2 percent for mixed pine stands to a high of 95.1 percent for jack pine plantations. Most of the mapping errors involved a confusion among the individual pine species (red pine, jack pine, and pine mixtures). Accuracies of the combined pine classes, were 93 and 77 percent. The swamp conifer type had consistently low interpretation accuracies in both test areas. As a single broad category, coniferous woodland was interpreted with an accuracy of 90 and 81 percent for the two sites.

Additional research has been aimed at developing automated techniques (i.e. computer classification of Landsat MSS digital data) for the identification and characterization of coniferous forest types in the northern part of Michigan's Lower Peninsula. Landsat-3 MSS data, acquired on February 26, 1979 (E-30358-15471), were used in an analysis of the same test sites referenced above. At this date, there was an average of 58.4 cm of snow on the ground as reported by the 17 weather stations in the area. Almost all of this snowfall occurred prior to several days before the Landsat overpass. As a result, virtually all non-forest cover types, including inland lakes, exhibited the spectral response of snow. Although the hardwood forests were leafless, their extensive mass of trunks and branches substantially altered the reflectance of the underlying snowpack. The coniferous forests, mostly red pine, jack pine, pine mixtures, and swamp conifers (primarly northern white-cedar), provided the only green-foliage reflectance in the entire scene.

A variety of "standard" classifiers (e.g. unsupervised clustering, minimum distance to the mean, and maximum likelihood) were evaluated in terms of their accuracy for disciminating among coniferous forest types (Table 3). Subsequent analysis of the digital brightness values led to the development of spectral response curve models (Hudson and Lusch, 1984). These models predict the multi-band brightness values corresponding to mixtures of various cover types on the basis of their spatial extent in the instantaneous field of view of the MSS instrument.

The Wexford and Crawford county test sites were classified using a two-band linear combination (BV6-BV5, BV6) with class thresholds determined by the response curve models. The resulting accuracies (Tables 4 and 5) exceeded those obtained by any of the "standard" classifiers, but, except for an improved discrimination of pine mixtures, were less than those achieved by visual interpretation procedures. Current research is attempting to quantify the ability of the spectral response curve models to provide a measure of stocking levels within the coniferous forest types.

The results from these three projects clearly indicate that, in addition to the more traditional mid-growing season (summer) coverage, Landsat data acquired during the fall, winter, or spring (i.e. leaf-off) can provide valuable forest resource information. Managers of any ecosystem in which conifers are an important component should not overlook the utility of these leaf-off, remotely sensed data.

ACKNOWLEDGEMENTS

This research was supported by a National Aeronautics and Space Administration grant, NASA NGL 23-ØØ4-Ø83, to Michigan State University, Center for Remote Sensing.

REFERENCES

Aldred, A.H., and F. W. Kippen. 1967. Plot Volumes from Large-Scale 7Ø mm Air Photograhs. Forest Science 13(4): 419-426.

Aldred, A.H., and J.J. Lowe. 1978. Application of Large-Scale Photos to a Forest Inventory in Alberta. Information Report FMR-X1Ø7, Canadian Forestry Service, Forest Management Institute, Ottawa, Ontario.

Antill, P.A. 1982. The National High-Altitude Photographic Data Base. Technical Papers, 1982 ACSM-ASP Fall Convention, American Congress on Surveying and Mapping and American Society of Photogrammetry, Falls Church, Virginia, pp. 28-37b.

Avery, T.E. 197Ø. Photointerpretation for Land Managers. Kodak Publication M-76, Eastman Kodak Co., Rochester, New York. 26p.

Avery, T.E., and G.L. Berlin. 1985. Interpretation of Aerial Photographs. Burgess Publishing Co., Minneapolis, Minnesota. 554p.

Avery, T.E., and M.P. Meyer. 1962. Contracting for Forest Aerial Photography in the United States. U.S.D.A. Forest Service, Lake States Forest Experiment Station, Station Paper No. 96. 37p.

Braun, E.L. 1950. Deciduous Forests of Eastern North America. Blakiston, Phildelphia. 596p.

Bryant, E., A.G. Dodge, Jr., and S.D. Warren. 1980. Landsat for Practical Forest Type Mapping: A Test Case. Photogrammetric Engineering and Remote Sensing 46(12): 1575-1584.

Colwell, R.N., and L.E. Marcus. 1961. Determining the Specifications for Special Purpose Photography. Photogrammetric Engineering 27(4): 618-626.

Craunt, H.R. 1968. Planning and Operation of a Color Aerial Photographic Mission. In Smith, John T., Jr., and Abraham Anson (eds.). Manual of Color Aerial Photography. American Society of Photogrammetry, Falls Church, Virginia. 550p.

Franklin, K.L., W.D. Hudson, and C.W. Ramm. 1983. Landsat Imagery for Identifying Coniferous Forest Types in Michigan. Research Report 448, Michigan State University Agricultural Experiment Station, East Lansing, Michigan. 7p.

Goba, N., S. Pala, and J. Narraway. 1982. An Instruction Manual on the Assessment of Regeneration Success by Aerial Survey. Ontario Centre for Remote Sensing, Ministry of Natural Resources. 49p.

Hill, J.M. and D.L. Evans. 1982. Bottomland Forest Community Delineation with Color-Infrared Aerial Photographs. LSU Forestry Notes #138, Agricultural Experiment Station, Louisiana State University, Baton Rouge, Louisiana.

Hudson, W.D. 1984. Interpreting Michigan Forest Cover Types From Color Infrared Aerial Photographs. Research Report 452, Michigan State University Agricultural Experiment Station, East Lansing, Michigan. 27p.

Hudson, W.D., and D.P. Lusch. 1984. Spectral Response Curve Models Applied to Forest Cover-Type Discrimination. Tenth International Symposium on Machine Processing of Remotely Sensed Data. Purdue University, West Lafayette, Indiana. pp. 175-179.

Karteris, M.A., W.R. Enslin, and J. Thiede. 1981. Area Estimation of Forestlands in Southwestern Michigan from Landsat Imagery, Second Eastern Regional Remote Sensing Applications Conference. NASA Conference Publication 2198, pp. 147-155.

Kirby, C.L. 1980. A Camera and Interpretation System for Assessment of Forest Regeneration. Information Report NOR-X-221, Northern Forest Research Centre, Canadian Forestry Service, Edmonton, Ablerta.

Lusch, D.P. 1981. Diazo Processing of Landsat Imagery: A Low-Cost Instructional Technique. CORSE-81, The 1981 Conference on Remote Sensing Education. NASA Conference Publication 2197, pp. 128-132.

Mead, R. and M. Meyer. 1977. Landsat Digital Data Application to Forest Vegetation and Land Use Classification in Minnesota. Fourth Annual Symposium on Machine Processing of Remotely Sensed Data. Purdue Univ., West Lafayette, Indiana. pp. 270-279.

Newman, C.S., and W.A. Shain. 1976. Aerial Photography As A Tool for Determining Detailed Forest Land Use Information. In Shahrokhi, F. (ed.). Remote Sensing of Earth Resources, Volume V. Space Institute, The University of Tennessee, Tullahoma, Tennessee, pp. 263-275.

Nielson, U., A.H. Aldred, and D.A. MacLeod. 1979. A Forest Inventory in the Yukon Using Large-Scale Photo Sampling Techniques. Information Report FMR-X-121, Canadian Forestry Service, Forest Management Institute, Ottawa, Ontario.

Roller, N.E.G., and L. Visser. 1980. Accuracy of Landsat Forest Cover Type Mapping In the Lake States Region of the U.S. Proceedings, Fourteenth International Symposium on Remote Sensing of Environment, San Jose, Costa Rica. pp. 1511-1520.

Sayn-Wittgenstein, L. 1961. Phenological Aids to Species Identification on Air Photographs. Technical Note No. 104, Forest Research Branch, Canada Department of Forestry, Ottawa, Ontario.

Sayn-Wittgenstein, L. 1967. The Best Season for Aerial Photography. Transactions of the Second International Symposium on Photo-Interpretation, Paris.

Sayn-Wittgenstein, L. 1978. Recognition of Tree Species on Aerial Photographs. Information Report FMR-X-118, Forest Management Institute, Canadian Forestry Service, Ottawa, Ontario.

Swanson, L.W. 1964. Aerial Photography and Photogrammetry in the Coast and Geodetic Survey. Delivered at the Congress of the International Society of Photogrammetry at Lisbon, Portugal.

Zsilinszky, V.G. 1966. Photographic Interpretation of Tree Species in Ontario. Ontario Department of Lands and Forests. 86 p.

Table 1. Landsat Classification Performance, Visual Interpretation Data, Wexford County Test Site

Known Cover Type	Number of Sample Points Classified as: Red Pine	Jack Pine	Pine Mixtures	Swamp Conifers	Non-conifer	Total	Percent[1] Correct
Red Pine	774	8	130	26	103	1041	74.4
Jack Pine	31	391	17	10	48	497	78.7
Pine Mixtures	71	2	87	9	13	187	46.5
Swamp Conifers	16	2	0	143	36	197	72.6
Non-conifer	57	3	36	89	2590	2775	93.3
Total	949	411	270	277	2790	4697	
Percent[2] Correct	81.6	95.1	32.2	51.6	92.8		**84.8**[3]

Table 2. Landsat Classification Performance, Visual Interpretation Data, Crawford County Test Site

Known Cover Type	Number of Sample Points Classified as: Red Pine	Jack Pine	Pine Mixtures	Swamp Conifers	Non-conifer	Total	Percent[1] Correct
Red Pine	23	38	0	1	16	78	29.5
Jack Pine	9	1500	18	18	33	1578	95.1
Pine Mixtures	0	23	33	11	10	77	42.9
Swamp Conifers	0	125	1	398	19	543	73.3
Non-conifer	2	301	1	222	307	833	36.9
Total	34	1987	53	650	385	3109	
Percent[2] Correct	67.6	75.5	62.3	61.2	79.9		**72.7**[3]

[1]considering only omission errors

[2]considering only commission errors

[3]overall classification accuracy; ratio of the sum of diagonal values to the total number of points

Table 3. Landsat Classification Performance Using "Standard" Classifiers

	Overall Classification Accuracy (%)	
	Wexford County	Crawford County
Default Cluster	77.6	64.8
Cluster with smaller radius	78.8	64.6
Level-Sliced default clusters	79.3	64.2
Level Sliced cluster with smaller radius	80.1	64.8
Minimum distance	79.3	65.2
Maximum Likelihood	79.4	65.1

Table 4. Landsat Classification Performance, (BV6-BV5, BV6) data, Wexford County Test Site

Known Cover Type	Number of Pixels Classified as -- Red Pine	Jack Pine	Pine Mixtures	Swamp Conifers	Non-Conifer	Total	Percent[1] Correct
Red Pine	4497	177	35	--	665	5367	83.8
Jack Pine	358	733	37	--	552	1680	43.6
Pine Mixtures	811	199	56	--	145	1211	4.6
Swamp Conifers	130	133	7	--	68	338	0.0
Non-conifer	757	405	5	--	13292	14459	91.9
Total	6553	1647	133	--	14722	23055	
Percent[2] Correct	68.6	44.5	42.1	0.0	90.3		**80.6**[3]

Table 5. Landsat Classification Performance, (BV6-BV5, BV6) data, Crawford County Test Site

Known Cover Type	Number of Pixels Classified as -- Red Pine	Jack Pine	Pine Mixtures	Swamp Conifers	Non-Conifer	Total	Percent[1] Correct
Red Pine	10	259	1	29	36	335	3.0
Jack Pine	5	4795	0	200	969	5965	80.3
Pine Mixtures	3	545	2	34	146	730	0.3
Swamp Conifers	2	769	0	585	60	1416	4.3
Non-conifer	2	1892	3	36	4229	6159	68.7
Total	22	8257	3	884	5440	14605	
Percent[2] Correct	45.4	58.0	66.7	66.2	77.7		**65.8**[3]

[1]considering only omission errors

[2]considering only commission errors

[3]overall classification accuracy; ratio of the sum of diagonal values to the total number of points

THE EVALUATION OF THEMATIC MAPPER DATA FOR RANGE MANAGEMENT APPLICATIONS IN WESTERN CANADA

K.P.B. Thomson
Canada Centre for Remote Sensing
717 Belfast Road
Ottawa, Ontario
K1A 0Y7

C. Gosselin
Intera Technologies Limited
Ottawa, Ontario

B.W. Adams
Alberta Energy and Natural Resources
Lethbridge, Alberta

I. Sutherland
Alberta Remote Sensing Center
Edmonton, Alberta

Previous studies (Brown et al., 1983) have shown that single-date LANDSAT MSS data were unable to provide reliable information concerning rangeland condition in the fescue grasslands. However, simulated Thematic Mapper (TM) imagery obtained from airborne data in 1981 and 1982 indicated that the TM short-infrared spectral region added significant information covering the density of the total biomass (green and brown) (Thomson et al., 1984).

A project to evaluate Thematic Mapper data for range management applications was started by the Canada Centre for Remote Sensing (CCRS) in 1984. The experiment is a joint effort between CCRS, the Alberta Department of Energy and Natural Resources, and the Alberta Remote Sensing Center.

The study area was located at the Waldron ranch in southwest Alberta. The foothills-type rough fescue grassland found at this location is subdivided into many individual fields with moderate slopes. Comparisons can easily be made between fields where grazing management has produced differences in vegetation cover.

Fifteen test sites were sampled on four dates during the grazing season. These dates coincided with LANDSAT overpasses, including June 6, July 7, August 8 and September 10. It should be borne in mind when considering the results that the 1984 growing season was abnormally dry.

The analysis of the TM spectral data, enhanced TM image data and the ground data showed that the short-infrared reflectance (TM 5 and TM 7) increases as the brown biomass increases. As expected, the near infrared (TM 4) reflectance shows an increase as the green vegetation component increases. This increase corresponds to the green-up period in the spring. The green vegetation pushes through and masks the litter, causing a decrease in the short-infrared reflectance (TM 5 and TM 7). In the July to August period the green grass senesces and the amount of litter increases. This causes a reduction in the near infrared (TM 4) reflectance with a corresponding increase in the short-infrared reflectance as the brown biomass or litter increases.

Green-up is detected earlier by TM 4 in the low biomass fields than in the high biomass fields. This is believed to be due to the large amounts of litter, which, in the high biomass fields, would effectively hide the emerging green vegetation. Enhanced colour composite images combining three of the four bands TM 3, TM 4, TM 5 and TM 7 provide significant information on range utilization. In particular, early green-up low-litter sites can be identified from high-litter late green-up sites when using one image from the beginning of the growing season and one from early July.

NOTE: This is an extended abstract. The full text illustrated with colour plates will be published in a scientific journal in 1986.

REFERENCES:

Brown, R.J., F.J. Ahern, K.P.B. Thomson, K. Staenz, J. Cihlar, C.M. Pearce and S.G. Klumph, 1983. Alberta rangeland assessment using remotely sensed data. Canada Centre for Remote Sensing, Research Report 83-1, 128 p.

Thomson, K.P.B., F.J. Ahern, R.J. Brown, C.M. Pearce, S. Hoyles, G. Fedosejevs, 1984. New remote sensing techniques for monitoring the fescue grasslands of Alberta. Proceedings of the 8th Canadian Symposium on Remote Sensing, Montreal, 1983, pp. 543-558.

COLOR INFRARED AERIAL VIDEO: APPLICATIONS IN NATURAL RESOURCE MANAGEMENT

Douglas Meisner
Remote Sensing Laboratory
University of Minnesota
St. Paul, Minnesota 55108

ABSTRACT

The University of Minnesota Remote Sensing Laboratory has developed a video camera which generates color infrared imagery. Advantages and disadvantages of the use of the camera in remote sensing applications are discussed. Several trial applications in forestry planned for this flight season are described. Issues relating to acquisition and interpretation of aerial video data are also discussed.

INTRODUCTION

Interest in the use of video imagery in remote sensing application has greatly increased in the past few years. This interest has resulted in the organization of an ad-hoc committee on video technology in the American Society for Photogrammetry and Remote Sensing, and a newsletter on video remote sensing has been published for several years (Nixon, ed., 1984-85).

During 1983, the Remote Sensing Laboratory at the University of Minnesota developed a multispectral video camera specifically for remote sensing application. E. Coyote Enterprises (Mineral Wells, Texas) has been licensed by the University of Minnesota to manufacture the camera, and is marketing it under the trade name Biovision. This camera uses an optical beamsplitter behind the lens to generate images simultaneously in three spectral bands: green (0.5-0.6 um); red (0.6-0.7 um); and near infrared (0.7-1.1 um). The three video signals are combined electronically to provide a standard (NTSC) color composite video signal using the color scheme of color infrared film and Landsat color composites: green is displayed as blue; red is displayed as green, and near infrared is displayed as red. The result is a video image equivalent to a color infrared photograph. Because the camera generates a standard color composite video signal, normal video equipment can be used to display or record the imagery. Similarly, standard video cassettes will store the data.

The size of the camera is approximately 4x8x12 inches, making it easy to use in a variety of mounting configurations. It weighs about 8 lbs. (3.5 kg). We use a 5-inch screen portable color monitor (JVC model TM-22U) to view the imagery during flight and when field checking. This screen size seems to be a good compromise between image size and monitor compactness. We record with a VHS-format portable video cassette recorder (Pentax model 1100). This unit has a five-head design for high quality freeze-frame capability and includes a six digit character generator for on-screen annotation of the video data. The recorder measures 10x10x3 inches and weighs under 10 lbs. The audio input for the VCR is plugged into an aircraft intercom system, providing a continuous audio-track annotation on the tape. All three pieces of equipment are powered by a rechargeable lead-acid battery pack (Gel-Cell model no. PP12120) using cigarette lighter-type power cords. This battery provides 3-4 hours of power. In aircraft having 12 volt systems (or in cars during field checking) the system can be operated directly via the cigarette lighter, and 24-12 volt converters are available for 24 volt aircraft.

This paper is intended to provide an introduction to the concept of color infrared video remote sensing, including advantages and disadvantages. It is hoped that this will help resource managers attending the conference to consider whether aerial video might be useful to them.

ADVANTAGES AND DISADVANTAGES

The Biovision camera was originally developed for projects in developing countries, where difficulties in obtaining color infrared film and processing has hampered efforts to use small format color infrared aerial photography (Meyer et al., 1983). In many countries, color infrared film must be obtained and processed abroad, adding weeks or months and considerable trouble to the analysis effort. The intention was to provide instant turnaround color infrared imagery. As the system was tested, it became apparent that a variety of additional advantages were also provided. These are:

Simplified acquisition. The availability of a live image during acquisition makes video more error proof than photography; if the system is set up wrong, the operator will be immediately aware of it. The live image also allows interactive exposure setting (always somewhat of a guessing game with color infrared film) and aids navigation by allowing the operator to be more sure that a particular site was covered.

Improved field checking. Post-flight field checking can be performed immediately with imagery in hand. This feature is important when dealing with changeable phenomena. With photography, the choice is between checking without imagery or delayed checking with the possibility of change having taken place.

Increased timeliness of data. Even with the availability of film and processing in this country, film processing can delay interpretation by several days to a week. For time-critical applications this can be a significant advantage.

Low media cost. In applications requiring coverage of large areas, video offers significant cost savings. A standard video tape cassette costs less than $10.00, one-half to one-third the cost of film and processing for a 36 exposure roll of color infrared film. Yet the tape provides two hours of coverage, compared with typically one to five minutes for the roll of film. In addition, the tape may be reused if the data do not need to be archived.

Potential for automated interpretation. The electronic format of video data makes it amenable to computer processing. A number of video digitizer products are available for converting aerial video imagery to digital data. This could be useful both for enhancement prior to manual analysis, or for automated computer classification.

Some disadvantages of video data are:

Low resolution relative to photography. Whereas even small format photography provides about 720 pixels across the field of view (at 20 line pairs per mm), video data generally provide only between 250-350 pixels. This translates to between 1-30 feet per pixel at typical video imagery scales.

As a result, video interpretation generally cannot be based on texture or shape details, and must depend more on spectral differences. In this way, the video image interpretation more in common with Landsat data analysis. Compared with satellite imagery, however, pixel sizes of 1-30 feet provide very good resolution.

Lack of stereo coverage. While end-lapped images can be easily generated from the continuous video coverage, the lack of detailed resolution at most scales limits the utility of stereo viewing.

Higher equipment cost. The cost of the full Biovision system ($12-13,000) is high compared with small format camera systems. While this may be a problem for the infrequent user, high volume users may find that the lower cost of tape vs. film will more than offset the higher initial equipment cost.

FLIGHT PLANNING CONSIDERATIONS

The potential video user should be aware of some points which affect the flight planning process. Photointerpreters frequently think in terms of image scale, but this parameter is not particularly meaningful in video. Whereas an image format size is specified for the Biovision camera (6.6 x 8.8mm, standard for 2/3-inch video tubes), from which a "contact scale" can be determined, there is no first generation, hard copy image corresponding to it. As in the case of satellite scanner data, the raw imagery has no absolute scale associated with it. A more appropriate measure for such electronic imagery is the ground dimensions of the pixel. Since the video signal is analog, there is not a set number of pixels in the image. At the March 1985 meeting of ASPRS, the members of the ad-hoc Committee on Video Technology decided that a value of 256 pixels per image width represented a reasonable (conservative) figure.

The 6.6 x 8.8mm image format size is necessary for computing aircraft altitude vs. coverage during flight planning. Since the 8.8mm dimension is about 1/4 the size of a 35mm frame, lens focal lengths must also be 1/4 the size. Thus, a 12.5mm lens provides normal coverage, an 8.5mm lens gives wide angle coverage, and a 25mm lens gives telephoto coverage. A sample flight planning table is included at the end of this paper. This table also lists the effective pixel sizes, based on 256 pixels per image width.

An intervalometer is not required when using video, since images are continuously recorded every 1/60 second. While this represents an excessive endlap factor of over 99%, the low cost of video storage makes it unimportant.

The relatively long exposure time of 1/60 second can cause blur due to forward image motion. Assuming image motion should be less than one image line per 1/60 second, the maximum aircraft speed for blur-free imagery can be computed for each pixel size or swath-width of coverage (Meisner, 1985). This value is listed on the right side of the flight planning table included at the end of this paper.

These figures show that forward image motion will not be a problem unless width of coverage is below 1000 feet (four-foot pixel size). In most applications, this is not a problem, since narrow coverage widths are inconvenient for two reasons: first, an unreasonably large number of flight lines are required for full coverage of an area; and second, it becomes difficult to locate oneself in the imagery when a narrow area is covered. Unlike photography, where a photomosaic can be constructed from many large scale photographs to allow orientation, video interpretation requires recognizable features in most image frames. These problems tend to limit the use of large scale/narrow coverage aerial video.

However, certain applications do not have such difficulties with narrow coverage. When full-area coverage is not required, large scale video can be very useful. Examples are coverage of linear features such as a narrow stream course, shoreline, or power line right of way; or a transect sample across an area. In these cases, image motion blur will be the limiting factor on image scale, requiring use of a shuttered camera. The Biovision camera is not currently available in a shuttered version.

INTERPRETATION PROCEDURES

Because video data are generally displayed on a video monitor screen, standard photointerpretation tools are not always directly usable. While hard copy video imagery can be generated (by simply photographing still-frame images), this eliminates the substantial cost advantage of video. Anticipated new products for making low cost color video hard copy may change this in the next 1-2 years.

Currently, however, most manual interpretation of video imagery is done using the "soft copy" video display. This is usually done with the still-frame feature of the video cassette recorder. High quality recorders provide reasonably noise-free images in this mode. Four head or five head machines often give superior still-frame results. Most video recorders will automatically stop after about five minutes in still-frame mode, under the assumption that the user may have forgotten and left the machine. This guards against possible tape or tape head damage, since the still-frame function requires that the tape track for the single frame be re-read 60 times a second.

While five minutes is sufficient for simple interpretations, longer still-frame analysis would require use of a digital freeze frame unit. The Robot Model 1200C (manufactured by Robot Research, San Diego, CA) converts the video image into three 256x240 pixel arrays (one for each spectral band) with 64 grey levels per pixel and color. This unit can display a single image indefinitely. It also allows viewing each channel separately as a black and white image, has a computer interface and a modem for transmission of imagery over phone lines to other Robot 1200C units. The system costs about $1300. One such unit has been in use at the University of Minnesota for the past year. In most applications, however, the still-frame function on the recorder is sufficient, and is more frequently used.

Actual interpretation procedures will vary between applications. In cases where positional accuracy is not critical, information can be visually transferred from the monitor screen to a map. In more critical applications, a Bausch and Lomb Zoom Transfer Scope can be modified to allow optical superimposition of a map and the video screen (Meisner and Lindstrom, 1985).

As mentioned previously, the electronic format of video data is amenable to digital analysis. A variety of low cost video digitizers are available, though most operate in black and white only. This includes remote sensing oriented systems, such as the LMS from Measuronics, VIP from Spectral Data, and ERDAS from Earth Resources Data Analysis Systems, as well as general purpose digitizer boards designed for microcomputers (Wilson, 1985).

Several systems do handle color or multispectral video, however. The Robot 1200C mentioned above can be interfaced to a computer via two parallel data lines. This interface is somewhat awkward (due to the need for two parallel ports on the computer), but the multiple uses of the unit and its low cost make it an interesting unit. The Innovion PGS provides a higher resolution image of 512x480 pixels x 24 bits per pixel (8 bits or 256 grey shades per spectral channel). It includes a serial computer interface and is priced at $3500.

Both of these systems have the advantage of being standalone units that can be used to display still frame imagery for manual analysis as well as to transfer the video image into a computer. As external units, however, they have slower data transfer rates, requiring several minutes to transfer an image to the computer. Faster data transfer rates are obtained by using internal circuit boards that interface directly to the computer data bus. AT&T manufactures a color video digitizer board which plugs into IBM PC-compatible microcomputers. This $1000 board digitizes the image into 256x240 pixels x 15 bits per pixel (5 bits or 32 shades per spectral channel).

Area measurement is frequently required in manual photointerpretation projects. Equipment commonly used (dot grids, polar planimeters, electronic digitizing tablets) is not readily adaptable to direct use on video monitor displays. Use of such equipment requires that a hard copy image be made, or that interpreted data be transferred to a map base.

On the other hand, image analyzers are area measurement tools which are video based. The more advanced systems, such as the Measuronics LMS, include computer programs for map registration. Although these systems are rather expensive (about $50,000), many are currently available in land management offices (especially federal agencies in the western U.S.). These systems use closed circuit T.V. cameras to provide video images of aerial photographs. Once in electronic format, the operator can locate areas of similar tonal values and instantly compute the area covered by those values. Interactive controls are provided to fine tune the analysis, allowing areas to be manually included or excluded from the tonal value class. These systems can process aerial video data in the same manner as the closed circuit T.V. input from aerial photos.

Currently available image analyzers are limited to processing black and white video imagery. With the availability of low cost color video digitizer boards, it is hoped that this limitation will not continue for long. Operator interaction with the system will need to be more complex if color is added, but the benefits in additional interpretability should be extensive.

APPLICATIONS IN FORESTRY

To date, the majority of the test flights with the University of Minnesota Biovision camera have involved agricultural applications. During the 1985 flight season, however, we are planning to try several forestry-oriented applications. Hopefully, preliminary results of these tests will be available when this paper is presented at the conference. Prospective applications are described below.

Aspen stand delineation. Aspen has become an economically important tree type in the upper midwest, but was not considered important when current inventories were made. As a result, aspen stands were not specifically mapped. Considerable work has been done recently using 35mm small format photography to map the stands (Hamann et al., 1984). Aspen stands can be easily differentiated if flights are carefully timed in early spring, due to early leaf-out of aspen, or fall, due to early leaf color change. Since at these times the differentiation is based on broad-area spectral differences rather than individual crown shape or texture, the application of aerial video seems to have promise. Planned test flights last spring were cancelled due to weather problems, but we intend to try again this fall.

Forest fire damage assessment. Rapid turnaround time is helpful for mapping damaged areas during the fire or immediately post-fire. This application was tested during August, 1984, over a large fire near Helena, Montana. While the infrared channel of the camera is not a thermal detector, the camera does provide excellent smoke penetration. Clear imagery was acquired at 6,000 feet above ground through smoke thick enough that the ground was not visible from the plane. The imagery would have been more useful if a wider field of view had been possible. This can be a problem when operating in non-pressurized aircraft over high-altitude terrain.

Forest disease mapping. The aerial sketchmapping procedure commonly used in forest pest management is extremely difficult work and is not as accurate as users would like. Test flights are planned for this summer to evaluate whether color infrared video could replace or supplement aerial sketchmapping. Several disease problems will be investigated, including oak wilt and spruce budworm. It is expected that the disease problems will have different degrees of success. In general, diseases will need to affect full stands rather than isolated crowns for the video data to be useful.

CONCLUSIONS

Color infrared aerial video is a new tool with unique characteristics for the remote sensing user. At this stage, few applications in forest and range management have been tried. Applications believed to have potential are time-critical data needs, which take particular advantage of the instant turnaround nature of video, and small scale mapping needs, where the lower resolution of video should not be a problem. In the case of small scale mapping, the interpretation needs to be based more heavily on spectral differences than on detailed spatial patterns such as individual crown shape or texture. To this end, the additional spectral information provided by a color infrared camera should prove beneficial.

ACKNOWLEDGEMENTS

The research reported in this paper was supported by the University of Minnesota, College of Forestry and Agricultural Experiment Station under Project MN-42-038.

Mention of trade names is intended for the readers benefit only and is not meant to imply endorsement by the University of Minnesota.

REFERENCES

Hamann, R., W. Johnson, N. Moody, and M. Meyer. 1984. "Small Scale 35mm Aerial Photography for Aspen Delineation." University of Minnesota Remote Sensing Laboratory Report No. 84-3. 25pp.

Meisner, D. E., and O. M. Lindstrom. 1985. "Design and Operation of a Color Infrared Aerial Video System." Photogrammetric Engineering and Remote Sensing, Vol. 51, No. 5, May 1985, pp. 555-560.

Meisner, D. E. 1985. "Fundamentals of Airborne Video Remote Sensing." Remote Sensing of Environment, in press.

Meyer, M. P., D. E. Meisner, W. L. Johnson, and O. M. Lindstrom. 1983. "Small Aircraft 35mm Photography and Video Imagery Applications to Developing Country Resource Surveys." University of Minnesota Remote Sensing Laboratory Report No. 83-1. 12pp.

Nixon, P. R., editor. 1984-85. "Video Technology Newsletter." Weslaco, Texas: USDA Agricultural Research Service, quarterly.

Wilson, A. C. 1985. "Image Processing with Personal Computers." Electronic Imaging, Vol. 4, No. 3, March 1985, pp. 44-51.

AERIAL VIDEO COVERAGE TABLE
(2/3-inch video format)

University of Minnesota Remote Sensing Laboratory

Coverage		Effective Pixel Size[1]	Altitude Required: Lens Focal Length (mm)					Maximum Aircraft Speed[2]
Width	Length		6.0	8.5	12.5	16.0	25.0	
(feet)		(feet)	(feet above ground)					(mph)
250	188	1	170	241	355	455	710	32
500	375	2	341	483	710	909	1,420	63
750	563	3	511	724	1,065	1,364	2,131	95
1,000	750	4	682	966	1,420	1,818	2,841	126
1,250	938	5	852	1,207	1,776	2,273	3,551	158
1,500	1,125	6	1,023	1,449	2,131	2,727	4,261	189
1,750	1,313	7	1,193	1,690	2,486	3,182	4,972	221
2,000	1,500	8	1,364	1,932	2,841	3,636	5,682	252
2,250	1,688	9	1,534	2,173	3,196	4,091	6,392	284
2,500	1,875	10	1,705	2,415	3,551	4,545	7,102	315
2,750	2,063	11	1,875	2,656	3,906	5,000	7,813	347
3,000	2,250	12	2,045	2,898	4,261	5,455	8,523	379
3,250	2,438	13	2,216	3,139	4,616	5,909	9,233	410
3,500	2,625	14	2,386	3,381	4,972	6,364	9,943	442
3,750	2,813	15	2,557	3,622	5,327	6,818	10,653	473
4,000	3,000	16	2,727	3,864	5,682	7,273	11,364	505
4,250	3,188	17	2,898	4,105	6,037	7,727	12,074	536
4,500	3,375	18	3,068	4,347	6,392	8,182	12,784	568
4,750	3,563	19	3,239	4,588	6,747	8,636	13,494	599
5,000	3,750	20	3,409	4,830	7,102	9,091	14,205	631
5,250	3,938	21	3,580	5,071	7,457	9,545	14,915	662
5,500	4,125	21	3,750	5,313	7,813	10,000	15,625	694
5,750	4,313	22	3,920	5,554	8,168	10,455	16,335	726
6,000	4,500	23	4,091	5,795	8,523	10,909	17,045	757
6,250	4,688	24	4,261	6,037	8,878	11,364	17,756	789
6,500	4,875	25	4,432	6,278	9,233	11,818	18,466	820
6,750	5,063	26	4,602	6,520	9,588	12,273	19,176	852
7,000	5,250	27	4,773	6,761	9,943	12,727	19,886	883
7,250	5,438	28	4,943	7,003	10,298	13,182	20,597	915
7,500	5,625	29	5,114	7,244	10,653	13,636	21,307	946
7,750	5,813	30	5,284	7,486	11,009	14,091	22,017	978
8,000	6,000	31	5,455	7,727	11,364	14,545	22,727	1,009
8,250	6,188	32	5,625	7,969	11,719	15,000	23,438	1,041
8,500	6,375	33	5,795	8,210	12,074	15,455	24,148	1,073
8,750	6,563	34	5,966	8,452	12,429	15,909	24,858	1,104
9,000	6,750	35	6,136	8,693	12,784	16,364	25,568	1,136
9,250	6,938	36	6,307	8,935	13,139	16,818	26,278	1,167
9,500	7,125	37	6,477	9,176	13,494	17,273	26,989	1,199
9,750	7,313	38	6,648	9,418	13,849	17,727	27,699	1,230
10,000	7,500	39	6,818	9,659	14,205	18,182	28,409	1,262
10,250	7,688	40	6,989	9,901	14,560	18,636	29,119	1,293
10,500	7,875	41	7,159	10,142	14,915	19,091	29,830	1,325
10,750	8,063	42	7,330	10,384	15,270	19,545	30,540	1,356

[1]Effective pixel size based on 256 pixels per format width.

[2]To avoid forward image motion blur if not using shuttered camera or forward image compensation.

LAND USE/LAND COVER MAPPING

Moderator: W. Miller

INTERIM PROGRAM FOR LAND COVER MAPPING IN ALASKA UTILIZING LANDSAT DIGITAL DATA*

Mark Shasby**
Technicolor Government Services, Inc.
U.S. Geological Survey
EROS Field Office
4230 University Drive
Anchorage, Alaska 99508-4664

David Carneggie
U.S. Geological Survey
EROS Field Office
4230 University Drive
Anchorage, Alaska 99508-4664

Leonard Gaydos
U.S. Geological Survey
Geographic Investigations Office
Ames Research Center 240-8
Moffett Field, California 94035

Katherine Fitzpatrick-Lins
U.S. Geological Survey
Office of Geographic and Cartographic Research
521 National Center
Reston, Virginia 22092

Donald Lauer
U.S. Geological Survey
Technique Development and Applications Branch
EROS Data Center
Sioux Falls, South Dakota 57198

Vincent Ambrosia
Susan Benjamin
Technicolor Government Services, Inc.
USGS/Geographic Investigations Office
Ames Research Center 240-8
Moffett Field, California 94035

ABSTRACT

The enactment of the Alaska National Interest Lands Conservation Act (ANILCA) in 1980 imposed mandates on all major land management agencies in Alaska to prepare comprehensive resource and managment plans to assess wildlife habitat, oil and gas exploration and development, wild and scenic rivers, land disposals, timber production, and archeological and cultural resources. To meet these objectives, the U.S. Geological Survey (USGS) has embarked

*Publication authorized by the Director, U.S. Geological Survey, on July 8, 1985.

**Technicolor Government Services, Inc., Anchorage, Alaska. Work performed under U.S. Geological Survey contract no. 14-08-0001-20129.

on a program to classify land cover for the entire State of Alaska using Landsat digital data. The USGS, in cooperation with other agencies, has completed Landsat-derived land use and land cover classification of 115 million acres for the State of Alaska. With this work as a substantial foundation, the USGS has prepared a comprehensive plan for classifying the remaining areas of Alaska. The development of this program will lead to a complete interim land use and land cover classification system for Alaska and provide for the dissemination of map products, statistics, and acreage summaries for all areas of Alaska at 1:250,000 scale. It also allows for the dissemination of Landsat digital data for those areas.

INTRODUCTION

The National Mapping Division (NMD) of the U.S. Geological Survey is responsible for producing and maintaining base maps and related thematic maps for the Nation. The Division carries out its mission by conducting a variety of national topographic mapping programs and a national land use and land cover mapping program. The latter is a congressionally mandated program which, to this point, has concentrated primarily on the 48 contiguous States and Hawaii. The program produces land use and land cover data (primarily at 1:250,000 scale) in graphic and digital form in response to Federal and State needs for consistent systematic information for land resource planning and management.

In Alaska, conventional land use and land cover mapping by the NMD has been limited to a single 1:250,000-scale quadrangle (Valdez). This is due primarily to (a) the limited availability of source material (a problem which essentially has been eliminated by the high-altitude aerial photography program active in Alaska since 1978), (b) the diversity of requirements for land cover maps by different agencies, (c) agency commitments to other resource mapping programs, and (d) the necessity for training air photointerpreters to apply the nationwide land use and land cover classification system to Alaska's predominantly wildland environment.

The enactment of the Alaska National Interest Lands Conservation Act (ANILCA) in December 1980, in particular, has imposed mandates on all major land management agencies in Alaska to prepare comprehensive resource and management plans which assess wildlife habitat, oil and gas exploration and development, wild and scenic rivers, land disposal, timber production, and archeological and cultural resources. To meet relatively short deadlines, each agency has adopted its own approach to land cover data collection, mapping, and compilation. Several agencies such as the Bureau of Land Management, U.S. Fish and Wildlife Service, U.S. Forest Service, and Alaska Department of Natural Resources are using digital Landsat data to produce land cover classifications. Others, such as the Alaska Department of Natural Resources, Alaska Department of Fish and Game, the Soil Conservation

Service, and the U.S. Forest Service are using both satellite images and aerial photographs to produce land cover and interpreted terrain maps. Although there is little consistency among the map legends produced by the various agencies, all agencies are using the same ground-based system (Viereck and others, 1980) to describe plant communities.

Objectives of the Interim Mapping Program

In Alaska, if conventional methods were used (namely manual interpretation of high-altitude aerial photographs) to delineate Level II land use and land cover categories (Anderson and others, 1976), standard land use and land cover program map products for the entire State could not be produced practically in the near term.

As an alternative, the Division has implemented an interim land cover mapping program to produce standard land cover maps (1:250,000 scale) and related products in Alaska. The Division's Earth Resources Observation Systems (EROS) Field Office in Anchorage, Alaska, and Geographic Investigations Office in Menlo Park, California, together with cooperating agencies (Bureau of Land Management, Alaska Department of Natural Resources, U.S. Fish and Wildlife Service, U.S. Forest Service, Army Corps of Engineers, and North Slope Borough) have produced land cover classifications from digital Landsat data for nearly one-half of the State. The interim mapping program includes reformatting existing land cover classifications to a common statewide map legend, completing the land cover classifications for Alaska, and producting paper maps, statistics, and computer-compatible tapes (CCT's) containing the land cover data for each of the 153 1:250,000-scale quadrangles in the State.

BACKGROUND

The USGS has been extracting land cover information from Landsat multispectral scanner (MSS) data since 1973 as part of its continuing research into the capabilities and applications of various types of remotely sensed data. The synoptic coverage, availability, and digital format of these spectral reflectance data showed early promise as economical alternatives to aerial photographs, especially where such conventional sources were lacking, as a tool for studying remote areas, and as input to digital geographic information systems.

During the first few years of this research, land cover maps were constructed by computer-assisted classification of Landsat MSS data for a number of areas. These research efforts provided analysts with an early appreciation of the relative advantages and disadvantages of this technology for land cover mapping. Apparent advantages included the availability of data for the entire land area of the country, resolution of data near the 1-acre level, and the digital format of the data allowing economical computer manipulation. Disadvantages included early problems with geometric and radiometric fidelity, lack of optimum coverage for specific seasons of the year, an inability to duplicate many of the categories being used in the nationwide land use and land cover mapping program, and a lack of success in extending spectral classes from one Landsat scene to another.

Today, many of the initial disadvantages have been overcome and Landsat data are being integrated into comprehensive digital spatial data bases with other types of map, map-like, and tabular data that are available in digital form or that are easily transferred to digital form. Advanced computer-assisted spatial analysis techniques are being applied to these data bases for deriving land cover as well as other types of resource information.

In 1977 a specific need for land cover data of a large remote region was expressed when the Bureau of Land Management (BLM) and the USGS assumed the responsibility for compiling a resource inventory and a land use study for the 97,000 square-kilometer National Petroleum Reserve in Alaska (NPRA). Based on the foregoing discussion of perceived advantages and disadvantages, it was decided to test the hypothesis that computer analysis of Landsat digital data could produce an inventory and map of land cover consistent with planning needs. A digital data base for the NPRA was created by the USGS Western Mapping Center's Geographic Investigations Office in which Landsat data were categorized into 10 vegetation classes in the form of 50-m cells tied to a Universal Transverse Mercator (UTM) map projection. These data were then registered with digital terrain data by BLM and have been used in assessing potential impacts of mineral exploration.

In 1980, the EROS Data Center expanded its remote sensing technology transfer program for the Department of the Interior and cooperating government agencies by opening a field office in Anchorage, Alaska. The mission of the field office was to meet the growing need by Alaskan scientists and resource managers to compile resource information. Field office personnel provide training in manual and automated analysis of remotely sensed data and make available an advanced digital image-processing system to government resource management agencies in Alaska. Field office personnel have worked cooperatively during the past 5 years with several Federal and State resource management agencies to produce vegetation and land cover classifications for approximately 135 million acres of Alaska. Classifications for an additional 50-60 million acres are in progress.

METHODOLOGY

The methodology for producing land cover classifications in Alaska employs digital Landsat and terrain data in a digital data base approach. The terrain data are arc-second digital elevation models (DEM's). Likewise, Landsat multispectral scanner (MSS) data are on computer-compatible tapes. Other source materials, like maps showing administrative boundaries or interpretations made from aerial photographs (for example, identification of urban, built up, or settled areas), are usually in map or map-like form, and are transformed into digital formats before they are entered into the data base.

For this program a data base with a grid-cell format is used for making land cover classifications. Raster-based data sets (such as elevation and Landsat data) and image processing equipment are readily available and multiple data sets are easily integrated and analyzed.

The procedures for preprocessing the Landsat data and making land cover classifications are summarized in table 1.

The post-classification procedures employ a variety of data categories to improve the separation of land cover classes. For example, digitized physiographic information can be used to segregate specific areas, and cluster statistics can be derived within these areas. In some cases, Landsat winter scenes registered to the data base are used to separate land cover classes that have similar spectral signatures on a spring or summer scene. Digital terrain data can be used in a similar way to select areas based on combinations of elevation, slope, and aspect. The accuracy of the Landsat classification results is improved when other types of digital data are used in the classification process.

An important aspect of the Landsat classification procedure is to ensure consistency among adjacent Landsat scenes. In a few cases, areas may need to be reclassified following establishment of new spectral classes. In other cases, misclassifications, such as misclassifying shadows on mountain slopes as water (a situation caused by the spectral similarity of water and shadow), are corrected by defining geographic regions where certain classifications cannot occur or by using digital elevation data limitations.

IMPLEMENTATION

To meet the objectives of an interim land cover mapping program, a three-phase approach is proposed which utilizes land cover classifications and terrain data from previous, current, and potential cooperative classification projects. A separate strategy for completing the mapping is proposed for the remaining areas of the State, where classifications do not exist or are not planned. All existing classifications will be integrated and reformatted using a common land cover map legend.

Phase I of the three-phase approach is underway and consists of identifying existing digital classifications that are suitable for use in the interim mapping program. The initial stages of this first phase are being addressed as a research effort designed to identify and evaluate all the different steps involved in converting the existing land cover maps from their agency-specific scheme to the interim scheme and onto standardized paper map products (six 1:250,000-scale quadrangles with existing classifications have been identified for conversion during this initial effort). Key elements of the research stage are the development of an acceptable standardized map legend and of an evaluation process following production of the maps involving State and Federal agencies.

Table 1. Land cover classification procedure used on Landsat data of Alaska

Analysis Procedures	Task Examples
Screen and preprocess Landsat data	Fix bad scan lines, perform radiometric destriping, and mosaic adjacent scenes.
Register Landsat and ancillary data to map base	Select and digitize control points, generate transformation, resample, and register digital data base.
Identify area(s) to be classified	Digitize study area boundaries and create appropriate digital masks.
Stratify raw Landsat data	Identify environmentally similar areas, locate training blocks within similar types, and develop training statistics.
Perform preliminary classification	Select decision rules, classify training blocks, and produce preliminary classification map.
Field verification	Visit training blocks in the field and describe areas associated with each spectral class.
Final classification	Use field and laboratory data to edit cluster statistics, reaggregate data clusters. Use post-classification procedures to make final map product. Adjacent map products are mosaicked and formatted to cover 1° by 3° map quadrangle.

Note: In some areas north of the 68th parallel, the raw Landsat data are not registered to the map base; instead, the classification results are resampled to 50-meter grid cells and registered to a Universal Transverse Mercator map projection.

The development of a common classification scheme for Alaska for the interim series is necessary to ensure consistency in form, resolution, and content from quadrangle to quadrangle and throughout the State. A series of interagency workshops attended by Federal and State personnel experienced in the use of Landsat data for mapping vegetation and land cover have been conducted with the sole purpose being the development of a classification scheme for the interim program. The workshop objectives were to identify a scheme that adequately addresses the information needs of the National Mapping Division's Land Cover Mapping Program and to provide a legend useful to the participating agencies. The classification legend, comprised of 17 land use/land cover types, that resulted from these workshops is presented in table 2. At present, it represents a combination of the Anderson and others (1976) Level II scheme and the Viereck and Dyrness (1980) vegetation classification system, which is the most widely used system for describing vegetation communities in Alaska. Establishing a link between the interim scheme and the Viereck system is a critical element in ensuring utility of the interim map products.

To begin, six interim land cover maps will be produced as a research project. Development of the six map products will be shared by the EROS Field Office in Anchorage and the Geographic Investigations Office (GIO) at the Western Mapping Center in Menlo Park, California. GIO will prepare data sets for the Mt. Michelson and Meade River quadrangles, while the EROS Field Office will work on the Arctic, Valdez, Fairbanks, and Dillingham quadrangles. As part of the mapping effort, area statistics for each township and range will be generated and presented on the back of each paper map product.

Following the development of the interim products for the six quadrangles and a review of the map products and their associated digital data bases by Federal and State agencies, Phase I of the Interim Program will move into a production mode, pending final approval by National Mapping Division. The existing land cover classifications which may be included in the proposed program come from the following projects or areas and are shown in figure 1:

- National Petroleum Reserve of Alaska (NPRA), and the Arctic Coastal Plain of the Arctic National Wildlife Refuge. These land cover classifications were produced by the Western Mapping enter's Geographic Investigations Office in conjunction with various cooperators.

- Kenai, Togiak, Tetlin, Yukon Flats, Kanuti, and Arctic National Wildlife Refuges (NWR). These classifications were produced cooperatively by EROS Field Office and USFWS personnel.

- Bristol Bay Subregion. These classifications were produced by EROS Field Office and Alaska Department of Natural Resources personnel for ADNR and the U.S. Fish and Wildlife Service.

Table 2. Proposed interim digital Landsat classification for use in Alaska

Proposed USGS Alaska Land Cover Classification Scheme		Corresponding Viereck et al.
Level I	**Level II**	**Level III Categories**
I. Forest (forest canopy cover of one-third or more)	**A.** Needleleaf forest (over two-thirds of tree cover contributed by needleleaf species)	Closed needleleaf forest Open needleleaf forest
	B. Broadleaf forest (over two-thirds of tree cover contributed by broadleaf species)	Closed broadleaf forest Open broadleaf forest
	C. Mixed forest (broadleaf or needleleaf species contribute one-third to two-thirds of the tree cover)	Closed mixed forest Open mixed forest
II. Shrub (forest canopy cover of less than one-third and shrub canopy cover of one-third or more)	**A.** Shrublands	Closed tall shrub scrub Closed low shrub scrub Open low shrub scrub Neddleleaf woodland Broadleaf woodland
	B. Dwarf shrublands and related communities (rarely exceeding 50 cm in height)	Closed low shrub scrub Open low shrub scrub Closed dwarf shrub scrub Open dwarf shrub scrub Needleleaf woodland
III. Herbaceous (vegetation with 5% or more of vascular and non-vascular [mosses and lichens] cover and less than one-third cover of woody plants)	**A.** Dry or moist	Dry graminoid herbaceous Mesic graminoid herbaceous Dry forb herbaceous Mesic forb herbaceous Needleleaf woodland Broadleaf woodland
	B. Wet	Wet graminoid herbaceous Wet forb herbaceous
	C. Very wet-aquatic	Aquatic herbaceous
	D. Mosses	Mosses Needleleaf woodland Broadleaf woodland
	E. Lichens	Lichens Needleleaf woodland Broadleaf woodland
IV. Agriculture		The Viereck et al. classification system is strictly for vegetation and does not apply to land use or non-vegetated lands.
V. Urban		
VI. Barren	**A.** Sparse vegetation **B.** Non-vegetated (rock, soil)	
VII. Water		
VIII. Ice and snow		

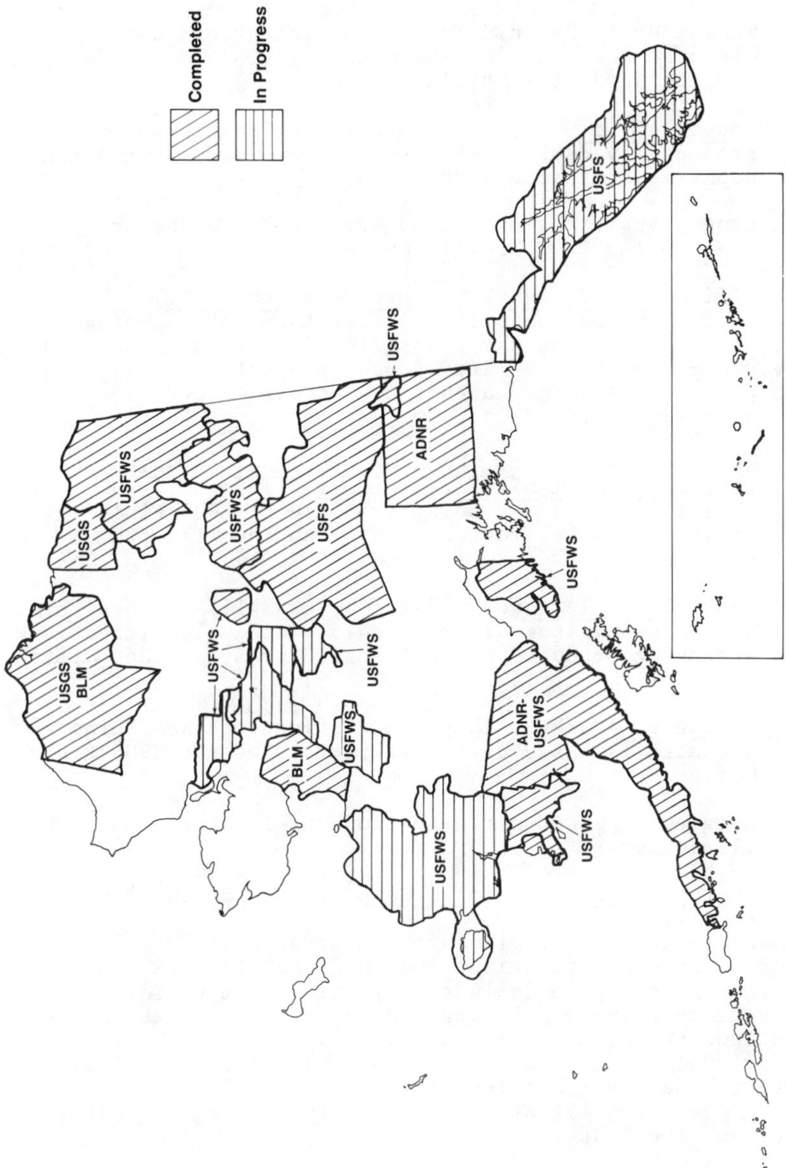

Figure 1. Current status of land cover classifications in Alaska.

- Nulato Hills. These classifications were produced at the EROS Field Office by the Bureau of Land Management.

- Tanana River Basin. These classifications were produced cooperatively between the U.S. Forest Service and the EROS Field Office.

- Copper River Project. These classifications were produced by EROS Data Center personnel for the Alaska Department of Natural Resources.

Existing classifications which are usable in the program total approximately 135 million acres.

Classifications Currently in Progress (Phase II)
Phase II involves immediate intervention into ongoing projects to reformat the classifications to a common land cover map legend as a part of each project's final output. The projects in this phase of the interim program are shown in figure 1 and include:

- Yukon Delta, Kanuti, Koyukuk, Nowitna, Selawik, and Innoko National Wildlife Refuges. These classifications are being produced by EROS Field Office personnel for the U.S. Fish and Wildlife Service.

- Central Arctic quadrangles (for example, Beechey Point and Sagavanirktok River quads). These classifications are being produced by Geographic Investigations Office personnel.

- Southeast Alaska. These classifications are being produced by the U.S. Forest Service at the EROS Field Office.

The area encompassed by these projects is approximately 75 million acres.

Strategy to Complete Interim Mapping Program in Alaska (Phase III)
Phase III involves implementing a strategy to complete the land cover classifications for the remaining areas of the State. Such a plan includes (a) entering into interagency agreements with cooperating Federal and State agencies to continue producing classifications which meet their individual mapping objectives and (b) embarking on an internal National Mapping Division program to complete land cover classifications for the remaining unclassified areas of the State.

Preparation of Uniform Map Products
One of the major factors guiding the development of the interim land cover map series is the need for consistency in the maps throughout Alaska. The basis for maintaining uniformity in this proposed program is (1) the establishment of the common land cover classification legend previously discussed and, (2) minimization of classification differences among Landsat scenes.

Minimizing scene differences is easier when adjacent Landsat scenes differ from each other by only a few days; but when adjacent scenes are acquired from different years and months representing different phenological stages, edge-matching is challenging. Under these latter conditions, a modest amount of discontinuity may occur within and between quadrangles upon completion of the interim mapping program. Where discontinuity problems are deemed unacceptable, reclassification may be required using Landsat scenes from the same season.

Beginning with Phase II projects, attention will shift toward both the common classification system and data continuity. Aggregation of spectral classes, use of digital terrain data, and editing procedures will be utilized to achieve the highest possible level of data continuity within these projects as they are being completed.

Phase I projects will pose more of a challenge. Classification results are variable, as each project was done at a different time, using different techniques, for different applications. It is anticipated, though, that nearly all of the Phase I project results can be incorporated into the program by re-aggregating spectral classes, editing certain problem classes by defining strata based on geographic area or terrain characteristics, and in some cases, re-clustering and re-classification. Specific solutions to data continuity problems will be proposed and carried out on a case-by-case basis.

The strategy that has been implemented is to define the common classification system at an early date and apply it to current and future projects (Phases II and III). Past projects (Phase I) will be examined with respect to the new classification system and data continuity. Specific procedures for reformatting each Phase I project will be developed.

PRODUCTS

L-series land cover maps at 1:250,000 scale and the land cover classifications in digital tape format will be available to the public and disseminated by the Mapping Division's National Cartographic Information Centers (NCIC). Digital data from this interim program will be maintained at the EROS Field Office in Anchorage.

A wide variety of products, other than those described above as standard products, are either available directly from the Alaskan data base or can be derived through additional processing of specific files within the data base. For example, geometrically corrected Landsat data will be available for the entire State. Landsat MSS scenes will have been screened and preprocessed whereby "bad" data lines are fixed and radiometrical striping is corrected. The preprocessed scenes will have been registered to a map base by means of selecting and digitizing control points, generating a transformation, and resampling and registering the data. It should be noted that Landsat MSS image maps at 1:250,000 scale could easily be made from these geometrically-corrected Landsat scenes since many of the processing steps for making image maps will have been completed. Thus, image maps could be made for all of Alaska at a minimal cost. During the process of making land cover classifications, a spectral cluster classification image is created for each Landsat scene that captures the diversity in spectral reflectance for the area within the scene. As many as 40-60 spectral classes are included in any single scene. Since many users wish to pool spectral classes in a manner that is specific to their own resource inventory or management problem, these spectral classification images for all of Alaska will be archived and available to the user community. The standards to which these other related products are made, as well as suitable descriptions of format, will be documented and made available to the public.

SUMMARY

The interim mapping program, with the objective of producing land cover maps and digital data bases for 153 quadrangles covering the State of Alaska, represents one of the largest multi-agency resource mapping efforts ever undertaken in the United States. The program is consistent with the National Mapping Division's responsibilities as they relate to its national land use and land cover mapping program. The program also assists cooperating State and Federal agencies in developing information they need to carry out their own resource management and planning mandates. Past and ongoing mapping efforts utilizing Landsat data will be the primary source of materials for carrying out the program. This will result in considerable cost savings as the necessary digital data bases and field reference data have already been developed. Future cooperative mapping projects will continue to provide the cooperating agencies with the specific thematic maps they desire, while the data bases will be constructed to facilitate development of the quadrangle-based maps for the interim program.

Completion of a consistent and comprehensive standardized land cover map series represents only one part of a series of products which will be derived through completion of the program. The digital data base approach also will yield quadrangle-formatted and registered Landsat MSS summer and winter data, detailed classified MSS data sets, elevation, slope, aspect and other various ancillary data types utilized in the mapping effort. This data base provides a very flexible source of information capable of being used by interested agencies for their own specific applications.

REFERENCES

Anderson, J. R., Hardy, E. E., Roach, J. T., and Witmer, R. E., 1976, A land use and land cover classification system for use with remote sensor data: U.S. Geological Survey Professional Paper 964, 28 p.

Viereck, L. A. and Dyrness, C. T., 1980, A preliminary classification system for vegetation of Alaska: U.S. Department of Agriculture Forest Service General Technical Report PNW-106, 38 p.

MAPPING RANGELAND VEGETATION USING LANDSAT MSS DIGITAL DATA FOR RESOURCE MANAGEMENT PLANNING

William R. Rush
Bureau of Land Management
3948 Development Avenue
Boise, Idaho 83705

Stephen M. Howard
Idaho Dept. of Water Resources, IIAF
450 West State Street
Boise, Idaho 83702

William D. Harrison
Soil Conservation Service
304 North 8th Street, Room 345
Boise, Idaho 83702

ABSTRACT

The objective of this project was to produce usable map products showing present rangeland vegetative communities based on Landsat MSS digital analysis and appropriate ancillary data. These products have provided the Bureau of Land Management, Cascade Resource Area (CRA) staff with a base inventory for the CRA Resource Management Plan and Environmental Impact Statement (RMP/EIS). They have also provided the Soil Conservation Service (SCS) with an inventory of private rangeland within BLM grazing allotments. Of particular interest was the ability of Landsat spectral data to distinguish non-native homogeneous stands of medusahead grass (*Elymus caputmedusae*) and cheatgrass brome (*Bromus tectorum*) from native plant communities. This project represents a cooperative interagency effort to generate uniform vegetation cover data at least possible cost.

INTRODUCTION

The Bureau of Land Management (BLM) resource inventory process is structured in a manner that meets legislative and policy requirements. It provides decisionmakers with the information necessary to make sound land-use and resource management decisions. Inventory data provides the factual basis for development of Resource Management Plans (RMP's). The Federal Land Policy and Management Act of 1976 states that inventories be maintained, on a continuing basis, of all public lands. These inventories are to be kept current to reflect changes in conditions and to identify new and emerging resource values (FLPMA 1976).

Traditional inventories in the Boise District BLM were conducted by a Resource Inventory Team (RIT). This team consisted of up to twelve resource specialists and was responsible for all inventory work associated with RMP's. The team concept was an effective but costly method of accomplishing the task. Costs for vegetation inventories alone averaged $.15/acre.* Faced with decreasing budgets and

---

*Average costs from 1977-1981 as reported by Boise District Resource Inventory Team Leader.

personnel, the Resource Inventory Team disbanded in 1982. Alternative methods to accomplish resource inventories were needed.

A rangeland inventory need existed for the BLM's Cascade Resource Area (CRA) in preparation for their Resource Management Plan. Limited rangeland inventory data existed for the CRA in 1982. A very scattered land ownership pattern and limited access contributed to this lack of current inventory. Some lands in the resource area had never even been visited by BLM personnel.

The use of remote sensing techniques was chosen by the CRA manager as an alternative method for gathering vegetation inventory data. The use of Landsat MSS digital data was available as a tool to accomplish this task. Of particular concern to the CRA manager was the continued proliferation of non-native homogeneous stands of medusahead grass (*Elymus caputmedusae*) and cheatgrass brome (*Bromus tectorum*). These two annual grass species encompass the majority of poor condition rangeland in the CRA. The ability of Landsat MSS digital data to distinguish these non-native homogeneous stands from native plant communities was imperative in this inventory effort.

USER REQUIREMENTS

The production of present rangeland vegetation maps based on Landsat MSS digital data was the basic requirement to meet the vegetation inventory needs for the CRA Resource Management Plan. In addition, the USDA Soil Conservation Service (SCS) was interested in using data gathered on private rangelands within BLM allotments for developing conservation plans. The information was planned for use in future SCS-BLM coordinated rangeland planning efforts.

Specialized image processing equipment was needed to process Landsat MSS data. The BLM's Branch of Technology Applications Assistance, at the Denver Service Center, was capable of processing the necessary data. However, due to an existing work schedule and limited personnel, this facility was unable to provide these services in the required timeframe. An alternate remote sensing facility was needed. With the limited funding available for this project, it was necessary that an alternate facility meet the following criteria:

1. Low cost. The contracted facility should benefit from the results and be willing to absorb certain processing costs associated with the project, a true cost share effort.

2. Close proximity. Due to restricted travel funds available for this project, the contracted facility must be close to both the project area and the Boise District Office.

3. Personnel interaction. Past experience has shown that the close interaction of contracted remote sensing facilities and BLM resource personnel is essential for the most effective use of the data.

4. Computer compatibility. The alternate facility must have a computer hardware/software package capable of interacting with the BLM's Interactive Digital Image Manipulation System (IDIMS) at Denver. This was required for hard copy output.

5. Interagency cooperation. The project was designed to be a cooperative effort and learning experience for all participants. The alternate facility must provide necessary training to BLM personnel to assist in data processing, therefore reducing costs.

IMAGE ANALYSIS FACILITY

The only facility meeting these requirements was the State of Idaho, Department of Water Resources, Idaho Image Analysis Facility (IIAF). It is a complete image processing facility, established by executive order, to service Landsat MSS user needs in the State of Idaho.

The IIAF is located in Boise, Idaho, close to the BLM Boise District Office and the project area. Travel costs were eliminated enabling BLM personnel to participate in data processing on a routine basis. This participation provided a better understanding of the complexity of the project and a more effective use of the output products.

The IIAF supports a suitable hardware/software computer package (VICAR/IBIS) which is compatible with BLM's IDIMS at the Denver Service Center. Computer compatible tapes were produced for hardcopy output. BLM personnel were able to work directly with IIAF personnel for data output. The IIAF was willing to provide all necessary training in the use of the (VICAR/IBIS) image analysis software and the interactive digital image display device. The primary emphasis of the IIAF is technology applications through development and application of operational remote sensing management and assessment programs.

The required data generation would normally cost about $.06/acre to process. The IIAF provided the required data to the BLM for about $.02/acre. The $.04/acre difference was absorbed by the IIAF to enhance its data processing capabilities for future project work. This facility is non-profit by directive.

DESCRIPTION OF THE PROJECT AREA

The CRA supports a diverse environment for a variety of vegetation types. A combination of man's actions, fire history, climate, elevation, aspect, and soil types account for this diversity.

The resource area stretches approximately 120 miles from north to south and 70 miles from east to west. The temperature/moisture regimes associated with this large area account for a variety of vegetation zones. The vegetation zones represented in the CRA include the following:

1. Sage/Grass Zone
2. Wheatgrass/Bluegrass Zone
3. Ponderosa Pine Zone
4. Douglas-Fir Zone
5. Spruce/Fir Zone

Because of the size and diversity of the CRA, it was decided that a sample area would be selected for processing in order to evaluate its usefulness over the entire area. This sample area, or Phase I, encompassed 508,000 acres. Native vegetation is primarily big sagebrush/perennial grass. Average annual precipitation is about 12 inches. About 75% of the soils are formed in residuum, alluvium, and colluvium from tertiary flow basalt and about 25% are formed in residuum, alluvium and colluvium from Pliocene stream and lake deposits. Elevations range from about 2400 feet to 6100 feet.

Of the 508,000 acres, about 350,000 acres are BLM administered public lands. The remaining 158,000 acres are private and state lands.

The limitations on fiscal year budgeting necessitated breaking the project into three distinct phases. Phase 2 and Phase 3, consisting of an additional 500,000 acres of public and private rangeland, were added to the project area after the successful completion of Phase I. The area of each phase was no larger than could easily be field checked in one field season. The Phase 3 classification was budgeted entirely by the SCS.

PROCEDURES

This project represents the operational use of Landsat MSS digital data. The classification techniques and inventory methodology developed as a result of the BLM-NASA Cooperative ASVT projects in Alaska, Arizona, and Idaho were used.

Environmental Stratification

Using geologic maps (1:250,000), the project area was stratified to separate soils derived from igneous and sedimentary parent materials. Agricultural lands were manually interpreted on National High Altitude Aerial Photography (NHAP) and transferred onto 1:100,000 scale maps using a Zoom Transfer Scope (ZTS). The strata were digitized and rasterized into image masks and were used to remove riparian confusion in the classification. Classification output for each geologic strata were treated separately, and the agricultural stratum was embedded into the final classification.

Administrative Stratification

BLM personnel compiled grazing allotment information onto the 1:100,000 scale Surface Management Map Series. Grazing allotment boundaries and BLM administered lands were digitized and converted to raster format. Image masks of grazing allotments by ownerships were created and used in an overlay process with the classified Landsat data. Tabular reports of landcover acreage for BLM and non-BLM lands by grazing allotment were then generated.

Classification of Landsat Data

A guided clustering methodology was employed to generate training statistics. BLM and SCS personnel familiar with the CRA selected training sites using NHAP and a 1:250,000 scale Landsat FCC (path:45 row:29 date:July 17 1980 Id #:82201317514XO). The Landsat MSS data for each training area (50x50 pixels) was clustered and classified into 10-12 spectral classes. The spectral classes were color coded, displayed, and photographed on a cathode ray tube (CRT). Display photographs and NHAP photography were registered using the ZTS. The delineated photos were taken into the field and each spectral class was labeled and ground checked for:

1. Physiography
2. Elevation
3. Slope %
4. Parent Material
5. Stratum
6. Aspect
7. Species Composition %
8. Range Site
9. Range Condition
10. Surface Stones %
11. Bare Soil %
12. Rock Outcrop %

A master statistics file was compiled from the individual training area statistics files. Spectral plots (band 5 vs. 7) for each training area were overlaid on a light table and the measurement space filled with as many distinct (Swain-Fu distance) spectral classes as possible. In this way a master statistics file containing 57 spectral classes representing 15 vegetation cover types was assembled (number of spectral classes and cover types varied for each phase). The project area was classified using a maximum likelihood decision rule.

The classified data were then resampled and registered to a 57 meter UTM grid. The agricultural stratum was embedded into the classified data and the spectral classes combined into the final output classes.

Continuity Evaluation

Due to time and funding limitations, a statistically designed accuracy assessment of the spectral data could not be conducted. An alternative action to assess the accuracy of the data was developed in the form of a continuity evaluation. This proved most useful in determining the reliability of individual spectral classes within each stratum and for developing land cover descriptions.

Random 40 acre sites were selected for individual spectral classes in each stratum. Each spectral class was evaluated at least twice within each stratum. The sites were transferred to 1:24,000 USGS quads for field verification.

A special "field" form was developed to record information at each site. Land surface feature data recorded included present vegetation, surface soil conditions, rock outcrop or stoniness, slope, aspect, and the probable presence of shadow. Care was taken to ensure that each spectral class was precisely located on the ground prior to documentation.

This data was used to evaluate the composition of each spectral class. As was expected, some spectral classes were very uniform in terms of land cover composition while others showed a wide range in composition. This data was also used to develop land cover descriptions for each spectral class. The actual composition of each description was based on the field data. A range-in-characteristics was developed indicating what could be expected in terms of vegetation and land surface composition within each spectral class.

The documentation of the variability within each spectral class served as the accuracy assessment for the project. Those classes having a wide diversity in land cover feature data were judged to be less accurate than those having a narrower diversity. Those land cover units having a wide diversity in composition were isolated for further analysis.

Output Products

Computer compatible tapes (CCT's) produced through IIAF were sent to the BLM's Denver Service Center. Through the BLM's Interactive Digital Image Manipulation System (IDIMS), hardcopy color maps were produced for the entire project area. In this form, the interpreted data was conveniently input to a geo-based information system. The data was smoothed and output at 1:100,000 scale to correspond to BLM planning base maps. These computer generated maps were used to compile the final rangeland vegetation and condition maps for the CRA Resource Management Plan.

The output products generated in the project include the following:

1. Tabular reports listing authorized grazing allotments showing acreage of vegetation cover types within BLM and non-BLM lands. This simplicity in cover type area determination is an additional cost advantage over manual map interpretation.

2. Color-coded maps (Applicon) of the CRA at 1:100,000 scale showing vegetation cover types.

3. Symbol mylar map overlays (Versatec) for the 7.5 minute topographic maps of the project area at 1:24,000 scale depicting vegetation cover types.

SUMMARY

Map products showing present rangeland vegetation based on Landsat MSS digital analysis have provided the CRA with a base inventory for resource management planning. These computer generated maps were used to compile the final rangeland vegetation and condition maps for the CRA Resource Management Plan and Environmental Impact Statement.

Landsat MSS data was able to easily distinguish and quantify non-native homogeneous stands of medusahead grass (*Elymus caputmedusae*) and cheatgrass brome (*Bromus tectorum*) from other native plant communities. This study has established baseline information for monitoring the changes in the distribution of these two species.

Interagency cooperation reduced traditional inventory costs of about $.15/acre to about $.06/acre. Through the use of a local remote sensing facility, IIAF, uniform vegetation cover data was generated at least cost. This facility, non-profit by law, absorbed some processing costs associated with the project. With a close proximity to the Boise District Office and project area, travel costs were virtually eliminated. The facility provided training to BLM and SCS personnel to assist in data processing which allowed close interaction of remote sensing specialists and professional resource specialists. This interaction is essential for the most effective use of the output products. Computer compatibility between IIAF and the BLM's Denver Service Center allowed the transfer of CCT's for the production of hardcopy color (Applicon) and mylar symbol (Versatec) maps.

A continuity evaluation was used to assess the accuracy of the image classification. This proved most useful in determining the reliability of the data and for developing land cover class descriptions.

The documentation of the variability within each spectral class served as the accuracy assessment for the project. The successful results of the continuity evaluation required close involvement between the remote sensing analyst, the image data, the field specialists, and the users of the data. The quality of the image classification can only be as good as the quality of the training data and accuracy evaluation used to perform the classification.

The entire project represents a cooperative effort by three agencies to generate uniform vegetation cover data at the least possible cost. The sharing of digital data bases and cooperative efforts are necessary cost saving mechanisms which should be addressed in future Bureau of Land Management remote sensing projects.

REFERENCES

Public Law 94-579, Federal Land Policy and Management Act of 1976, 43 USC 1711 Sec. 201a

Bureau of Land Management, United States Dept. of Interior 1982, Idaho Resource Management Plan Guidebook, pp. 2-13

Bureau of Land Management, United States Dept. of Interior 1983, #1734 Inventory and Monitoring Coordination: Reference Guide To The Inventory Coordination System

Pettinger, Lawrence R. 1982, Digital Classification of Landsat Data for Vegetation and Land-Cover Mapping in the Blackfoot River Watershed, Southeastern Idaho, Geological Survey Professional Paper 1219

THE USE OF LANDSAT "RANGELAND ENHANCEMENTS" TO MONITOR PUBLIC RANGELANDS IN SOUTHERN ALBERTA

S. G. Klumph and B. W. Adams
Alberta Energy and Natural Resources
Lethbridge, Alberta T1J 2J8

ABSTRACT

"Rangeland enhancements" are LANDSAT images produced by a special enhancement technique which was developed jointly by federal and provincial agencies. Public Lands Division of Alberta Energy and Natural Resources uses this type of imagery to identify range site types and monitor range use on mixed prairie range areas of Southeastern Alberta. Enhancements of spring and fall rangeland scenes are produced at the Alberta Remote Sensing Centre of Alberta Environment, on a computerized image analysis system. Field staff use the satellite images in conjunction with field checks to identify and map patterns of range use and to plan corrective management where necessary. The images have proven particularly useful for locating potential watering developments, relocating salt licks and cross fences, monitoring burn areas as well as the conversion of native range to pasture or cropland. The Canadian Centre for Remote Sensing is continuing to assist the provincial agencies in developing enhancement applications for other rangeland types in Alberta.

INTRODUCTION

The Public Lands Division of Alberta Energy and Natural Resources is responsible for the administration and management of public grazing lands in the settlement areas of the province. Public grazing lands in Alberta are administered as grazing leases or permits, or as government operated community pastures known as grazing reserves. There are 2.5 million acres of mixed grass prairie in Southeastern Alberta, to which this paper pertains, which are held under some form of public grazing disposition. Field staff carry out range inspections of these lands on a regular basis to evaluate grazing management and make adjustments to stocking levels and livestock distribution. Given the acreage of lease land that Public Lands must administer, a thorough assessment of these lands on a regular basis is time consuming and expensive by conventional means.

The traditional approach has been to conduct ground reconnaissance with the aid of black and white panchromatic photographs. Although conventional aerial photographs are quite useful in this process, they give no indication of current forage utilization or the status of ground biomass at a particular point in time. Conventional photographs are only acquired periodically, are expensive and many are required for assessing each grazing disposition. Regardless of the amount of time spent conducting a ground survey, it is difficult to accurately map differences in range use and condition. Actual boundaries between grazing-use levels or condition classes can only be roughly estimated.

The need for resource management agencies, such as Public Lands, to obtain more efficient techniques to monitor native rangelands is made essential by the growing competition for the use of these lands. The trend towards intensified grazing management and the increase in recreational activity, A.T.V. use, man-caused wildfires as well as oil and gas exploration and development can exert a powerful influence on the range resource. These problems place additional demands on field staff, diverting their energies from their principal task of monitoring the agricultural uses made of crown land.

Prior to the advent of rangeland enhancements, a variety of studies had been carried out in Alberta to apply remote sensing technology to rangeland applications (Intera 1975, Intera 1977, Jaques and Cihlar 1979, Intera 1979 and Pearce 1980). Results from these studies with false colour infrared aerial photography and LANDSAT MSS imagery indicated that estimation of species composition and biomass could not be made reliably. Estimation of these parameters were complicated by soil backgrounds, plant phenology and spectral reflectance characteristics of green and brown biomass (Thomson et. al., 1983). A significant development in these studies was change in the emphasis from the estimation of green biomass to that of total grassland biomass where brown vegetation was a significant stand component. Green biomass could not be reliably estimated since further increases in reflectance from the green component diminishes beyond threshold levels and also because of the masking effect of brown vegetation (Thomson et. al. 1980, Brown et. al. 1983). In spite of these complications, much qualitative information was, nonetheless, available from the various image products.

RANGELAND ENHANCEMENTS AND THEIR ACQUISITION

The rangeland enhancements currently used by Alberta Public Lands for monitoring grazing management on mixed prairie ranges are the product of a three-year joint study, conducted by the Canadian Centre for Remote Sensing, Alberta Energy and Natural Resources and Alberta Environment. This study resulted in the development of a special LANDSAT image from which the rapid, low-cost monitoring of range resources can be made. The final product is a colour-coded image requiring no special equipment and a minimum of training to interpret. The details of these studies have been described by Ahern et. al. 1981, Brown et. al. 1983 and Pearce et. al. 1984.

The operational production of computer-enhanced LANDSAT images for assessment of mixed prairie rangeland has been outlined by Sutherland and MacCallum (1985). Rangeland enhancements can now be ordered as would conventional aerial photography. Alberta Energy and Natural Resources handles ordering and distribution, while the Alberta Remote Sensing Centre of Alberta Environment produces the enhancements on a Dipix Aries III image analysis system.

The Alberta Remote Sensing Centre routinely acquires data tapes of cloud-free coverage of the region (six scenes) for two dates, one in May or June and another in August or September. The spring date enables the manager to assess spring carry over, green up conditions and the supply of surface water at the beginning of the season while the fall image is to assess the effects of the past summer's grazing. Computer compatible tapes are ordered from the satellite receiving station at Prince Albert, Saskatchewan. Once orders are submitted, images will be produced from cloud-free scenes to a maximum enlargement of 1:140,000 (2.5 in/mi).

INTERPRETATION AND OPERATIONAL USE OF RANGELAND ENHANCEMENTS

A rangeland enhancement is an 8" x 10" colour print with an overall blue-green cast. Differences in dark to light tones indicate a shift from heavy to light plant cover on the soil surface. Ungrazed or lightly grazed mixed prairie will have substantially more surface litter than will moderately to heavily grazed grassland. Although, to be technically precise, range condition is determined from the percent composition of desirable species on a given range site and is independent of forage utilization, it tends to be positively correlated with litter abundance. Therefore, dark blue-green tones tend to indicate sites that are in better condition than those with lighter tones. The limitations of this basic generalization will be discussed later. Since a "digital stretch" technique is applied to maximize contrasts in the mixed prairie ground cover, other cover types (dryland crops, irrigation, summer fallow, stubble and water) exhibit either a saturated or subdued appearance.

Further interpretation of the enhancements is similar to that of colour infrared photography. Green growth adds a red hue to the blue-green background on spring images, whereas red is largely lacking in autumn images when vegetation has begun to senesce. Areas that are moderately to heavily grazed have less litter to mask green growth in the spring. Overgrazed sites and moist depressional swales will exhibit a bright red appearance. Pink hues tend to appear where broad leaved plants are prevalent.

Public Lands field staff order imagery from the autumn previous for the current season's ground work. The imagery is no substitute for field checks and the first step is to ground truth each image enhancement to a field base. Once the major levels of range use are identified within each field the boundaries of these types can be mapped with a fair degree of accuracy. A major advantage afforded by these enhancements is that field staff are able to pinpoint problem areas that require checking without having to systematically search the entire lease and can overview large areas rapidly and inexpensively. When problem areas are identified, corrective management can be worked out with the rancher and implemented. Enhancements have now been used successfully to plan and develop new fence locations and dugouts to improve livestock distribution and to serve as an overall base for the range management plan.

In addition to monitoring the pattern and intensity of grazing, rangeland enhancements have been utilized for a variety of other purposes. Rangeland burned by wild or man caused fires have been delineated and assessed on enhancements. Areas burned during the growing season usually require two to five years of rest from grazing until plant vigour is restored. The eventual recovery of a burned area may be judged complete when the affected area regains the darker blue-green characteristic of the surrounding unburned grassland. Likewise, the extent of the effects of a particularly devestating hailstorm was mapped along the Canada-U.S. border in 1983. A mosaic composed of enhancements revealed the track of the storm and three other similarly affected areas in northern Montana.

LIMITATIONS OF RANGELAND ENHANCEMENTS

Estimating range condition on winter ranges must be done with caution. With winter grazing, mixed prairie vegetation may be utilized to a higher percentage since plants have put down fall reserves and are dormant and the ground is frozen. Consequently, although there may be a high percentage of desirable species present indicating that the stand is in good condition, there will tend to be less litter present than would be present on spring and summer grazed ranges. Because of this, there may be a tendency to under rate range condition on winter range. Winter ranges should, therefore, always be assessed with fall imagery.

Pearce (1985) has stressed the importance of accounting for seasonal variation in precipitation. Normal grass growth and litter accumulation is necessary so that there is sufficient above-ground biomass to detect differences in range use. In this regard, the value of field reconnaissance and intimate knowledge of growing conditions at the time the imagery was acquired, cannot be over emphasized.

Otherwise, it is impossible to determine whether apparent differences in plant cover and plant biomass are the result of grazing, site or climatic variation. For example, range site, range condition and utilization can easily be confused if images are interpreted without adequate field checks. Solonetzic sites in good to excellent condition tend to exhibit the same appearance on images as more productive sites which are moderately to heavily grazed.

Pearce (1985) also noted a number of technical problems in using rangeland enhancements for archival studies. The major problem was that of registration between images from different satellites due to variation in orbit and orientation. Because of this, accurate comparisons between different years, are at times difficult to make if there is no clear point of reference. Also, in certain images there appeared an excessive background redness which did not in fact correlate with vigorous green vegetation on the ground and could not be explained.

FUTURE OF RANGELAND ENHANCEMENTS

Rangeland enhancements have been described as "a highly technical data type reduced to base essentials and [delivered] into the hands of people in the field" (Sutherland and MacCallum 1985). Although resource management agencies have been the principal users of these products to date, further support of this technology is required until the ranching public become more familiar with this useful tool.

The prospect of producing rangeland enhancements with data from the French satellite known as SPOT, will improve spacial resolution (10 m) and also add to the spectral characteristics of the existing system. However, SPOT data may not be in operational use for a number of years.

Although use of the present system is restricted to the mixed prairie, the Canada Centre for Remote Sensing is continuing to support provincial agencies in developing applications for other rangeland areas of Alberta. Preliminary evaluations show that LANDSAT TM data has value for providing range resource information in the fescue grasslands of the Alberta foothills (Thomson et. al. 1985).

REFERENCES

Brown, R. J. et. al. 1983, Alberta Rangeland Assessment Using Remotely Sensed Data, Research Report 83-1 Canada Centre for Remote Sensing Ottawa, 128 pp

Intera 1975, Remote Sensing of Rangelands in Western Canada, DSS Contract Number OSZ4-0413, Intera Environmental Consultants Limited, Calgary, Alberta, 104 pp

Intera 1977, Rangeland Remote Sensing Pilot Project, DSS Contract Number OSZ 76-00183, Intera Environmental Consultants Limited, Calgary, Alberta, 183 pp

Intera Environmental Consultants Limited 1979, Remote Sensing for Estimation of Rangeland Productivity and Condition at Onefour Research Substation Alberta, Report submitted to Canada Agriculture, Lethbridge Research Station, Calgary, Alberta, 51 pp

Jaques, D., J. Cihlar 1979, Estimation of Vegetation Biomass in Bunchgrass Rangelands of Alberta Using LANDSAT Imagery: Remote Sensing of Earth Resources, Volume VII, University of Tennessee Space Institute, Tullahoma, Tennessee, Edited by F. Shahrokhi (in press)

Pearce, C. M. 1980, An Evaluation of Satellite-Generated Information for Inclusion in Rangeland Monitoring Programs in Alberta, M. Sc. Thesis Dept. of Geography, University of Calgary, Alberta, 206 pp

Pearce, C. M. et. al. 1983, Monitoring Rangelands in the Mixedgrass Prairies of Southern Alberta with Rangeland Enhanced LANDSAT Imagery: A Users Guide, Publication No. 83-2 Alberta Remote Sensing Centre, Edmonton, Alberta

Pearce, C. M. 1985, An Evaluation of Enhanced Archival Landsat Imagery For Mixed Prairie Rangelands in Alberta. Research Report for the Alberta Remote Sensing Centre, Edmonton, Alberta, 46 pp

Sutherland, I. and F. MacCallum 1985, Operational Production of Computer Enhanced LANDSAT Images for Rangeland Assessment in Alberta, (unpublished) presented at the Canadian Institute of Surveying Meeting May 1985, Edmonton, Alberta

Thomson, K. P. B. et. al. 1985, The Evaluation of Thematic Mapper Data for Range Management Applications in Western Canada, PECORA 10 Symposium Fort Collins, Colorado

CURRENT FOREST MAPPING TECHNIQUES IN SOUTHEAST ASIA[1]

Lawrence Fox III
Humboldt State University
Arcata, CA 95521

ABSTRACT

Remote sensing procedures such as interpretation of aerial photographs, satellite image interpretation, and digital analysis of satellite data have been used to various degrees for forest mapping in Southeast Asia since 1950. Aerial photographs were the primary tool employed until the launch of Landsat in 1972. The traditional strategy for forest mapping is based on medium scale aerial photography with Landsat imagery used only on an experimental basis. Bangladesh, China, India, Iran, Indonesia, Malaysia, Sri Lanka, and Nepal take this approach. Since the establishment of a consistent, Landsat MSS data base in 1979-1980, visual interpretation of single band, MSS imagery has been the most commonly employed technique for forest mapping with satellite data. Foreign receiving stations in Southeast Asia have been able to produce high quality black-and-white single band imagery since 1980. The quality of color composites is still low primarily because advanced film writers have not been incorporated into image processing labs. The satellite approach to mapping is based on visual interpretation of Landsat imagery as the operational method for country-wide analysis with digital analysis in the research and development stages. The countries of Pakistan, Thailand, and Vietnam use this approach.

TRADITIONAL MAPPING STRATEGY

The traditional strategy, as practiced by eight of the eleven countries discussed here, is based on interpretation of aerial photographs for classification of vegetation cover and land-use. Landsat imagery is used in parallel to aerial photography for limited areas as an experiment for comparative purposes. An interesting variation to this technique is under way in Sri Lanka, where Landsat imagery is being used as a cartographic base map for displaying photo interpreted vegetation cover and land-use.

---

[1]Much of the material summarized here was presented at the ESCAP Regional Workshop on Remote Sensing Applications for Vegetation Mapping in Colombo, Sri Lanka, November 16-18, 1983. Proceedings were edited by M. U. Chaudhury of the UNDP/ESCAP Regional Remote Sensing Program and printed by the Survey Department of Sri Lanka, 1983, 107 p.

Bangladesh

Remote sensing began in Bangladesh in 1948 when the country was flown for the first time with panchromatic film at 1:30,000 scale. The first forest resource inventories were produced from these photos in 1952. Recent forest mapping and inventory projects include: area estimation and mapping of the Sundarban mangrove forest in the southwestern section of the country, area estimation and mapping of the Sal forest (*Shorea robusta*) near Dhaka, monitoring the Coastal Afforestation Project of the Bangladesh Forestry Department and stratification mapping of homestead forest land from edge-enhanced Landsat data (M. A. H. Pramanik).[2] The mapping of the Sal forest near Dhaka serves as a representative example of remote sensing procedures in Bangladesh. The objective of this project was to develop a forest community classification and land-use classification system for monitoring the impact of agricultural development in the Sal forest region. Stereo interpretation of black-and-white and color infrared photography allowed the identification of four classes of Sal forest (40-60% crown closure, 50-80% crown closure, seedling to pole stands, and new plantations) three classes of agriculture and a village class. Digital enhancement of MSS Landsat data failed to achieve the detail of the vegetation classification interpretable from the aerial photographs (J. Islam).

China

Aerial photographs were used in forest resource inventories as early as 1953 in China and in 1958 more than 1/3 of the entire country was photographed at 1:25,000 scale. The most recent work also at 1:25,000 scale employs color, color infrared and black-and-white infrared film, in addition to panchromatic photography. The photography has been used to construct forest type maps while volume measurements are obtained from ground surveys. Recently the Forestry Department has experimented with more advanced uses for the photos such as stand measurements, aerial stand volume estimation and site factor determination by point sampling.

Experimentation with Landsat began in 1977. Generalized forest maps of Tibet, an autonomous region of China, were produced at 1:500,000 scale by visual interpretation. An experimental survey of land status was accomplished throughout the country in 1980 and 1981. The goal was to illustrate the distribution of land area in agriculture, forest and rangeland in each province. Results compared favorably with existing maps (Table 1). A stratified sampling plan is recommended for the future with Landsat imagery used to define broad associations and sampling within Landsat strata to describe the character of each strata (L. Liuyu and L. Wenrui).

---

[2] The date has been omitted from this reference and from all references to the ESCAP Regional Workshop on Remote Sensing Applications for Vegetation Mapping, 1983, since several citations are from this workshop. See Reference section.

Table 1. Accuracy of Landsat Interpretation in China

The percent accuracy of Landsat image interpretation and existing land-use maps (Southern segment only) determined by comparison with 1:10,000 scale aerial photography.

	Southern segment		Northern segment
Land Type	Landsat image percent correct	Land-use map percent correct	Landsat image percent correct
Cultivated Land	96.0	93.5	93.8
Forest Land	95.7	39.7	84.0
Range Land	80.6	44.4	96.7
City, Town, Industrial and Transportation Land	20.7	73.0	79.0
Water Area	58.0	72.0	73.0

India

The Forest Survey of India has prepared forest cover maps and land-use maps at 1:50,000 scale since 1965. Photos have also been used in planning and executing forest inventories during this same period. Based on these inventories, the Forest Survey has recommended the development of 30 industrial plants including pulp and paper, plywood, sawmills and fiber board operations. The Indian Institute of Remote Sensing has also carried out numerous projects using 1:10,000 scale and 1:25,000 scale aerial photos.

Landsat data has been used with visual interpretation and computer aided interpretation for forest mapping on an experimental basis since 1977. In one research project, eight forest features including cover types, fire damage and clear cutting, were evaluated with visual interpretation and with computer classification. Five features were identifiable with visual while only four were mapped reliably with computer classification. Visual interpretation of computer enhanced images is proving to be a very cost effective technique (P. S. Roy). An example of a map made by visual interpretation of Landsat imagery is included as Figure 1.

Iran

General land cover mapping in Iran using 1:50,000 scale aerial photography is in progress. The landscape is being classified into major uses such as: irrigated annual crops and orchard, dry farming, range and pasture, forest, urban area, etc. More detailed forest mapping is also under way

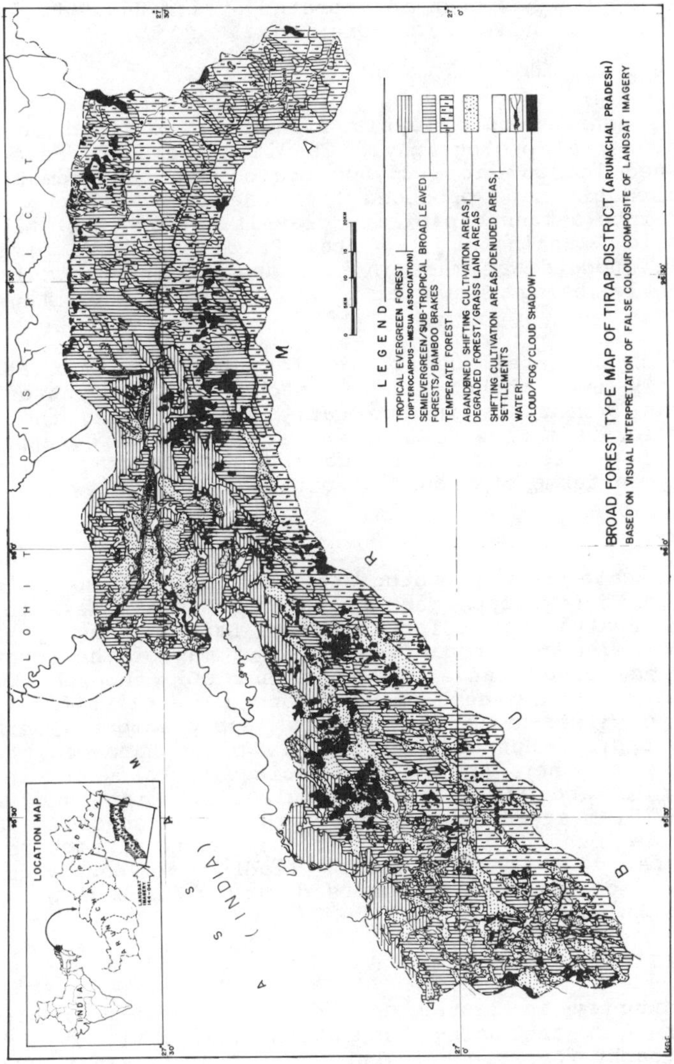

Figure 1. An example of a forest cover map made from visual interpretation of Landsat imagery. Indian Institute of Remote Sensing.

by the Ministry of Agriculture using aerial photographs (M. H. M. Shojaie). Landsat imagery has been used experimentally with promising results. Single band and false color imagery has been interpreted at 1:1,000,000, 1:500,000 and 1:250,000 scale (J. Sadighian).

Malaysia

Remote sensing in Malaysia is mainly confined to conventional aerial photography. The first complete coverage was obtained in 1967 at 1:25,000 scale with panchromatic film. The thematic maps produced from these photos were consolidated into a land capability classification (LCC) map by the Economic Planning Unit of the Prime Ministers Office. A partial update was made in 1975 and Penninsular Malaysia is currently being flown at 1:40,000 scale for the second update.

Research on satellite applications has been initiated only recently even though Landsat technology was introduced to Malayan Scientists in the early 70's. Among the reasons cited for this slow development are: poor topographic maps, lack of professional training and poor image quality especially in terms of ground resolution (K. Y. Siew).

Nepal

Large scale aerial photographs are being used to prepare detailed forest type maps on an operational basis in Nepal. Visual analysis of color composite Landsat imagery has been used to produce a broad forest cover map of the country and selected areas have been classified through computer analysis. In a recent project supervised classification was used to differentiate five forest associations (*Quercus* and other *spp.*, *Pinus* with hardwoods, mixed hardwood, hardwood *Sal*. and other *spp.*, and deciduous hardwood including *Dalbergia* and *Acacia*). The classification was verified using 70 mm aerial photography and ground checks. The mapping was found to be 90% accurate in spite of high relief in some parts of the watershed leading to deeply shadowed terrain sometimes being confused with water bodies (K. B. Malla).

Indonesia

This country is listed here in the traditional group although they are using Landsat in combination with aerial photography for forest inventory in a multi-stage procedure. Satellite imagery is interpreted at 1:1,000,000 or 1:250,000 scale to obtain reconnaissance level information concerning ground coverage and forest type classification based on topographic condition. Small scale aerial photographs (1:100,000) are then interpreted within forested areas identified on Landsat to obtain forest type maps and finally large scale photographs (1:20,000) are consulted for information on specific trees and groups of trees, crown diameter, and average tree height. Inventory methods commonly include stand volume per hectare estimates directly from the photographs (Y.L. Rombe). An example of a local, aerial stand volume table is included as Table 2.

Table 2. Local Aerial Volume Table of Agathus Lorantifolia Salisb of Age Class 35.

Crown Area Percent	Volume in Cubic Meters Per Hectare
10	112.40
20	163.66
30	214.93
40	266.19
50	317.45
60	368.72
70	419.98
80	471.24
90	522.51
100	577.77

Sri Lanka

This country is using panchromatic aerial photography (1:40,000) for a country-wide survey of forest cover, land-use and general vegetation cover. They have employed Landsat data in quite a unique way however. Landsat MSS band five black-and-white imagery is enlarged to 1:100,000 scale. This image is geometrically corrected and used as the base map for the photo interpretive work. The country's 1:63,360 scale topographic maps were determined to be inadequate for a planimetric base and it was also felt that transferring photographically interpreted land cover and land-use categories to these topographic maps would be difficult. The transfer of photo interpreted information to the Landsat image base is facilitated by the appearance of natural features (e.g., distinct ecotones, rivers, steep mountain formations) on the Landsat image as well as on the photography. Also, the geometrically corrected Landsat image was felt to be more accurate, planimetrically, than the National map base (C. Nanayakkara).

SATELLITE MAPPING STRATEGY

The satellite mapping strategy is practiced by three of the eleven countries discussed here. This approach relies on the visual interpretation of Landsat imagery in an operational program to map the forest resources of a nation.

Pakistan

This country is listed here in the satellite section because information on mangrove forests was obtained from Landsat for the entire mangrove area of the nation. However, this mapping does represent a one-time effort and it is probably not appropriate to assume that Landsat is completely operational in Pakistan. Vegetation maps at 1:250,000 scale were prepared for 1.5 million acres in the Indus delta region of the country. Interpretation of Landsat data was initiated with the demarcation of the study area on MSS band seven prints. For the identification of different land-use classes, Diazo prints (false color composites of MSS band 4, 5, 7 at 1:250,000) were obtained for three dates: December 21, 1973; December 14, 1976 and December 4, 1978. No significant changes in the size or shape of the mangrove forests were observed on the imagery. The total area of the tidal zone of the Indus deltaic region was estimated at 1.5 million acres of which approximately 0.65 million acres (44%) were in mangrove vegetation. The remaining area was determined to be in: sparse or no vegetation (22%); sand (7%); and water (27%). On the basis of separable differences in color on the Diazo prints, mangrove was further classified into dense vegetation (0.13 million acres) and normal vegetation (0.52 million acres). These results indicated to government officials that Landsat data could be effectively used to measure the area under mangrove vegetation as well as other classes of land cover (Miraz, et. al.).

Thailand

Medium scale aerial photography (1:25,000 and 1:40,000) has been used for forest type mapping in Thailand since 1955. In addition, the Thai Forestry Department has used this photography for aerial estimates of timber volume and productivity classes. The standard method of Landsat data analysis for forest mapping is visual interpretation of black-and-white band five and band seven imagery at 1:250,000 scale. The entire country was mapped in 1978 and 1982 using these techniques. The forested area of the country decreased by 18,624 square km during this four-year period, a loss of 10.6% from the 1978 forested area (175,224 square km). While the Thai Forestry Department has made a major, operational committment to Landsat visual interpretation, Thai officials sense that computer classification would be more cost effective and possibly provide a bridge between the reconnaissance level of Landsat visual interpretation and the very detailed information available from aerial photographs. Research is under way in computer classification techniques of Thematic Mapper and, in the future, SPOT imagery in order to find the optimum mix of Landsat and airphoto techniques (S. Tikumpomvarokas).

Vietnam

The forested area of Vietnam occupies approximately 2/3 of the country. A large portion has been degraded by shifting cultivation, excessive cutting and other forms of forest removal. Realizing that a complete country-wide mapping effort was needed rapidly for one point in time to cope with this dynamic situation, Vietnamese officials have depended completely on a satellite image data base. A preliminary vegetation map of the entire country was completed in 1983 based principally on Landsat imagery, space photography from the KATE-140 and multiband photography (MKF-6). Imagery was obtained between 1979-1982.

Interpreters indicate that optimum results are achieved when a complex approach is taken in which all available geographical data are interpreted and used in the decision logic. The Vietnamese are taking a terrain analysis approach to vegetation mapping. Expected vegetation types in light of terrain variables, inferred micro-climate, and ground experience weigh heavily in the image interpretation. A minimum type size of 100 hectares has been adopted for Landsat imagery at 1:500,000 scale although interpreters report that forest areas of 20 to 25 hectares can be observed on the imagery. Landsat images have been enlarged to 1:160,000 with an additive color viewer to delineate the boundaries of small forest areas quite clearly. The landscape has been discriminated into nine classes including:

1. Rich broad-leaved evergreen forest.
2. Medium to poor broad-leaved forest.
3. Pine bamboo forest.
4. Mixed bamboo and broad-leaved forest.
5. Pine forest.
6. Mangrove forest.
7. Woodland, rich and medium to poor land.
8. Bare land, grassland and brush land.
9. Agricultural land and other land.

Vietnamese officials plan to experiment with digital processing of satellite data and expect digital processing to prove more effective for forestry investigation than visual analysis.

CONCLUSION

It seems apparent from reviewing the remote sensing experience of these developing nations that some of them are using a Landsat data base much more extensively than the developed nations such as the United States or countries in Europe. In many cases, the absence of an institutional system for mapping based on aerial photography and already in place may have facilitated the switch to a satellite based mapping effort.

REFERENCES

References below are from the Proceedings of the ESCAP Regional Workshop on Remote Sensing Applications for Vegetation Mapping, Colombo, Sri Lanka, November 16-18, 1983. UNDP/ESCAP Regional Remote Sensing Program, Bangkok, Thailand. (RAS/81/034/C/01/53). 107 p.

Huy, C. Q. Application of remote sensing data in compiling vegetation maps of Vietnam. p.100-102

Islam, J. Vegetation mapping through remote sensing. p.50-52

Liuyu, L. and L. Wenrui, An application of remote sensing techniques in the survey of land status and forest resources. p. 53-59

Malla, K. B. Applications of remote sensing to forest type mapping in Nepal. p.85-88

Mirza, M. I., et. al. Identification and area estimation of mangrove vegetation. p. 89-92

Nanayakkara, C. Remote sensing applications in Sri Lanka. p. 103-107

Pramanik, M. A. H. Vegetation mapping activities in Bangladesh using remote sensing techniques. p. 49

Rombe, Y. L. The use of remote sensing for forest inventory in Indonesia. p. 66-73

Roy, P. S. Remote sensing of vegetation - an Indian experience. p. 60-65

Sadighian, I. Vegetation mapping of the Jaz-Murian watershed, S.E. Iran, using Landsat Imagery. p. 74-75

Shojaie, M. Remote sensing applications to vegetation mapping in Iran. p. 76-77

Siew, K. Y. Vegetation mapping using remote sensing in Malaysia - a country report. p. 78-84

Tikumpomvarokas, S. Forest cover classification and mapping methods used in Thailand. p. 93

TROPICAL DEFORESTATION AND THE IMPLICATIONS OF MICROCOMPUTER-BASED IMAGE PROCESSING TECHNOLOGY CASE STUDY INDONESIA

Sean C. Ahearn
Ralph W. Kiefer
Thomas M. Lillesand

Environmental Remote Sensing Center (ERSC)
University of Wisconsin-Madison
1225 West Dayton Street
Madison, WI 53706

BIOGRAPHICAL SKETCHES

Sean C. Ahearn is a Ph.D. student in the University of Wisconsin-Madison Environmental Monitoring Program. He received a B.S. from the College of Environmental Science and Forestry at Syracuse (NY) in 1978 and an M.S. degree in Environmental Monitoring from the University of Wisconsin-Madison in 1982.

Ralph W. Kiefer has been a faculty member at the University of Wisconsin-Madison since 1962. He has B.S., M.S. and Ph.D. degrees from Cornell University. At the U.W., he is Professor of Civil and Environmental Engineering and the Institute for Environmental Studies and is Chairman of the Environmental Monitoring Program, a graduate program in environmental remote sensing.

Thomas M. Lillesand has B.S., M.S. and Ph.D. degrees from the University of Wisconsin-Madison. He is a Professor of Environmental Studies, Forestry, and Civil and Environmental Engineering. Prior to coming to the UW in 1982, he was a faculty member at SUNY-Syracuse and the University of Minnesota. He is Director of the Environmental Remote Sensing Center (ERSC).

ABSTRACT

Tropical deforestation is perceived by many to be one of the most important natural resource issues of the century. Until recently the data necessary for monitoring this phenomenon have been scarce or nonexistent. With the development of Landsat MSS receiving stations in Thailand, Indonesia, China, Australia, and Brazil, this data gap is beginning to close. The ability of these stations to receive Thematic Mapper (TM) and SPOT data in the future will provide an even higher level of image detail over tropical areas.

In the past, the question as to who would do the analysis of remotely sensed data for deforestation monitoring was confined to a limited number of centralized facilities with large computers and extensive support equipment for image analysis. With the advent of microcomputer technology this need no longer be true.

A case in point is the Institut Pertanian Bogor (IPB) Remote Sensing Center in Bogor, Indonesia. The center is the result of a cooperative project between the IPB and the University of Wisconsin-Madison, supported by the United States Agency for International Development. It is intended to act as a national training, research and policy center for the monitoring and management of Indonesia's natural resources.

The center's lab consists of an IBM-PC/XT microcomputer-based image processing system that is capable of reading and analyzing Landsat MSS, AVHRR and other satellite data tapes. The software includes a wide range of digital image processing programs for image restoration, enhancement, and classification, as well as basic geographic information system operations.

With the change from expensive and difficult to maintain minicomputers, to affordable, relatively resilient microcomputer-based image processing systems, the ability to tackle deforestation monitoring problems is present at the national and regional levels. This places the analysis of imagery in the hands of the people who know the resources of their country best and in turn can generate a greater within country commitment to policy promulgation and implementation in the area of deforestation.

We believe that the development of national training, research and policy centers throughout the tropics is necessary for the proper management and monitoring of tropical rain forests. Integration of the national centers under the umbrella of an international network for tropical forest monitoring and policy analysis will ensure a coordinated approach to the management of the tropical environment.

INTRODUCTION

Various segments of the scientific community have been concerned about the threat of deforestation and its implications for the world for more than twenty years. While many studies have contributed to the understanding of the problem, only recently has deforestation been seen in the context of its effect on the global system (NASA, 1984). One of the biggest obstacles has been the lack of reliable quantitative information about the extent and rate of deforestation. Even to this day, an accurate estimate of the magnitude of deforestation eludes us. With the greater resolution afforded by the Landsat Thematic Mapper (TM) and the SPOT satellite system, estimation of the extent and rate of tropical deforestation will be less of a problem. Emphasis will begin to shift from the search for even generalized data to the site specific determination of the rate and extent of deforestation, and to questions such as who will do the monitoring, what policy should be implemented to abate intolerable rates of deforestation, and who will be responsible for this abatement.

Past barriers to the monitoring and analysis of tropical areas have been lessened by technological advancements in satellite systems, microcomputers and geographic information systems. What is required now is the structural development of an international network for the monitoring, management and policy analysis of the world's tropical natural resources.

This paper describes the salient causes and implications of deforestation, the impact of the technological development of microcomputers on deforestion monitoring, a framework for an institutional structure for establishing an international network for tropical deforestion assessment, and the description of a project in Indonesia that represents one model that could comprise the beginning of this network. It also highlights a second generation microcomputer system under development at the Unviversity of Wisconsin-Madison based on the IBM PC AT.

CAUSES OF DEFORESTATION

The causes of deforestation can be viewed from two vantages, the indirect causes and the direct causes. The indirect causes are population growth, a country's national debt and the need for foreign exchange, a development mentality, the scarcity and cost of alternative fuel sources, and cultural practices. These are the socio-economic factors that result in various forms of deforestation. The direct causes can be as diverse as shifting cultivation, commercial logging operations, firewood gathering, cattle ranching, crop monocultures, tree plantations, road building, transmigration and hydropower development.

IMPLICATIONS OF DEFORESTATION

Deforestation can manifest its impact locally, regionally and even globally. At the local level, it can lead to soil infertility, soil erosion, and loss of productive land. Regionally, it can lead to damaging floods, changes in rainfall patterns, depletion of groundwater and loss of sustainable use of forest lands. Globally it may induce a disruption of the carbon dioxide balance, the reduction of species diversity and cause global climatic changes (Talbot 1981, Meyers 1980). What is worse, in the absence of hard data and valid models, we do not really know what the global implications of deforestation might be under alternative future global development scenarios.

DEFORESTATION: WHO IS RESPONSIBLE?

Given the ramifications of deforestation both known and unknown, it is important to determine who will be responsible for its abatement. To expect the tropical countries of the Amazon, Africa and Southeast Asia to take full responsibility is unrealistic. Many of these countries are wallowing in billion dollar debts and must pursue any means possible to capture foreign exchange. They will not pay the opportunity costs of not cutting the forests so that the rest of the world can preserve its genetic supply or keep the carbon cycle in balance.

It is clearly up to the international community to share in the opportunity cost of preserving the world's tropical forests. What is necessary is a full-scale effort by this community to set up the institutional structure and rationale for the monitoring, management and policy formulation for tropical rain forests.

THE IMPACT OF TECHNOLOGY

Increase in the accessibility of information is the primary contribution of microcomputer technology to a greater understanding and control of deforestation. With decentralization of data analysis, no longer are expensive, difficult to maintain, minicomputers needed for remote sensing and natural resource analysis. Relatively resilient, inexpensive microcomputers are able to do much of the work that was traditionally done by minicomputers. This makes the data analysis process available to the regional and local levels of government.

The ability of regional and local governments to analyze data is complemented by the recent availability of Landsat TM data and the forthcoming availability of SPOT data. These data, which have a resolution of 30 meters and 10/20 meters respectively, provide a level of detail that is suitable for local and regional analysis.

The effect of having site-specific data that can be analyzed at the regional level will be to enable better management of developed lands and increased monitoring of lands undergoing change. This will give more options for resource development and insure the sustainability of land that is already under development. It will also facilitate the identification of particularly ecologically sensitive areas.

INTERNATIONAL NETWORK FOR TROPICAL MONITORING, MANAGEMENT AND POLICY (INET)

Several movements are now underway to increase the understanding of the global ecosystem through modeling of the world's climatological, hydrological and biogeochemical cycles. A chief participant in these activities is NASA (NASA, 1984). Given that the greatest rate of change in the global environment is occuring in the third world countries, integration of these countries into the scientific analysis of global ecosystems is of paramount importance. Establishing a network of third world countries that ties into NASA's Earth Observing System (EOS) project will insure that the project is truly of long term global significance.

What has been frequently overlooked by the participants in the purely scientific global projects in the past, is the connection between the scientific conclusions that are generated by such projects and the implementation of policy based on those recommendations. Actions that are altering the global environment are happening at the national, regional and local levels. Thus, in order to implement policy based on scientific observations from the global studies, a network must be formed that extends from the international level down through the national level to the regional and local levels. There are no institutional mechanisms at the global level to change the global environment. Actions must be taken at lower levels that can propagate upward to affect the global system.

This requires an international network that not only looks at the scientific questions concerned with the global environment, but also at policy formulation as directed by the answer to those scientific questions.

International Body - Global Trends and Policy Advisor

What is envisioned is an international body that is concerned with global monitoring and policy advising to the national governments. Ideally, such a body could be integrated into, or run parallel with, the Earth Observing System project. It would have similar scientific objectives as EOS, but in addition would explore the policy issues that surface in response to future global scientific observations as well as those already identified (Office of Technology Assessment 1984). The body would also be involved in coordinating programs that promote the sustainablility of the world's tropical resources.

National Centers - Training, Research, & Policy Formulation

National centers would focus on training, research and policy formulation. They would interact with the international body through policy formulation in response to the international body's recommendations, and through the passage of site specific data compiled from the regional levels of government, to the international body. Their emphasis would be on increasing the productivity and sustainability of developed lands and on deciding how best to manage undeveloped lands. National centers would be involved in various types of research in the areas of remote sensing, geographic information systems, and systems modeling. They would also act as training centers for personnel at the regional and central level.

Regional Centers - Policy Implementation

The primary concern of the regional centers would be in the implementation of policy promulgated at the national level. They would interact with the national center through the transmission of policy directives from the national to regional level. They would also provide detailed information to the national level for compilation and monitoring of national trends.

Requirements for an International Network

There are certain requirements that are essential if an International Network for Tropical Monitoring, Management and Policy (INET) is to become a reality. These are: system compatibility between the international, national and regional levels; a data linkage between these levels; a well trained interdisciplinary staff at the international, national and regional levels; and a commitment by the international funding organizations to establish such a network.

System compatibility and data linkage between the regional, national and international levels is critical to the formation of a cohesive unit. Having a single package for a geographic information system/image processing system will ensure communication and data transfer between the three levels and thus a coordination that would otherwise be lost.

Without a well trained staff at the regional, national and international levels there is no formula for the success of the network. The technology transfer and training during the IPB/UW remote sensing/GIS project discussed below is presented as one model that could be used to accomplish the training and development of expertise.

The international funding agencies are clearly major driving forces behind development in the tropical countries. The coordination of these agencies and support for an international network for tropical monitoring, management and policy is a necessary ingredient for the realization of this goal.

THE INDONESIAN EXAMPLE

The Center for Remote Sensing and Geographic Informations Systems at the Institut Pertanian Bogor (IPB) in Bogor, Indonesia (Ahearn et al., 1985), represents but one model for a national component within the international network described above.

This Center was set up during a cooperative project (funded by the U.S. Agency for International Development) between the Institut Pertanian Bogor and the University of Wisconsin-Madison during the period running from January 1983 to July 1985. The center is intended to act as a national training, research and policy institute for the monitoring and management of Indonsesia's natural resources. It is interdisciplinary in character and is staffed by IPB faculty from soil science, statistics, agrometeorology, forestry, and agronomy. Personnel from agricultural economics, natural resource planning, and rural sociology will be integrated in the future.

The original design for the IPB image processing system hardware is shown in Figure 1. The computer system now consists of one IBM PC microcomputer, six IBM PC/XT microcomputers, two dot matrix printers, two Gould FD5000 high resolution graphics terminals, an 800/1600 Bpi tape drive, and a 160 MB hard disk. The software includes a wide range of digital image processing programs for image restoration, enhancement, and classification, as well as basic geographic information system operations.

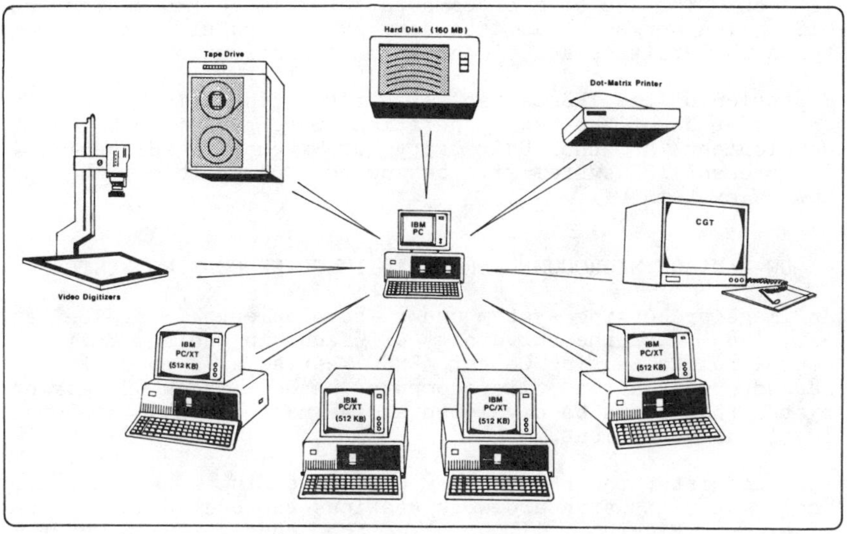

FIGURE 1 - IPB IMAGE PROCESSING SYSTEM HARDWARE

The primary emphasis of the IPB center is in four areas:

1) development of a strong curriculum and research capability in remote sensing, geographic information systems, systems modeling (in agriculture, ecology, forestry, hydrology, and meteorology), agricultural economics, rural sociology and natural resource planning.

2) training government personnel from regional centers.

3) technology transfer to regional centers.

4) developing links to the international community with the prospects of developing an international network for monitoring, management and policy of tropical countries.

While the Center is in its early stages of development it is in a good position to assume the role as the Indonesian national link in the network described above. It is already actively engaged in research in the areas of crop yield forcasting, vegetative mapping and transmigration site selection.

A model for rice yield forcasting using a combination of remotely sensed data and ground information is also under development at the Center. The Center's scientists are also using AVHRR data, received by the Indonesian ground station LAPAN, to develop a vegetative index map for the country.

The IPB Remote Sensing and GIS Center has also been involved in training activities. A two-week workshop was given at IPB on "The Use of Microcomputers for Image Processing and GIS." The workshop was attended by 24 participants from three universities and six government institutions.

Expansion of the IPB Center's laboratory is anticipated with the addition of the next generation micro system now under development at the University of Wisconsin-Madison (UW). The present IPB system will be upwardly compatible with the new UW system.

UW-MADISON MICROCOMPUTER-BASED IMAGE PROCESSING SYSTEM

An image processing system under development (Scarpace, et al., 1985) at the University of Wisconsin-Madison Environmental Remote Sensing Center (ERSC) serves as a model for a second generation microcomputer-based image processing system that could be developed at the national and regional levels of government.

When complete, the ERSC system will have nine work stations, including eight standard work stations and one advanced work station. Figure 2 illustrates the configuration of a standard work station. It is built around an enhanced IBM PC AT, with an 80287 math co-processor and one Megabyte of RAM memory.

The work station has both an IBM monochrome monitor and an IBM Professional Graphics Display. The resolution of the professional graphics display is 640 columns by 480 rows. A lookup table on the graphics card allows 256 colors (out of a palette of 4096 colors) to be shown simultaneously. This is sufficient to provide a photograph-like color image display.

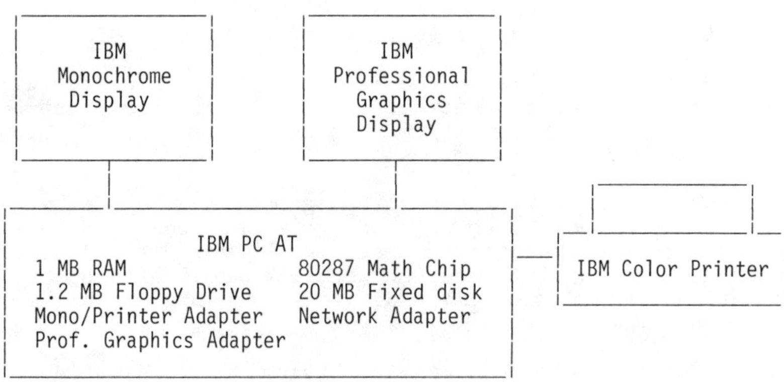

FIGURE 2 - STANDARD WORK STATION

One station will be configured as an advanced work station (see Figure 3). It will control a Gould DeAnza model FD5000 image processor (the same system as used at the IPB in Bogor). The FD5000 has a color monitor that displays 512 rows by 512 columns, 28 bits deep, and has a feedback processor. In addition a video camera is connected to the FD5000 as a data input device.

The advanced work station will also have additional storage provided by an IOMEGA Bernoulli Box and by an 800/1600 bpi model TC-1054 tape drive from Innovative Data Technology, Inc. A GTCO digitizing tablet and Houston Instruments drum plotter are connected with this work station to facilitate the input and output of line graphics.

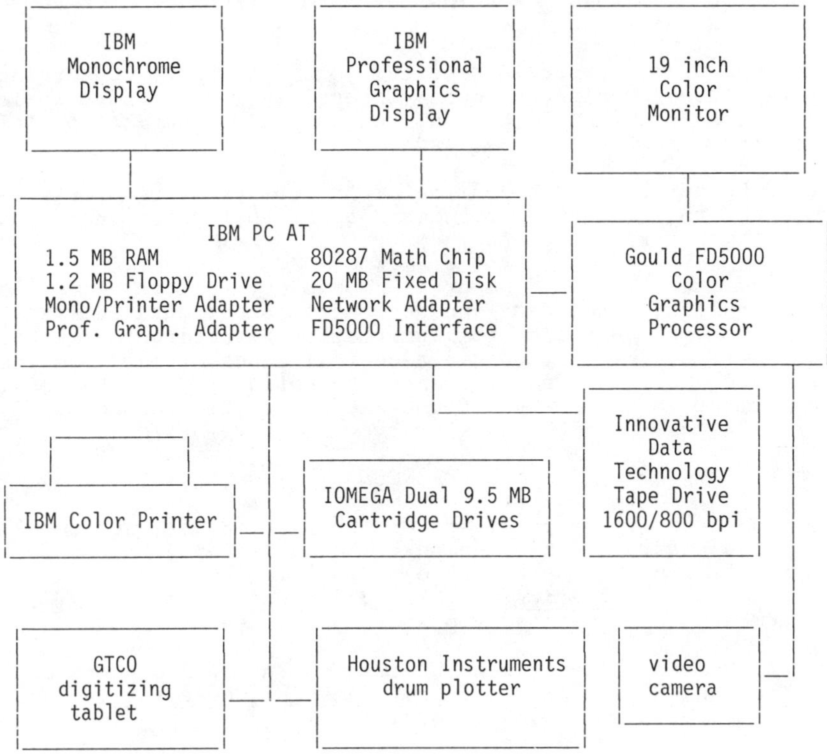

FIGURE 3 - ADVANCED WORK STATION

All eleven work stations will be linked together by an IBM PC Network. The network will enable the rapid exchange of images and other data among the work stations. It is likely that the advanced work station will be connected with as many as three other local AT networks that will be devoted to photogrammetric, cartographic, and geoinformation applications.

The ability to network systems within a center and between centers demonstrates an important capability for linkage between the international, national and regional levels of analysis. This would ensure a smooth passage of data through and between the levels of the International Network discussed above.

CONCLUSION

Interdisciplinary training, research and policy centers can and should be established in the tropical countries at the national levels. The center being established at the Institut Pertanian Bogor is but one model for such centers.

An International Network for Tropical Monitoring, Management and Policy could form a nexus for integration into the Earth Observing System.

The long-term, overall benefits of the Earth Observing System are contingent upon integrating the third world countries into the collection, analysis and compilation of ground and remotely sensed data.

Clear recognition must be made of the policy issues attendant to the scientific recommendations that arise from such global monitoring efforts.

The advent of microcomputer technology allows for the decentralization of data and customization of data analysis. A bottom-up approach to information gathering for global monitoring is clearly within technical reach. What remains is the development of an international institutional commitment to fully exploit the nature of this technology.

REFERENCES

Ahearn, Sean C., Ralph W. Kiefer, Larry L. Seidl, and Uup S. Wiradisastra, "A Microcomputer-Based Training Center for Remote Sensing and Resource Analysis: The Indonesian Example", Proceedings of the 1985 Machine Processing of Remotely Sensed Data Symposium, pp. 53-58.

Myers, Norman, "The Present Status and Future Prospects of Tropical Moist Forests", Environmental Conservation, Vol. 7, No. 2, Summer 1980., pp. 101-114.

NASA, "Earth Observing System", Technical Memorandum 86129, NASA Goddard Space Flight Center Greenbelt, Maryland 20771 1984.

Scarpace, Frank L., Ralph W. Kiefer, Thomas M. Lillesand, and Peter R. Weiler, "An IBM PC/AT Microcomputer-Based Image Processing Laboratory", Technical Papers of the American Society for Photogrammetry and Remote Sensing, ACSM-ASPRS Fall Technical Meeting, September 1985.

Talbot, James J., "What's Happening to the World's Forest Resources", Proceedings of the 1981 Machine Processing of Remotely Sensed Data Symposium, p. 587-592, 1981.

Technologies to Sustain Tropical Forest Resources, Office of Technology Assessment (OTA-F-214), U.S. Congress, Washington, D.C., March 1984.

GEOGRAPHIC INFORMATION SYSTEMS APPLICATIONS

Moderator: D. Hunter

UPDATING RANGE SURVEYS USING A GEOGRAPHIC INFORMATION SYSTEM*

Jeffery C. Eidenshink**
Technicolor Government Services, Inc.
EROS Data Center
Sioux Falls, South Dakota 57198

David C. Sjaastad***
Cheyenne River Agency
Bureau of Indian Affairs
Box 833
Eagle Butte, South Dakota 57625

ABSTRACT

A spatial data base was developed for the Fort Berthold Indian Reservation in North Dakota to demonstrate the use of a geographic information system for natural resource management. A key component of the digital data base was a detailed soil survey. Range site boundaries were derived by aggregating soil mapping units on the basis of interpretations from the published soil surveys. The problem was that no range survey data including stocking rates and condition class were available for the soil-survey-derived range sites. The only available range survey was a survey that was based on range site delineations from a Missouri River Basin Inventory (MRBI) program conducted in 1961, prior to the availability of soil survey data. The MRBI survey information was correlated with the revised range sites derived from the soil surveys to derive stocking rates and condition classes. A 1:24,000-scale map which included revised range site boundaries and management boundaries was produced for each township within the reservation. All polygons on the maps were uniquely labeled to permit discrimination of the potential differences in range site stocking rates due to grazing management practices. The maps and the associated polygon attribute data were merged with maps of the MRBI transect information to derive the stocking rate and condition class for each revised range site within a management unit. The resulting data base components were a map of revised range site boundaries within management units and an associated tabular file containing attribute information including stocking rate and condition class.

*Publication authorized by the Director, U.S. Geological Survey, on June 17, 1985.

**Technicolor Government Services, Inc. Work performed under U.S. Geological Survey contract no. 14-08-0001-20129.

***Formerly with Shive's Rangeland and Environmental Consultants, now with Cheyenne Agency, Bureau of Indian Affairs.

INTRODUCTION

A range survey was completed in 1982 for the Bureau of Indian Affairs (BIA) Fort Berthold Indian Reservation in central North Dakota. The size of the reservation is approximately 500,000 acres. This is the first reservation-wide survey since the Missouri River Basin Inventory (MRBI) was completed in 1961. The 1982 survey used the same range site information developed in 1961 to establish field survey transects, but the survey resulted in a new determination of stocking rates and condition classes for the MRBI range sites. An independent effort was underway at the same time to develop a computerized resource data base for natural resource management on the Fort Berthold Indian Reservation. The data base was to be used within a geographic information system for storage, retrieval, and analysis of resource information in support of land management activities.

One of the major activities is management of the rangeland within a range leasing program called the Range Permit Payment System (U.S. Department of the Interior, 1972). The program involves aggregating small individual ownership parcels, generally less than 320 acres into larger units called range units and leasing them for livestock grazing. Participating landowners receive a payment in proportion portion to the carrying capacity of their land. One of the major information needs is to know the stocking rates of the range sites within each ownership parcel. Ultimately, a total carrying capacity is calculated for each parcel and for each range unit.

The data base includes modern soil surveys information which could be aggregated to dominant range sites (U.S. Department of Agriculture, 1979). However, no stocking rate and condition information was available for the revised range sites. The original MRBI survey was completed far in advance of published detailed soil surveys; thus, the range sites in the MRBI did not conform to those derived from soil mapping units. A major data entry effort would have been required to computerize the MRBI range site/range survey data. The BIA decided to transform the field information collected during the 1982 range survey to derive stocking rates and condition class information for the revised range sites derived from soils data.

This paper describes the use of a geographic information system to combine baseline information about revised range sites with the field-collected data to update the range survey at the Fort Berthold Indian Reservation.

THE DATA BASE

The Fort Berthold Reservation data base contains several layers of information: soils, ownership, range unit boundaries, public land survey data, and land use. All layers were geometrically registered to 50-meter grid cells in a Universal Transverse Mercator grid projection, with each grid cell representing about 0.6 acres.

The data layers used were soils and range unit boundaries. The soil survey boundaries were digitized from Soil Conservation Service published soil survey maps. There were over 10,000 separate soil delineations representing 184 different soil mapping units. The range unit boundaries were digitized from base maps provided by the BIA and included 90 range units within the reservation. Generally, range units are contiguous land parcels; however, in some cases they are not. For this reason, another layer of information called management units was derived from the range unit data. The management unit layer comprises unique numbers assigned to individual range unit polygons within the reservation. Therefore, if a range unit is three separate polygons, the range unit would be composed of management units 1, 2, and 3. The importance of the management unit designation will be evident later.

PROCEDURES

Range Site Classification

In recent published detailed soil surveys, a table identifies the range site for each soil mapping unit. This information was used to reclassify the 184 individual soil mapping units in the data base to 16 possible range site classes. The range site classes at Fort Berthold are: silty, overflow, shallow, sandy, wet meadow, wetland, saline lowland, closed depression, sands, thin upland, clayey, claypan, thin sands, very shallow, thin claypan, and not classified. The result of the reclassification was a data layer comprising approximately 3,300 range site polygons of various sizes.

Unique Range Site Labeling

Each range site delineation was uniquely labeled to discriminate it from others. The discrimination of separate range sites, regardless of range site classification, is necessary because each range site could have different stocking rates and condition classes as a result of different management practices. After the range site labeling process, three characteristics were known about each range site: site number or label, site class, and size.

Area Filtering

The size information for each range site was used to eliminate polygons smaller than 40 acres. They are generally considered by the BIA to be unmanageable. Range sites smaller than 40 acres were essentially absorbed by the larger range sites surrounding them. These procedures resulted in a final range site layer comprising approximately 1,100 unique revised range site delineations which could be discriminated by a site number.

Management Unit Determination

Range site stocking rates and condition are affected by management practices. The most common delineations of boundaries between management practices are fencelines where individual owners have determined the grazing practices. It is difficult to identify all fences between land parcels, yet it is relatively easy to identify fences or boundaries of major land parcels. These boundaries are the range unit boundaries. The range unit boundaries were most important because the BIA has the responsibility of managing the land within the range units but only serves an advisory role on land outside the range units.

The large blocks of land not in a range unit were also assigned as a management unit. Since these blocks were extremely large, they were subdivided into smaller management units. Township and range boundaries from the public land survey data were used to complete the subdivision of large tracks of land into more meaningful management units. As a result of the subdivision, it was possible to treat large blocks of land on a township basis to better account for variations in management practices.

Overlay Analysis

At this point, five major data layers are being used: range sites greater than 40 acres, range sites where each polygon is uniquely labeled, range units, management units derived from range units, and township and range lines. Geographic information systems have a capability called overlay or intersection that can combine multiple layers spatially. One output of the intersection function is a tabular file which identifies all unique combinations of the five data layers, especially the range sites within a management unit.

The tabular file created by the intersection function is stored within a data base management system (DBMS) within the geographic information system. The DBMS provides several methods of tabular data manipulation including sorting, merging, query, statistics, mathematical expressions, editing, and data entry (DBD Products, Inc., 1983). Table 1 is a portion of the tabular data file, which contains the numeric values of each of the data layers. Each case or row in the file represents a unique combination of the five major layers. The township and range number (14993) is Township 149, Range 93. The variable "RANGE UNIT" is the range unit number. "MANAGEMENT UNIT" is the management unit polygon label. "SITE NUMBER" is the unique range site delineation label. "SITE NAME" is the range site class descriptor. "SITE ACRES" represents the area of the unique combination.

In table 1, the second case listed in Table 1 indicates that for range site number 254, in Range Unit 4, Management Unit 89, the range site class is thin upland, and it is 247.73 acres in size. There are also several more range sites within Range Unit 4 (65, 255, 268, 269).

Table 1. Facsimile of a tabular data file that contains unique combinations of five data layers.

TOWNSHIP--14993

RANGE UNIT	MANAGEMENT UNIT	SITE NUMBER	SITE NAME	SITE ACRES
4	89	65	SHALLOW	872.33
4	89	254	THINUP	247.73
4	89	255	SILTY	24.09
4	89	268	SILTY	4.32
4	89	269	THINUP	33.97
26	169	65	SHALLOW	323.72
26	169	262	SANDY	63.63
26	169	288	THINUP	33.97
26	241	302	SANDY	4.32
26	241	324	THINCLAY	40.77
26	241	65	SHALLOW	62.39
26	241	262	SANDY	5.56
26	241	335	THINUP	45.09
213	231	388	SANDY	51.27
213	231	379	SHALLOW	41.39
213	231	377	SANDY	46.95

Spatial Identification

The tabular data provided all the information pertaining to the range sites except the spatial location. The spatial location is, of course, necessary in order to correlate the information collected in the field with the revised range sites. The spatial information was produced by plotting two data layers; the range site boundaries and management unit boundaries. A 1:24,000-scale line plot of each township was produced that would overlay directly with the maps of range survey information that had been collected in the field. The polygon label or site number for each range site delineation was included on the plots. The management unit boundaries were plotted directly over the range site data. The management unit boundaries were plotted using bolder or darker lines, and the management unit numbers were plotted using bolder numbers in order to distinguish them from range site numbers. Figure 1 is a subset of the township plot. The area represented in figure 1 is the data for Range Unit 4 in table 1.

Using both the tabular and spatial information, a specific range site can be located and its size and type determined. Forty township plots and associated tabular data were delivered to the BIA.

Determination of Stocking Rates and Condition

The information used by the BIA for determination of stocking rates and condition classes included detailed field transect reports and MRBI range site maps. Each report contained complete listings of occurring plant species, percent species composition (by weight), total site production (dry weight), recommended stocking rates, and comments on site condition and trend. Each transect

T149N R093W

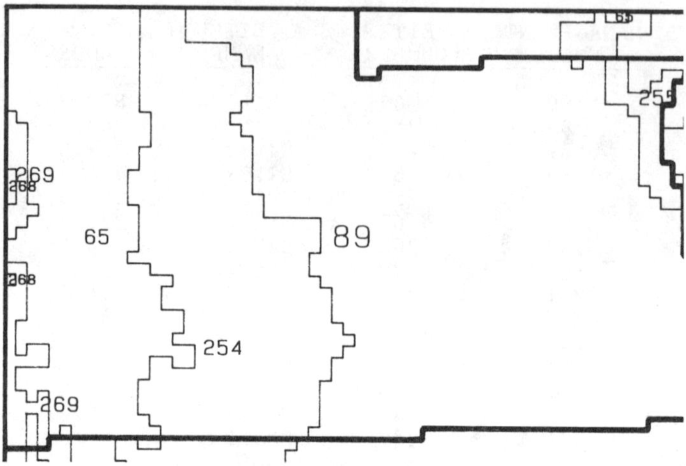

Figure 1. Subset of 1:24,000-scale township plot of range sites and management units, emphasizing Range Unit 4, Management Unit 89.

location was pin pricked on an aerial photograph and the direction was given on each report. One MRBI range site map was made for each township on the Fort Berthold Reservation. Each map consisted of a base 7.5-minute USGS topographic map and a clear acetate range site overlay. Each overlay contained range site polygons with symbols, percent condition, and stocking rate shown in acres per Animal Unit Month (Ac/AUM).

The revised range sites were assigned a condition and stocking rate unless it was determined from field notes, aerial photographs and USGS topographic maps that the site consisted entirely of water or was previously non-surveyed land (1982 survey). The detailed field data collected enabled decisions to be made concerning the revised range sites that did not correspond directly to the MRBI range sites.

Where several revised range sites made up one 1982 MRBI range site, in most cases the 1982 transects for each site crossed over and sampled parts of each range site. The plant composition data collected in each field report were used to determine new separate site conditions (in percent) and stocking rates (Ac/AUM).

Finally, there were cases in which several MRBI range sites were combined to make up one revised range site. The conversion was accomplished by taking the area-weighted mean of each old site to derive a new condition and stocking rate. For example, two silty range sites and a sandy range site make up one revised silty range site. First, a new percent condition would be calculated for the sandy site utilizing the species composition list. Acreage of each MRBI site lying within the site would then be determined, and the total acreage of the site would be determined. Then, a percentage of each MRBI site would be calculated and a weighted average of each would be taken.

The stocking rate and condition class for each revised range site within a management unit were entered into the DBMS and merged with the existing tabular information. Table 2 shows a facsimile of the tabular data including stocking rate and condition class for the same information listed in table 1.

There were cases in which the MRBI survey classified a range site as one type and the revised range sites classified it differently. For example, in one case the MRBI survey classified a range site as sandy, whereas the revised range site was classified as silty. This required a re-rating of the species list, with the percent composition, to derive a new percent condition based on criteria used to rate silty sites.

Table 2. Facsimile of the tabular data file including stocking rate and condition class for the same township depicted in table 1.

TOWNSHIP--14993

RANGE UNIT	MGT. UNIT	SITE NUMBER	SITE NAME	STOCK RATE	COND.	SITE ACRES
4	89	65	SHALLOW	2.8	55	872.33
4	89	254	THINUP	2.3	75	247.73
4	89	255	SILTY	2.2	65	24.09
4	89	268	SILTY	1.8	65	4.32
4	89	269	THINUP	2.7	55	33.97
26	169	65	SHALLOW	3.5	60	323.72
26	169	262	SANDY	2.2	60	63.63
26	169	288	THINUP	2.2	60	33.97
26	241	302	SANDY	2.2	60	4.32
26	241	324	THINCLAY	2.2	60	40.77
26	241	65	SHALLOW	2.2	60	62.39
26	241	262	SANDY	2.2	60	5.56
26	241	335	THINUP	2.2	65	45.09
213	231	388	SANDY	3.1	55	51.27
213	231	379	SHALLOW	2.7	55	41.39
213	231	377	SANDY	1.8	60	46.95

CONCLUSION

A computerized resource data base and a geographic information system were used to take independent range survey field information and to integrate it with the revised range sites derived from a soil survey to produce an up-to-date range site/range survey data base. The procedures used eliminated the need to digitize a very complex out-of-date range site map. The adaptation of soil-survey-based range sites enables the BIA to take advantage of revised stocking rate guidelines provided by the Soil Conservation Service (U.S. Department of Agriculture, 1984). In fact, in 1984 the Soil Conservation Service revised the recommended stocking rates for North Dakota. The revision essentially resulted in a 25-percent increase in the carrying capacity of most range sites. In less than two days, all 100 range site stocking rates were adjusted and new carrying capacities for the range units were calculated. Without the use of a computerized geographic information system, the process would have taken weeks.

Using geographic information system spatial analysis capabilities and tabular data handling, the carrying capacity of every ownership parcel within a range unit and thus the payment due each landowner participating in the Range Permit Payment System could be determined. Any revisions or other pertinent information provided by the Soil Conservation Service or by the BIA range scientists could be easily integrated into the data base.

REFERENCES

DBD Products, Inc., 1983, Relational data base--fact or fancy?: Rockville Centre, New York, DPD Products, Inc., 179 p.

U.S. Department of Agriculture, 1979, Soil Survey of McLean County, North Dakota: Soil Conservation Service, p. 79-108.

U.S. Department of Agriculture, 1984, Soil Conservation Service, North Dakota: Technical Guide Notice ND-35, p. 4.

U.S. Department of the Interior, 1972, Range Permit Payment System: Bureau of Indian Affairs, 57 p.

COST EFFICIENCIES OF SURFACE COVER MAPPING FOR GEOGRAPHIC INFORMATION SYSTEM APPLICATIONS

Samuel C. Williamson
U.S. Fish and Wildlife Service
2627 Redwing Road
Fort Collins, CO 80526

Ivo E. Lindauer
Department of Biological Sciences
University of Northern Colorado
Greeley, CO 80639

and

Patricia DePlazes
Department of Biological Sciences
University of Northern Colorado
Greeley, CO 80639

ABSTRACT

A project was initiated in the fall of 1983 to quantitatively assess the accuracy of vegetation cover maps produced by three remote sensing techniques: machine classified Landsat (MCL) digital multispectral scanner (MSS) data, color composite Landsat (CCL) MSS images, and medium scale color infrared (MSCIR) aerial photography. Information from the three techniques was compared to ground reference data gathered in the Piceance Basin Planning Area of northwest Colorado.

Assessments were made using a site-specific random sampling design and analyses based on classification error matrices. Several comparisons were made to compensate for differences among the vegetation classification systems used with each remote sensing technique. The heterogenous mixture of upland shrubland, sage shrubland, aspen woodland, and pinyon-juniper woodland was difficult to map accurately, particularly with Landsat imagery. Using the preferred classification system (without overlap of cover types) MCL accuracy was 26%, CCL accuracy was 27%, and MSCIR accuracy was 58%. Color composite Landsat had the lowest total mapping costs at .062 cents/acre, followed by MCL at 4.1 cents/acre, and MSCIR at 8.2 cents/acre. Based on an accuracy:cost ratio analysis, the most cost-efficient procedure for mapping large expanses of rangelands would use a combination of MSCIR aerial photography for heterogenous or high interest areas, and small scale aerial photography or CCL imagery for homogeneous and low interest areas where lower accuracy is acceptable.

DEVELOPING A RESOURCE MANAGEMENT DATA BASE FOR THE OKANOGAN NATIONAL FOREST FROM MULTISPECTRAL IMAGERY AND THE USE OF GIS

G. O. Klock
G. O. Klock & Associates
2113 Sunrise Circle
Wenatchee, WA 98801

P. Gum
Okanogan National Forest
U.S. Forest Service
Okanogan, WA 98840

L.E. Jordan III
ERDAS, Inc.
430 Tenth St.NW
Atlanta, GA 30318.

ABSTRACT

The vegetation and overstory crown closure for approximately three million acres of the Okanogan National Forest, adjacent forest and range lands have been derived from Defense Mapping Agency topographic information and classified Landsat multispectral scanner (MSS) imagery to form a database for fire fuels management. This database, includes vegetation specie, conifer size class, crown closure, average elevation, aspect and slope classes. Summary statistics are available for each 0.6 acre for fire management. This database forms the framework for a Geographic Information System (GIS) to be used in all resource management activities throughout the Okanogan Forest. Additional information (e.g. land ownership, wildlife habitat) is being added to the database through tablet digitization and video digitizing. Individual files of the various data layers can be displayed in color for interpretation, subjected to overlay, matrix, and proximity analyses to make composite data layers, or output to geo-referenced color maps for field use. This GIS capability, on an ERDAS IBM PC/XT microcomputer, provides the Okanogan Forest with a detailed level of information management that can easily be kept current and accurate.

INTRODUCTION

At the "1984 International Symposium on Inventorying Forest and Other Vegetation of the High Latitude and High Altitude Region", a presentation was made describing Phase I of a U. S. Forest Service project which utilized Landsat MSS imagery for forest vegetation inventory (Klock et al. 1984). Phase I involved the site specific "test" classification and stratification of two (2) 7.5 minute USGS quadrangles, followed by a field verification and validation to determine if Landsat imagery could be used with sufficient accuracy to construct a fire fuel inventory similiar to that developed by Consentino et al. (1981) and Salazar (1982). Phase II involved the fire fuels classification of the remaining ninety-two (92) 7.5 minute USGS quadrangles in the project. After briefly reviewing work completed in Phase I, this presentation focuses on the techniques and methodology used to complete Phase II of the project. Opportunities for using and supplementing the project database in GIS by resource managers are also discussed.

PHASE 1

The primary objective of the Okanogan National Forest project was to classify and stratify the vegetative fire fuels on a variety of ecotypes throughout the Forest and adjoining lands. This vegetative fuels inventory was to be produced, based on satellite data, in a form suitable for use by the Forest computerized fire management strategy system. The end product was a listing of vegetative fuels by 0.6 acre mapping units (50m x 50m pixels) with summarization by GEOLOC grid units (Renslow 1979) for approximately three million acres of forest and range land. Each GEOLOC unit, as defined by latitude and longitude, is approximately 80 acres.

The Okanogan National Forest, located in North Central Washington, is administered by the United States Forest Service. Physiographically, the study area is encompassed by a western extension of the Rocky Mountains, the Okanogan Highlands, and the eastern and western slopes of the North Cascade Range. Both alpine and continental glaciers have played a major role in shaping the landforms of the Forest. Elevations range from 1,100 feet to greater than 8,000 feet. The climate of the Forest varies from pseudo steppe conditions east of the Cascade Range to a cool, moist maritime environment with heavy snowfall along the crest of the Cascades. Both climatic conditions often are found within a relatively short distance within the Forest. During the winter months, the normal eastward movement of maritime air is occasionally disrupted by cold arctic air moving down north-south trending valleys. Annual precipitation varies from 10 inches at the lowest elevation to 160 inches or more in the North Cascade Range. Geologically, the study area is composed of a variety of igneous, metamorphic and sedimentary rocks. The combination of diverse climatic and geologic conditions and a varied topography results in a wide range of forest vegetation species. Predominant coniferous overstory species are: ponderosa pine (*Pinus ponderosa*), Douglas-fir (*Pseudotsuga menziesii*), lodgepole pine (*Pinus contorta*), western larch (*Larix occidentalis*), subalpine fir (*Abies lasiocarpa*), and Engelmann spruce (*Picea engelmannii*), western hemlock (*Tsuga heterophylla*), and Pacific silver fir (*Abies amabilis*) (Williams and Lillybridge 1983).

The Okanogan National Forest vegetation mapping project required classification of Landsat MSS imagery to delineate all individual conifer species which comprised more than 60 percent of the stand. When two or more species represented 25-40 percent or 40-60 percent of the stand, all specie associations were categorized as separate classes. With the exception of lodgepole pine, each specie class was stratified into three size classes: seedlings less than 10 feet high, poles 4-14 inches in diameter, and "mature" trees larger than 14 inches diameter. Due to the natural distribution of stem diameters in non-plantation conifers a stand was considered to be "mature" if, within the sample site, at least one stem in ten exceeded 14 inches diameter breast height (d.b.h.). Lodgepole pine stands were stratified using four types:

(1) old growth decadent stands characterized by large amounts of downfall with little or no reproduction;
(2) healthy, vigorous stands characterized by limited downfall, reproduction in open or fringe areas, the presence of other species in the understory, and no obvious signs of stunted growth;
(3) seedlings characterized by large amounts of reproduction and generally vigorous in appearance; and,
(4) "dog-hair" stands which are severely stunted, often decadent, with an obvious lack of vigor.

Each class of conifers was stratified utilizing four levels of crown closure (1) 0%, (2) 1-30%, (3) 31-70% and (4) 71-100%. Since crown closure estimates are somewhat judgmental, classes 2, 3, and 4 were derived by applying specified mapping criteria consistent with fire control management objectives. A stand was identified as class 2 if individual trees could be identified on a 1:24,000 scale aerial photograph, class 3 if continuous forest floor could be observed within the stand, and class 4 when observations of the forest floor were discontinuous.

Accuracy standards for project mapping were tested area wide by field and aerial photograph verification of each specie, crown closure, and cover type. In Phase I, standards were required to meet or exceed an aggregate 80 percent correct classification of primary sampling unit, an 80 acre GEOLOC grid cell. A confidence level of at least 85 percent was required. Site-specific accuracy standards for each secondary sampling unit, 2.4 acre or four (4) pixel area, were required to meet 70 percent correct classification with confidence level to exceed 85 percent.

In the initial project plan, it was assumed that standard forest inventory techniques could be used to collect data for identifying training classes for correlation with individual species, stem size, and crown closure. It was also assumed that the available forest inventory data such as that used by Sugarbaker et al. (1980) in the Okanogan region, would provide the necessary information to identify training classes for the supervised classification of the Landsat imagery. Early in the project, however, it became obvious that the necessary project accuracy could not be met by relying on the combination of available forest inventory information and techniques used previously to construct Landsat forest vegetation inventories. In order to meet the requirements of Phase I, a new approach to gathering field ground-truth data became necessary.

To meet project accuracy an inventory of stands to be used as "training" sites was made for the Phase I Test Quadrangles. Field ground data as well as aerial photographs were used to select these "training" sites which were larger than 10 acres, preferably greater than 40 acres, and contained the required various vegetative fuels by mapping classes. It was particularly useful that the technician selecting the "training" sites understood ecosystem dynamics as well as site/vegetation relationships. Each training unit location was carefully identified on USGS 7.5 minute quadrangle maps. Then coordinates of each "training" site were interactively delineated on the ERDAS image processing system display.

Through the USGS EROS Data Center a search of available Landsat scenes revealed an extremely good quality September 1981 scene that covered most of the Okanogan National Forest. This scene was particularly useful because the aerial photos available for the project were taken approximately one month earlier. The raw Landsat MSS data was pre-processed to remove the effects of Earth rotation, skew, mirror velocity, and atmospheric effects such as haze. This provided corrected "clean raw" Landsat data. The raw data for the Landsat scene can be displayed on the ERDAS interactive color image processing system in a false color mode (3 channel overlay) simulating color infrared photography.

Using the image processing system joystick, 24 ground control points were selected at equal intervals throughout the two (2) Phase I Test Quadrangles. Initially, small water bodies were used for ground control points. These points, however, did not meet the ground control accuracy required for the project. Thus, road material "rock pits," road junctions, and prominent geologic features were utilized providing a more acceptable providing ground accuracy within +/-50 meters.

Nearest-neighbor algorithms were employed to perform the geometric correction of the Landsat MSS data. The data were "rectified" to the UTM coordinate system based on a 50 meter pixel size. A digital mosaic was created for areas split between boundaries.

The field-collected training samples delineated interactively on the image display were used in conjunction with a supervised maximum likelihood algorithm to perform a forest vegetative fuels classification. To "fine tune" the supervised classification new interactive training samples were used in areas of potential misclassification. Twenty-four classes of vegetation were identified in the two (2) Phase I Test Quadrangles.

In Phase I it became apparent that the use of topographic stratification to support the supervised maximum likelihood classification (similiar to that used by Hoffer et al., 1979) would significantly increase the mapping accuracy of the vegetative fuel classes. It was also observed that spectral data alone could not be used to identify the small difference spectral reflectance value differences among similar species such as ponderosa, lodgepole, and whitebark pine (Pinus albicaulis). However these species do fill an ecologic niche in the forest landscape with locational differences characterized by aspect and elevation thereby creating the moisture and temperature environment favorable to a particular pine specie. There is generally minimal overlap in location between the occupied environmental zones in which the respective pine specie is dominant. Thus, ecologic models were developed for the Okanogan National Forest that reflect the expected location of dominance by each tree specie. During the supervised classification of the Landsat MSS data, these ecologic models were used to "mask" aspect and elevation zones in which a particular specie is unlikely to be present.

A test of the Phase I project accuracy showed that the crown closure classification ranged from 90 to 94 percent correct on both the primary and secondary sample units. The vegetative fuel classificaton accuracy ranged from 92.5 to 94 percent correct on the 80 acre primary sampling units, and as was expected, dropped to 82 to 84 percent correct on the secondary or individual pixel sampling units. Project accuracy was sufficient to proceed with Phase II of the project.

Phase II

Phase II required the vegetative fuels and crown closure mapping for an additional ninety-two (92) 7.5 minute quadrangles with the assumption that all procedures acceptable in Phase I would be applicable. The majority of the computer image processing and spectral classification procedures developed in Phase I were used successfully in Phase II. There were, however, several basic assumptions in the original project requirements that did not remain valid for a supervised classification of a large mountainous region.

In developing the project, the two (2) Phase I Test Quadrangles were chosen specifically because they appeared to reflect the greatest range of different vegetative fuel classes. The Tiffany Mtn. NW Quadrangle represented the upper range of elevations found in the Forest (5,000 to 8,000 feet), while the Loup Loup NW Quadrangle represented lower elevations (3,000 to 6,000 feet). For these two (2) quadrangles 24 different fuel classes were identified, and it was assumed that these classes would represent a majority of classes within the forest. This was not the case as evidenced by the final project output of 142 different vegetative fuel classes for the Forest. Thus, a major effort was needed to identify all the additional training sites required.

It was also assumed that after the signature of a respective fuels class was identified from a training site, this signature could be used to classify respective fuels class throughout the entire Landsat scene. Again, this was not the case because the spectral reflectance value for the same vegetative fuel class was markedly different within the various geographic regions. A review of the problem showed that the spectral reflectance values for a given class remained quite similiar within uniform areas that could be identified as biogeoclimatic zones. It was reasonable to assume spectral differences among zones for the same specie because within the Landsat scene precipitation ranges from 10 to 160 inches per year and elevations range from 1,100 to more than 8,000 feet. Thus, the forest was divided into eight areas or biogeoclimatic zones, and training sites for each vegetative fuels class within the zone were identified and used for developing signatures. Often it was difficult to find a modal vegetative stand that represented the entire range of spectral features for a given vegetative fuels class. Thus, to compensate for large variability within a fuels class, multiple spectral signatures were often used to classify a specific vegetative fuels class. After classification, the output files were merged to represent a single fuels class.

The original project assumed that if the Landsat classification was sufficiently accurate in Phase I, this standard would be maintained in Phase II. Based on the addition of a large number of mapping classes and the observed biogeoclimatic variability factor, this assumption did not appear valid. Therefore, a method was developed to check the accuracy and to validate at least one (1) and possibly two (2) quadrangles in each biogeoclimatic zone. This approach provided a thorough accuracy verification on approximately 15 percent of the total project area.

One challenging area in Phase II was the definition, by spectral classification, of vegetative fuel classes which contained significant amounts of three (3) plant species in a common plant association. Each plant species had a markedly different spectral reflectance value when observed in pure stands. An example is a fuels class that contains 1-25 percent ponderosa pine, 25+ percent bitterbrush (*Purshia tridentata*) and 25+ percent grasslands. The spectral range of the signature representing a training site with this specie composition is quite large, and when used for classifying the Landsat scene introduced significant error. For spatial definition of respective mapping classes in stands represented by three or more spectral dominant species, this project experience indicates the following: MSS data classified to the dominant specie and refined with ancillary data provides greater accuracy than use of an individual training signature.

DATABASE FOR GIS

The original project was developed to establish a database for a computerized fire management strategy system. A further review of the information being generated pointed out its value for use in a Forest-wide Geographic Information System (GIS). The outputs of Phases I and II (vegetative fuel classes, crown closure, average aspect, slope, and elevation with summary statistics on each Ø.6 acre pixel) provided the framework for a GIS database that could be supplemented with other ancillary data to support not only fire management, but all resource management activities in the Forest. The Okanogan National Forest purchased an ERDAS Integrated Image Processing/Geographic Information System to analyze, display, and provide color hardcopy of this database. An additional ten layers of management data are now being prepared for use by the Okanogan National Forest via tablet-digitizing, video digitzing, the MSS data used previously.

One of the key features of the ERDAS image processing system is the capability to analyze the original data layers and create "synthetic" maps by use of overlay, weighted sums, matrix, proximity analysis, and additional analysis functions. For example, areas of high solar radiation intensity, where mature stands are now present, can be identified. Regeneration may be difficult in areas of high solar intensity. Thus, this information could be reflected in a stand silvicultural prescription.

At present, G. O. Klock & Associates is developing two additional layers of information for the Okanogan GIS system that require the database information created by the Phase II project. The two layers are the current and potential timber productivity of National Forest lands on twenty (2Ø) 7.5 minute quadrangles. Coupling the analytic capablitiy of the GIS system with field plot data provides the spatial distribution of vegetation and abiotic factor levels that control current and potential timber productivity on these lands.

Data layers in the GIS system can be visually inspected on the Okanogan National Forest`s ERDAS IBM PC-XT low cost Image Processing/Geographic Information System. Individual files of the various data layers (vegetation, crown closure, aspect, etc.) can be displayed for visual interpretation, overlayed to make composite data layers, or output to scaled geo-referenced color maps for field use. Maps often are produced at the 1:24,ØØØ scale so that they may be used at the same scale as the aerial resource photos. These tools provide the resource managers of the forest an opportunity to develop inexpensive current color maps for many activities including planning and project monitoring programs. Through the use of reclassification and the input of digitized information, the database can easily be kept current and accurate.

SUMMARY

The vegetative fuels classes and crown closure for ninety-four (94) 7.5 minute USGS quadrangles were successfully inventoried for input into a fire management strategy system by use of Landsat MSS imagery for the Okanogan National Forest and adjoining forest and range lands in Washington State. Many new techniques, that have not been used extensively in previous forest land inventory studies, were introduced in the project to improve the vegetation classification accuracy. It is apparent that the accuracy of the Landsat classification depends on the capabilities of the image processing system as well as the ability of the field technician to "read the land" and relate ecosystem features to spectral reflectance values collected by imagery.

Information collected in the Okanogan National Forest vegetative fuel class inventory, based on using Landsat MSS and DMA topographic data, provides an excellent framework for a low-cost GIS that can be used by fire management as well as other resource managers. Aside from the basic data layers provided in this project, additional layers of resource information such as land ownership, wildlife habitat, soils, and hydrology can be input through tablet digitizing for continual support of Forest-wide planning and project monitoring programs.

REFERENCES

1. Consentino, M. J., C. E. Woodcock, and J. Franklin. 1981. **U. S. Forest Service Vegetative Fuels Research - Final Report.** Geography Remote Sensing Unit, University of California, Santa Barbara.

2. Hoffer, R. W., M. D. Fleming, L. A. Bartolucci, S. M. Davis, and R. F. Nelson. 1979. **Digital Processing of Landsat MSS and Topographic Data to Improve Capabilities for Computerized Mapping of Forest Cover Types.** Annual Report on NASA Contract NAS 9-15508, Laboratory for Applications of Remote Sensing, Purdue University, West Lafayette, Indiana. 159 pp.

3. Klock, G. O., L. E. Jordan III, and P. Gum. 1984. A Forest Vegetation Inventory with Landsat Imagery for use with Fire Management Strategies in Washington State. **Presented Paper, International Symposium on Inventorying Forest and Other Vegetation of the High Latitude and High Altitude Regions,** Fairbanks, Alaska, JULY. p. 182-187.

4. Renslow, M. S. 1979. **Geo-Referencing/Locator Grid and Map Numbering System.** (unpublished) Pacific Southwest Region, USDA Forest Service, 24 pp.

5. Sugarbaker, L. J., T. W. D. Gregg, E. W. Barthmaier, R. B. Scott, M. D. Fleming, and R. M. Hoffer. 1980. **Forest Cover Type Mapping in Eastern Washington with Landsat and Digital Terrain Data.** Final Report, Pac. Northwest Reg. Comm. by State of Washington Dept. of Nat. Res., 40 pp.

6. Williams, C. K., and R. R. Lillybridge. 1983. **Forest Plant Associations of the Okanogan National Forest.** USDA Forest Service, Pacific Northwest Report. R6-Ecol-132b-1983. 140 pp.

GEOGRAPHIC INFORMATION SYSTEM AND REMOTE SENSING APPLICATIONS IN RURAL ALASKA SUBSISTENCE USE PROTECTION

Kevin Gerard Meyer
Bureau of Land Management
4700 E. 72nd Avenue
Anchorage, Alaska 99507

ABSTRACT

The protection of rural Alaskan subsistence use emerged as an important management issue for federal land managers with the passage of the Alaska National Interest Lands Conservation Act of 1980 (ANILCA). Provisions of Title 8 of the Act, Subsistence Management and Use, require that traditional subsistence use by rural residents of the State, both Native and non-Native, receive preferential consideration in land use planning and environmental review. The Bureau of Land Management, in its interpretation of ANILCA, has developed specific guidelines to protect the "supply side" components of subsistence use. These components are: 1) wildlife and plant species utilized in subsistence; 2) their habitat; 3) traditional access to subsistence resources; and 4) protection from significant increases in competitive use by non-subsistence users. The framework for an operational subsistence use protection management program has been developed by the author which incorporates geographic information system capabilities as a base for data management. The program utilizes remote sensing and the USDI Fish and Wildlife Service's Habitat Evaluation Procedures for data input and analysis. Major components of the management system include: 1) documentation of traditional subsistence use; 2) evaluation of subsistence resources within subsistence use study areas; 3) conflict identification and impact analysis; and 4) administrative decision making and documentation.

INTRODUCTION

The objectives of this paper are two-fold: 1) to introduce the subsistence use protection program to data acquisition and information system technicians, and 2) to initiate a discussion with resource managers of the potential applications of geographic information systems and remote sensing in rural Alaska subsistence use protection. This paper briefly reviews ANILCA's legal mandates for subsistence use protection, the Bureau of Land Management's administrative response to ANILCA Title 8 requirements, and potential applications of geographic information and remote sensing systems to an proposed subsistence use protection management program.

ADDRESSING THE SUBSISTENCE ISSUE

The Alaska National Interest Lands Conservation Act of 1980 (ANILCA), (P.L. 96-487), was the first piece of federal legislation to specifically address the subsistence issue. Title 8 of the Act, Subsistence Management and Use, states that: "the continuation of the opportunity for subsistence uses by rural residents of Alaska, including both Natives and non-Natives, on the public lands and by

Alaskan Natives on their Native lands is essential to their physical, economic, traditional, and Native cultural existence." ANILCA provides a legal interpretation of the term subsistence by defining its associated activities. Section 803 defines subsistence use as: "The customary and traditional uses by rural Alaskan residents of wild renewable resources for the direct personal or family consumption as food, shelter, fuel, clothing, tools or transportation, for the making and selling of handicraft articles out of the nonedible by-products of fish and wildlife resources taken for personal or family consumption and for customary trade." Congress recognized that protecting the opportunity for a subsistence way of life required both the protection of the wild renewable resources upon which the subsistence user draws; and the protection of existing subsistence use--as a land use activity. Individual sections within Title 8 address these two components of the protection problem. Of these, Section 810 most directly affects the day-to-day operations of federal land management agencies as it requires the integration of subsistence protection into land use planning and environmental review.

The Bureau of Land Management (BLM), a Department of the Interior agency, is authorized under the Federal Land Policy and Management Act of 1976, (P.L. 94-579), to provide multiple-use management of approximately 100 million acres of federal public lands within the State of Alaska. With the passage of ANILCA, subsistence use protection was elevated to a primary management concern for the Bureau. BLM identifies that ANILCA's Subsection 810(a), depending upon circumstances, requires up to four steps. These steps are: 1) an evaluation of the affect a proposed action will have on subsistence uses and needs; 2) a finding of whether or not a proposed action will have a significant restriction on subsistence use; 3) a notice and hearing if an action is found to have a significant restriction; and 4) a three part determination. BLM's 810 process places a strong emphasis on the protection of the "supply-side" components of subsistence resource use; while sociocultural and socioeconomic aspects are addressed within the National Environmental Policy Act (NEPA), (P.L. 91-190), environmental review process.

COMPONENTS OF AN OPERATIONAL SUBSISTENCE USE PROTECTION MANAGEMENT PROGRAM

In response to BLM's administrative interpretation of ANILCA Title 8 requirements, the author has proposed a framework for an operational subsistence use protection program. The major components of the framework are outlined in Table 1.

Current Subsistence Use Documentation.

The first component of a subsistence use protection management program is the identification and description of current subsistence use activities by their geographic location. An initial step in the development of a subsistence use data base is the establishment of study area boundaries. Boundaries should be defined utilizing information provided by Alaska Department of Fish and Game (ADFG)-Subsistence Division. Study areas should include subsistence use areas and adjoining areas which contribute to, or support wildlife species availability within use areas. Subsistence use information is typically collected by ADFG-Subsistence Division and displayed on USGS topographic series maps. The map base makes it ideally suited for GIS entry. This information, when formulated into GIS data base overlays provides the foundation for an operational subsistence use protection management system.

TABLE I
COMPONENTS OF AN OPERATIONAL SUBSISTENCE USE PROTECTION MANAGEMENT PROGRAM

1) Identify and describe current subsistence use activities by their geographic location.

DATA NEEDS: Locations of use, seasons of activity, species utilized, means of access, intensity of use, importance of use, relative productivity of use areas.
DATA SOURCES: Alaska Department of Fish & Game-Subsistence Division, interviews with subsistence users, documented history.
TECH APPS:[1] Data input, storage, manipulation and output.

2) Locate and evaluate subsistence resources within subsistence use areas.

DATA NEEDS: Species populations, species distributions, population dynamics, habitat evaluation, access assessment, levels of competitive use.
DATA SOURCES: Alaska Dept. of Fish and Game-Game Division, Wildlife census and population sampling, USFWS's Habitat Evaluation Procedures, subsistence use data and interviews, topographic maps, hunting records.
TECH APPS: Species census and population estimates, biotelemetry, habitat mapping, access assessment, data storage, organization manipulation and output.

3) Identify geographic conflicts and conduct impact assessment of proposed land uses within subsistence use areas.

DATA NEEDS: Identification of characteristics of proposed land use, identification of geographic overlaps within subsistence use areas, impact analysis: impact identification, mitigation, identification of unavoidable impacts, and determination of significance.
DATA SOURCES Proponents of a proposed land use, standard NEPA compliance methodologies, USFWS Habitat Evaluation Procedures.
TECH APPS. Conflict overlay, impact type sorting, mitigation modeling, data management.

4) Make administrative decisions and complete ANILCA Title 8 procedural requirements and documentation.

DATA NEEDS: Identification of the affected environment, subsistence use characteristics, subsistence resource conditions.
DATA SOURCES: Current subsistence use documentation, subsistence resource evaluation, project plans, conflict and impact analysis, public participation.
TECH APPS: Evaluation documentation, outputs for public review, maps for field operations.

[1] TECH APPS = Potential technical applications of remote sensing and/or geographic information systems.

The subsistence use data base should include the following: a) the geographic extent of specific subsistence activities (hunting, fishing, trapping, gathering) and species sought; b) the season(s) in which the activity takes place; c) secondary uses which occur in the course of that activity, e.g. the taking of small game of fishing while big game hunting; d) means of human access to and from the subsistence use area; e) the intensity of use in the area, perhaps expressed as hunter days per season or cycles of use; f) the relative productivity of the area, e.g. the relative success of obtaining subsistence resources while in the area; and g) the relative importance of the area to the individual subsistence users, perhaps expressed along the lines of traditionally assigned subsistence use areas. Each of these information components should be documented as a data layer within the GIS system. Data may be collated by individual user, village or region depending upon detail and complexity. Typically, ADFG's subsistence use data does not detail the intensity, relative productivity or importance of subsistence use areas. Consequently, a second round of subsistence interviews, aided by GIS produced map products of subsistence use areas and evaluated wildlife habitat, may be necessary to improve the information base.

Subsistence Resource Evaluation.

Once the subsistence use characteristics of a study area are defined, an evaluation of the subsistence resources with the area is developed utilizing a GIS system for documentation and manipulation of collected data. The evaluation process begins by collecting existing information on species, habitats and access characteristics. Generalized wildlife population information is available from ADFG which identifies the known distribution of many subsistence species but does not define their specific habitats. Information on population numbers and trends are often available in the form of ADFG departmental reports or by interview with area biologists. Small game populations may be indirectly assessed through the evaluation of their habitat. The USDI Fish and Wildlife Service has been actively modeling habitat requirements under their Habitat Evaluation Procedures (HEP) program. HEP models, coupled with seasonal and cyclic fluctuation information, can provide data on the relative abundance of small game species and associated furbearing predators. In some instances, remote sensing or traditional inventories may be necessary to update old data, fill data gaps, or provide detailed information on species habitats. Potential remote sensing applications include standard film and thermal infrared photography interpretation, Landsat computer-aided analysis, and for direct species census work, the possible application of multispectral linear array instruments (Wyatt 1977). Radio tracking may also be employed to follow migration and estimate the availability of subsistence species for particular geographic areas.

An evaluation of the distribution of gathered resources (berries, craft materials, building materials, fire wood and herbs) may be facilitated by analyzing the biologic and geographic character of known gathering sites. Field measurement in known sites will allow the development of gathering site model descriptions. A GIS data base containing land cover and terrain data can then be used to identify potential gathering sites and field inventories may be used for verification.

Identifying and rating big game habitat within subsistence use study areas is the second major element of the subsistence resource evaluation procedure. Konkel et al (1980) has compiled 29 preliminary bird and mammal HEP habitat index models for major species within the

State. In areas where detailed landcover mapping is available, these models can be utilized; but unfortunately, landcover information is not available for many wildland subsistence areas. To respond to this problem, a system to integrate remote sensing to HEP is beginning to evolve. Armbruster and Farmer (1981) demonstrated success in applying aerial photography mapping techniques to HEP while developing a sandhill crane habitat suitability model along the Platte River in central Nebraska. Payne and Long (1984) provide guidance on translating relatively detailed HEP vegetation mapping requirements to versions amenable to other forms of remote sensing derived data. Early mapping efforts provide insight on the potential applications Landsat systems. George et al (1977) applied remote sensing procedures to map reindeer range on the Seward Peninsula of western Alaska. His efforts showed the suitability of aerial photography for habitat mapping. Meyer and Spencer (1983), among others, have demonstrated that Landsat computer-aided landcover mapping can provide relatively detailed landcover information in mixed tundra, tiaga and boreal forest ecosystems. Their efforts identified 68 detailed computer landcover classes. From these, 16 generalized landcover classes were created which corresponded to Level III of the Suggested Classification for Alaska Vegetation, (Viereck et al. 1979). Konkel et al. (1980), in his work, recommends the application of Level II vegetation information to the Alaskan HEP models. Therefore, Landsat derived data for the Nulato Hills appears to meet HEP requirements. In addition, early versions of HEP habitat models have been applied to Landsat derived landcover data. The USFWS, in support of the 33 million acre Bristol Bay cooperative planning effort, conducted an exploratory mapping program utilizing Landsat computer-aided classification techniques, aerial photography and ground sampling. USFWS biologists utilized the derived landcover information and HEP to evaluate the relative quality of wildlife habitat for this regional planning effort.

The third component of the subsistence resource evaluation process focuses on methods of human access to subsistence resources. The evaluation of access is partially dependent upon information provided by subsistence users and partially on the analysis capabilities of a GIS. Subsistence users can provide valuable information on the location of trails and traplines, the limits of navigability of streams and rivers, and locations of snowmachine and dog sled trails. Topographic maps can supplement that information by providing data on the locations of historic trails and major lakes which support floatplane operations. A GIS can also be used to evaluate traveling characteristics off of main trails and water access points. Don Hinrichsen, BLM's Area Manager for the Peninsula Resource Area in Alaska, working with Dr. Page Spencer, BLM's Alaska State Office Remote Sensing Coordinator, have used GIS to evaluate overland travel characteristics in the Kvichak area of the Bristol Bay Regions of the State. Their technique applies weighted value assignments, for various slope and vegetation types, to overlay grid cells. The weighted values represent the relative resistance of overland travel with increasingly higher values assigned to more difficult travel conditions.

Defining the level of existing resource use is the final component of the resource evaluation procedure. Because information on the intensity and importance of subsistence use areas have not been a component of ADFG subsistence use reports, this data must be obtained from a second round of interviews with subsistence users, and from data contained in sport hunting records.

Conflict Identification and Impact Assessment.

The first step in the conflict analysis and impact identification process is the identification of the characteristics of a proposed land use. Information should include: season(s) of activity, surface occupancy requirements, number of personnel involved, duration of the project, types of equipment, air and overland travel support requirements, surface disturbing activities, fuel and waste handling methods, etc. With data on the proposed action, a GIS analyst is able to identify the geographic conflicts between the proposed use and subsistence. This process simply identifies whether the proposed action occurs within, across or adjacent to a subsistence use study area. If there is no geographic overlap, and subsistence use activities are not likely to be indirectly affected by the action, the resource manager is free to find that the action presents no significant restrictions to subsistence use. If however, there is geographic overlap, or subsistence may be indirectly impacted by watershed or airshed impacts, disturbances to wildlife from air support activities, or potential disruptions to migration patterns, than a subsistence impact assessment must be conducted.

The first step in the impact analysis process is to assess the season of conflict. If proposed activities occur outside of subsistence use periods, they may not have a significant affect on subsistence use if they do not impact species, habitat, access or competition. Under these circumstances, a resource manager may find that there will be no significant restriction to subsistence use <u>if</u> the activity permit has specific seasonal restrictions, surface stipulations, and subsistence protection requirements. In the case where seasonal restrictions are not adequate to prevent potential restrictions, the impact analysis procedure moves to a second phase. Here, potential subsistence use restrictions are identified, impacts are evaluated, mitigation measures are developed, alternatives are identified and assessed, and unavoidable impacts are listed. These operations are intended to provide the land manager/decision maker with a clear analysis of potential subsistence impacts from the proposed action. Once conflicts are identified, the analyst lists the specific impacts associated with them. Impacts should be divided into four groups: species oriented impacts, habitat impacts, access impacts, and increases in competition. This list of impacts provides the framework for the identification of mitigation measures and alternatives. The GIS provides the base for sorting and responding to the identified impacts.

Species impacts can be divided into two groups, direct and indirect impacts. Direct impacts are the loss of individual animals from a population. For wildlife, this occurs primarily from water related fish kills, the burial of small mammals during construction operations, and hunting, fishing and trapping losses. Indirect impacts include the displacement of wildlife populations, and habitat loss. Mitigation for direct species impacts should center on protecting fish producing streams and lakes; locating activities away from productive or important habitats; and restricting sport hunting, fishing and trapping activities by workers of permitted actions. Mitigation for wildlife displacement should focus on routing surface support and aircraft operations away from wildlife concentration areas and subsistence use areas; avoiding surface activities within wildlife concentration areas, subsistence use areas or along migration routes; restricting on-the-ground activities during critical subsistence use periods; and designing and conducting operations in a manner to minimize noise, smoke and human presence.

Habitat impacts are associated with the on-the-ground activities of a proposed action. Overland travel, surface construction, water quality and quantity degradation, human disturbance, and the introduction of exotic species can all lead to detrimental alteration or loss of valuable wild habitat. Mitigation should focus on locating activities away from productive habitats, reducing the aerial extent and duration of surface occupancy, limiting the severity of surface impacts, and requiring habitat rehabilitation after impact. Standard surface protection requirements, such as those recommended by the USFW's National Resource Protection and Petroleum Development in Alaska Handbook (Hanley et al, 1981), and HEP outputs can be adapted to control habitat resource impacts.

Impacts to access are most likely to occur from the degradation or obstruction of traditional access. River access can be restricted by the construction of dams, hydraulic alteration of channels, increased sedimentation, or the blockage of channels with debris. Surface access can be affected by the construction of roads or other facilities across trails, or by surface disturbance which affects surface travel conditions. Mitigation should focus on preventing impacts to rivers or streams, avoiding traditional subsistence trails and traplines, and controlling surface impacts.

Assessing the impacts from increased competition is the final component of the impact assessment process. Competition from individuals associated with permitted actions, and general public use of access improvements are the primary sources of conflict. Restricting hunting, fishing and trapping by the employees of permitted operations; and the removal, by rehabilitation, or the restriction of public use of access roads, trails or airstrips are recommended stipulations to control increases in competitive use.

Administrative Decision Making and Documentation.

With the evaluation procedures completed, the resource manager/decision maker is in a position to make an administrative decision on the proposed action and complete the documentation requirements of ANILCA Section 810. A key component in the decision making process is determining the significance of potential impacts to subsistence use. The determination of significance is dependent upon many factors. These include the characteristics of subsistence use, the condition of subsistence resources, and the type and severity of unavoidable impacts. BLM's policy is that a potential for a significant restriction to subsistence use exists whenever there is likely to be: a) a reduction in the harvestable resources used for subsistence purposes; b) a reduction in the availability of resources caused by an alteration in their destruction, migration, or location; and c) a limitation on the access of subsistence users to harvestable resources.

Determining the level of significance for species impacts is dependent upon: 1) the levels of subsistence utilization; 2) the populations of utilized species; and 3) the dynamics of those populations. Clearly then, the finding of significance is dependent upon the availability of this information. Two scenarios exist: (1) There is inadequate information on subsistence use characteristic. In this instance, any direct impact to species by the activities of a proposed action, or increases in competitive use has the potential of creating a significant restriction to subsistence use. To eliminate this potential, management should maintain wildlife populations at their steady state. Steady state being defined as the existing population level fluctuating within natural system dynamics, i.e, the carrying capacity of available habitat, reproduction rates, climatic and cyclic influences, predation, and current subsistence use. (2) Specific information on subsistence use exists, or resources managers and subsistence users identify an upper level of subsistence use. In this case, a resource manager may allow some loss of species without causing a significant restriction _if_ available species numbers are maintained above known or established subsistence use demands and a "buffer". The size of the buffer will vary with species population numbers, population dynamics, and the level and importance of subsistence use. Buffer values should be established in consultation with wildlife experts, subsistence users and ADFG.

Defining levels of significance to habitat impacts is primarily dependent on whether habitat is a limiting factor for species populations. If habitat is a limiting factor and subsistence demand for that species is near its replacement rate, then a steady state of habitat quality (perhaps defined by HEP) must be maintained to prevent potential significant impacts. Where habitat is not the limiting factor, i.e. where competition, predation, disease, weather, or cyclic influences limit population, some degree of habitat loss may be allowed without a significant impact to subsistence use. A secondary consideration is the location of habitat impact. Determining significance based on the location of habitat loss is dependent upon the access characteristics of the study area and subsistence use patterns. If impact occurs in an area readily accessible to subsistence users, then mitigating habitat improvements must occur in areas equally accessible. By the same token, if impact occurs within an individual's subsistence use area, mitigation should be located within that same area. If impact occurs in areas remote and not readily accessible to subsistence users, habitat improvements may be located anywhere within the study area.

Defining the level of significance for impacts to human access must be tempered by a review of the existing access patterns within an impacted area. There are two types of access concerns: linear access associated with trails, traplines, rivers or streams; and general overland access off of linear features and point access sources (airstrips and lakes). In the first case, a potential significant restriction would be total loss of a traditional source of access. Significant restrictions may also arise from partial losses of access or a major decrease in the quality of trails or rivers. A case by case review of impacts is necessary to determine the level of significance in such instances. General overland access may be affected by surface disturbing activities and the construction of pipelines or surface facilities which block or restrict surface travel. Again, identifying the level of significance is dependent upon the review of the degree of impact, its location, and traditional use patterns. Experience has shown that mitigation measures can control extensive surface degradation in most situations, and the siting and design of facilities can usually accommodate traditional access. In addition, use of access roads, trails or other improvements can be restricted by management agencies through the permitting process.

Determining the level of significance for competitive use of wildlife, is closely related to the direct species impact analysis presented above. Competition increases, especially those facilitated by the use of project developed access, have a much greater potential of direct impacts to species than most other project associated activities. By insuring that adequate levels of resources are available to meet subsistence demands, resource managers may facilitate increases in sport, recreation or commercial utilization without creating the potential for significant restrictions. Increases in competitive use should be carefully monitored to ensure that changing subsistence use conditions, or other factors do not lead to significant subsistence impacts.

Once the resource manager/decision maker has made a finding on the significance or potential unavoidable impacts, a decision must be made and the rationale for that decision documented. In the decision making process, the decision maker must weigh the potential subsistence impacts against social, political, and environmental considerations, and make a determination of what is in the best interest of the public as a whole. If the decision is to proceed with the proposed action which poses a significant restriction, the agency must notify the ADFG Habitat and Subsistence Divisions, the appropriate Regional and Local Subsistence Advisory Councils and give notice and hold public hearing(s) in the vicinity of the affected area. Products generated during the evaluation procedure should be used to provide information to the State, Regional Advisory Councils, local Advisory Committees, and the public. The GIS can readily produce map products at a variety of scales and formats to meet information transfer needs. For the State and advisory groups, maps on the specific locations of conflict and the characteristics of subsistence use and subsistence resources along with a well documented analysis procedure will aid their review and comment. For public hearings, GIS generated map products can be particularly helpful in displaying subsistence use patterns; the locations and quality of subsistence resources; the location of potential impact sites; and presenting alternatives to the proposed action.

SUMMARY

In 1980, the Alaska National Interest Lands Conservation Act's Title 8, Subsistence Management and Use, legitimizing Native cultural rights when it declared subsistence use of federal public lands was to receive preferential consideration in land use planning and environmental review. The Bureau of Land Management has interpreted the provisions of Title 8 and provided guidance to its land managers on the assessment of potential impacts to the "supply side" components of subsistence use. This paper provides a framework for an operational subsistence use protection program. It defines four major elements of an operational program: 1) identification and description of current subsistence use activities by their geographic location; 2) location and evaluation of subsistence resources within subsistence use areas. 3) identification of geographic conflicts and impact analysis of proposed land uses; and 4) administrative decision making and the completion of ANILCA Title 8 procedural requirements and documentation.

RECOMMENDATIONS

The operational system proposed by this paper, incorporated with the program development efforts of other resource specialists, managers and system technicians, has the potential of meeting the "letter of the law" expressed in ANILCA for the protection of rural Alaskan subsistence use. The proposed operational system is generic in its makeup, it does not detail every decision pathway or hardware system interface necessary to respond to the subsistence use protection problem and therefore, it is not recommended as a final format for a subsistence use protection program. The paper is simply intended to stimulate communication between resource specialists, managers, technicians and subsistence users on the potential applications of a systematized management program--a program which insures the protection of current subsistence use while providing a management base for continued subsistence use and resource production in the future. It is hoped that such communication will lead to the development of an operational management system which embraces new technology, meets the needs of resource managers, and provides a foundation to meet the "spirit of the law" expressed in ANILCA--the protection of traditional rural Alaskan culture values.

BIBLIOGRAPHY

Armbruster, M.J., A.H. Farmer. 1981. Draft Sandhill Crane Habitat Suitability Model. U.S. Fish and Wildlife Service, Western Energy and Land Use Team, Fort Collins, CO. 80526

George, T.H., W.J. Stringer, and J.S. Baldridge. 1977. Reindeer Range Inventory in Western Alaska from Computer-aided Digital Classification of Landsat Data. Proc. of the 11th Int. Symp. on Remote Sensing of Envir. 1:671-682

Hanley, P.T., J.E. Hemming, and J.W. Morsell, 1981. Natural Resource Protection and Petroleum Development in Alaska, Office of Biological Services, USFWS, USDI. Washington, D.C. 20240

Konkel, G.W., L.C. Shea, 1980. Terrestrial Habitat Evaluation Criteria Handbook-Alaska. USDI Fish and Wildlife Service, 1011 E. Tudor Rd. Anchorage, Alaska 99503

Meyer, K.G., P. Spencer, 1983. User's Guide-Nulato Hills Digital Landcover and Terrain Data Base. USDI Bureau of Land Management, Anchorage District Office, 4700 E. 72nd Ave., Anchorage, Alaska 99507

U.S. Fish and Wildlife Service, 1980. Habitat Evaluation Procedures ESM 102, 103, Division of Tech. Services. USFWS, Dept. of Interior Wash.

Payne, B.S., K.S. Long, 1984. "Airborne Sensor Potential for Habitat Evaluation Procedures (HEP)," Technical Report EL-84- U.S. Army Engineer Waterways Experiment Station, CE, Vicksburg, Miss.

Viereck, L.A., C.T. Dyrness, 1979. A Suggested Classification for Alaskan Vegetation. USDA-Forest Service, Institute of Northern Forestry, Fairbanks, Alaska 99701

Wyatt, C.L., D.R. Anderson, and R.K. Harshbarger, 1984. Design, Development and Tests of Two Prototype Instruments for Remote Sensing of Mule Deer. Space Dynamics Laboratory, Utah State University, Logan UT 84322

CHANGE DETECTION AND MONITORING

Moderator: J. Merchant

A PERSPECTIVE ON USING REMOTE SENSING FOR DETECTING AND MONITORING CHANGE IN RENEWABLE RESOURCES

Lee F. Werth and Fred T. Batson
Bureau of Land Management
Branch of Remote Sensing
Building 50, Denver Federal Center
P.O. Box 25047
Denver, Colorado 80225-0047

ABSTRACT

Reduced funding and manpower will require increased efficiency for change detection and monitoring of public land resources. The vast acreages administered by public agencies for range, wildlife, recreation, forestry, and watershed, mandate increased emphasis on remote sensing in combination with ground surveys to monitor resource status and change. The type of remote sensing system required will depend on the level of detail desired and the resource extent. Remote sensing will never replace field monitoring, but it can make the on-the-ground efforts more efficient. The key questions determining the utility of a remote sensing monitoring system are: 1) are the changes indicated real (valid) and 2) what relevance do they have to resource management decisions? To effectively use remote sensing technology, there is need for closer communication and understanding between resource managers and remote sensing specialists. Each must make further attempts to understand the other's requirements and capabilities. This is especially important since the types of data provided via remote sensing systems cannot duplicate some of the traditional field gathered data that resource management personnel are comfortable dealing with.

This paper attempts to facilitate communication between remote sensing specialists and resource agencies (users) for resource change detection and monitoring using remote sensing. The papers that follow in this session will discuss specific applications of remote sensing techniques.

Austere federal budgets now and in the future will require greater efficiency in change detection and monitoring of public land resources. These funding limitations combined with the vast acreages administered by public agencies for range, wildlife, recreation, forestry, and watershed, mandate increased emphasis on remote sensing in combination with on-the-ground data gathering to monitor resource status and change. The amount of time spent gathering field data, out of economic necessity, will need to be decreased but not eliminated. Remote sensing is one tool in combination with selected field visits that could increase resource monitoring efficiency. The type of remote sensing system required will depend on the level of detail desired and the extent of the resource. For example, satellite systems may be adequate for monitoring gross changes over large areas. However, large scale aerial photography or ground visitation may be more appropriate for site specific monitoring of key plant communities within livestock grazing allotments.

Remote sensing may provide a means of quantifying vegetation change or it may just signal that a change has occurred. Historical photography compared with recent aerial photography may indicate that the amount of sagebrush cover in an area has increased by 20 percent or aspen has decreased by 10 percent. On the other hand, high altitude photography and satellite sensors may show that a change has occurred but not exactly what the change is. Field efforts could then be focused on those areas where change has occurred.

Resource managers need to detect subtle changes in vegetation. For example, most resource managers already have maps showing where forests have been clearcut. However, they need to know where succession is altering forest and range species composition. Range managers need to know if trends in resource values are up, down, or staying static. If the change between two dates is 50 percent or greater (depending on the vegetation attribute), they will probably already have discovered this on the ground and the change may have gone too far (Eshelman 1985). Yet a change of 20 percent in vegetation cover detected by remote sensing would be quite useful. Generally, we monitor to determine the effects of our management actions. The management plans have specific objectives. Monitoring must be sensitive enough to determine whether or not the area is responding to the management (positive or negative). We can then make adjustments in midcourse if necessary.

Possible remote sensing approaches for change detection and monitoring are manual interpretation of aerial photography or satellite imagery and digital image processing. Aerial photography will provide greater detail because of better spatial resolution but photo comparisons may not be available for the desired time interval. Imagery from satellites is more readily available because of repetitive coverage but decreased resolution will limit the vegetation parameters that can be detected. Manual interpretation (especially if a 3-dimensional view is available) provides more clues or characteristics than digital processing (tone, texture, pattern, shape, size, shadow, and site) to use in convergence of evidence and the final interpretation decision. With digital image processing, the primary characteristic used in change detection is tone; however, some algorithms are able to incorporate the spectral reflectance of adjacent pixels rather than just one pixel at a time. The likelihood of making a false change detection call with manually interpreted aerial photographs is minimal. However, misregistration and differences in plant phenology between two or more dates can result in error. With digitally processed remote sensing data a false alarm could result from a variety of factors, i.e., misregistration, phenology differences, atmospheric differences, terrain and sun angle. The human interpreter can compensate for atmospheric differences, terrain and sun angle. Resource managers need to have some assurance that an indicated change is indeed real or they will quickly lose confidence in presumed changes that cannot be verified in the field.

Various techniques have been developed to examine the digital reflectance data from satellite sensors. The usual approach of each technique is to extract as much information as possible from the sensors that record reflected light split into various wavelengths or bands. Although some of the techniques will be discussed in the papers that follow, it will be helpful to give a brief explanation of some of the more common approaches. Tasseled cap or the Kauth-Thomas transformation is a mathematical procedure that separates multiband data into the components of greenness, brightness, and yellowness (Kauth and Thomas 1976). Greenness represents vegetation biomass in a scene, brightness the amount of bare soil, and yellowness the atmospheric haze. For change detection purposes, tasseled cap is combined with BLOB clustering algorithm and change vector analysis (Colwell et al. 1980). Principal components analysis, not specifically designed for remote sensing vegetation analysis, reduces multiband data into different components or axis where the first component contains the maximum spectral variability in the scene. Some consider the first component similar to greenness and the second component to brightness of the tasseled cap transformation (Jenson and Waltz 1979, and Kauth and Thomas 1976). Albedo is the percent of incoming radiation reflected from the ground and albedo differencing is comparing the albedo for two or more dates to see if it increased or decreased (Robinove et al. 1981). Another common technique is ratioing whereby the band or bands of one date are divided by the band of another date. Normalized difference is similar to ratioing but the difference and sum of the same two channels are ratioed against each other.

Resource monitoring entails taking the pulse of the rangeland or forest site to determine whether management objectives are being met. Changes in greenness, brightness, and albedo mean something to a remote sensing specialist but probably do not have any significance to a resource manager. The resource manager needs these terms translated into something he is familiar with or he will just shrug his shoulders and say, "So what." For example, detected change in greenness may signify that vegetation cover has increased. However, does this cover change relate to a highly palatable forage species or a noxious weed? Broad generalities and unknown terms are not useful, but specific statements about the range condition on a specific grazing allotment are.

Traditionally, resource managers have been taught by colleges and universities to think in terms of very site specific vegetation parameters such as density, frequency, cover, and production. These parameters are very appropriate for field investigations but are difficult if not impossible to obtain from satellites and high flying aircraft although the density and cover of some species could be measured from large scale aerial photographs. The detailed vegetation parameters are important for management but may not be for more generalized resource management plans. In fact, maybe too much detail is gathered that later has to be synthesized for general planning purposes. There is some truth to the statement that data gathered in millimeters are later applied in kilometers to management plans.

What is all this suggesting? Simply, there must be a meeting ground or halfway point between resource managers and remote sensing specialists if remote sensing is to play a larger operational role in change detection and monitoring activities. Resource managers will be very leery of a new monitoring method until they somehow see that it can help them and it can be easily implemented. Also, the remote sensing community must be careful with high technology buzzwords that sound good academically but have little practical meaning and cannot be demonstrated. Resource managers must determine their true data requirements <u>before</u> monitoring and not ask for the number of seedlings when all they really need is the dominant species of a plant community. The ultimate test of any new technique is whether it is adopted and used in management activities.

How can this polarity between resource managers and remote sensing specialists be reduced? The solution lies partly in providing resource managers with a good remote sensing background. Preferably this should be done at colleges and universities and should be required in their curriculums. Another option is at agency training sessions and workshops where realistic management data needs are clearly expressed and remote sensing systems are designed to satisfy these requirements. Another part of the solution may be to reexamine the traditional vegetation parameters and see if new ones can be derived that will meet management needs and are within the capabilities of remote sensing.

The remote sensing community has been guilty of overselling some kinds of technology and not emphasizing others enough. Therefore, it is not surprising that there are managers reluctant to use any remote sensing data especially if it is derived from satellite sensors. We must be very straightforward and honest with resource managers about the capabilities and limitations of various remote sensing systems. If we gain their trust and confidence, it will be easier to implement remote sensing systems that really work. Remote sensing will never be a panacea and replace ground monitoring but can be an important tool to increase management efficiency if used judiciously and incorporated with a geographic information system for rapid update and resource analysis.

REFERENCES

Colwell, J., G. Davis, and F. Thomson, 1980. Detection and Measurement of Changes in the Production and Quality of Renewable Resources. Final Report, NFAP, RRI, AgRISTARS Rep. No. 145-300-4-F. 144 pp.

Eshelman, K. 1985. Personal communication, Denver, Colorado.

Jenson, S.K., and F.A. Waltz. 1979. Principal Components Analysis and Canonical Analysis in Remote Sensing in Proceedings of the American Society of Photogrammetry, Washington, D.C. p. 337-347.

Kauth, R.J., and G.S. Thomas. 1976. The Tasseled Cap -- A Graphic Description of the Spectral-Temporal Development of Agricultural Crops as Seen by Landsat, in Proceedings of Second Symposium on Machine Processing of Remote Sensing Data, LARS, Purdue University, p. 4B 41-51.

Robinove, C.J., P.S. Chavez, D. Gehring, and R. Holmgren. 1981. Arid Land Monitoring Using Landsat Albedo Difference Images. Remote Sensing of Environment 11: 133-156.

CHANGE DETECTION IN RANGELAND ENVIRONMENTS USING LANDSAT MSS DATA - A QUANTITATIVE APPROACH*

David C. Johnston and Robert H. Haas**
Technicolor Government Services, Inc.
EROS Data Center
Sioux Falls, South Dakota 57198

ABSTRACT

A range forage utilization study on the Crow Creek Indian Reservation in central South Dakota provided the opportunity to use Landsat multispectral scanner (MSS) data for examining range condition trends. A procedure was developed to compare change in spectral reflectance over time for polygon areas, defined by resource type within management units. A t-test was used to evaluate changes in brightness and greenness within pastures between September 27, 1978, and September 18, 1983. The first principal component transformation from four-band MSS images for both dates was used as a measure of brightness. Greenness was measured using the second principal component transformation for both dates. Examination of the brightness data showed that the assumptions required for a valid t-test were met. The greenness data violated the assumption of independence between dates and was not used for trend comparisons. The t-values calculated from each polygon were coded into three groups: (1) those indicating significant brightness decrease, (2) those indicating significant brightness increase, and (3) those indicating no significant brightness change. Significance was determined at the 5-percent level. These results were formatted into an image, which is a preliminary product for evaluating range condition trends over a 5-year period.

INTRODUCTION

A cooperative project between the EROS Data Center (EDC) and the Bureau of Indian Affairs (BIA) permitted study of the possibility of predicting range forage utilization on the Crow Creek Indian Reservation in central South Dakota (fig. 1), using transformed Landsat multispectral scanner (MSS) data. One aspect of that project was to determine the feasibility of using spectral change over time to monitor range condition trends on the reservation, where trend was defined as change in vegetation quantity and quality between two dates.

*Publication authorized by the Director, U.S. Geological Survey, on July 1, 1985.

**Technicolor Government Service, Inc. Work performed under U.S. Geological Survey contract no. 14-08-0001-20129.

The forage utilization project evaluated measures of scene brightness and greenness for predicting utilization at the end of the growing season (Haas and Johnston, unpublished data, 1984). These same measures were used in the change detection study. Scene brightness was found to be correlated with such ground characteristics as amount and color of bare soil, amount of litter, and range condition, while scene greenness was found to be correlated with the amount of green biomass present. Several measures of both brightness and greenness are possible; a decision was made to use the first principal component image (PC1) of a scene as the measure of brightness and the second principal component image (PC2) of a scene as the measure of greenness.

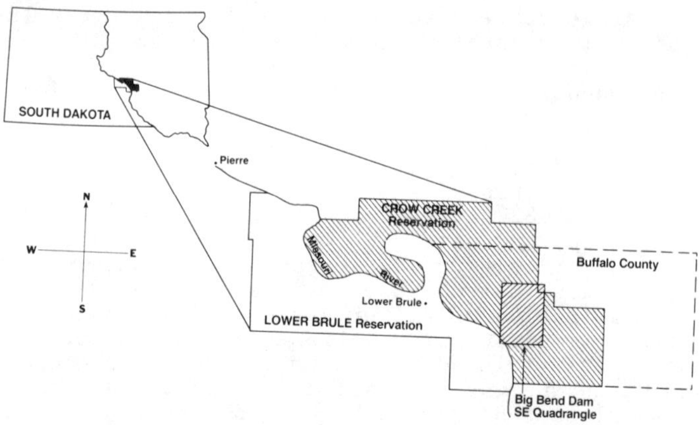

Figure 1. Approximate location of the Crow Creek and Lower Brule Indian Reservation Study Site in South Dakota.

There are several methods of studying spectral change. One method is discussed by Arndt (1983), who calculated albedo from two temporal Landsat images of the same area and studied the difference between the two images. Arndt used brightness thresholds, developed from histograms of the difference image, to detect areas of probable mining trespass on Bureau of Land Management (BLM) administered lands. His approach records change on a pixel-by-pixel basis.

Another method of monitoring change is to consider the average spectral change within identifiable areas of interest. This method has the advantage of allowing change to be measured relative to the spectral variation in these areas so that the statistical significance of the change may be inferred. Significant change implies that some set of factors other than random variation is operating. This method for measuring average change over homogeneous areas of land within pastures was evaluated in the present study.

MATHEMATICAL AND STATISTICAL CONSIDERATIONS

Principal Component Transformations

Principal component analysis is a mathematical technique which transforms a multivariate data set into another data set which has a simpler statistical structure. A discussion of the general mechanics of principal component analysis is given by Jenson and Waltz (1979), while Morrison (Chapter 8, 1976) gives a more detailed theoretical account. Most statistical software packages include principal component analysis routines.

Two basic problems encountered with multivariate data sets are that people generally can not visualize relationships between variables when more than three variables are involved, and multivariate data sets usually contain correlated variables. Principal component analysis simplifies a multivariate data set by generating a new set of variates--the principal components which are uncorrelated; thus, the total variation in the system can be expressed as the sum of the variances of the principal components. The proportion of the variance accounted for by each component can thus be easily calculated, and since a few of the components usually account for a large proportion of the variation, these few uncorrelated variables can be used in lieu of the original data with little loss of information. The principal components of a data set are linear combinations of the original variables formed from the normalized eigenvectors of the data's covariance matrix. Also, the eigenvalues of the covariance matrix are the variances of the principal components.

A four-band Landsat MSS scene contains four highly correlated bands. The principal component transformation yields four new uncorrelated bands in which the first two usually account for at least 95 percent of the spectral variation within the scene. The first principal component usually contains all positive coefficients; therefore, PC1 may be interpreted as scene brightness. When the coefficients of the second principal component are such that the first two coefficients, corresponding to the visible portion of the spectrum, are negative and the last two coefficients, corresponding to the infrared portion of the spectrum, are positive, PC2 may be interpreted as scene greenness. This association represents the contrast between the visible and near infrared wavelengths caused by the absorption of red band radiation by chlorophyll and the high reflectance of near-infrared radiation by growing plants.

Statistical Development

Student's t-test is often used to test for differences between the means of two populations. To conduct a valid t-test three assumptions must be met: (1) the data in both groups must be normally distributed, (2) the two groups must be statistically independent, and (3) the variances of the two groups must be equal. When samples of equal sizes are taken from both groups, the calculated t-statistic is given by the following equation:

$$t_c = \frac{(n)^{1/2} \quad (\bar{X}_1 - \bar{X}_2)}{(S_1^2 + S_2^2)^{1/2}} . \qquad (1)$$

where

n = number of observations in each group

$\bar{X}_1$, S_1^2 = mean and variance of group one

$\bar{X}_2$, S_2^2 = mean and variance of group two

The hypothesis tested with equation (1) is that there is no difference between group means. Large absolute values of t_c cause rejection of this hypothesis. The rule is to reject the hypothesis if the absolute value of t_c is greater than a corresponding entry in a t-table for some probability level and 2n-2 degrees of freedom. If the hypothesis is rejected, the conclusion is that the groups have different means. The absolute magnitude of t_c is a function of both the magnitude of the mean difference and the magnitude of the sum of the variances. The smaller this sum, the more sensitive the test is to differences in the means.

PROCEDURES

Development of the Spatial Data Base

Landsat MSS images for September 27, 1978, and September 18, 1983, scene identification numbers 83020616445 and 84041316531, respectively, were obtained and registered to a 50-meter Universal Transverse Mercator (UTM) grid. Sub-scenes (1,000 by 2,000 pixels) which included the Crow Creek and Lower Brule Reservations, were copied into the data base, and transformed into principal component images. Only the PC1 and PC2 images were kept in the data base. The effect of resampling during image registration had little effect upon the principal component transformation.

The eigenvalues and eigenvectors, as well as the percentage of variation accounted for by each principal component, are shown in table 1. To illustrate how the eigenvectors generate the principal component images consider PC1 for September 27, 1978. All values of MSS band one are multiplied by 0.189; all values of MSS band two are multiplied by 0.355, and so forth for the remaining two bands. Addition of these four band multiples results in the PC1 image. The second eigenvector for September 18, 1983, does not fit a strict definition of greenness given above; however, the third coefficient is small and has little influence on the interpretation. Examples of brightness and greenness images are given for September, 1981 (fig. 2). On rangeland areas at this season, brightness and greenness tend to be complementary. The lighter areas of the brightness image are darker on the greenness image, indicating little or no standing biomass and no green biomass. The high green biomass areas along the streams, which are very light on the greenness image, are darker on the brightness image.

Table 1. Statistics related to the principal component analyses of the MSS data of Crow Creek and Lower Brule study site for September 27, 1978, and September 18, 1983.

	September 27, 1978			September 18, 1983			
	Eigenvalues	% variation		Eigenvalues	% variation		
PC1	145.10	89.5		137.09	94.9		
PC2	13.17	8.1		3.29	2.3		
PC3	2.14	1.3		2.23	1.6		
PC4	1.83	1.1		1.70	1.2		
Eigenvectors				Eigenvectors			
PC1	PC2	PC3	PC4	PC1	PC2	PC3	PC4
0.189	-0.522	0.684	-0.473	0.060	-0.444	0.780	0.437
0.355	-0.710	0.260	0.550	0.193	-0.851	-0.488	-0.020
0.606	0.046	-0.537	-0.585	0.634	-0.028	0.327	-0.700
0.686	0.470	0.420	0.362	0.746	0.279	-0.214	0.564

Rangeland on the Crow Creek Reservation is managed by pasture units. Most of the pastures are large heterogeneous aggregations of grassland plant communities and of range sites, ranging from loamy to dense clay. In an attempt to reduce variation to a manageable level, areas within pastures were stratified by range sites, producing more homogeneous areas within pasture units.

A soil series map was available only for the Buffalo County portion of the Crow Creek Reservation. This map was digitized and entered into the data base. The soils classes were aggregated into the eight dominant range site classes listed in table 2. A ninth class composed of water and a few very minor range sites was partitioned out, and excluded from the analyses. Eleven 7.5-minute quad sheets containing the boundaries of all pastures in Buffalo County were digitized and entered into the data base. The pasture and range site images were then merged, and finally a statistical-strata mask file was created where each polygon in the strata file was a unique range site within pasture combination.

All greenness and brightness values for both scenes were extracted from each polygon in the strata file. Since that volume of data could not be analyzed as a unit, the analysis was conducted for one 7.5-minute quadrangle, Big Bend Dam SE.

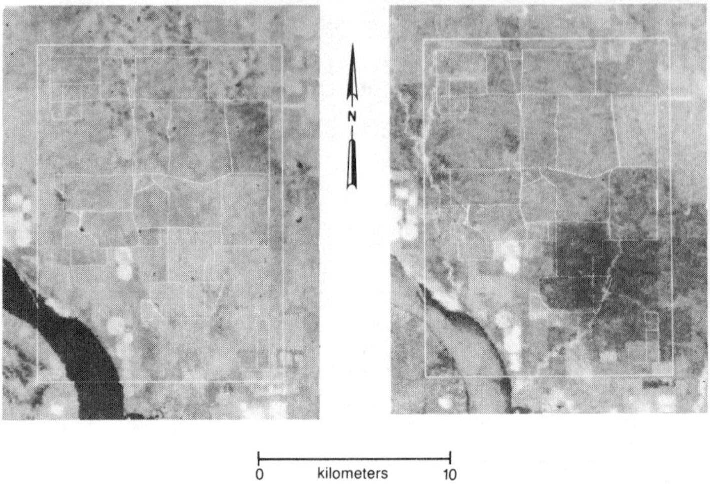

Figure 2.--PC1 (left) and PC2 (right) images corresponding to September, 1981, Landsat MSS data for the Big Bend Dam SE quadrangle, South Dakota. Pasture boundaries are shown in white.

Figure 2 shows the pasture boundaries in that quadrangle. Since a given range site could occur over several unique areas within a pasture, the extracted greenness and brightness values were amalgamated over polygons into range-site-within-pasture classes; there were 121 such classes on the quad. Greenness and brightness means and variances were then calculated for each class for both dates.

Table 2. Dominant range sites and percent coverage of Buffalo County, S. Dak.

Dominant Range Site	Percent Coverage
Claypan	2
Clayey	38
Dense Clay	17
Overflow	4
Silty	12
Shallow Clay	21
Thin Claypan	4
Thin Upland	2
Total	100

Statistical Analyses

Since measurements of the same features at different times tend to be autocorrelated, a time series analysis which removes the autocorrelation is the preferred analysis technique for evaluating change over time; however, when only two dates are under consideration, time series analysis may not be appropriate.

Student's t-test may be used in these two-sample situations, if the test assumptions are valid. Therefore, before conducting the t-tests, the assumptions were checked for validity. Histograms of the data showed that neither greenness nor brightness was normally distributed, but since most of the range-site-within-pasture classes contained at least 30 pixels, the central limit theorem was used to assume normality of the means. The assumption of homogeneous variance was valid. Table 1 shows that while the PC1 variances (the eigenvalues) are very similar, the PC2 variances are quite different; however, the stratification procedure resulted in very nearly identical variances for both variables at each date for a given range-site-within-pasture class.

The assumption of independence between dates was difficult to substantiate. If two variables are independent, they are also uncorrelated, but uncorrelated variables are independent only for some distributions, the normal distribution being an example. Correlation coefficients were calculated for PC1 means between dates and for PC2 means between dates. The correlation between dates for PC1 was insignificant at the 5-percent level, but the correlation between PC2 means was significant. Since the central limit theorem allowed the assumption of normality, the PC1 means between dates were assumed to be independent because they were uncorrelated. Greenness change was excluded from consideration because of the significant correlation coefficient.

Since the brightness data met the assumptions, the t-tests were conducted. Equation (1) was applied to each of the 121 range-site-within-pasture classes, and each value was compared to the appropriate entry in a two-tailed t-table at the 5-percent probability level. The brightness data for September 27, 1978, were considered group one. The order of the means in equation (1) causes negative values to imply an increase in brightness between dates, while a positive value implies a decrease in brightness.

Once the tests were completed, each class was mapped (fig. 3) depending upon whether there was significant brightness decrease, significant brightness increase, or no significant brightness change. Table 3 summarizes brightness change of selected pastures within Big Bend Dam SE quadrangle. These pastures were selected for presentation because they illustrate the range of spectral change over the quadrangle, and they illustrate typical grazing conditions. Note the relatively small mean differences, which are significant. This is due primarily to the range site stratification reducing within-class spectral variation.

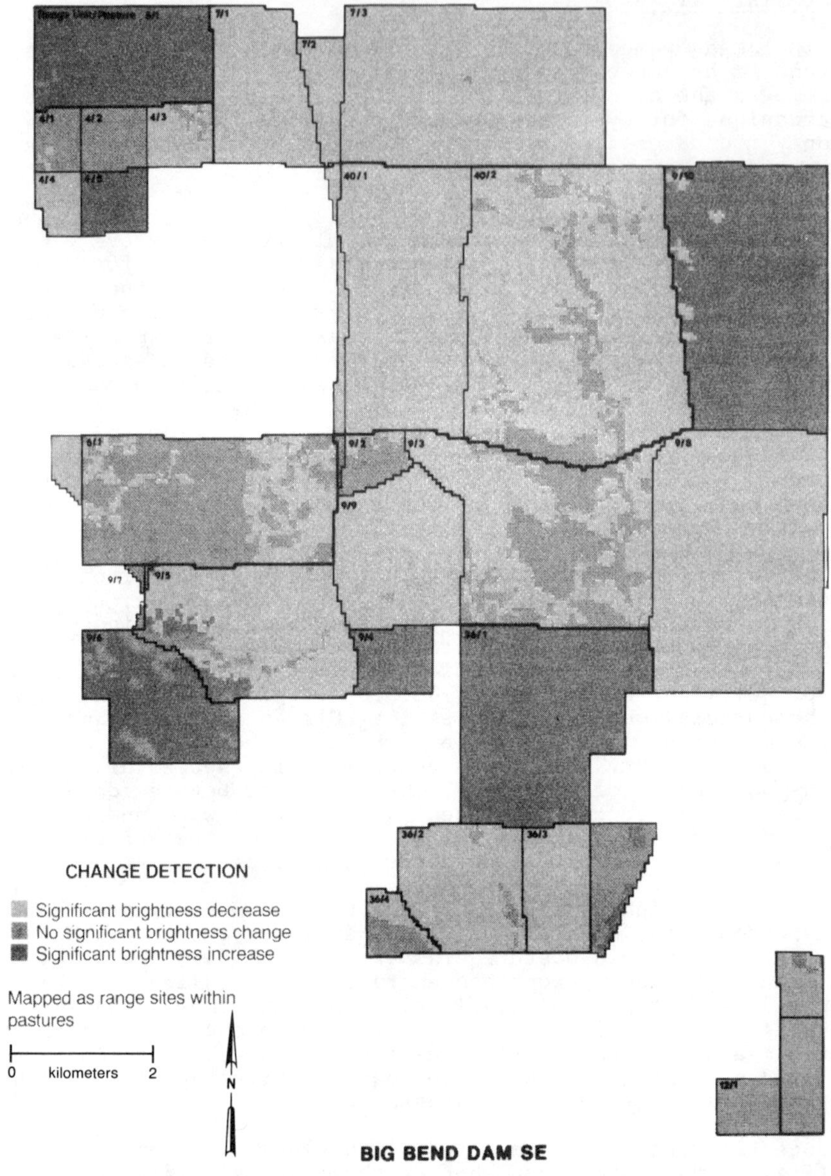

Figure 3. Brightness change from September 27, 1978, to September 18, 1983, Big Bend Dam SE quadrangle, South Dakota, summarized by range sites within pastures.

Table 3. Summary of brightness change of selected pastures from Big Bend Dam SE quadrangle, South Dakota, between September 27, 1978, and September 18, 1983.

Range Unit	Pasture	Range Site	Acres	Mean Difference
4	1	Clayey	1.9	1.000
		Dense Clay	39.5	-0.937
		Overflow	6.2	2.900*
		Shallow Clay	69.9	-0.580
		Pasture Mean		-0.492
	4	Dense Clay	45.7	3.324*
		Shallow Clay	63.0	1.490*
		Pasture Mean		2.261*
5	1	Clayey	202.6	-0.782*
		Dense Clay	166.8	-1.737*
		Overflow	7.4	-1.750
		Shallow Clay	310.8	-3.196*
		Thin Claypan	6.2	-2.900*
		Pasture Mean		-2.143*
9	10	Claypan	8.6	-6.714*
		Clayey	339.2	-1.802*
		Dense Clay	991.0	-1.226
		Shallow Clay	87.7	-3.366*
		Thin Claypan	36.5	5.000*
		Pasture Mean		-1.409*
36	3	Clayey	83.4	1.251*
		Dense Clay	52.5	0.976*
		Shallow Clay	189.7	0.420*
		Pasture Mean		0.722*

* Indicates significance at the 5-percent probability level.

DISCUSSION

Since very little field data were available for September, 1978, only general interpretations can be made to relate spectral change to vegetation change. Spectral changes are obviously related to grazing management practices because of the association between change and the pasture boundaries. As an example, no significant brightness changes are noted in pasture 4/1, except for the overflow range site areas which show a significant brightness decrease. This pasture probably had only light grazing use prior to the September, 1983, Landsat overpass. Reduced grazing pressure in 1983 may account for the vegetation improvement on the better overflow sites.

The grazing histories of pastures 9/10 and 36/3 are better documented than the other pastures. Pasture 9/10 shows significant brightness increase from 1978 to 1983 on all range sites except thin claypan. This pasture was only slightly grazed during 1978 but in 1983 it was heavily grazed from June through September. The increase in brightness reflects an increase in litter and bare soil. While the 1983 grazing was not destructive, the brightness increase shows that the pasture was heavily grazed.

Pasture 36/3 displays significant brightness decrease across all range sites. Most of the spectral change within the 5-year period took place between 1981 and 1983. Figure 2 shows that pasture 36/3 was extremely bright and non-green in September, 1981, indicating extremely poor vegetation conditions. By September, 1983, the vegetation conditions were much improved, due to limited grazing and abundant moisture during the 1982 and 1983 growing seasons.

This study demonstrates a method of using Landsat MSS data to monitor spectral change over time within a statistical framework, using a raster data base and geographic information system (GIS). This approach to change detection is a very efficient way of documenting trends. With adequate field data the approach provides the land manager with a means to flag problem or potential problem areas. The method may be used in any agricultural or renewable natural resource setting; however, the assumptions of the t-test must be validated if this method is to be used effectively. If the assumption of independence is not valid, conducting the test may result in misleading or incorrect conclusions. Under such conditions, a time series analysis, or some type of qualitative assessment of change should be employed.

REFERENCES

Arndt, R. E., 1983, The use of Landsat albedo difference algorithm for land use and land cover change: in Conference on Renewable Resource Inventories for Monitoring Changes and Trends, Proceedings, Oregon State University, College of Forestry, p. 79-85.

Jenson, S. K. and Waltz, F. A., 1979, Principal components analysis and cononical analysis in remote sensing, in American Society of Photogrammetry Annual Meeting, 45th, Washington D. C., 1979, Proceedings: Falls Church, Virginia, American Society of Photogrammetry, v. I, p. 337-348.

Morrison, D. F., 1976, Multivariate statistical methods, McGraw-Hill Book Company, New York, New York, 2nd edition, 415 p.

ASSESSING DYNAMIC FORAGE CONDITIONS IN INDIVIDUAL RANCH PASTURES USING THEMATIC MAPPER IMAGERY AND AN IBM PERSONAL COMPUTER

L.D. Miller, Y.K. Yang, M. Unverferth, and T. Cheng
Nebraska Remote Sensing Center
University of Nebraska
Lincoln, Nebraska 68588-0517

GENERAL CONCEPTS

Frequent, timely assessment of the mean and spatial distribution of estimates of the green canopy biomass density or green forage within each of his pastures can provide the ranch manager with valuable, quantitative information for his livestock management decisions. Landsat Thematic Mapper (TM) images are of sufficient ground resolution to provide a detailed map of the spatial distribution of the green forage within each pasture. Transformation of selected spectral bands of the TM image into appropriate, calibrated green canopy density maps can readily be completed using one of the established vegetation indexes such as the Transformed Vegetation Index (TVI). Ground sampling by clipping test plots at reference positions within pastures provides a basis for calibrating these green forage maps in tons per acre. The mean, variance, and distribution of the green forage can then be computed for each irregular pasture specified.

Practical application of these general procedures for improved, on-site management of individual pastures requires that appropriate image analysis techniques be available to the rancher. These techniques must be usable by him on his on-site microcomputer and interface to the computer ranch management decision models currently employed. A microcomputer image processing system has been constructed which meets these criteria. It can perform all the necessary image analysis functions (Miller, et al., 1984).

The IBM PC is readily available to rural ranchers and is being used for maintaining livestock records, cash flow analysis, determining feed mixes, and many other related functions. Extending its application into image processing requires the addition of a new display subsystem for portraying the Thematic Mapper and other images in an appropriate size, format, and color balance to support the recognition of individual fields, ground sample locations, fence boundaries, etc. The TM imagery is easily accessed on the standard 5 inch floppy disks or from 8 inch disks with an inexpensive additional 8 inch disk drive.

Almost all ranchers utilize the U.S. Geological Survey's (USGS) 7.5 minute topographic maps in their operations if they are available. It was therefore convenient to design the microcomputer based image processing system to operate upon subareas of the original TM image which are called image maps and which accurately correspond to these USGS topographic maps. Each 7.5 minute TM image map can be conveniently stored on floppy disks and accurately related to the current fence boundaries which can be annotated on the corresponding topographic maps. These 7.5 minute image maps can be displayed in high quality natural color or color infrared on the auxiliary display screen and a mouse-controlled cursor used to draw a complex polygon around each pasture (Miller, et al., 1984). Each of these pasture boundaries can be saved on disk and superimposed upon successive, new image map(s) acquired for the ranch.

The transformation of the appropriate spectral bands for each .25 acre picture element within a specified pasture can be performed to estimate its relative level of green forage. If corresponding ground calibration data has been collected the relationship between this transformation and tons of green forage per acre can be established using the microcomputer techniques. The relative green forage level for each cell can then be converted to tons per acre using this calibration curve. Either the relative or calibrated green forage levels can then be displayed as image maps using colors to code their ranges and suppressing all other areas and features except the selected pasture or pastures. Green forage maps are easily prepared and have a geometry matching the original topographic map upon which the fence boundaries were annotated. If an appropriate plotter is available these pasture by pasture forage maps can be plotted to scale and overlaid directly upon the corresponding topographic map. This estimate of green forage for each quarter acre cell can also be integrated or summed over the irregular pasture area and a tabular output prepared summarizing conditions in each pasture such as the total, mean, and variance of the green forage in each pasture, the confidence limits of their calibration, and even a graphic histogram of their distribution within each field (Cheng, T.D., 1985).

ASSESSMENTS OF LARGER MANAGEMENT UNITS

The image map utilized in the examples was extracted from the 24 September 1982 TM image to accurately match the Hagan Lake N.E. 7.5 minute USGS topographic map. The boundaries and internal features of this image map correspond to the associated topographic sheet to within 1 picture element (Yang, Y.K., 1985). This inherent geographic calibration allows images of various dates to be co-registered to each other, existing cultural features, and management unit boundaries. The image map of 7 TM spectral bands was extracted to floppy disk from the EROS source tape and geo-referenced using only the IBM PC and its associated peripherals, including a standard open reel tape drive.

TM bands 3 and 4 were used as a basis for computing the Transformed Vegetation Index (TVI) for the image map (figure 1). A color infrared image map of the same area was then displayed on the monitor and used as a basis for

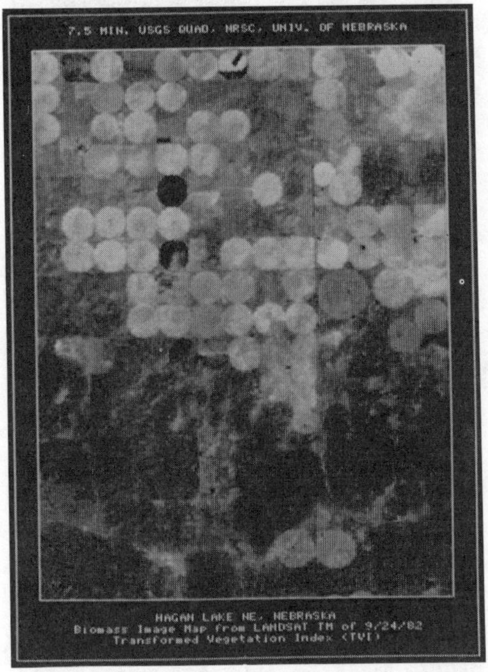

Figure 1. Transformed Vegetation Index (TVI) of 7.5' TM image map.

the direct photointerpretation of the location of cultivated agricultural lands within the image map. A mouse-controlled cursor and associated software was then used to define these areas for suppression from the TVI version of the image map (figure 2). The resulting image map of the biomass distribution within the remaining area is then displayed to the range manager with appropriate color codes (figure 3) and summation statistics.

The 7.5 minute image map is the most suitable window of Thematic Mapper imagery for local on-site microcomputer analysis in a range, agricultural, or other natural resource management context. However, since the picture elements within each image map are accurately geographically labeled, concatenation of several 7.5 minute image maps and their derivative biomass assessments can be readily assembled to determine the status of an irregularly bounded large ranch or other geographic management unit. Similarly, a small subportion of an image map such as an individual pasture, can be easily located, outlined in geographic terms, and repeatedly extracted from each new corresponding 7.5 minute TM image map and its derivative products. Thus, the inherent geocalibration available in the use of the image map concept allows multitemporal image or biomass spatial information to be assembled at no additional effort beyond that involved in calibration of

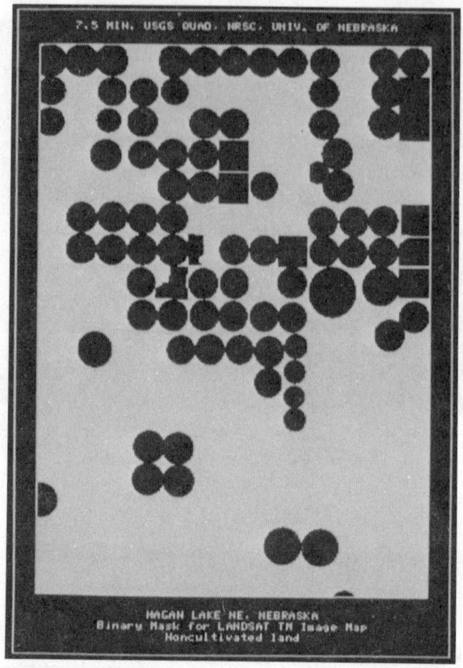

Figure 2. Binary mask to suppress cultivated fields from 7.5' TM image map.

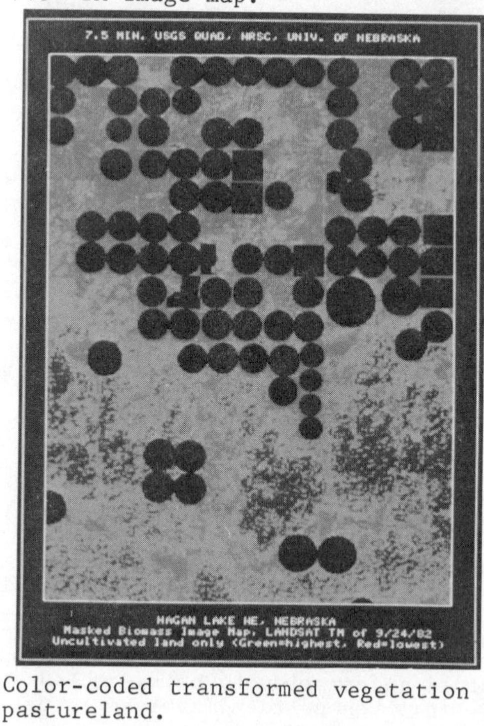

Figure 3. Color-coded transformed vegetation index of pastureland.

each individual image map to a common topographic map. The reassembly of this multitemporal image, associated canopy biomass map or other vegetation condition assessments into new geometric areas larger than the 7.5 minute map (a large ranch) or significantly smaller (a pasture), is easily done on the microcomputer from a stored file of polygon boundaries.

ASSESSMENTS OF INDIVIDUAL PASTURES

A mouse cursor pointing device is very inexpensive and easy to use by a rancher to outline the boundaries of his individual pastures upon a color infrared image map displayed on the IBM PC system. Using his intimate knowledge of his area, his pasture fencing, general pasture use, and so on, he is able to directly annotate the boundaries of his individual management units upon a high quality rendition of the color infrared image map (figure 4).

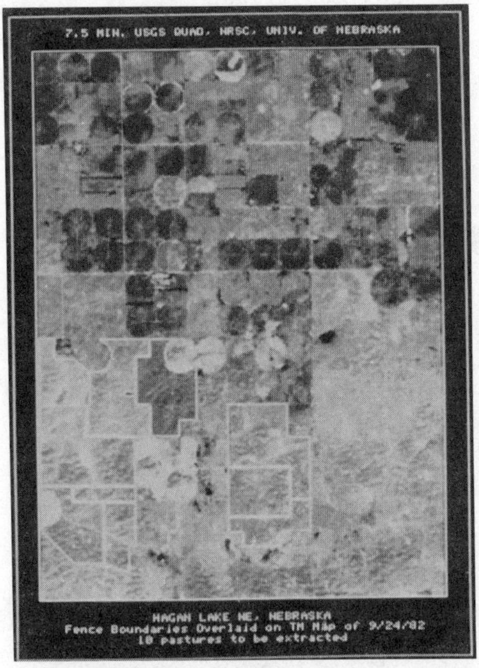

Figure 4. Color infrared TM 7.5' image map with pastures annotated.

Occasionally TM image maps of two different dates may be required to accurately locate these boundaries. This on-site photointerpretive approach is similar to that used to suppress the cultivated agricultural areas as described above. This method of locating the features of interest employs all the resident, incipient knowledge of the end user that could never be available at any higher level. For example, the rancher-owner knows that he has no gate, or is not maintaining the fence between two enclosures and that they must be outlined and managed as one fenced pasture. However, this hybrid photointerpretation and computer analysis approach also necessitates that the display of the colored image map by the microcomputer be of sufficient size and of a very good color balance and rendition to enable the use of these accurate direct photointerpretation procedures. A simple color coding or even a good black and white rendition of the image map will not suffice.

Accurate pasture boundaries can be mapped and stored in this fashion for subsequent reuse on each new image map or multidate pasture condition assessment. This boundary definition process is not repeated unless new fencing is added, property or leases acquired or dropped, or other modification made to the areas managed. The effects of short term area or spatial management decisions such as the opening or closing of gates between adjacent pastures is easily handled by specifying the adjustment of the existing boundary polygon files.

The fence boundary files saved above can be instantly recalled and accurately overlaid upon any new image map or derivative product. For example, the transformed vegetation index map of the 7.5 minute Hagan Lake (figure 1) can be overlaid with "John Hancock's" stored pasture boundaries (figure 5).

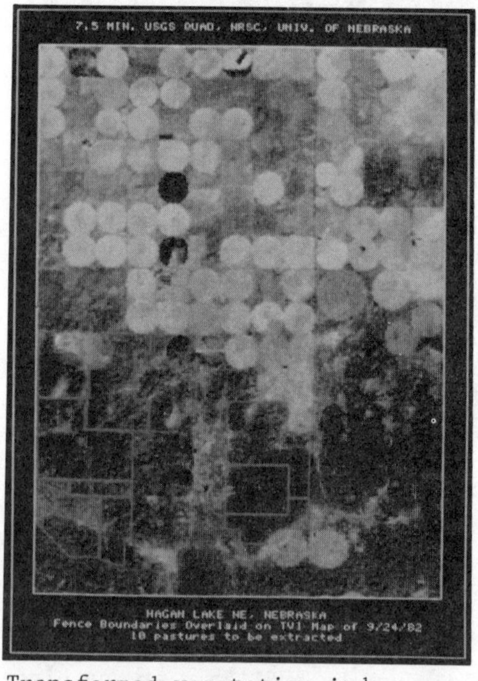

Figure 5. Transformed vegetation index map with pastures annotated.

Since John Hancock is not interested in the balance of the area of this image map which he is not managing, he can request that it be suppressed. The inherent geographic location of his fields within the image map is still preserved enabling subsequent overlay of information extracted from earlier dates of TM assessments, ground control information, digitized soil property maps, etc. The relative distribution of the green canopy biomass within each of his pastures can now be color-coded to clearly display the status and distribution of the growing portion of available forage within each of his pastures (figure 6). He may also summarize the current conditions of all his pastures by requesting the microcomputer to print a composite table or draw a graph of the mean and variance of the biomass with each pasture.

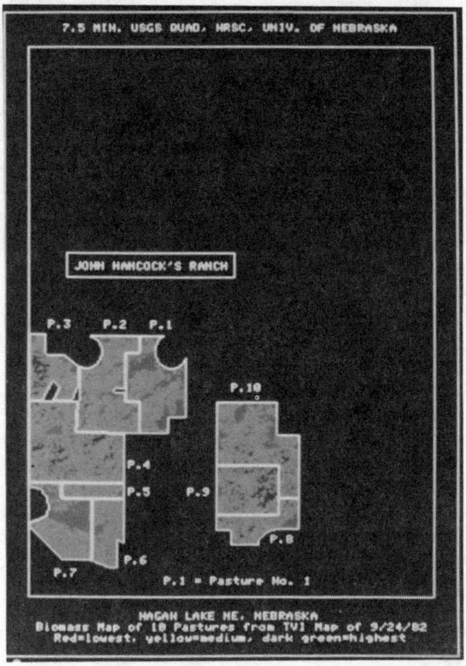

Figure 6. Color-coded biomass of individual pastures.

Should John Hancock require a more quantitative assessment of the pasture conditions he can collect the necessary ground control information within his pastures at a time near the satellite's overpass by clipping test plots, ocular estimates, or similar techniques. This information can then be input to his microcomputer, together with the image estimates for the corresponding picture elements, to form a regression equation between satellite and ground measurements. This then provides a calibration curve with which every one of the .25 acre transformed vegetation index assessments of his pasture can be converted to tons per acre of green biomass together with a measure of the confidence of this calibration (figure 7). After calibration, these cells can be summed by pasture and presented in the form of a graph or table presenting the green forage condition in tons per acre, its range of variation, and the statistical confidence of these measurements.

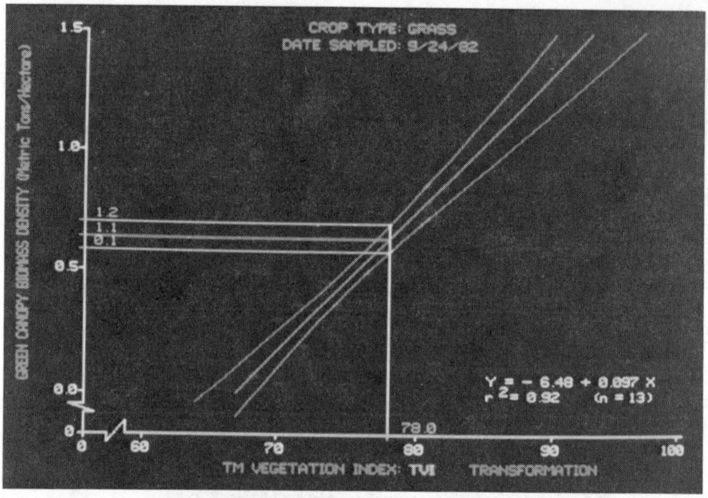

Figure 7. Sample biomass calibration curve.

MICROCOMPUTER SPEED AND COST CONSIDERATIONS

A few general comments on the current costs and timing of the processes outlined are in order. These are based on current IBM PC microcomputer technology. Suitable, good quality, 7.5 minute TM natural color or color infrared image maps can be displayed with innovative new software techniques on specialized color display boards which use 256 colors selected from a much larger palette of colors. A typical low end display board of this type and an appropriate analog color monitor can now be purchased as an add-on to the IBM PC series (PC, XT, and AT) for \$1,500. Projections for the evolution of the PC graphics market indicate that this display board and monitor combination will fall below \$1,000 in the next two years. The addition of a mouse pointing device (\$100), an auxiliary arithmetic processing chip (\$100), and appropriate software would provide a complete system capable of analyzing 7.5 minute TM image maps as demonstrated in this report. Current access to 7.5 minute TM image maps is more complicated. Service bureaus will prepare 7.5 minute image maps in appropriate floppy disk formats from user supplied TM tapes. An add-on package consisting of a good quality, standard open-reel tape drive, IBM PC interface card, and general tape access software is currently \$4,000. This package is capable of reading the TM image tapes distributed by EROS. However, not only is this still an expensive add-on for the personal user, but the process of extracting geographically calibrated 7.5 minute image maps from TM scenes is still rather complicated and involved.

The time to complete the various image processing activities outlined is becoming a rather inconsequential portion of the effort. It depends directly upon which end of the IBM PC series is used. At the low PC model end with the least costly, but suitable display boards, the process of reading of a typical Nebraska 7.5 minute TM image map from a floppy disk, color balancing it, and displaying it would take 2 to 5 minutes depending upon the user's familiarity with the approach (382 by 488 cells times 3 bands). The recolor balance and redisplay of the same image, with this floppy disk based system, would require the same steps and time. At the upper end with an AT and the most advanced color display boards it would require 20 seconds to accomplish this task from an image resident on hard disk and 5 seconds from a RAM disk. A recolor balancing of the image displayed, regardless of its storage media, is instantaneous or more practically, as fast as the user can provide new inputs.

Computing the TVI with a floppy disk-based PC system would take about 15 minutes for the whole image map or less than 1 minute for a single pasture. Using the AT system, the TVI of the whole image map can be completed in a couple of minutes and the individual pastures in a few seconds. Outlining pasture boundaries with a mouse on a good quality display is quite easy compared with earlier joystick or track ball procedures. It can be a one time effort of a few minutes if the pasture boundaries are easily detected on the display. The process of identifying the desired boundaries by photointerpretation determines the length of time involved as the mouse can be used to rapidly outline these boundaries once they have been interpreted.

CONCLUSIONS

The capability for quantitatively assessing the spatial variation in plant canopy and surface soil conditions as a function of time within individual ranch pastures has been demonstrated. The Thematic Mapper supplies imagery of suitable spectral bands and ground resolution. Currently available microcomputers, already being employed in some ranch operations for other purposes, can do all the analysis necessary to extract and portray the desired information from this imagery with the addition of appropriate hardware and software add-ons. These periodic assessments of conditions for each pasture, such as green canopy biomass (as illustrated here), near surface soil moisture, canopy moisture stress, etc., can then be passed to management models which can also be run on the same microcomputer. The utilization of these techniques at the ranch or other local management level will provide a synergistic relationship between this new technical information and the wealth of local "ground control" and experience with each individual management unit.

No significant technical hurdles appear in this general approach. Images are already moved in real time from their source to the individual ranch in the form of entertainment television, often by direct downlink to a rancher's private dish. It has already been shown that digital information can be encrypted into standard video, stored, moved, and decoded back to its original digital form at no information loss. Technically, all the TM imagery collected during a day could be moved out to subscribers during the early morning hours via these TV links. Thus, the remaining discussion on the worth of schemes for the on-site use of multitemporal satellite imagery and microcomputer image processing in ranch, agricultural, and forest management is purely socio-economic. Does anyone need the incremental benefits provided and can the costs of the image acquisitions and transmission be paid for with this benefit?

REFERENCES

Cheng, T.D. 1985. Canopy Biomass Measurement of Individual Agricultural Fields with LANDSAT Imagery. Ph.D. Thesis, Dept. of Forest Science, Texas A&M Univ., College Station, Texas. 199 pp.

Miller, L.D., Y.K. Yang, T. Cheng, M. Unverferth, M.G. Kim, B. Elliot, and K. Wills. 1984. A Table-Top, Microcomputer Approach to the Management, Analysis and Display of Geographic and Image Data Using a Map-Oriented, Geo-Referenced Framework. IEEE 1984 Proc. of the Pecora 9 Spatial Information Technologies for Remote Sensing Today and Tomorrow. IEEE Computer Society, Silver Springs, Maryland. pp. 331-340.

Miller, L.D., Y.K. Yang, T. Cheng, M. Unverferth, and K. Wills. 1984. A Table-Top, Microcomputer Approach to the Management, Analysis and Display of Geographic and Image Data Using a Map-Oriented, Geo-Referenced Framework. Proceedings of the Ninth IBM University Study Conference, IBM Academic Information Systems, Special Studies Department, Milford, Connecticut. pp. 318-332 (13 color plates).

Yang, Y.K. 1985. Microcomputer Image Analysis of LANDSAT Satellite Thematic Mapper Images, Airphotos, and Related Maps. Ph.D. Thesis, Dept. of Forest Science, Texas A&M Univ., College Station, Texas. 350 pp.

EVALUATION OF A LAYERED APPROACH FOR CLASSIFYING MULTITEMPORAL LANDSAT MSS DATA

D. Fabian Lozano-Garcia and Roger M. Hoffer
Department of Forestry and Natural Resources
Purdue University
West Lafayette, Indiana 47907

ABSTRACT

Two different methods for analysis of multitemporal MSS data were compared in this study. The first involved the use of a single stage maximum likelihood classifier. The second approach made use of a hierarchical classifier (Layered classifier) which consist of a sequence of classification steps, each of which can incorporate a different wavelength band or combination of wavelength bands. In addition, at each step in the classification process, only one spectral class or group of classes are separated from all other classes in the data. Thus, the classification procedure is more efficient since only a relatively small number of wavelength bands and a small portion of the data is involved in each decision step.

Landsat-1 MSS data were obtained over the Hoosier National Forest in June 1973 and February 1974. After performing a digital registration of these two data sets, a series of four specific classifications were compared. The first consisted of a standard single stage maximum likelihood classification using an eight channel training statistics deck. The second utilized the four best channels of the eight available, using a single stage maximum likelihood classifier. The third classification involved the Layered classifier and the same eight channel training statistics. In the fourth approach each of the two dates were used independently to develop two sets of training statistics which were then used as input for the Layered classifier algorithm.

The results indicate that the Layered classifier is a more effective and efficient approach for the analysis of multitemporal data sets. The classification accuracies were relatively high for all four classifications but the Layered classifier required only one third of the CPU time used in the single stage classifications.

INTRODUCTION

Classification of multispectral scanner data has been performed traditionally with single stage classification algorithms. One characteristic of these algorithms is that they involve the categorization of the data using only one of the many available sets of spectral channels; (usually all the available channels are used). The large number of computations required and the sensitivity of the classification algorithm to category variance are the primary limitations of the single-stage techniques.

For some remote sensing applications, a classification technique based on the layered or decision tree approach can be utilized. In this method, the multispectral scanner data is classified through a hierarchical decision procedure (Figure 1) in which the analyst goes through a series of decision nodes, and at each node he/she selects the best set of spectral channels that separate a class or group of classes from certain other cover types. This separation is made step by step, until all classes have been separated. The actual classification of each pixel is performed using any of the available per-point algorithms.

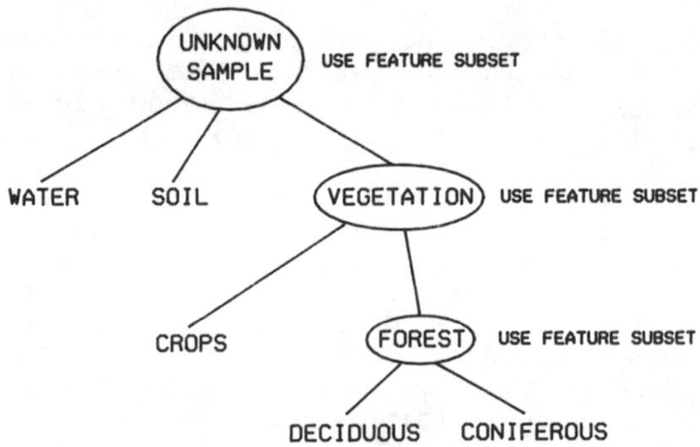

FIGURE 1. HYPOTETICAL HIERARCHICAL DECISION TREE.

Two of the major limitations of the single stage classifiers (that are not present in the Layered classifier) are:

1) Only one of the many possible combinations of wavebands (features) is used in the single stage classification. However, there may be some subsets which are more effective than the entire set of channels, or some subsets may be more effective for the discrimination of a particular spectral class.

2) In the standard single stage classification, every data sample is tested against all classes. This characteristic leads to very low efficiency.

Hierarchical classifications have not been used commonly in the analysis of remotely sensed data. One of the earliest applications of this scheme was the work of Bartolucci et al. (1973) in mapping water temperatures. The authors used this approach to separate water from all other cover types in the first stage of the decision tree, using the best set of channels to accomplish this separation. The second step was to classify the water temperatures using the available thermal channel. They found this approach to be superior to the use of the single stage classification procedure. Hoffer et al. (1979) used the Layered classifier in a study of combined multispectral scanner and digital topographic data. Their approach was to separate cover types using the spectral data in the first stage. The next stage of the classification involved the utilization of topographic data to divide the major cover types into individual forest cover types, and to remove some misclassification errors due to shadow effects. The results showed an improvement of approximately sixteen percent as compared to the use of spectral data alone using a single stage classification.

The Layered classifier has been applied to a multitemporal data set by Landgrebe (1976), who classified a set of agricultural areas. Weismiller et al. (1977) also utilized this technique in a change detection procedure. Hixson et al. (1980) compared this classifier against six others. In general, their conclusions indicate that the time required by the analyst to design the decision tree can be significant. Also, they concluded that the Layered classifier is well suited to handle multitemporal MSS data sets.

STUDY AREA

The study area is located in the Hoosier National Forest and the Brown County State Forest, south-east of Bloomington, Indiana. The forest of the area is classified according to Petty and Jackson (1966) as:

WESTERN MESOPHYTIC ASSOCIATION.-- In this community, ten to twenty species frequently share dominance in the crown cover and exert their controlling influence on the forest community. The mixed forest usually occurs in ravines and on the cooler slopes, whereas oak or oak-hickory forests cover the drier slopes and ridges.

OAK-HICKORY ASSOCIATION (Quercus-Carya).-- The oak-hickory forests are usually found occupying south-facing and west-facing slopes. In general, moisture content of the soil is consistently lower than in the oak-hickory type than the mesophytic mixed forest.

The study area is largely forested, mostly by tulip poplar, oak, maple, hickory, ash, walnut and sycamore. Small stands of pine are scattered in the study area. The steep slopes and heavily dissected topography have discouraged the extensive clearing of this area for agriculture, although selective logging has altered the composition of most stands. On the valley bottoms in the western, south-western and east central portion of the area the forest has been substituted for crops and pasture. Some portions of suburban Bloomington occur in the north-western corner of the area. Monroe Reservoir, Lemon Lake, Yellowood Lake and Grandview Lake are the major water bodies in the area.

AVAILABLE DATA

Seven dates of registered multispectral scanner images of the Landsat-1 satellite (Table 1) were available for a portion of the Hoosier National Forest including Monroe Reservoir and the surrounding areas. The data sets from June 1973 and February 1974 were selected due to their high quality, minimum cloud cover, and available ground reference data.

Table 1. Available Landsat MSS data sets.

Scene ID	Date	Season
1285-16001	May 4, 1973	Spring
1320-15541	June 8, 1973	Late spring
1392-15531	August 19, 1973	Summer
1411-15584	September 7, 1973	Late summer
1482-15514	November 17, 1973	Late fall
1572-15493	February 15, 1974	Winter
1591-15550	March 8, 1974	Late winter

ANALYSIS PROCEDURES USED

Table 2 shows the four different classification techniques utilized in this work. They can be divided into single stage and Layered classifiers. In the single stage approach two methods were tested -- one with all eight channels and the other with the four best channels. The Layered approach was first tested using the same training statistics generated for the eight channel, single stage approach. The second layered classification involved generating two sets of independent training statistics -- one for each date.

Table 2. Characteristics of classifications being compared.

	Classification Number			
	1	2	3	4
Classification Technique	Single Stage	Single Stage	Layered Stage	Layered Stage
Statistics used for training	All 8λ	"Best" 4λ	Combined Dates (All 8λ)	Separate Dates (4λ for June; 4λ for Feb.)
Algorithm used for classification	GML	GML	GML	GML

<u>Multitemporal/Multispectral Single Stage Classifications</u>
The first analysis conducted in this study involved

classifications of a two date, eight channel data set. The dates selected were June of 1973 and February of 1974.

Training statistics were generated using a "multi-cluster blocks" approach, similar to the one described by Fleming et al. (1975).

Two different classifications were performed with this data set. The first used all the channels of both dates -- Multitemporal/Multispectral 8 channels (Figure 2). In the second classification, four channels were selected based upon the minimum and average transformed divergence value (Figure 3). The selected channels were:

Table 3. Selected wavebands for the single-stage four wavebands classification.

Waveband	Date	Spectral Region
0.6-0.7 μm	June	Visible (red)
0.7-0.8 μm	June	Reflective IR
0.6-0.7 μm	February	Visible (red)
0.7-0.8 μm	February	Reflective IR

It is important to point out that the analysis sequence resulted in one channel in each major portion of the spectrum (i.e., visible and reflective infrared) covered by Landsat-MSS was selected for this classification.

<u>Layered Classifications</u>

One of the characteristics of the layered classifier is that it permits the analyst to optimize the use of certain spectral bands in the separation of a class or group of classes. In a multitemporal/multispectral classification, this algorithm also permits the use of the best season (represented by a set of spectral channels) for the identification and separation of cover types.

The selection of the classes that will constitute a particular node, and the set of spectral channels to be used to separate this node were based on the statistical distance between training classes. This distance was obtained by calculating the transformed divergence values of the training classes for all possible combinations of spectral channels. The best set of features to be used in each particular node was also defined using the separability information, based on a threshold of Dt=1750.

The first approach used with this classifier involved using the same 8 channel training statistics developed for the single stage Multitemporal/Multispectral classification (Figure 4). The second approach used with the layered classifier was the development of training statistics independently for each of the two dates which were then used as part of the input to the classifier (Figure 5).

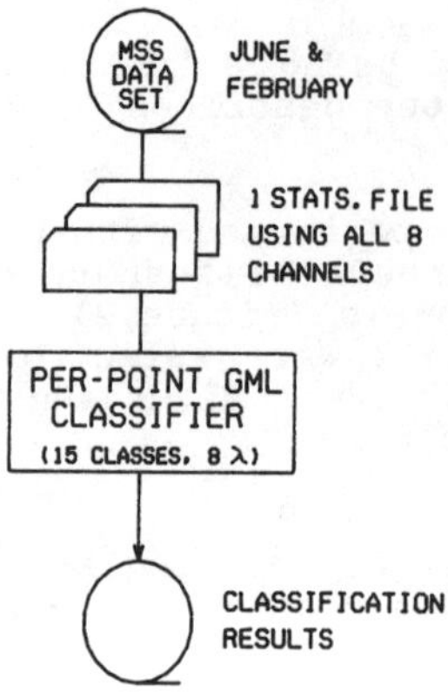

FIGURE 2. MULTITEMPORAL/MULTISPECTRAL SINGLE STAGE CLASSIFICATION, 8 CHANNELS.

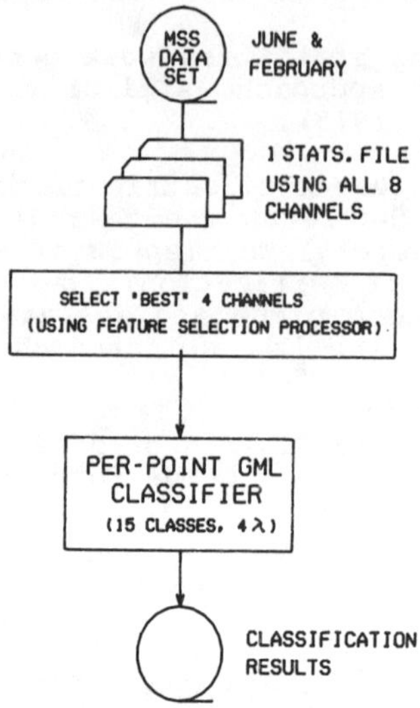

FIGURE 3. MULTITEMPORAL/MULTISPECTRAL SINGLE STAGE CLASSIFICATION, 4 CHANNELS.

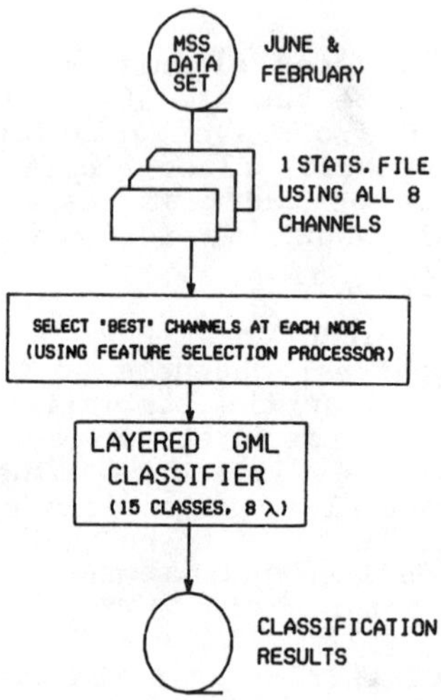

FIGURE 4. LAYERED CLASSIFIER, ONE SET OF 8 CHANNELS.

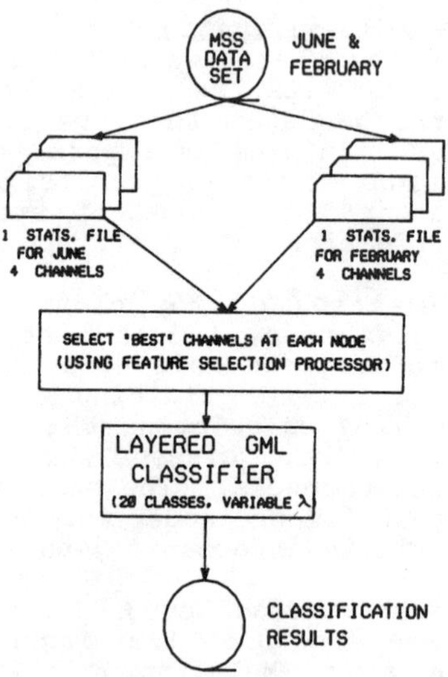

FIGURE 5. LAYERED CLASSIFIER, TWO SETS OF 4 CHANNELS.

Evaluation Methods

To determine the accuracy of the classification of MSS data, a set of statistically valid test fields was developed. The standard color IR composite was displayed on the Comtal Vision One/20 and a test grid with dimensions of 50 lines by 50 columns was selected as the basis for the sampling procedure, over the area where aerial photographs were available. The cell to be analyzed was selected at random. Each cell of 2,500 pixels was subdivided into four quadrants of 25 lines by 25 columns (625 pixels). One quadrant was selected at random, and the biggest, most homogeneous field corresponding to each specific cover type present in that block was identified. The minimum number of observations (pixels) used for the evaluation of any particular cover type was 100 pixels (as suggested by Landgrebe 1976).

A statistical evaluation was carried out using one-factor analysis of variance, based upon the arcsine transformation of the performance values. (This transformation was necessary because of the nature of the results -- a proportion dealing with binomial data, i.e., pixels are identified correctly or incorrectly). To determine if there were significant differences between the performance values of cover types or classifications, a Newman-Keuls Range test was performed at an alfa level of 0.1.

The criterion used for determining the cost effectiveness of the classification results was based on the amount of computer CPU time (Central Processing Unit) used to perform each classification. This was considered the most objective and accurate way to compare and evaluate the cost of each classification scheme. Because the analyst become increasingly familiar with the characteristics of the data during the sequence of analysis, it was believed that the "analyst time" required to develop the training statistics would be biased.

RESULTS

Total CPU time required for each classification and the overall and average performances are shown in Table 4.

The Multitemporal/Multispectral single stage classifications (both eight channels and the four best channels) provided results showing detailed informational classes. For the deciduous forest, a class representing forest in shadow was identified. Bare soils were differentiated into two groups: those that are subject to flooding and those that are not. Two distinctive classes of water (deep water and shallow water) were recognized in Monroe Lake.

Figure 6 shows the classification performance by class. For the single stage classifications, both overall performance and the performances by class were very good, except for the class "Pasture" which had an accuracy of 31.5% for the 8 channel classification and 32% for the 4 channel classification. Due to confusion between the pasture and soil classes, 58% of the pasture test pixels were classified as soils in the 8 channel classification, and 50% of the pasture test pixels of the 4 channel classification were assigned to the soils classes.

Table 4. CPU time and overall classification performance.

	CPU TIME	No. of Spectral Classes	Overall Performance	Average Performance
Multitemporal/ Multispectral 8 Channels	77.3 min	14	91.0%	83.0%
Multitemporal/ Multispectral 4 Channels	25.4 min	14	90.8%	83.2%
Layered 1 set of 8 Channels	16.5 min	14	90.8%	82.3%
Layered 2 sets of 4 Channels	9.1 min	20	91.8%	89.9%

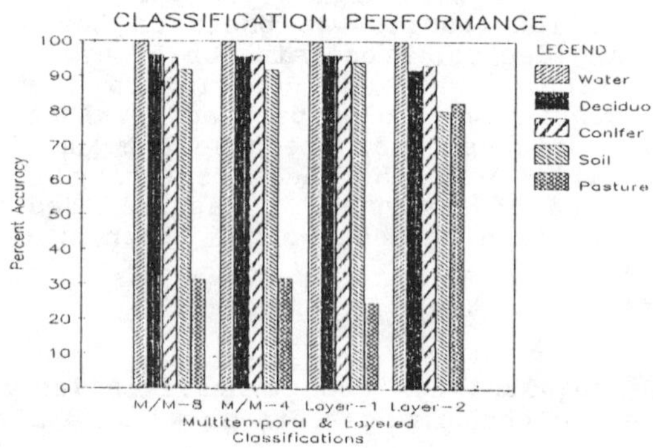

Figure 6. Classification Performance by Class.

The third classification using the layered technique (i.e., training statistics based on one set of 8 channels) showed no significant difference from the results obtained in the single stage Multitemporal/Multispectral classifications. A small decrease in the percent accuracy for the pasture class was found, but this difference was not statistically significant. The primary difficulty in classification of pasture was again due to confusion with the soil class. However, the CPU time required was only 60% (10 minutes less) of that required for the single stage Multitemporal/ Multispectral 4 channel classification.

The fourth classification (two date layered classification) had a classification accuracy of 91.8%, the best overall performance of all four classifications. Both deciduous and coniferous forest cover types had over 90% correct classification. Most of the misclassifications in each of these forest classes were actually due to confusion occurring between them rather than between forest and non-forest categories.

A more consistent classification was obtained for the non-forest classes -- over 80% as can be seen on Figure 6. The statistical comparisons of the percent correct classification for the five classes show three groups (Water, Forest and Soils-Grasslands) in which there were no significant differences in the classification accuracy. It was clear that the classification was improved due to the capabilities of the layered classifier, primarily because of this classification processor allows the analyst to select the best set of features to separate a class or group of classes.

As shown in Table 4, the overall classification performance varied only from 90.8% to 91.8%. However, in the first three classification results shown in Table 3, the average performance values indicate an important variability in relation to the overall performance values. This was caused by the very low performance of the pasture class.

Of the four classifications, the Multitemporal/Multispectral 8 channel classification required the highest amount of CPU time, followed by the Multitemporal/Multispectral 4 channel classification, then the Layered 1 set of 8 channels, and with the lowest CPU time of all four, the Layered 2 sets of 4 channels classification. Thus, based on both accuracy and CPU time, the Layered 2 sets of 4 channels approach was the best method.

SUMMARY AND CONCLUSIONS

The results of this research show the advantage of the Layered classification approach over the Multitemporal/ Multispectral classification approach in the analysis of Multitemporal MSS data.

The single stage Multitemporal/Multispectral approach provided an adequate classification accuracy. The weak points of this scheme are: the complexity involved in the development and interpretation of the training statistics, and also the CPU time required to perform the classification. The eight channel classification required 8.5 times more CPU time than the best layered classification, and even the four channel Multitemporal/Multispectral classification required 2.8 times more CPU time than the best layered classification. The selection of the four best channels of this data set indicates that one channel of each date for each of the available regions of the electromagnetic spectrum in the MSS data are required to perform an effective multitemporal classification.

The Layered classification procedure proved to be the best in terms of classification accuracy, for both the Layered 2 sets of 4 channels and Layered 1 set of 8 channels. Although percent accuracies in the Layered 2 sets of 4 channels for the individual forest cover types were slightly lower in relation to the Multitemporal/Multispectral single stage classifications, these differences were not statistically significant. The layered technique also provided more consistent results, since all accuracies were over 80%. The design of the decision tree for the classification is one of the most important and difficult tasks in this approach.

In summary, the combination of high classification accuracy, low CPU time required and the flexibility in handling multitemporal data sets makes the Layered classifier a very effective, efficient and useful tool in multitemporal analysis of remotely sensed data.

REFERENCES

Anderson, J.R., E.E. Hardy, J.T. Roach and R.E. Witmer. 1976, A Land Use and Land Cover Classification System for Use with Remote Sensor Data. U.S. Geological Survey Professional Paper 964. Washington, D.C.

Anderson, V.L. and R.A. McLean. 1974, Design of Experiments: A Realistic Approach. Marcel Dekker, Inc. New York. 418 pp.

Bartolucci-Castedo, L.A., R.M. Hoffer and T.R. West. 1973, Computer-aided Processing of Remotely Sensed Data for Temperature Mapping of Surface Water from Aircraft Altitudes. LARS Information Note 042373. LARS-Purdue University, West Lafayette, Indiana 47906. 143 pp.

Bartolucci, L.A., P.H. Swain and C. Wu. 1976, Selective Radiant Temperature Mapping Using Layered Classifier. IEEE Trans. on Geoscience Electronics GE 14(2):101-106.

Duda, R.O. and P.E. Hart. 1973, Pattern Classification and Scene Analysis. John Wiley & Sons. New York. 482 pp.

Fleming, M.D., J.S. Berkebile and R.M. Hoffer. 1975, Computer-aided Analysis of Landsat-1 MSS Data: A Comparison of Three Approaches, Including a "Modified Clustering" Approach. LARS Information Note 072475. LARS-Purdue University, West Lafayette, Indiana 47906.

Hixson, M., D. Scholz, N. Fulls and T. Akiyama. 1980, Evaluation of Several Schemes for Classification of Remotely Sensed Data. Photogrammetric Engineering and Remote Sensing 46(12):1547-1553.

Hoffer, R.M. 1979, Computer-aided Analysis Techniques for Mapping Earth Surface. LARS Technical Report 020179. LARS-Purdue University, West Lafayette, Indiana 47906. 20 pp.

Hong, J-K. and J. Lisaka. 1982, Coastal Environment Change Analysis by Landsat MSS Data. Remote Sensing of Environment 12(2):107-116.

Jensen, J.R. 1981, Urban Change Detection Mapping Using Landsat Digital Data. The American Cartographer 8(2):127-147.

Latty, R.S. and R.M. Hoffer. 1981, Waveband Evaluation of Proposed Thematic Mapper in Forest Cover Classification. LARS Technical Report 041581. LARS-Purdue University, West Lafayette, Indiana 47906. 12 pp.

Lindsay, A.A., D.V. Schmelz and S.A. Nichols. 1969, Natural Areas of Indiana and Their Preservation. Indiana Natural Areas Survey. Lafayette, Indiana. 594 pp.

Petty, R.O. and M.T. Jackson. 1966, Plant Communities. pp. 264-296. In Lindsey, A.A. (editor) Natural Features of Indiana. Indiana Academy of Science.

Swain, P.H. 1976, Layered Classifier Adapted to Multi-temporal Data Sets. LARS Final Report, NASA Contract NASA-14016. LARS-Purdue University, West Lafayette, Indiana 47906.

Swain, P.H. and H. Hauska. 1977, The Decision Tree Classifier: Design and Potential. IEEE Transactions on Geoscience Electronics GE 15(3):142-147.

Swain, P.H., C.L. Wu, D.A. Landgrebe and H. Hasuska. 1975, Layered Classification Techniques for Remote Sensing Applications. LARS Information Note 061275. LARS-Purdue University, West Lafayette, Indiana 47906. 11 pp.

Weismiller, R.A., S.J. Kristof, D.K. Scholz, P.E. Anuta and S.A. Momin. 1977, Change Detection in Coastal Zone Environments. Photogrammetric Engineering and Remote Sensing 43(12):1533-1539.

Williams, D.L. 1979, Canopy Closure in the Pine Forest of North Carolina. pp. 19-22. In Williams, D.L. and L.D. Miller. Monitoring Forest Canopy Alterations Around the World with Digital Analysis of Landsat Imagery. NASA, Goddard Space Flight Center.

Williams, D.L. and F. Yates. 1977, A Reduction in Ag./Residential Signature Conflict Using Principal Components Analysis of Landsat Temporal Data. Proceedings American Society of Photogrammetry Fall Technical Meeting. Little Rock, Arkansas. pp. 230-238.

Wu, C.L., D.A. Landgrebe and P.H. Swain. 1974, The Decision Tree Approach to Classification. LARS Information Note 090174. LARS-Purdue University, West Lafayette, Indiana 47906. 174 pp.

COMPARATIVE EVALUATION OF DIGITAL CHANGE DETECTION METHODS IN FORESTLAND AND RANGELAND ENVIRONMENTS USING LANDSAT MULTISPECTRAL SCANNER DATA

Gregory S. Burns
NASA/Earth Resources Laboratory
National Space Technology Laboratories, MS 39529

ABSTRACT

Procedures for digitally processing bi-temporal spectral data sets were tested in control areas in Louisiana, Kansas, and Arizona for their ability to identify and map areas of surface cover change. Location-specific, temporal radiometric differences between equivalent areas of co-registered spectral data pairs were characterized as measures of land conversion, and the results were compared with digitized land cover change maps for a reference standard. The maps were produced from aerophotography that temporally and spatially coincided with the ends of the test intervals determined by the satellite overpasses for each of the three MSS test pairs. The maps were processed into rasterized format along with georeferenced MSS data-derived results and placed into common multichannel data base files representative of each geographic area for performance testing. The two computer classification techniques used to identify and map changes were (1) a land cover post classification comparison (PCC) and (2) spectral change pattern analysis (SCPA) of a bi-temporal change classification. Unsupervised training routines were used to define statistical groups of spectral similarity (1) for the four bands from each date independently for the PCC technique and (2) for corresponding visible and near-infrared band pairs from each date for the SCPA technique. A numerical pattern recognition routine was then used to collapse these same channels of reflective variability into its respective spectral group and map each data cell into its proper ground location. Land cover changes were identified, and each cell was compared with its photo-identified counterpart to sum the agreement between the two data sources. Results indicated that better than 90% to 96% correspondence between data sources could be obtained over the three test areas.

INTRODUCTION

This paper presents areal measurements of regional land cover conversions in three study areas in Louisiana, Kansas, and Arizona. These measurements were made with two computer classification techniques using digitally co-registered sets of bi-temporal Landsat Multispectral Scanner (MSS) data for each test location. Additionally, parallel measurements of land cover conversion in these same areas were made by manual photointerpretation from concurrent pairs of aerophotography and are presented for comparison.

DESIGN, DATA REQUIREMENTS, AND DATA SELECTION

The intent of this investigation was to determine how well MSS data could be used to map regional land cover conversions by comparing results against a credible reference standard. To carry this out bi-temporal MSS data sets would be compared for location-specific spectral changes, characterized in terms of land covers, and contrasted with equivalent measures within a control area. The only reliable means for mapping land cover changes in retrospective over sizable areas has been the interpretation of dated pairs of aerophotos where available. This need for a control area to evaluate technique performance was but one of the restrictions that dictated data requirements and severely limited the choice and extent of possible test sites. To be defined as a control area, each test site required at least two seasonally equivalent, atmospherically clear frames of Landsat data with the highest radiometric quality ratings in all bands. The purpose of these requirements was to avoid spectral differences between dates that might be of equal magnitude but not associated with the targeted land cover changes. The common areas of these MSS test data pairs had to coincide both spatially and temporally with the intersection of aerophotographic pairs. The net extent of each test area was the intersection of these four contemporaneous data sets. Additionally, several test areas that offered a variety of environmental conditions as well as different scales and types of change were desired that spanned a five year period or more. So the national data archives at EROS Data Center were searched for candidate domestic areas undergoing known, wide scale land cover transformation, due to increases in human use pressure, for data satisfying all these requirements to establish test areas. After examining the possibilities, the data shown in Table 1 were selected for three areas that closely matched the design criteria. The locations of these areas are shown in Figure 1.

Table 1. Data Pairs and Acquisition Dates

	LANDSAT MSS		AEROPHOTOS	
AREA	START	END	START	END
LA	10 Oct 74	02 Oct 79	04 Oct 74	24 Oct 79
KS	17 Aug 72	13 Aug 78	18 Aug 72	xx May 77
AZ	25 Jun 73	17 Jun 78	14 Jun 73	07 Mar 78

xx - acquired over several days

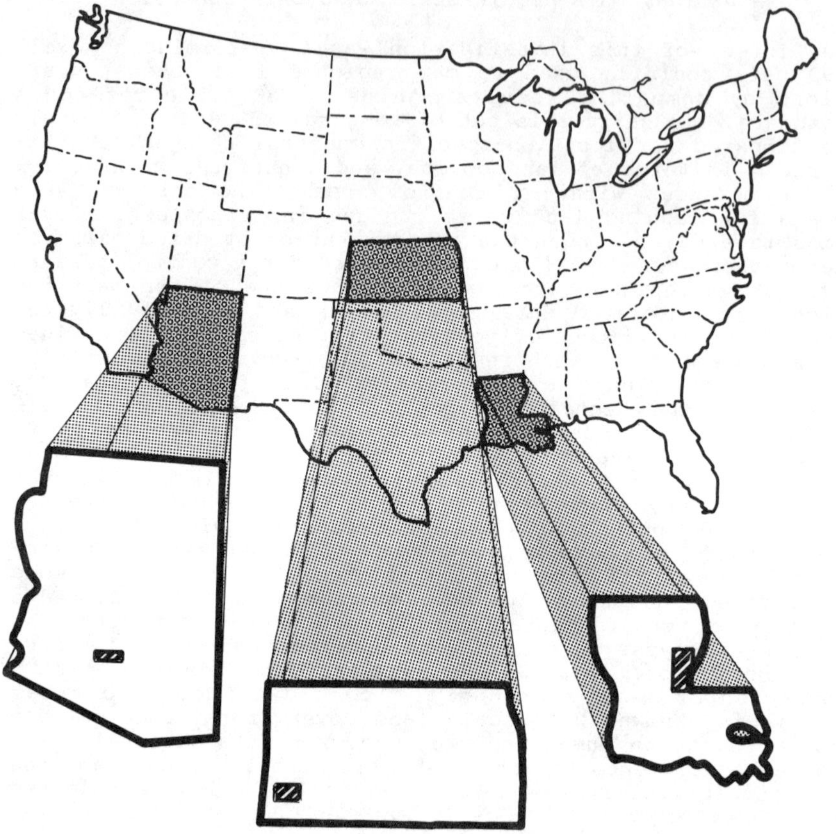

Figure 1. Locations of the Three Test Areas.

SITE DESCRIPTIONS

Louisiana

The intersection of the MSS test pairs with the aerophotographic verification pairs produced a half degree by one degree test area covering eight USGS 15-minute quadrangles that extend from 91.5° to 92° W. longitude and from 31° to 32° N. latitude. This area covers 5261 square kilometers of alluvial plain situated on the west side of the Mississippi River in east central Louisiana. This region was extensively forested with bottomland hardwood and forested wetland communities until the middle of this century. Large scale clearing, leveeing, and draining have occurred now that access is possible with new types of heavy machinery that can alter this previously impenetrable environment. Cropland and some pasture have replaced the previous land covers with some tracts as large as four miles by four miles being deforested during the five year study period from 1974 to 1979.

Kansas

A 15 by 30 minute test area was established in southwest Kansas, west of Garden City along the Arkansas River valley that covers 1220 square kilometers of eight 7.5 minute USGS quadrangles that were previously known as the Lakin and Deerfield 15-minute quadrangles. This area falls within the Central High Plains of the Great Plains and is very typical of this region physiographically. The temperate, semiarid rangelands are covered by native short grasses and forbes where uncultivated. Dryland farming has been the predominant farming practice except for areas along the river where water could be diverted for flood irrigation. The majority of land cover conversion has taken place upon the sand sage prairie communities. Pivotal irrigation technology has made it possible to irrigate this undulating aeolian terrain, and crops have replaced the native vegetation on these dry sandy soils. The scale of most of these changed areas is on the order of one-quarter mile across though both smaller and larger change areas were encountered over a six year period from 1972 to 1978.

Arizona

This arid, temperate test control area falls within a broad intermontane valley region that is within both the Sonoran Desert and Mexican Highland sections of the Basin and Range physiographic province. Drainage is poorly developed except for the Salt and Gila Rivers which drain snow melt from nearby mountains. This water supply, where harnessed, has spawned a locally huge, irrigated crop industry and allowed the city of Phoenix to grow to its present proportions. The extent of this test area was controlled by photocoverage for the second date. Its intersection with the three other data sets required to define this site was limited to parts of two east-west flightlines across the southern fringes of the Phoenix metropolitan area. The result was two separate polygonal control areas that cover 1186 square kilometers and parts of fifteen 7.5 minute USGS quadrangles. The flightline to the north dissected Tempe and Mesa and extended to just north of Chandler, while the second line was centered on a section of the Maricopa-Pinal county line just south of Chandler. The northerly flightline covers mixed suburban and irrigated agricultural land with remnants of desert rangeland. An increasingly urbanized Phoenix population has expanded into these areas replacing agriculture and range with urban cover in parcels as large as quarter-sections. The flightline to the south covers a more open area of rangeland where new cropland is the more dominant change type. Once again, contiguous areas as large as quarter-sections were the largest change areas encountered during the study interval between 1973 and 1978.

PROCEDURES

Computational Facility

All data processing was carried out with the NASA Earth Resources Laboratory (ERL) Analysis Software (ELAS) - based, Geographic Information System (GIS) package. The computational hardware utilized during the conduct of this project included a Perkin-Elmer MPS Super-minicomputer with four auxillary processor units operating under OS/32 version 7.2

with four local memory controllers, eight megabytes of main memory, and 300 megabytes of online disk storage. Other essential peripherals were used frequently and included image analysis stations, an X-Y digitizer, an array processor, and other standard hardware available at the ERL computer facility.

Preprocessing

Both co-registration to compare data between dates to produce change maps and geo-registration to compare these results with the UTM-mapped verification data were essential to this project. So co-registered, georeferenced, bitemporal, multiband source files were created for further processing as the necessary data components were compiled for each test area. To facilitate and streamline the production of these source files for each test area, data subsets from the first date were displayed on image display devices, and map control points were selected from the USGS base maps that coincided with the image coordinates. Mapping coefficients were developed relating the scanner coordinate system with that of the UTM, and the four bands of scanner data were resampled to conform with the map projection. Next image tie points were selected to define coordinates of the second date with those georeferenced coordinates of the first date and were modelled to relate the two coordinate systems. The four bands of data from the second date were transformed via the mapping equations to match those of the first. Spatial integration of both the MSS test data pairs and verification pairs was carried out with this procedure in the three test areas to decrease redundancy of these operations.

Ground Reference Data

Photo-verification pairs were all handled in the same manner. In each case the areas of commonality between data source pairs were examined for land cover changes between corresponding photos at each date. These polygonal areas were delineated and transferred onto a common base. Maps of the largest scale USGS topographic series available for each test area were used for this purpose. The land cover change maps were digitized with an X-Y digitizer to define the polygonal boundary information within the UTM coordinate projection. Each geographic area was processed into a multi-channel, gridded data file so that each 57 meter data cell represented either a change or no change area. This format allowed compatibility with georeferenced MSS-derived data products for performance testing of corresponding ground areas.

Methods Used to Map Changes

As stated earlier the intent of this investigation was to identify and map types of land cover changes. To accomplish this objective requires taking advantage of the geometric as well as the spectral relationships preserved with scanner data to locate the areal distribution of, to identify the types of, and quantify the areas of these temporal changes. Weismiller (1977), Jensen (1981), and Burns (1983) among other investigators have shown this capability in their reviews of change mapping methods and determined that different techniques should be tailored to purpose. The two classification methods selected for these tests were capable of providing all of these informational components, and are particularly suited for regional change mapping of this type. They are very similar in the way that they produce land cover change maps and vary more by the sequence in which the data is processed and results interpreted rather than the processing algorithms themselves.

Post classification comparison requires reduction of data at both ends of an interval of time into a user-defined thematic classification scheme. After this, data must be co-registered so that data cells representative of equivalent ground areas can be compared for a change in classification. In most cases with multispectral data this classification is derived through automated spectral pattern recognition techniques. These test cases used an unsupervised sampling routine, called SRCH, to develop training statistics for a maximum likelihood classifier. SRCH adds a beneficial spatial bias to the sampling by collecting statistics through a 3x3 cell window from "homogenous" fields as measured by an allowable range of standard deviations in each channel. This process served to use the broader, more abundant land cover types for training statistics to describe the area surface cover spectrally. A maximum likelihood algorithm was then used to map the multiband spectral response of each area into one of the statistically defined spectral groups. In this case, the areas represented by each spectral group were assigned to one of three regional land cover categories through spatial, spectral, and statistical interpretation. The resulting land cover classifications at each date consisted of either forest, agriculture, or water in the Louisiana area; either agriculture, range, or water in the Kansas area; and either agriculture, range, or urban in the Arizona area. Each cell area was compared for a descriptive change, and the change/no change status was recorded for each location. Amounts and types of change measured with this method are shown in Table 2 for the three test areas.

Spectral change pattern analysis (SCPA) requires co-registration initially so that clusters can be developed from equivalent ground areas from coresponding bands from each date. The bitemporal statistical signatures that are defined represent ground areas with similar spectral characteristics between dates which can be identified as patterns of change. For these tests visible and near infrared band pairs (MSS bands 5 and 7 from each date) were used as the source for signature development. The unsupervised training procedure, SRCH, was used to define statistical boundaries of classes of similar spectral response for both dates. The spectral variability measures of each of these bands were mapped by their probability densities into these bitemporal composite classes via a maximum likelihood algorithm. The resulting classification map represented areas that have similar spectral response characteristics of both dates. Each of these classes which represent both dates were temporally sliced to show either equal response in both dates (no change) or a change in spectral response between dates. Change classes were identified, and land cover conditions were determined by examining the spectral means of each class at both points in time, plotting the signature means of each band pair at each date, and analyzing the resulting change image map. The amounts and types of change that were identified by this method are shown for each test area in Table 2.

Table 2. Area of Land Cover Conversions by Method and Test Area.

		Method of Determination					
		PCC		SCPA		AEROPHOTOS	
TEST AREA	TYPE OF CHANGE	Amount (km2)	% of Test Area	Amount (km2)	% of Test Area	Amount (km2)	% of Test Area
LA	F-A	686.6	13.0	625.7	11.9	538.2	10.2
KS	R-A	261.5	21.4	205.2	16.8	232.5	19.06
AZ	A-U R-A R-U	158.9	13.4	100.8	8.5	28.6	2.4

Types of Change:
F-A = Forest to Agriculture
A-U = Agriculture to Urban
R-A = Range to Agriculture
R-U = Range to Urban

Results

Rather than only quantify and identify the types of change measured with these methods, it was desirable to evaluate how accurately their distribution was mapped relative to another data source. The results of each of these methods were spatially integrated with its photo-derived counterpart in the geographic data base file for each test area. A simple arithmetic operation between corresponding cell locations in each channel produced a reference output map detailing how each change map spatially coincided with the verification map. The outcome of this operation identified (1) areas that both data sources agreed there was not change, (2) areas they agreed there was change, (3) areas of change identified with MSS only (commission, false positive, disagreement, and (4) areas of change identified with aerophotos only (omission, false negative, disagreement). These results are shown in Table 3 in terms of their respective percentages of the total area of each test site.

Table 3. Agreement of MSS-Derived with Photo-Derived Land Cover Change Areas.

	Technique	Agreement % of Total Area			Disagreement % of Total Area	
		NO CHANGE	CHANGE	TOTAL	COMMISSION	OMISSION
LA	PCC	85.8	9.1	94.9	3.9	1.1
	SCPA	87.0	9.1	96.1	2.8	1.1
KS	PCC	76.3	16.7	93.0	4.7	2.3
	SCPA	79.5	15.4	94.9	1.4	3.7
AZ	PCC	85.2	0.7	85.9	12.6	1.5
	SCPA	89.9	0.8	90.7	7.7	1.6

REFERENCES

Burns, G.S. 1983, Land Cover Change Monitoring Within the East Central Louisiana Study Site -- A Case for Large Area Surveys with Landsat Multispectral Scanner Data, 36 pgs., Report #E84-10031, NASA-TM-85469, NSTL/ERL-221 NTIS, Springfield, VA.

Jensen, J.R. 1981, Urban Change Detection Mapping Using Landsat Digital Data: The American Cartographer, Vol. 8, No. 2, pp. 127-147.

Weismiller, R. A., et al. 1977, Change Detection in Coastal Zone Environments: Photogrammetric Engineering and Remote Sensing, Vol. 43, No. 12, Dec. 1977, pp. 1533-1539.

VEGETATION DAMAGE ASSESSMENT

Moderator: P. Murtha

INTERPRETATION OF LARGE-SCALE COLOR-IR PHOTOGRAPHS FOR BARK BEETLE INCIPIENT ATTACK DETECTION

Peter A. Murtha
Faculty of Forestry
University of British Columbia
Vancouver, B.C. Canada V6T 1W5

ABSTRACT

Recent studies into the operational application of large-scale (1:1000 - 1:2000) color-infrared (CIR) aerial photographs for the detection of early stages of bark beetle attack have led to the description of the Christmas tree effect" for successfully attacked trees. The description is effective for both spruce beetle [*Dendroctonus rufipennis* Kby.] attacked Engelmann spruce [*Picea engelmannii* Perry] and mountain pine beetle [*Dendroctonus ponderosae* Hopk.] on lodgepole pine [*Pinus contorta* Dougl.]. Healthy or unsuccesfully attacked trees present a variegated magenta pattern on CIR photos which is caused by natural reflectance differences among foliage age classes. When a tree has been successfuly attacked (killed), one of the first foliage symptoms is the loss of the foliage age class spectral reflectance differences, even though the foliage still remains visaully green. Similar to cut Christmas trees indoors, all foliage on the attacked tree becomes similar in reflectance patterns and the natural variegations are lost. Thus there is a definite but subtle way in which green attacked trees can be photo interpreted.

Key words : Air photo interpretation; spruce beetle; mountain pine beetle; color-infrared; vegetation damage.

INTRODUCTION

The remote detection of green-foliaged, bark beetle successfully-attacked (killed) conifers has eluded foresters, entomologist, and remote sensors for decades (Gimbarzeusky, 1984; Henninger and Hildebrandt, 1980). Remote detection methods have primarily used aerial sketch-mapping, and have concentrated on locating trees with faded foliage, red crowns or defoliated trees (Harris *et al*., 1983). Outbreaks have been considered difficult to detect since the foliage does not fade or turn color until a year after attack (Furniss and Carolin, 1977). Aerial sketch mapping cannot be used to diagnose new attacks. Ground probes are used to locate and identify new attack when trees still maintain a full compliment of green foliage through detection of brown boring dust around the base of the tree for spruce beetle (*Dendroctonus rufipennis* Kby.), or pitch tubes for mountain pine beetle (*Dendroctonus ponderosae*, Hopk.), or by removing bark to search for feeding galleries for both. Early detection of beetle attack would make forest management and beetle control more efficient. Early detection means finding green or current attack [successfully attacked and killed tress] as soon as possible after attack and before the foliage fades, turns red brown or drops off. The purpose of this paper is to describe a photo interpretation technique that can be used to identify green attack. Two case studies are used to support the interpretations through independently collected ground reference data.

BACKGROUND

Bark Beetles : The Entomological View

The life cycles of both the spruce beetle and mountain pine beetle have been well documented (Furniss and Carolin, 1977; Johnson and Lyon, 1976), and the entomological viewpoint of the impacts on conifers have been well described (Davidson and Prentice, 1967). Survey techniques have been developed based on these impact impressions. Safranyik *et al.* (1982) described the mountain pine beetle (MPB) impact as follows:

> "Green trees are attacked by the mountain pine beetle in mid-summer. The first noticeable foliage color change usually occurs in spring of the following year. Occassionally, following an early attack and a hot, dry summer, crowns will begin to fade in the fall. The foliage changes from green to yellow, to yellow brown, to red brown, and then drops off the tree. In June and July of the year after attack, trees are usually yellow green or yellow brown."

They follow this foliage description changes with the statement that:

> "This change in foliage permits detection from the air, lookout towers or other vantage points. However, trees killed by any agent, including beetles, usually show the same sequence of color changes and therefore a ground survey must be done."

According to Furniss and Carolin (1977), after a spruce has been attacked by the spruce beetle (SB) and dies, the foliage fades during the summer, and/or through the following winter. [It does not turn red brown as do lodgepole pine (*Pinus contorta* Dougl.) attacked by mountain pine beetle]. The foliage first falls from the branch ends and the outer portions of the crown since it is exposed to the elements and is more readily knocked off. Defoliated trees have been called "grey" attack.

Wood and van Sickle (1983), and Wood *et al.* (1985) categorized both mountain pine beetle and spruce beetle attack into the following five categories for insect survey and report summary purposes:

1. Healthy - no beetle attack;
2. Current - attack presumed successful in the year of the survey; the tree foliage is still green. This stage is sometimes called green attack; determined on the ground;
3. Red - for MPB -- successful attack occurrred in the year prior to the survey, and
   - for SB -- the attack occurred two years prior to

survey
- for both insects, the foliage has turned red brown, but this situation is rare for spruce beetle attacked spruce;

4. Grey - the trees are defoliated and were successfully attacked at least two years prior to the survey;
5. Partial or Strip Attack - Attack is not considered successful, the trees display visually green foliage, and are frequently re-attacked the following year. Strip attack is determined on the ground by removing some of the bark from the tree trunk.

A final category is also used by the ground survey crew to describe a tree which successfully overcomes the attack by the beetles by overwhelming it with excessive production of resins and pitch. When this happens, the beetle is described as "pitched out" and the attack is thought to be unsuccessful. The red and grey categories have been mapped by aerial sketch-mapping (Harris et al., 1983) or by photo interpretation (Klein, 1982), but the other categories (healthy, current or green, partial or strip, and pitch-out) have required data from ground surveys (Safranyik et al., 1982; Wood and van Sickle, 1983).

Bark Beetles : The Air Photo View and Healthy Trees

It is clearly evident from insect survey reports and research papers on photo interpretation of bark beetle attacks that the early stages of attack (which includes current, green, strip attack and pitch out) produce no obvious visual symptoms of attack, otherwise they would have most certainly been reported in the literature (Gimbarzevsky, 1984). Puritch (1981) after an analysis of the pertinent literature reported that there were many unsuccessful attempts in the 1960's and 1970's to detect the early stages of bark beetle attack, and concluded "...bark beetles...do not cause immediate changes to the appearance of the tree and consequently remain undetected through early infestation stages." The time period for the early infestation stages applies to that time period between the initial successful beetle attack and visual appearance of obviously faded foliage, which generally occurs almost a year later in the spring following the mid-summer attack. These faded crowns have been nick-named "faders" and generally preceed the red brown foliage phase of mountain pine beetle attack or loss of foliage after spruce beetle attack.

However, it is perceived by this author that one of the major problems in the application of aerial photographic and interpretation techniques to the evaluation of bark beetle impacts is that the small photographic scales (1:6000 and smaller) have permitted interpretation of only the more obvious faded, red, and grey attack categories. Spencer (1984) reported that larger scales (between 1:4000 and 1:2000) were requireed to interpret dieback in Australian eucalypt forests, that scales between 1:3000 and 1:2500 were required to make interpretations on medium-sized branches, and that a scale of 1:500 was required to assess leaf and twig patterns. Murtha (1983) noted that a 1:4000 photo-scale sufficed for counts of dead trees, but 1:1000 photos were required for the photo interpretation of subtle strain symptoms which included thin crowns (premature loss of older foliage), scattered clumps of yellowed or red brown foliage, and darker or lighter magenta hue variations of foliage in a crown on color infrared (CIR) photos. Because of these perceptions it was decided to formulate a scenario which would describe the crown condition, and sequence of symptoms possible for a tree successfully attacked (killed) by bark beetles.

The first part needed the description of a healthy conifer. (If a healthy tree is not described, how is an attacked tree to be described?) On CIR photo, the average magenta-hued trees were considered to be normal. Darker (IIIOa) and lighter (IIIOb) magenta-hued variations (Murtha and McLean, 1981) were shown to be vigor variations of the normal tree. [The coding for the darker and lighter variations is after a key to forest "damage types" presented by Murtha [1972, 1978]. These darker and lighter variations are noted best at large scales (1:1000) and even at 1:4000, misinterpretations can be expected (Murtha 1983).

After the examination of 624 ground-checked spruce (Murtha 1985) and over 200 ground-checked pine (Hobbs and Murtha 1984) followed on multitemporal CIR photographs from 1981 to 1984, a second obvious pattern for a healthy tree has emerged. On both normal (healthy) non-beetle-attacked spruce and pine, there are hue variations along the branch. I have called this pattern a variegated magenta (VM) pattern for CIR photos, variegated green (VG) for normal color (NC) photos.

Sometimes the variegated patterns are seen associated with generally darker (IIIOa) or lighter (IIIOb) magenta-hued crowns. Hence for healthy or unsuccessfully attacked trees there are three "damage types" designations, VM, VM/IIIOa, and VM/IIIOb (Table 1). For the CIR large-scale aerial photographs, the unattacked trees displayed a range of tones within the magenta hue along each branch. Careful observation is needed to preclude possible confusion with contained shadows in the tree crown. The foliage on the branch tips is a light magenta while the inner foliage is a darker magenta. On some trees, as many as 3 steps of magenta tones can be seen (Figure 1). Some conifers maintain several years worth of foliage, and each year as the foliage grows older, it becomes darker, or less reflective at all wavelengths. Steiner and Gutermann (1966) documented this fact for pine and spruce (Figure 2). The different age classes of foliage produce the variegations along each of the branches. The visual impact is a salt and pepper appearance to the foliage of the tree crown.

Because of the natural stress of poor sites, some trees suffer premature loss of older foliage, -- that is, they may only maintain two live foliage age classes. The oldest age class dies in the third year, turns red brown and eventually drops off (Figure 1). This fact causes another pattern:

Figure 1. Because of spectral reflectance differences among foliage age classes, on CIR photo, different hue tones are evident in the foliage along a branch of an unattacked tree.

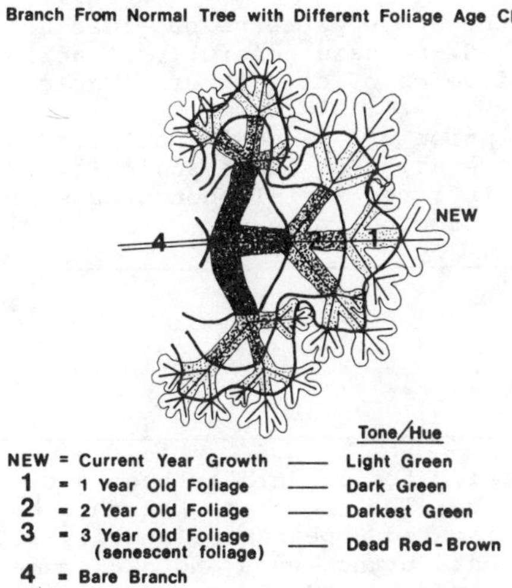

Figure 2. Steiner and Gutermann (1966) measured spectral reflectance differences between new and old foliage for both pine and spruce. As foliage ages, spectral reflectance drops at all wavelengths. [Diagram based on data presented by Steiner and Gutermann (1966)].

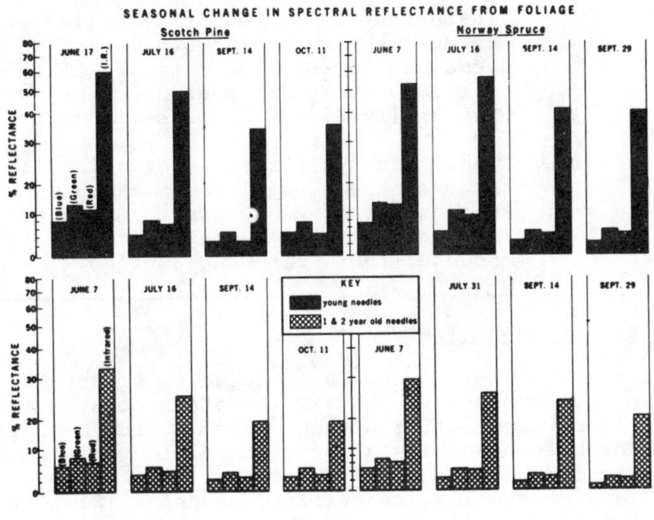

Foliage age class	Tone and Hue
current	- light magenta
2nd year foliage	- medium dark magenta
3rd year foliage (dead and red brown)	- straw yellow

Recognition of the above patterns provides another clue to a normal, unattacked, or unsuccessfully attacked tree, which is under the influence of site stress (Table 1).

Table 1. Description of "damage types" for the photo interpretation of large-scale (1:2,000) color-infrared aerial photographs of unattacked spruce or pine.

Photo Interpretation Category (Damage Type)	Description
VM	"Variegated magenta"; these are <u>unattacked</u> trees. The crown and branches have a mottled variegated appearance, usually with light-pink magenta branch ends and darker magenta toward the tree trunk. Some very light straw hues may be present on the inner part of the crown (Figure 1). The overall crown hue may be lighter or darker than other trees, but note the tonal variation. Do not confuse contained crown shadows as part of the tonal variation which occurs along the illuminated branch and is caused by spectral reflectance differences from foliage in different age classes.
VM/IIIOa	These are "VM" trees as described above, however the overall crown magenta hue is somewhat darker than normal. [This is a similar pattern as that described by Murtha and McLean (1981) for Douglas-fir].
VM/IIIOb	These are "VM" trees as described above however the overall crown magenta hue is light magenta to somewhat pink. [This too is a similar pattern as that described by Murtha and McLean (1981) for Douglas-fir].

<u>Beetle Attacked Trees</u>

For bark beetle, successfully-attacked trees, two patterns have been observed. The first pattern occurs within a couple of weeks after attack. When a tree has been killed by the beetle and its associated <u>Ceratocystis</u> sp. fungus, the foliage dies, dries, chlorophyll breaks down and the foliage eventually losses its greenness. Similar to a cut Christmas tree in a house, the foliage along each branch becomes the same green tone since spectal reflectance differences among the different foliage age-classes are lost

(Figure 3). The variegated tonal pattern is lost, and the absence of this pattern can be interpreted on large scale CIR photographs. The non-variegated pattern is most evident in the upper tree crown and is virtually impossible to see from the ground. When compared to unattacked trees the foliage usually appears darker, and somewhat tannish. These attacked trees have been designated as IIIOa-T (darker magenta-tannish)(Table 2). The tan hue is not present in all cases, but the lack of the variegated pattern is essential to observe.

Figure 3. Bark beetle killed (successful) green attack causes all foliage to attain perceptually the same tone within weeks after attack. The tonal variations (Figure 1) have been lost.

Bark Beetle Successful Green Attack (Tree has been killed)

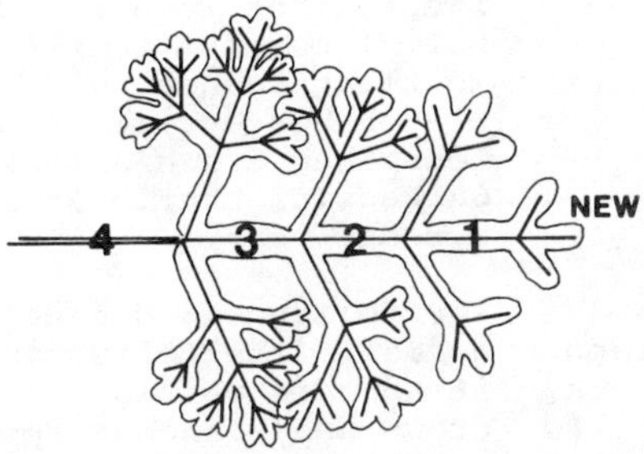

NEW = Foliage Growth in Year of Attack

New, 1, 2, and 3 year old foliage have the same tone and hue.

Normal senescence of 3 year old foliage does not occur.

Given sufficient time and dry weather conditions successfully attacked trees start to glow like a "halo". The loss of greenness, the dryness, and the lack of spectral variation among foliage age-classes produce light-toned, highly reflective, mono-hued tree crowns which are highly reflective. These successfully attacked trees have been designated IIIOb-H (lighter magenta-halo) (Table 2). Even though these trees stand out with marked contrast from other unattacked trees, they can be misjudged on the ground without destructive tree sampling. Murtha (1985) reported that during a study of spruce beetle attack, and using the damage types as described in Tables 1 and 2, photo interpretation had suggested 68 (93.1%) of 73 trees had been successfully attacked and therefore killed. This conclusion was not corroborated by the 1983 field data which indicated that the beetle had been successfully "pitched-out" and the trees would survive. The photo interpretation was confirmed a year later during the following field season when the trees were found dead.

The variegated magenta theory was also applied to the mountain pine beetle attached stand used in the Hobbs and Murtha (1984) study. On the 1:2000 CIR photos, 188 pine were re-examined by photo interpretation. The results are presented in Table 3.

Table 2. Description of damage types for the photo interpretation of large-scale (1:2,000) color-infrared aerial photographs of bark beetle attacked pine or spruce.

Photo Interpretation Category (Damage Types) [with beetle attack status]	Description
IIIOa-T [Current, green or Strip Attack]	The foliage is decidedly darker than normal magenta to tan-magenta. Tonal mottling is not evident. Foliage on each branch has a similar hue and tone. Contained crown shadows will be present. Successful beetle attack occurred in current year.
IIIOb-H [Current, green or Strip attack]	The foliage is decidedly lighter than normal. The foliage may be so light that it appears whitish, or off-white, and the crown and branches appear to have a "Halo" appearance. Tonal mottling is not evident. [To the ground observer these trees still display visually green foliage, but it appears totally different on air photos.] Successful beetle attack occurred in current or previous year.
IID [Grey attack] Spruce beetle attack	The crown has lost dead foliage from branch tips, and some usually light toned foliage remains. Various degrees of defoliation may be evident but the inner branches will be obscured because of the residual foliage. This is the early part of the grey-phase classification of beetle attacked trees. Beetle attack is most likely two years old.
IIIG [Red attack] Pine	The foliage has turned red brown and is seen as yellow or straw yellow on CIR photographs. MPB attack occurred in the previous year.
IB [Grey (dead)]	The crown is defoliated, the twigs and branches are grey. The tree has been dead for more than 2 years.

For the unattacked trees, 72/95 (75.8%) displayed a variegated magenta pattern in the foliage. For the green attack trees 63/93 (67.3%) fell into the IIIOa-T or IIIOb-H categories. Unattacked trees which fell into either of these

these categories were commission errors, in that these trees displayed symptoms common to attacked trees, but the ground data said no attack. For the total population commission errors amounted to 12.2% (23/188).

Table 3. Photo interpretation vs. beetle attack status from ground data of a mountain pine beetle infested pine stand. Aerial photography Aug 18/81; ground check May 25/82. Numbers in Table are from interpretation of the air photos.

Photo Interpretation Damage Type		Beetle Attack Status				Total Trees
		Unattacked		Green Attack[1]		
		No.	%	No.	%	
Unattacked	VM	53	72/188=38%	25		78
	VM/IIIOa	2	or	2	30/188=15.9%	4
	VM/IIIOb	17	72/95=53.8%	3		20
Attacked	IIIOa-T	9	23/188=12.2%	26	63/188=33.5%	35
	IIIOb-H	14		37		51
		95	50.5%	93	49.5%	188

1 Green attack was verified on May 25, 1982 through examination for pitch tubes and boring dust. No bark was removed, but by Sept. 1982 the trees had red crowns.

Trees which were described as having a variegated magenta pattern, and as being green attack, are misses or omission errors. For the total population omission errors were 15.9% (30/188). [Omission errors can be expected to be high since the time required for crown foliar symptoms to develop after successful attack is generally unknown, and reported to be as long as one year (Safranyik *et al*. 1982)].

Management Implications

The management implications are obviously open to discussion and debate; but it is suggested here, that with proper application large-scale photos and photo interpretation can be used to help manage beetle infested stands. Firstly, they can be used as a sampling tool to monitor forest stands. The correct sampling procedures need to be developed. The large scale-photos should be used in conjunction with smaller scale, overview photos which would clearly show red, grey and old attack. Once suspect areas have been identified, ground probes can be used to collect further beetle population data and large-scale photos can be obtained and used to assess tree condition without destructive sampling. This is the second and most important implication in that tree condition and hence successful beetle attack may be easier to interpret from the large scale photos than it is (without destructive tree sampling) on the ground (Murtha 1985). As forest management progresses into single-tree "silvicultural" management for beetle infested stands, assessment of condition becomes more critical. With large sums of money being spent to remove single infested trees, it seems that interpretation results of large-scale CIR photo should become part of the management decision.

Conclusion

Recent results from the photo interpretation of large scale (+1:2000) color-infrared photos for early stages of bark beetle attack have indicated that:

1. A variegated magenta pattern indicative of tonal variation on the foliage means a normal or unattacked tree;
2. Successfully attacked trees have non-variegated mono-toned crowns which at first are dark magenta to tannish (IIIOa-T);
3. Foliage on successfully attacked trees dries out and becomes highly near-infrared reflectives while it is still visually green. Such tree crowns grow like a halo (IIIOb-H) and stand out in marked contrast to unattacked trees.

Acknowledgements

The author would like to thank his funding sources, the British Columbia Science Council for the spruce beetle study, and the Natural Sciences and Engineering Research Council of Canada for the mountain pine beetle study.

LITERATURE CITED

Davidson, A.G. and R.M. Prentice. 1967. Important forest insects and diseases of mutual concern to Canada, the United States and Mexico. Can. Dept. Forestry and Rural Develop., Queens Printer, Ottawa, Ont., 248 p.

Furniss, R.L. and V.M. Carolin. 1977. Western forest insects. USDA. For. Serv. Miscell. Public #1339.

Gimbarzevsky, P. 1984. Remote sensing in forest damage detection and appraised: Selected annotated bibliography. Can. For. Serv.; Pacific For. Res. Centre, Victoria, B.C., Inform. Rep. BC-X-253, 55 p.

Harris, J.W.E., A.F. Dawson and R.G. Brown. 1983. Evaluation of mountain pine beetle damage using aerial photography taken with a hand-held 70-mm camera, Gold Bridge-Clinton, B.C., 1981. Can. Forestry Serv., Pac. Forest Res. Centre, Victoria, B.C. Inform. Rep. BC-X-245, 15 p.

Henninger, J. and G. Hildebrandt. 1980. Bibliography of publications on damage assessment in forestry and agriculture by remote sensing techniques. 2 ed.Abt. Luftbildmessung und-interpretation, Univ. Freiburg, 280 p.

Hobbs, A.J. and P.A. Murtha. 1984. Visual interpretation of four scales of aerial photography for early detection of mountain pine beetle infestation. p. 433-444. In "Renewable Resources Management, Applications of Remote Sensing". Amer. Soc. of Photogrammetry, Full Church VA. 774 p.

Johnson, W.T. and H.H. Lyon (ed.) 1976. Insects that feed on trees and shrubs. Cornell Univ. Press, Ithaca, N.Y., 463 p.

Klein, W.H. 1982. Estimating bark beetle-killed lodgepole pine with high attitude panoramic photography. Photogram. Eng. and Remote Sens. 48(5):733-737.

Murtha, P.A. 1972. A guide to air photo interpretation of forest damage in Canada. Can. Forest. Serv. Public #1292, 63 p.

Murtha, P.A. 1978. Remote sensing and vegetation damage: a theory for detection and assessment. Photogramm. Eng. and Remote Sens. 44(9):1147-1158.

Murtha, P.A. 1983. Some air-photo scale effects on Douglas-fir damage type interpretation. Photogramm. Eng. and Remote Sens. 49(3):327-335.

Murtha, P.A. 1985. Photo interpretation of spruce beetle-attacked spruce. Can. Jour. Remote Sens. (In Press - July, 1985).

Murtha, P.A., and J.A. McLean. 1981. Extravisual damage detection? Defining the standard normal tree. Photogramm. Eng. and Remote Sens. 47(4):515-522.

Puritch, G.S. 1981. Nonvisual remote sensing of trees affected by stress. Can. For. Serv., Pac. For. Res. Centre, Victoria, B.C., Forest Tech. Rep. #30. 38 p.

Safranyik, L., D.M. Shrimpton, and H.S. Whitney. 1982. Management of lodgepole pine to reduce losses from the mountain pine beetle. Can. For. Serv., Pac. Forest Res. Centre, Victoria, B.C., Forest. Tech. Rep. No. 1, 25 p.

Spencer, R.D. 1984. Detecting dieback in native eucalypt forest using aerial photographs. Austral. Forest. 47(4):243-249.

Wood, C.S. and G.A. van Sickle. 1983. Forest insect and disease conditions, British Columbia and Yukon. 1982. Can. For. Serv., Pac. For. Res. Center, Victoria, B.C., BC-X-239, 31 p.

Wood, C.S., G.A. Van Sickle, and T.L. Shore. 1985. Forest insect and disease conditions, British Columbia and Yukon. Can. For. Serv., Pac. Forest Res. Centre, Victoria, B.C., BC-X-259, 32 p.

THE USE OF LANDSAT MSS DIGITAL DATA TO DETECT MORTALITY OF LODGEPOLE PINE CAUSED BY THE MOUNTAIN PINE BEETLE

John A. Brockhaus, Heather M. Cheshire, Siamak Khorram, and William H. Klein

School of Forest Resources
North Carolina State University
Raleigh, NC 27695-8001

Forest Service, USDA
Methods Application Group
Denver, CO

ABSTRACT

In 1981 the United States Forest Service (USFS) Methods Applications Group initiated a pilot project to determine the utility of Landsat multispectral (MSS) digital data for monitoring the intensity and spatial extent of lodgepole pine mortality caused by the mountain pine beetle, *Dendroctunus ponderosa*. Portions of the Ashton and Park Island Ranger Districts within the Targhee National Forest, located in eastern Idaho, were selected as test sites for this project. Average reflectance values in each MSS waveband were obtained for twenty-nine sample plots which represented a range of lodgepole pine mortality conditions. MSS digital data were clustered for each plot generating a set of spectral signatures for three mortality classes which were based upon percent dead merchantable timber volume: (1) 0-34.5%; (2) 35-66.5%; and, (3) 67-100%. Signatures were applied to the MSS digital data for the study site through a table look-up, minimum distance multispectral classifier which classified each pixel within the study area into one of the three mortality classes. Results of this classification process seem to correspond well to ground conditions described by USFS personnel. Analysis of classification of the 29 sample plots indicated that 22 of the 29 plots (76%) were correctly classified into their original mortality class.

INTRODUCTION

During the past twenty years the mountain pine beetle, *Dendroctonus ponderosa*, has caused widespread mortality of lodgepole pine forests of the northern Rockies. Particularly damaging outbreaks have occurred in portions of northern Utah, western Wyoming, eastern Idaho, western Montana and Canada. In Utah, Wyoming, and Idaho the outbreaks have receded but in northern Montana and Canada, extremely damaging outbreaks persist (Kohler et al. 1981). In the wake of these outbreaks are several million acres of severly depleted lodgepole pine forests, many areas accruing losses of more than 100 trees per acre and upwards of 50 percent of merchantable volume (Klein 1976).

The long lasting effects of this intense and widespread damage on management and the future of these depleted forests is uncertain. The forest manager must not only make decisions affecting the utilization of this dead lodgepole but must also look into the future to the time when these stands again become merchantable and/or beetle fodder (Wellner 1978). The beetles depredation has made existing management plans obsolete, causing them to be updated or completely revised, and resulting in new plans of action (Carter 1978). These plans are based on stand inventory data, and the location, extent and intensity of beetle-killed lodgepole pine is a critical part of that data base. Currently there are methods available for inventorying annual mountain pine beetle induced mortality of lodgepole pine using multistage sampling techniques (Klein 1979; Bennet et al. 1979), but no method currently exists for efficient monitoring of a large forested region.

Landsat multispectral imagery has the advantage over conventional aerial photographic techniques in that it produces almost simultaneous imagery of very large areas, provides repetitive coverage, and can be digitally processed and analyzed. The most serious shortcoming of MSS data has been its relatively poor resolution, slightly more than one acre. However, considering the magnitude of dead lodgepole pine throughout the West and considering advances made in Landsat digital interpretation technology, it shows promise as being an ideal inventory system for mountain pine beetle-killed lodgepole pine over large areas. Landsat analysis could be an effective, highly efficient, and relatively inexpensive tool for assessing lodgepole pine mortality and identifying problem areas.

Unfortunately, in forestry applications, Landsat utilized data has had more failures than successes, and most, if not all of these failures, resulted from undue expectations, i.e., extending its potential beyond its practical capabilities (Sharp 1979). Landsat is a passive system and its multiband sensors are totally dependent on terrestrial solar reflection; its utility being that each of the myriad objects on the earth surface has its own unique reflectance. For example, very large objects having uniform surfaces, such as water or large tracts of timber, reflect uniformly and can be consistently separated from other objects having different or variable reflectance. Lodgepole pine is a relatively uniform cover type and the intensive bark beetle caused mortality within the stands should afford good reflectance contrast from a healthy stand, and may have a unique reflectance signature.

METHODOLOGY

Study Area

Portions of the Ashton and Park Island Ranger Districts within the Targhee National Forest in eastern Idaho were selected as test sites for this study. The area sustained a damaging mountain pine beetle outbreak during the 1960's which had collapsed by 1971. In 1971, ground truth plots were established as part of a long term study to measure total tree mortality and to monitor the depleted forests recuperation rate (Klein 1976).

Field Data Collection and Analysis

USFS personnel identified 53 lodgepole pine stands to be used as sample plots for quantifying the degree of mortality caused by the mountain pine beetle. These stands were selected so as to represent a wide range of mortality conditions. Examination of the plot data indicated that the field data for three plots had been improperly coded and were therefore not used for further analysis. Twenty-one of the remaining 50 plots were found to lie outside of the boundaries of the selected Landsat scene. A decision was made to utilize the existing Landsat scene along with the remaining 29 plots which occurred within that scene rather than purchasing an additional scene which included the other 21 plots. Data collected at each stand included the following: (1) stand size in acres; (2) number of trees within the stand; (3) number of dead trees within the stand; (4) merchantable timber volume within the stand; and, (5) dead merchantable timber volume within the stand (Table 1). From this data two additional variables were generated; percent dead trees, and percent dead merchantable timber volume. Analysis of the available data indicated that the variable percent dead merchantable timber volume provided the best results when attempting to group the 29 sample plots into three or more mortality classes. Investigation of histogram plots of the percent dead merchantable timber volume data lead to the selection of three mortality categories: (1) 0-34.5% dead merchantable volume; (2) 35-66.5% dead merchantable volume; and, (3) 67%+ dead merchantable volume. Data for the remaining variables did not provide the separation between mortality categories which was obtained when using the percent dead merchantable timber volume data. Each sample lodgepole pine stand was then assigned to one of the three mortality classes.

Remotely Sensed Data Acquisition and Analysis

A Landsat MSS digital data set obtained during July, 1980 had been selected and purchased by the USFS. As discussed earlier this scene did not provide full coverage of the study area. Within this scene a 512 x 512 window which encompassed the remaining portion of the study area was identified and stored for further analysis. Data for cover types which were not of interest to this project, such as large agricultural blocks and other non-forest cover types, were masked out of the 512 x 512 window so as to restrict the data analysis to those areas representing lodgepole pine stands as much as possible.

Visual location of the sample plots was found to be difficult due to the relatively small size of the sample plots; average sample plot size was approximately 70 acres which represents less than 70 Landsat pixels. Since it was difficult to accurately locate plots of this size

Plot	Acres	Total Trees	Dead Trees	% Dead Trees	Total Volume	Dead Volume	% Dead Volume
1	68	431	156	36.2	9448	3341	35.4
2	72	1736	415	23.9	6017	2006	33.3
3	85	1019	383	37.6	6847	4147	60.6
4	37	1372	730	53.2	11484	4524	39.4
5	78	958	608	63.5	8188	4450	54.4
6	140	1721	27	1.6	7277	4682	64.3
7	53	1206	43	3.6	8383	1429	17.0
8	66	655	18	2.7	7898	6578	83.3
9	50	601	35	5.8	11252	6797	60.4
10	61	944	237	25.1	5938	3810	64.2
11	90	939	75	8.0	8413	2783	33.1
12	51	488	28	5.8	8494	6680	78.6
13	62	1658	49	2.9	5500	3095	56.3
14	72	361	28	7.8	10421	4646	44.6
15	77	1472	27	1.8	4078	2086	51.2
16	212	800	157	19.6	17129	12739	74.4
17	138	1159	49	4.2	9101	4558	50.1
18	158	465	68	14.6	3666	2184	59.6
19	180	1881	21	1.1	7245	3968	54.8
20	79	1298	112	8.6	5118	3700	72.3
21	88	1500	40	2.7	11840	6891	58.2
22	131	609	59	9.7	8901	6028	67.7
23	70	2786	145	5.2	9712	5748	59.2
24	97	2178	120	5.5	5460	2760	50.5
25	82	906	57	6.3	7650	5149	67.3
26	116	2843	261	9.2	15578	8078	51.8
27	73	3450	96	2.8	10288	8028	78.0
28	85	2152	104	4.8	6186	3293	53.2
29	109	635	36	5.7	4076	2927	71.8

Table 1. Field data collected for sample plots.

visually, an alternative technique was developed which related Landsat image x,y coordinates to Universal Transverse Mercator (UTM) easting and northing coorinates. Location of sample plots were delineated on United States Geological Survey (USGS) 7.5' orthophoto quadrangles, and UTM easting and northing coordinates were then calculated for the center of each of the 29 sample plots. A set of 34 control points were selected and used in developing calibration equations relating Landsat image x,y coordinates to UTM easting and northing coordinates. These equations were then used to locate the sample plots within the Landsat digital data by entering the UTM easting and northing coordinates for each sample plot center which were then transformed into Landsat image x,y coordinates. Sample plots were then located by navigating through the image until the desired image x,y coordinate was found.

As a sample plot was located, a block of data encompassing the sample plot and surrounding features which were readily identifiable on aerial photographs or the USGS orthophoto quadrangles would be extracted for further analysis. This data was then clustered, and the cluster means and standard deviations stored in a statistics file. Results of the clustering process were then displayed and the spatial distribution of the clusters compared to the patterns which could be observed on the photographs and/or the USGS orthophoto quadrangles. By comparing the clustered image to known ground conditions the identity of each cluster was determined. Only those clusters found to represent a lodgepole pine mortality class were kept, data clusters found to represent non-mortality cover types were deleted from the statistics file. Each sample plot was located and analyzed in this manner generating a statistics file comprised of 44 data clusters representing the three mortality classes. Spectral signature plots of the cluster means (Figure 1 a, b, c) show that there is little or no difference between the signatures within the visible portion of the electromagnetic spectrum. However, the two near infrared wavebands provide significant separation between spectral signatures representing the three lodgepole pine mortality classes. Sample data for cover types other than lodgepole pine mortality classes, such as; grass, hardwoods, brush, rivers, lakes, bare soil, and clouds, were clustered and added to the mortality statistics file. These spectral signatures then were applied to the study area digital data set through a table-look up, minimum distance multispectral classifier classifying each pixel within the study area into one of the lodgepole pine mortality or non-mortality classes.

RESULTS AND DISCUSSION

The determination of the reliability of the classification of an image is an integral part of any remote sensing study. Ideally a verification data set, independant of the sample data used to classify an image, should be selected to accomplish this task. As with many projects of this nature which have been conducted in the past, it was not possible to select an independant verification data set due to budget and time constraints. In an attempt to determine the reliability of the image classification in the absense of an independant verification data set it was decided to use the 29 sample plots to verify the results of the classification. By comparing the known mortality class of each sample plot to its predicted class, based upon the classification, a determination of the reliability of the classification was calculated. Twenty-two of the 29 sample plots were classified into the actual mortality class observed in the field, this gives an overall reliability of 76 percent. Analysis of the classification confusion matrix (Table 2) shows that the mis-classifications were never more than one mortality class apart. For example, a sample plot whose actual mortality class was 0-34.5% was never classified as being in the 67%+ mortality class. The majority of classification errors occurred when plots which were in the 35-66.5% mortality class were classified as being in one of the other two categories. The plots which were mis-classified in this manner were 1, 4, 6, and 10. The actual percent dead merchantable timber volume on these plots is near the upper or lower limits of the mortality class in question, therefore these classification errors are understandable. No attempt was made to determine the reliability of the classification of the non lodgepole pine

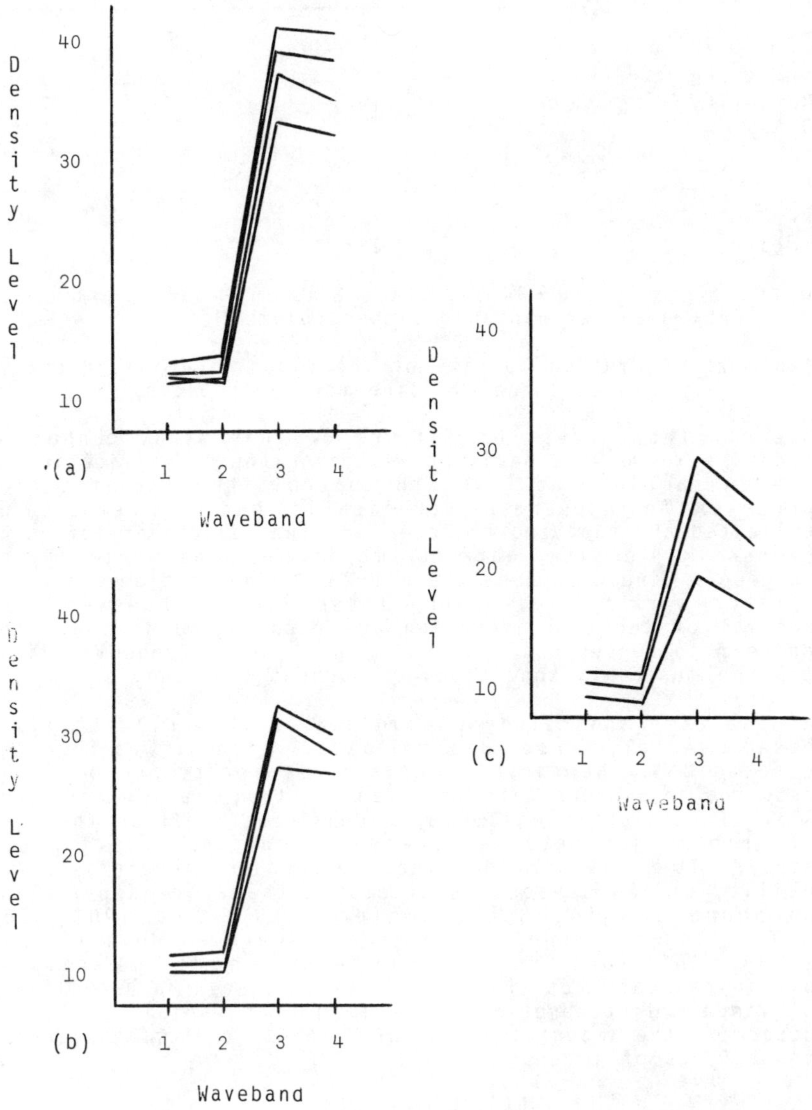

Figure 1. Spectral signatures for lodgepole pine mortality mortality classes: (a) 0 - 34.5% dead merchantable timber volume; (b) 35 - 66.5% dead merchantable timber volume; and, (c) 67%+ dead merchantable timber volume.

mortality classes.

		Actual Mortality Class			
		1	2	3	Total
	1	2	2	-	4
Predicted Mortality Class	2	1	14	2	17
	3	-	2	6	8
Total		3	18	8	29

where class 1 is 0 -34.5%, class 2 is 35 - 66.5%, and class 3 is 66%+ dead merchantable timber volume.

Table 2. Confusion matrix for the classification of the 29 sample lodgepole pine mortality plots.

Clearly the success or failure of this study cannot be judged soley on the basis of analyzing the classification of the 29 sample plots used to control the classification process. An independant verification data set needs to be collected and analyzed to determine the reliability of using Landsat MSS digital data for monitoring the mortality of lodgepole pine caused by the mountain pine beetle before any further work is done along these lines. However, the results of the classification are encouraging in that they do seem to correspond well with what USFS personnel believe are the mortality conditions throughout the study area.

At the time this study was initiated the only satellite based remotely sensed digital data available was from the Landsat MSS. New developments in sensor technology have resulted in sensors with improved spectral, radiometric, and spatial resolution which may significantly affect the type of information obtainable when conducting this type of study. It is recommended that researchers investigate the utility of new sensors, such as the Thematic Mapper (TM), Airborne Imaging Spectrometer (AIS), and SPOT, for monitoring the condition of forests as was done in this study. This work should not be restricted to the analysis of digital data but should also investigate the usefullness of visually interpreting these images at various scales to determine the amount and type of information obtainable from these types of imagery.

ACKNOWLEDGEMENTS

The authors would like to thank the USFS personnel responsible for the collection of the ground truth data and the delineation of plot locations on the aerial photography. The ground truth data provided by these individuals was indispensible to the completion of this project. This research was conducted using the image processing facilities located at the Computer Graphics Center, North Carolina State University, Raleigh, North Carolina. We would also like to thank Mr. Alan Roberts, formerly with the Computer Graphics Center now with the Research Triangle Institute, for the many hours spent developing special purpose software needed to complete this project.

REFERENCES

Bennett, D. D., W. E. Bousfield, M. D. McGregor and K. E. Gibson. 1980. Evaluation of multistage sampling techniques to measure lodgepole pine mortality caused by the mountain pine beetle in Montana, 1979. USDA For. Serv., Northern region, Rep. No. 80-13. 11 pp.

Carter, William W. Jr. 1978. Potential impacts of mountain pine beetle and their mitigation in lodgepole pine forests. Pages 27-6 In Theory and Practice of Mountain Pine Beetle Management in Lodgepole Pine Forests, Wash. State Univ., Pullman, WA. 212 pp.

Klein, William H. 1976. Measuring damage to lodgepole pine caused by the mountain pine beetle. Pages 35-41 In Current Topics in Forest Entomol. Selected papers for the XV Intl Congress of Entomol., Wash. D.C. 174 pp.

Klein, William H. 1979. Measuring mountain pine beetle-caused tree mortality with aerial photography. Pages 132-135 In Proceedings of the Society of American Foresters. Town Meeting Forestry--Issues for the 1980's. 320 pp.

Kohler, Steve, Jerald E. Dewey, Robert L. James, Kenneth E. Gobson and Lawrence E. Stripe. 1981. Forest insect and disease conditions in Montana, 1980. USDA For. Serv., Northern Region, Rep. No. 81-9. 33 pp.

Sharp, James M. 1979. Demonstrating the value of Landsat Data: A case of lowered expectations. Photogramm. Eng. 45:1487-93.

Wellner, Charles A. 1978. Management problems resulting from mountain pine beetles in lodgepole pine forests. Pages 9-15 In Theory and Practice of Mountain Pine Beetle Management in Lodgepole Pine Forests, Wash. State Univ., Pullman, WA. 212 pp.

SURVEYING SPRUCE BUDWORM DEFOLIATION WITH AN AIRBORNE PUSHBROOM SCANNER: A SUMMARY[1]

F.J. Ahern
Canada Centre for Remote Sensing
2464 Sheffield Road
Ottawa, Ontario
K1A 0Y7

W.J. Bennett
INTERA Technologies Limited
785 Carling Avenue, 4th Floor
Ottawa, Ontario
K1S 5H4

E.G. Kettela
Maritime Forest Research Centre
Canadian Forestry Service
P.O. Box 4000, College Hill
Fredericton, New Brunswick
E3B 5P7

INTRODUCTION

Insect damage is one of the most serious causes of forest depletion and growth loss in Canada, contributing to an estimated 88 million m^3 of lost wood volume from Canadian forests (Baskerville, 1980). Informed forest management requires yearly information on insect damage to plan countermeasures and to revise predictions of future wood supply. LANDSAT MSS data have not generally been satisfactory for insect damage assessment. In 1983 the Canada Centre for Remote Sensing carried out an assessment of the ability of airborne multispectral scanners (mechanical and pushbroom) to detect current-year damage to softwood stands by spruce budworm larvae.

In eastern Canada, where forestry is fundamental to the economic health of the region, forest managers are battling one of the worst forest insect problems in Canada, the spruce budworm [*Choristoneura fumiferana* (Clem.)]. Periodic outbreaks of the budworm (three in the 20th century) have resulted in widespread defoliation, leading to growth loss (both radially and in height) and mortality of spruce-fir stands in Ontario, Quebec and the Atlantic provinces.

---

1 Colour images are a key element of this work, providing a basis for the detection and recognition of spruce budworm damage and also for showing the role of spatial and radiometric resolution. Since these proceedings cannot include colour plates, this report is simply a summary of our work. A complete report will be submitted for publication in *Photogrammetric Engineering and Remote Sensing*.

To monitor the activity of the spruce budworm and to plan and assess spray programs, federal and provincial forestry agencies conduct annual damage surveys. Aerial sketch mapping surveys record the extent and severity of current-year defoliation, that is, damage to growth. The survey is carried out during a short period at the end of June/beginning of July, as soon as damage to current-year growth is complete. The damage gives the attacked trees a red-brown appearance.

Although conventional survey methods continue to supply foresters with useful information about the status of the spruce budworm, more reliable surveys related to individual stands could provide better timber forecasts (Erdle, 1985) when used as input to improved growth forecast models (McLean and Erdle, 1984).

This paper summarizes the results of a remote sensing project jointly undertaken by the Canada Centre for Remote Sensing (CCRS) and the New Brunswick Department of Natural Resources (NBDNR). The goal of the study was to investigate the potential of the high resolution MEIS-II pushbroom scanner (McColl et al., 1984) for detecting current-year spruce budworm defoliation.

Detecting current-year defoliation is a difficult remote sensing problem. Since it is only the remains of the damaged foliage that turn red-brown, defoliation is difficult to detect remotely. Secondly, the detectable red-brown stage of damage has a very short biowindow, lasting only two to three weeks. Finally, it is not easy to predict where an attack will occur, making project planning difficult. For these and other reasons related to data acquisition and resolution, LANDSAT data have not been useful for mapping current-year defoliation. Higher resolution airborne multispectral scanner data also proved to be ineffectual until 1981 when Leckie and Gougeon (1981) were successful in digitally classifying four levels of cumulative defoliation (severe, moderate, light, healthy). Leckie and Ostaff report (In preparation) some success in using airborne multispectral scanner data to classify current-year defoliation into four levels in softwood stands but more difficulty in mixedwood stands. The CCRS-NBDNR study reported here was the first aimed at detecting spruce budworm damage by visual analysis of enhanced data.

STUDY AREA

The study site is located north of Fredericton and extends for 25 km along the Miramichi River between Boiestown and Doaktown (NTS map sheet 21J) (figure 1). The terrain is typically hilly and, in some places, relatively steep, except for the floodplain areas along the river. Forest cover in the area is mixed softwood-hardwood.

DATA ACQUISITION

We obtained data over two test areas with a wide variety of sensors and scales (table 1). The flight with the Canadian Forestry Service Cessna 170 provided an initial reconnaissance of the study area and confirmed that the flight lines drawn up for the CCRS aircraft would cover both defoliated areas and areas lacking defoliation. This flight also aided the specification of flight lines for the 70 mm photography.

The areas accessible by road and by foot along the transmission line cut (figure 1) were visited to confirm the levels of defoliation damage which had been observed and photographed from the air. Close-up ground-based photography was also used to document the damage. Quantitative assessment of the degree of defoliation was not attempted because of the difficulty of obtaining a statistically significant sample; rather this was viewed as an initial evaluation of the MEIS-II and Daedalus 1260 scanners using enhanced images for qualitative visual assessment.

Table 1: Sensors

Sensor		Spectral Regions	Platform	Altitude	Data Obtained
35 mm camera 50 mm lens		blue, green, red (natural colour)	Cessna 170	3 000 ft agl*	29 June 83
70 mm camera 80 mm lens		blue, green, red (natural colour)	Cessna 172	4 000 ft agl	10 July 83
RC-10 camera 150 mm lens		blue, green, red (natural colour)	Falcon 50 CCRS aircraft C-GRSD	33 000 ft agl	12 July 83
Daedalus 1260 Multispectral scanner	3	0.45-0.50 μm	Falcon 50	33 000 ft agl	12 July 83
	4	0.50-0.55 μm			
	5	0.55-0.60 μm			
	7	0.63-0.70 μm			
	8	0.68-0.78 μm			
	9	0.77-0.90 μm			
	10	0.87-1.04 μm			
	11	1.55-1.75 μm			
MEIS-II Pushbroom scanner	1	0.43-0.46 μm	Falcon 50	33 000 ft agl	12 July 83
	2	0.51-0.54 μm			
	3	0.58-0.61 μm			
	4	0.66-0.69 μm			
	5	0.84-0.90 μm			

*agl = above ground level

DATA ANALYSIS AND INTERPRETATION

MEIS data analysis

The MEIS data for flight lines 4 and 5 were copied from high-density tape to computer-compatible tape and then to disk. Visual inspection of the images and histograms showed that band 1 (0.43-0.46 micrometres) had very little contrast and a low signal-to-noise ratio so this band was dropped from subsequent analysis.

The remaining four bands, corresponding to green, yellow, red, and near infrared portions of the spectrum were linearly contrast stretched and displayed on the Canada Centre for Remote Sensing Image Analysis System (CIAS, Goodenough, 1979).

A natural colour image produced from MEIS band 2 displayed as blue, band 3 displayed as green and band 4 displayed as red shows a green forest with brown areas of discolouration. Transparencies made from the RC-10 negatives show brown discolouration in corresponding places, but they are difficult to see because the contrast-reducing effects of atmospheric path radiance and attenuation are not removed from the photography as they are from the digital MEIS data. A comparison with the 70 mm photography shows that the areas of brown discolouration are indeed associated with spruce budworm defoliation.

MSS data analysis

Data from the Daedalus 1260 airborne multispectral scanner were also processed in an attempt to reveal spruce budworm defoliation. The Daedalus 1260 multispectral scanner obtains data over an angular range of ±37.5° of nadir. Flight lines 4 and 5 were flown within 31 degrees of the solar azimuth. With this geometry, the radiance of a given ground cover type will increase nearly symmetrically about nadir as a result of target bidirectional reflectance effects (Staenz et al, 1982) and variations in path radiance as a function of viewing angle (Turner et al, 1971). This resulting variation in radiance can be well fitted with a low order polynomial. Flight lines 4 and 5 were corrected for changes in radiance as a function of view angle with the program NEWSCN. Data from the individual bands were then contrast stretched to fill the available 0-255 range and a number of composite colour images were formed. Principal component transformations were also performed on the data and colour composite images were formed from the first three principal components.

None of these data treatments resulted in images that clearly showed spruce budworm defoliation distinctly from other sources of variation.

We examined our results in comparison with those reported by Leckie and Ostaff (In Preparation) to try to understand why the MSS was unsuccessful in our enhancement test, but moderately successful in their classification test. Two related reasons emerged:

1. The classification test was restricted to a relatively small number of classes while an enhancement processes the entire image. This reveals confusion classes that may have been omitted from the classification trial.

2. Leckie and Ostaff (In preparation) reported the best separation of defoliation classes with spectral bands in the near infrared region. We computed class separability measures that gave similar results, but found that variations in deciduous/conifer proportions and crown closure caused much larger variations in infrared reflectance, which confused the smaller changes caused by defoliation.

MEIS-MSS comparison

The MEIS and MSS data were processed further to determine what difference in characteristics enables the MEIS data to show the defoliation when the MSS data could not.

Three MSS bands (4, 5, 7) corresponding to the MEIS bands 2, 3 and 4 were radiometrically matched to the MEIS data. This was done through a linear transformation, which matched the lower limit and mean value of the histogram of each MSS band to the lower limit and mean value of the corresponding MEIS band. The resulting image still did not show brown discolouration corresponding to that visible in the MEIS image.

Next, to test whether the MEIS spatial resolution of 7 m compared to the MSS spatial resolution of 25 m played a significant role, the MEIS image was averaged with a sin x/x convolution with an effective size of 25 m between the innermost two zero crossings. The brown discolouration was more apparent in the resulting image because of the reduction in radiometric variation brought about by spatial averaging. Since the defoliation areas are generally larger than 25 m, a spatial resolution of 25 m appears adequate to detect defoliation. The additional resolution of MEIS is thus not crucial.

Finally, random noise was added to the MEIS data to make MEIS match MSS in terms of noise variance. This was done by finding an area, in a cloud shadow, where the signal was low and essentially constant and the image variance was assumed to be representative of sensor noise. The variance of the data in each displayed band of both sensors was measured. Then noise from a gaussian distribution was added to the 25 m resolution MEIS image to make its variance equal to that of the MSS image.

The additional noise made it more difficult to see the defoliated areas but they were still visible in the degraded MEIS image and not in the MSS image. Thus it seems that the better noise performance of MEIS relative to MSS explains part, but not all, of the success of MEIS.

Since the ability of MEIS to detect defoliation cannot be attributed entirely to its superior spatial or radiometric resolution compared to MSS, we turned our attention to the difference in spectral bands. Figure 2 shows two band intensities of five classes obtained from training areas in the MEIS image and figure 3 shows two band intensities of the same five classes obtained from the same training areas in the MSS image. Water, bright conifer, and dark conifer all have similar locations in both figures. With the MEIS bands (figure 2) the hardwood class is lower in the red band, and overlaps with bright conifer, and the defoliation class is separated from these, being brighter than both. In the MSS bands (figure 3), the defoliation class overlaps the other conifer classes, and the hardwood class is brighter than these. We have concluded that these differences are a result of the fact that the MEIS red band is very narrow and well positioned relative to the chlorophyll absorption minimum at 675 nm of vegetation reflectance. This spectral region is particularly sensitive to the difference between healthy and defoliated trees (Teillet et al, 1985). Thus it is concluded that the major factor for the success of the MEIS sensor in detecting spruce budworm defoliation is its narrow, well-positioned spectral bands, particularly the red band.

CONCLUSIONS

In this study we have shown that it is possible to create natural colour images showing moderate levels of current-year spruce budworm defoliation as a brown discolouration on a green forest background using data from an electro-optical pushbroom scanner. This demonstration suggests the possibility of a defoliation survey that would provide more accurate maps of current-year defoliation than those presently produced by aerial sketch mapping. This capability could improve spray program planning and effectiveness assessment, and ultimately improve the wood supply through better growth-loss predictions for harvest scheduling.

Successful detection of defoliation was achieved with the pushbroom (MEIS) scanner, but not with the mechanical (MSS) scanner. The mechanical scanner was unable to detect defoliation with either near-nadir viewing, or when viewing the affected forest with a geometry that looked at the sun-lit sides of the trees.

The difference between the MEIS and MSS performance was traced primarily to a difference between the widths and placements of their spectral bands, particularly the red band. This difference was substantiated through subsequent ground-based spectroscopic studies (Teillet et al., In preparation). The greater signal-to-noise ratio of MEIS compared to that of MSS was also shown to be important.

ACKNOWLEDGEMENTS

The authors would like to thank D. Carlin, T. Erdle, D. Leckie, D. Ostaff and P. Teillet for assistance and valuable discussions, and K. Thomson and J. Cihlar for their critical review of this work. We also thank D. Gareau, M. Trindade and C. Wade for assistance with the preparation of this paper for publication.

REFERENCES

Baskerville, G.L., 1980, Insects, Diseases and Forest Management: Proceedings, Canadian Forest Congress, The Forest Imperative, Toronto, Ontario, September 1980, pp. 52-54.

Erdle, T.A., 1985, Harvest Scheduling and its Impact on Wood Supply: Unpublished Report, New Brunswick Department of Natural Resources, Forest Management Branch, Fredericton, New Brunswick.

Goodenough, D.G., 1979, The Image Analysis System (CIAS) at the Canada Centre for Remote Sensing: Canadian Journal of Remote Sensing, Vol. 5, pp. 3-17.

Leckie, D.G. and Gougeon, F.A., 1981, Assessment of Spruce Budworm Defoliation Using Digital Airborne MSS Data: Proceedings, Seventh Canadian Symposium on Remote Sensing, Winnipeg, Manitoba, September 1981, pp. 190-196.

Leckie, D.G. and Ostoff, D.P., In Preparation. Assessment of Current Defoliation Caused by the Spruce Budworm, Choristoneura Fumiferana (Clem) Using Airborne Multispectral Scanner Data.

McColl, W.D., Neville, R.A. and Till, S.M., 1984, Multi-Detector Electro-Optical Imaging Scanner MEIS II: Proceedings, Eighth Canadian Symposium on Remote Sensing and Fourth Conference of L'Association québécoise de télédétection, Montreal, Quebec, May 1983, pp. 71-79.

McLean, D.A. and Erdle, T.A., 1984, A Method to Determine Effects of Spruce Budworm on Stand Yield and Wood Supply Projections for New Brunswick: The Forestry Chronicle, June 1984, pp. 167-173.

Staenz, K., Ahern, F.J. and Brown, R.J., 1981, The Influence of Illumination and Viewing Geometry on the Reference Factor of Agricultural Targets: Proceedings, Fifteenth International Symposium on Remote Sensing of Environment, Environmental Research Institute of Michigan, Ann Arbor, Michigan, pp. .

Teillet, P.M., Leckie, D.G., Ahern, F.J., Fedosejevs, G. and Ostaff, D.P., In preparation. Submitted to ISP Commission VIII for Presentation at the 1985 Symposium on Spectral Signatures of Objects in Remote Sensing.

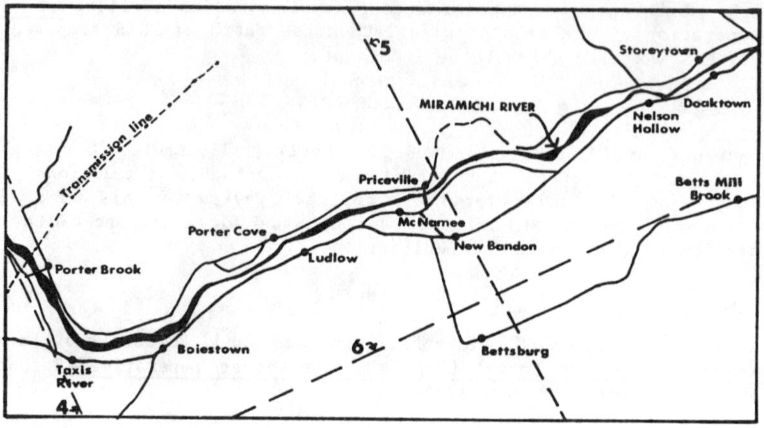

Figure 1. Study area in central New Brunswick showing flight lines 4, 5, and 6.

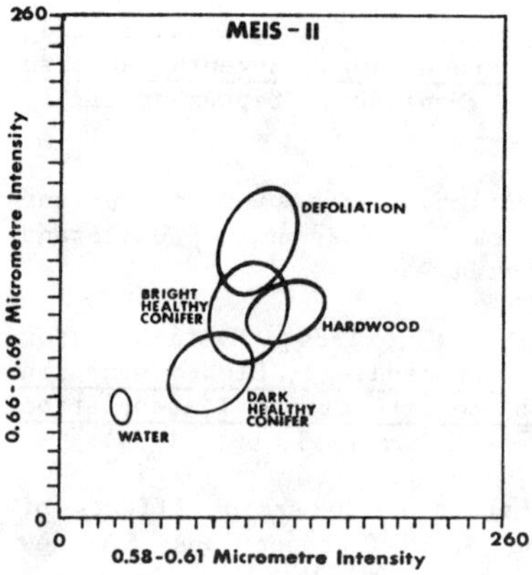

Figure 2. Equi-probability ellipses of MEIS-II bands 3 and 4 for five classes.

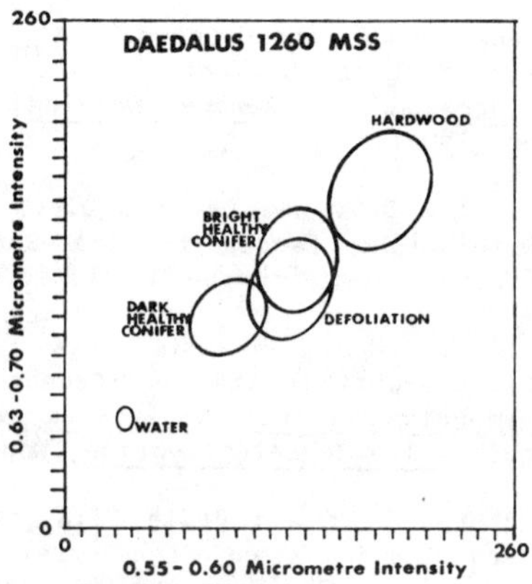

Figure 3. Equi-probability ellipses of MSS bands 5 and 7 for the same five classes shown in figure 2.

DIGITAL AIRBORNE AND SATELLITE DATA FOR EVALUATING SPRUCE BUDWORM DAMAGE IN QUEBEC

J. Beaubien
Laurentian Forest Research Centre
Canadian Forestry Service
Sainte-Foy, QC, Canada G1V 4C7

P. Laframboise
Ministère de l'Énergie et des Ressouces du Québec
Centre québécois de coordination de la télédétection
Sainte-Foy, QC, Canada G1N 4H9

ABSTRACT

Since 1970, northeastern North America has experienced a spruce budworm (*Choristoneura fumiferana* (Clem.)) outbreak that has had an important influence on management of Quebec's fir/spruce forests. Landsat MSS, high altitude CIR aerial photography, and medium scale photography have low capabilities for detecting spruce budworm damage level. The potential of higher spatial and/or spectral resolution digital data for detecting, mapping, and quantifying cumulative defoliation, needed for spraying protection programs and stand harvesting schedules, was studied. In November 1983 and August 1984, a Deadalus 1260 and a MEIS II (Multi-Detector Electro-Optical Imaging Scanner) collected multispectral digital data. Ground resolutions were 20 and 10 m, and 5.6 and 2.8 m. From these data, future SPOT multispectral and panchromatic imagery was simulated. A Landsat TM scene of May 18, 1984, was also acquired. Various enhancement techniques for discriminating damage levels were investigated. Principal component enhancements of SPOT simulated multispectral data gave the best results for producing imageries that can be used as an additional tool in the traditional sketch-mapping from small aircrafts. The amount of dead crowns or tops inside stands can be evaluated. The study continues to determine the optimum spectral band combinations from the sets of data for enhancing or classifying damage levels.

INTRODUCTION

Since 1970, northeastern North America has experienced a spruce budworm (*Choristoneura fumiferana* (Clem.)) outbreak that has an important influence on management of fir/spruce forests in Quebec. Spruce budworm, the most important destructive forest insect in Quebec, affects the valuable pulpwood species: balsam fir (*Abies balsamea* (L.)) and spruces, mainly white spruce (*Picea glauca* (Moench) Voss). The economic impact is important; tree mortality covers 12 million hectares, and, since early 1970, over 22 million hectares of forest were sprayed in an effort to control this pest.

Information is needed on the amount and location of budworm damage for planning stand management or spray strategies. Current methods, estimating damage over large areas, rely on aerial visual surveys and various types of ground surveys. These are the only methods operational. They are costly and time consuming, and partially fill present needs. Recent remote sensing tools may have the potential for a more rapid and accurate assessment.

Color and chiefly color infrared (CIR) aerial photographs were investigated for evaluating spruce budworm damage. Ashley (1976) concluded normal color, summer imagery, scales 1:15 840 and larger, was the best for evaluating current defoliation when the affected needles are brown and still held on branches by the budworm's webbing. Heavy cumulative (current and past) defoliation was identified on CIR, scales 1:31 680 and larger, whereas moderate and light defoliations were not detected at any scales ranging from 1:9 900 to 1:63 360. Beaubien (1975) also found that only very severe damage of one year feeding or more could be identified on high-altitude (1:60 000) CIR photographs. Color variations due to certain stand characteristics (density, age, site quality) prevented the detection of lighter damage.

Satellite imagery acquired on a systematic and repetitive basis is theorically a powerful tool to monitor changes in forest canopy. Landsat-MSS data with a resolution of 80 m have low capabilities for evaluating levels of insect damage. However, in coniferous stands, tree mortality was delineated over large areas of pure or nearly pure stands (Beaubien and Daus 1977).

Higher resolution data were studied through airborne MSS data. Leckie and Gougeon (1981), using 19 and 4.5 m data, concluded that classification of total defoliation levels was difficult in mixed stands and in stands of varying density, hardwood component and crown closure affect stand signatures. However, four (4) levels of total defoliation were classified in dense fir/spruce stands by a supervised classification of 4.5 m resolution data, with misclassification error of approximately 25 percent. Nelson _et al_. (1984) used airborne data to simulate Landsat TM data for delineating boreal forest cover types in Maine where coniferous stands were heavily damaged by spruce budworm. As data were obtained October 12, 1981, when leaf drop was beginning, mixedwood and conifer severe defoliation was confused with hardwoods. They suggested a data acquisition in August for an overall (cumulative defoliation) tree condition evaluation.

Remote sensing has not proved its effectiveness to replace traditional sketch-mapping from small aircrafts. Each year large areas of forest susceptible to defoliation by spruce budworm have to be surveyed to plan protection programs or stand harvesting schedules. The best known tools are medium to large scale aerial photographs, but their yearly acquisition is too costly. This is also true for airborne MSS data if their capability was proven. Satellite data have the operational capability to be used over large areas, probably as a supplementary tool to current methodology.

A study was initiated to determine the potential of higher spatial and/or spectral resolution digital data, particularly TM and future SPOT data, for detecting, mapping, and quantifying cumulative defoliation.

STUDY AREA

The study area is in the Pohenegamook Lake area, about 170 km NE of Quebec City and 40 km SE of the St. Lawrence River. This low relief, rolling area is covered by mixed coniferous stands: balsam fir, white spruce and black spruce (_Picea mariana_ (Mill.) B.S.P.), and also some pure, black spruce stands. Hardwood and mixedwood stands are abundant; white birch (_Betula papyrifera_ Marsh.) and trembling aspen (_Populus tremuloides_ Michx.) prodominate.

Spruce budworm has been in the area for more than ten years. Most balsam fir and white spruce that have not been sprayed are severely or completely defoliated. Yearly spraying preserved a few patches. Consequently the study turned out to be more an evaluation of the proportion of dead trees or dead tops inside stands, than an evaluation of defoliation levels.

DATA ACQUISITION

A Deadalus 1260 and a MEIS II (Multi-Detector Electro-Optical Imaging Scanner) collected multispectral digital data over the area in November 1983 and August 1984. Therefore data collected in two seasons, with and without leaves, could be studied. Flight altitudes of 8 000 m and 4 000 m gave ground resolutions of 20 and 10 m, and 5.6 and 2.8 m for both dates. Coincidental color IR aerial photographies were acquired on both flights. Future SPOT multispectral imagery was simulated from Deadalus data, panchromatic channel from MEIS-II data (Saint and Podaire 1982). Geometric corrections and resampling produced 10 m images, overlaying MSS and panchromatic data. A normalization was applied to attenuate the effect of scanning angles. A Landsat TM scene of May 18, 1984, was also acquired.

METHODOLOGY

The objective was to investigate the effectiveness of various enhancement techniques and to determine the optimum spectral band combinations from these sets of data for enhancing or digitally classifying levels of cumulative damage. Recent major problems with our image analysis system limited the projected analysis to the production of enhancements: stretches, band ratios, and mainly principal component transformations (PCT) on SPOT simulated multispectral data and a TM scene.

A first PCT general enhancement using Beaubien's (1984) method was produced from Deadalus autumn and summer raw data (bands 0.55-0.60, 0.77-0.90, 1.55-1.75 µm) and was used as source documents for two reconnaissance samplings from a helicopter, the first in summer and the second in autumn 1984. Each observation was accompanied by an oblique 35 mm color slide carefully located on the printed enhancements. Ground sample plots were also established. In spring 1985, a second PCT enhancement oriented only on conifers (healthy and damaged) was produced from SPOT simulated multispectral data of both acquisition dates. The imageries were used for a third reconnaissance flight and additional ground samplings. The ground-truth data enabled to set final enhanced images by modifying the histograms of the component digital values. Only the two first components were used, the third one being too dependent of the scanning varying angles. Using the Taylor (1974) color space, the first component was loaded on a "brightness" axis, the second on a red-green axis and the first was repeated on a blue-yellow axis.

A TM scene of May 1984 was also enhanced by PCT, studied and compared with preceeding results.

PRELIMINARY RESULTS

All enhancements were analyzed through ground-truth data and by interpreting the CIR aerial photographs. The following is a summary of the preliminary results:

- severe cumulative defoliation (1/3 of the upper crown and over) or tree mortality was observed on each set of data;
- only amount of dead crowns inside stands could be evaluated, not defoliation level;
- on winter (autumn) images, damages are over-estimated in some stands because of the high density of leafless hardwood branches;
- on summer images, damages tend to be under-estimated in mixed-stands, some dead crowns hiding behind hardwood leaves;
- principle component enhancements from SPOT MSS data, viewed through Taylor's color space, gave the closer picture of stand conditions;
- because of the prodominance of radiometric variation along scan lines on the third component, only the first two components were usable from SPOT data;
- the best image for conifer defoliation detection extracted from the TM scene was obtained by using the three first components viewed through the standard color space (red, green, blue), originating from spectral bands 2 (green), 3 (red), 4 (near infrared), and 5 (middle infrared 1). As suggested by Nelson _et al_. (1984), bands 1 (blue), 3, 5, and 7 (middle infrared 2) were tried but with less success. Other combinations have to be tested;
- SPOT summer data gave the best results. Dead tops in stands (conifers, mixed dominated by conifers or mixed dominated by hardwoods) could be evaluated according to three approximate percentages: 25-50 percent, 50-75 percent, and more than 75 percent; and,
- SPOT winter imagery was preferable to separate black spruce from balsam fir.

CONCLUSIONS

Current methods to quantify adequately spruce budworm damage must be improved. Satellite imagery with it's repetitive coverage should be seriously investigated as a possible tool to facilitate the task of forest managers. Because of their poor resolution, Landsat-MSS data have low capabilities for insect damage appraisal except for large areas of pure to nearly pure severely affected stands. Landsat-TM or future SPOT imagery may have the potential of producing valuable information.

Partial results are encouraging, at least as a valuable additional tool to the present eyeball evaluation. Principle component enhancements produced from SPOT simulator data allowed us to evaluate the percentage of dead conifer tops in forest stands according to three classes: 25-50 percent, 50-75 percent, and more than 75 percent. This is valuable information to guide a more detail survey or to plan a spraying program. Up to now the results from TM data analysis are not so conclusive; they locate the highest concentrations of dead tops. One should notice that the overall results allow only an approximate evaluation of the proportion of dead tops in the forest canopy, not an appreciation of defoliation levels among trees. In dense stands, dead tops and killed trees are impossible to differentiate from data of this resolution.

Data acquired in summer seem more convenient than those acquired in winter, hardwood naked branches have the effect of exagerating damage. On the other hand, hardwood leaves hide a certain amount of dead branches, this varies with stand characteristics.

Considering the common heterogeneity of forest cover and the numerous factors affecting it's spectral response, defoliation is an additional confusing factor not easy to characterize. It is felt that interpretation of digitally enhanced images is the only way of extracting the needed information on spruce budworm damages. This study will continue with the main objective to select better spectral band combinations to improved enhancements. Digital classification will be also tested.

Methodology based on satellite data will not give more than an approximate estimation of severe cumulative damage to conifers within stands. For current defoliation evaluation, aerial data must be acquired during a short period (2-3 weeks) when dead needles remain on branches. Presently, the chance of obtaining a satellite image in a so short period is slight. To obtain levels of total defoliation among conifer stems, data of much higher resolution will have to be acquired: in the order of 1 m for digital data. Those data will be convenient for sampling; they will be too costly to cover large areas as presently needed in Quebec. The principal anticipated limitation of this methodology will probably be the poor availability of satellite images for the desired areas and dates.

ACKNOWLEDGMENTS

The authors thank Mr. Georges Dorval of the "Service de la Protection du Ministère de l'Énergie et des Ressources du Québec" for his collaboration in this project.

REFERENCES

Ashley, M.D. 1976, Spruce Budworm Damage Evaluations Using Aerial Photography: Photogrammetric Engineering and Remote Sensing, Vol. 42, pp. 1265-1272.

Beaubien, J. 1975, High-Altitude Colour-IR Photographs for Evaluating Spruce Budworm Damage in Quebec: Proceedings 3rd Canadian Symposium on Remote Sensing, Canadian Aeronautics and Space Institute, Ottawa, Ontario, pp. 281-283.

Beaubien, J. and S.J. Daus 1977, Le traitement numérique des données de Landsat pour la cartographie: Proceedings 4th Canadian Symposium on Remote Sensing, Canadian Aeronautics and Space Institute, Ottawa, Ontario, pp. 19-26.

Beaubien, J. 1984, Une méthode de rehaussement d'images Landsat pour la classification du couvert végétal: Proceedings 8th Canadian Symposium on Remote Sensing, Canadian Aeronautics and Space Institute, Ottawa, Ontario, pp. 559-566.

Leckie, D.G. and F.A. Gougeon 1981, Assessment of Spruce Budworm Defoliation Using Digital Airborne MSS Data: Proceedings 7th Canadian Symposium on Remote Sensing, Canadian Aeronautics and Space Institute, Ottawa, Ontario, pp. 190-196.

Nelson, R.F., R.S. Latty and G. Mott 1984, Classifying Northern Forests Using Thematic Mapper Simulator Data: Photogrammetric Engineering and Remote Sensing, Vol. 50, pp. 607-617.

Saint, G. and A. Podaire 1982, Méthodes de simulation radiométrique et leur représentativité: Comptes rendus du colloque sur le système SPOT d'observation de la terre, Association québécoise de télédétection, Sainte-Foy, Québec.

Taylor, M.M. 1974, Principle component color display of ERTS imagery: Proceedings 2nd Canadian Symposium on Remote Sensing, Canadian Remote Sensing Society, Ottawa, Ontario, pp. 295-314.

USE OF COLOR INFRARED AERIAL PHOTOGRAPHY FOR ASSESSING RED SPRUCE MORTALITY AND DECLINE

Melvyn J. Weiss, Lew R. McCreery, and Imants Millers
USDA Forest Service, Forest Pest Management
P.O. Box 640, Durham, NH 03824

William M. Ciesla
USDA Forest Service, Forest Pest Management
3825 E. Mulberry, Ft. Collins, CO 80524

ABSTRACT

Red spruce decline and mortality in high mountain regions of the Northeastern United States has recently received considerable publicity. In 1984 a survey was conducted over portions of New York, New Hampshire, and Vermont by the Forest Pest Management staff, Northeastern Area, USDA Forest Service. The purpose was to provide data on the amount of red spruce decline and mortality. The survey method combined estimates from 1:8000 scale CIR aerial photos and ground surveys. Data were obtained for four survey regions: Two in New York, the Adirondack Mountains and the Tug Hill Plateau, and Vermont and New Hampshire. Primary sample units were randomly selected 8,000 acre blocks which were stratified into vegetation classes. Vegetation classes with a spruce-fir component were stratified into 3 mortality classes based on the proportion of spruce and balsam fir mortality. A series of 2.5 acre photo sample plots were established within these strata and counts of dead trees made. A subsample of photo plots were ground cruised. The survey provided estimates of the proportion of spruce-fir area in each of 3 mortality classes. The number and volume of dead trees were also estimated.

INTRODUCTION

Historical Records of Mortality

Extensive mortality of red spruce, *Picea rubens* Sarg., and balsam fir, *Abies balsamea* (L) Mill., has occurred in the Northeast at several times during the last two centuries. Historically, the most important causes of mortality have been fire, high winds, the spruce beetle, *Dendroctonus rufipennis* (Kirby), and the spruce budworm, *Choristoneura fumiferana* (Clem.). The spruce beetle killed vast numbers of trees at various times during the 19th century (Hopkins 1901) and has been a sporadic mortality factor in this century (U.S. Department of Agriculture 1985). The spruce budworm recently has caused significant mortality in northern New Hampshire, northeastern Vermont and Maine (Kucera and Taylor 1984).

While most historical records were concerned with these major mortality factors, other problems were also reported. For example, in the early 1950's red spruce mortality occurred on the Green Mountain National Forest in Vermont (Mook and Eno 1956). The mortality was attributed to the combination of old age and severe growing conditions. At about the same time, winter injury and drought led to mortality of spruce and balsam fir in the Adirondacks (Curry and Church 1952).

Through the 1960's, mortality of red spruce and balsam fir was reported at several locations in New Hampshire, New York and Vermont (Ford 1969, Ford and Murray 1966, Hadfield 1968, 1969a, 1969b, Siccama 1974, U.S. Department of Agriculture, Forest Service 1970, Vermont Department of Forests and Parks 1960, 1962, 1966). The major factors involved were considered to be drought and winter injury. Root disease was associated with mortality on some sites.

Reports of damage and mortality continued into the 1970's and 1980's (Allison 1975, Friedland and others 1984, Johnson and Siccama 1983, Siccama and others 1982, Snowden 1972, Sprugel 1976, U.S. Department of Agriculture, Forest Service 1973).

Recent Reports of Mortality and Justification for Survey

Red spruce and balsam fir mortality in New Hampshire, New York, and Vermont recently received increased attention following reports by Vogelmann (1982) and Johnson and Siccama (1983). Vogelmann (1982) reported on red spruce mortality on Camel's Hump in Vermont and discussed possible causes, including acid rain (Fig. 1). Johnson and Siccama (1983) reported that on seven study sites in New Hampshire, New York, and Vermont substantial reductions in basal area of some size classes of red spruce occurred between the mid-1960's and late 1970's. Reduction in balsam fir basal area occurred in some size classes at the same sites, but to a relatively lesser amount than red spruce. They also discussed possible causes, including drought and acid rain.

Under authority of the Cooperative Forestry Assistance Act of 1978, the USDA Forest Service has the responsibility to conduct surveys to detect and assess insect and disease damage, including damage from

Figure 1. Dead and dying red spruce trees (*Picea rubens*), Camel's Hump, VT, 1984

abiotic agents. This responsibility is carried out directly on Federal lands and through the State Foresters on State and private lands. With increased concern about the red spruce and balsam fir resource, the Northeastern Area, USDA Forest Service and the State forestry agencies of New Hampshire, New York, and Vermont decided in early 1984 to conduct a survey to determine the extent of tree mortality over the range of the spruce-fir type in the three States. A detailed Interim Report of the survey was released in March, 1985 (Weiss and others 1985). A summary of the methods and results are presented here.

Survey Objectives

The primary objectives of the survey were to determine:

1. The proportion of spruce-fir types classified with (a) light - less than 10 percent, (b) moderate - 10 to 30 percent and (c) heavy - more than 30 percent cumulative mortality.
2. The number and volume of dead trees.
3. Spruce regeneration stocking.

Cooperation

The survey was a cooperative effort between the USDA Forest Service, Northeastern Area; the New Hampshire Department of Resources and Economic Development, Division of Forests and Lands; the New York State Department of Environmental Conservation, Division of Lands and Forests; and the Vermont Department of Forests, Parks and Recreation, Division of Forests. The State agencies helped plan the survey and collected much of the field data. In addition, several other groups provided support. Forest Pest Management's Methods Application Group, Ft. Collins, CO, provided the basic survey design and flew the photo mission. The Nationwide Forestry Applications Program of the USDA Forest Service, Houston, TX, was responsible for photo interpretation. Some photo interpretation was also done by the Forest Pest Management staff, Durham Field Office. The Northeastern Forest Experiment Station participated on the steering committee for the survey. The White and Green Mountain National Forests participated on the steering committee and provided logistical support to the field crews.

METHODS

Survey Area

The survey area covered portions of the three northeastern states New Hampshire, New York and Vermont. The boundaries of the survey regions within each state were delineated in consultation with the State forestry agencies. Type maps were considered, and in general, the areas where spruce-fir type was likely to occur were included. In New York, two regions were selected: the Adirondack Mountains and Tug Hill Plateau (Fig. 2). One region was established in each of the other two states (Fig. 2). In New Hampshire, the survey region included the White Mountains and portions of the Western Highlands. In Vermont, the survey region included the Green Mountain Range and the Central Plateau. Parts of the spruce-fir forest in northern New Hampshire and northeastern Vermont were not included in the survey regions because of the presence of spruce budworm caused tree mortality. All ownerships were included in the survey regions, including the Green Mountain and White Mountain National Forests.

Timing

The first phase of the survey, acquisition and interpretation of aerial photos, extended from July to November, 1984. Ground data collection extended from September, 1984 to January, 1985.

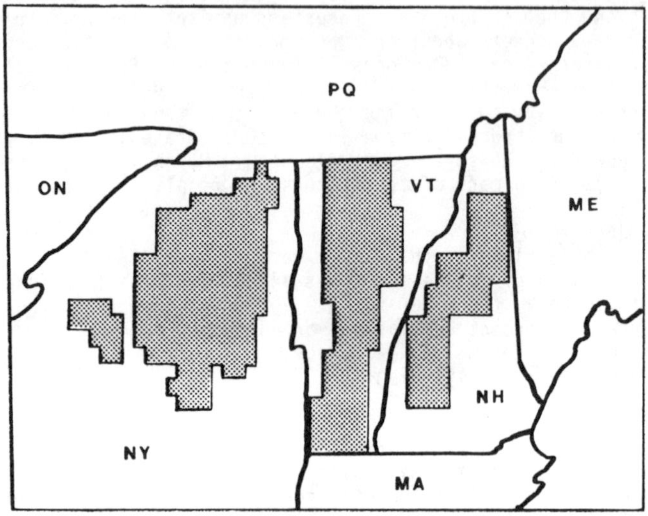

Figure 2.--Survey regions (shaded) in the Cooperative Survey of Red Spruce and Balsam Fir Decline and Mortality, 1984 (l. to r. Tug Hill, Adirondack, Vermont, New Hampshire)

Sample Blocks

The primary samples units were rectangular blocks, each 8,000 acres in size. The number of blocks to be photographed by region was as follows: Adirondack-20, Tug Hill-5, New Hampshire-15 and Vermont-15. USGS 7-1/2 minute topographic maps or their equivalent were chosen at random. Then, one quarter of each of these quadrangles was randomly selected for placement of the sample block. Each sample block was examined visually from an aircraft to determine if at least 100 acres of spruce-fir type were present. If not, another block was chosen at random and examined until the predetermined number of blocks with sufficient type were selected for aerial photography.

Aerial Photography

Complete photo coverage was obtained of each sample block. The aerial photo mission was flown between July 14 and August 27, 1984, using a USDA Forest Service Beechcraft Queen Air aircraft equipped with a Zeiss RMK 21/23 9-inch format camera with an 8-1/4 inch focal length lens. All sample blocks and flight lines were marked on USGS 7-1/2-, 7-1/2 x 15-, or 15- minute topographic maps to aid in aerial navigation. Each block was photographed with a forward overlap of 70 percent and a sidelap of 20 percent at a scale of about 1:8,000. Kodak Aerochrome Color Infrared (CIR) film (Type 2443) was used in combination with Wratten 12(minus blue) and antivignetting filters. CIR film was chosen to provide haze penetration, easy separation of conifers and hardwoods, and enhance conifer tree mortality and foliage loss. Approximately 56 photos were taken for each block.

Photo Interpretation

Photo interpretation guidelines were developed for the project and provided to the interpreters (Ciesla 1984).

*Mention of commercial products does not imply endorsement.

Indexing and Annotation. All photos were indexed and the principal point of every other photo frame was plotted on a 7-1/2 minute topographic sheet or equivalent. The effective area of each photo frame and photo scale were then determined.

Stratification. After annotation, photo interpreters, using Old Delft scanning stereoscopes, drew polygons around cover types and mortality classes directly on the CIR transparencies.

Each sample block was stratified into 5 cover types: Spruce-fir swamp - Low lying bogs or swamps at least 10 acres in size with more than 50 percent of the overstory in spruce or balsam fir; mixedwood - Areas at least 10 acres in size with between 25 and 50 percent of the overstory in spruce or balsam fir; spruce-fir slope - Areas at least 10 acres in size with more than 50 percent of the overstory in spruce or balsam fir (this is generally the high elevation forest, but may include spruce-fir "flats" at lower elevations); high elevation balsam fir - a subclass of the spruce-fir slope where over 90 percent of the stand is balsam fir with small crowns and average diameter of less than 5 inches; other - all other areas including lakes, fields, other forest cover types and alpine or tundra vegetation.

The 4 cover types containing spruce and fir were further stratified by the abundance of spruce and fir mortality: Light - less than 10 percent of standing spruce or fir trees dead; moderate - 10 to 30 percent of standing spruce or fir trees dead; heavy - more than 30 percent of standing spruce or fir trees dead. The mortality classes (light, moderate and heavy) do not necessarily indicate increasing rates of current mortality or the extent of any abnormal conditions. The assumption was that about 10 percent standing dead trees or less represent a normal forest condition. One to three percent annual mortality has been calculated from inventories of commercial forests in New Hampshire, New York and Vermont since the 1940's. Dead trees usually remain standing 5 years or longer. Therefore, 10 percent cumulative standing mortality may be the normal condition for commercial forests. However, many high elevation non-commercial stands were included in our survey, and in these stands, "normal" cumulative mortality may be considerably greater than 10 percent.

Map Transfer and Area Determination. After all photo frames within a sample block were stratified, polygons drawn on the photo frames were transferred to the USGS 7-1/2 minute topographic maps, using a Bausch and Lomb zoom transfer scope. Polygon areas were then measured using a Bausch and Lomb Resource Measurement System. The acreage in each stratum was determined in this way for each sample block.

Photo Plot Selection and Tree Counts. After all sample blocks in a region were stratified and strata acreages determined, photo plots were randomly selected within the mixedwood and spruce-fir slope cover types. Photo plots were not established in the spruce-fir swamp and high elevation balsam fir types. Each photo plot was a square marked on the photo estimated to represent 2.5 acres on the ground. The number of photo plots allocated by region was as follows: Adirondack-620, Tug Hill-44, New Hampshire-620, and Vermont-264. The number of photo plots in the Tug Hill and Vermont regions were fewer than originally planned because of the small amount of acreage in the two higher mortality classes (the sampling intensity was a minimum of 10 acres per photo plot). Photo interpreters counted all standing dead spruce and fir trees in each photo plot, using Old Delft scanning stereoscopes. Photo interpreters did not distinguish between spruce and fir. Dead trees were defined as those trees with foliage missing but most of the fine lateral branches present, or trees with all foliage red to brown in color (yellow on CIR film). Snags (those dead trees with only a few heavy limbs remaining) were not counted.

Photo Plot Ground Tree Counts

After photo interpretation was completed, photo plots were examined on the ground. All of the photo plots in the light mortality strata were examined on the ground. About 10 percent of the photo plots in the two higher mortality classes were randomly chosen for ground examination. In New Hampshire, a smaller percentage of photo plots was examined because of hazardous field conditions. The number of photo plots examined on the ground by region was as follows: Adirondacks-82, Tug Hill-17, New Hampshire-64, and Vermont-78. Field crews located photo plot corners as accurately as possible to correspond with the same trees and land area observed by photo interpreters. All standing dead spruce and fir (snags excluded) 5.0 inches diameter at breast height (dbh) and larger on the 2.5 acre plot were counted. For each dead tree, the species, dbh, total height to nearest 10 feet, and crown class were recorded.

Variable Radius Plots

In each of the 2.5 acre photo plots examined on the ground, five variable radius plots (10 BAF) were systematically located as follows: One in the center, and the other four about 2 chains away, one in each quadrant. The variable radius plots were used to sample site, stand and tree variables. Site data included elevation, aspect, slope, soil depth, soil drainage, and evidence of disturbance. Stand data included crown closure, and age. Tree data were collected for trees 5.0 inches dbh and larger. These data included species dbh, total height, crown class, and crown condition.

Regeneration Plots

At each variable radius plot, five milacre plots (3.7 foot radius) were established to determine the presence of regeneration. One milacre plot was located at the center and the other four 20 feet away, in each cardinal compass direction. The presence or absence of regeneration (at least one living tree 1 foot in height to 4.9 inches dbh) in each milacre plot was noted in four categories as follows: 1) No regeneration present; 2) spruce present (other species may be present); 3) spruce absent, fir present (other species may be present); 4) stocked with commercial species other than spruce or fir.

Statistical Procedures

A combination of statistical procedures was used.

Area Estimates. The sample blocks were chosen at random from a pool of blocks, without replacement of selected blocks.
A sequential sampling plan was used to assure that a predetermined number of sample blocks in each region had at least a minimum number of acres with spruce-fir type. The sample blocks without spruce-fir type were considered as non-type in calculations. The mean area of each stratum and the error of the mean was determined for each region from sample block acreages using a correction for the

sequential method*.

Mortality Estimates. A fixed number of photo plots were randomly selected in each stratum. Tree counts from ground examination were used to adjust photo interpreter counts using a double sampling with regression method (Wear and others 1966). The mean mortality per acre of each stratum was expanded to the total acreage in the region.

Regeneration Estimates. For regeneration estimates, milacre plots were systematically chosen at each variable radius plot. From these, the proportion of milacre plots stocked with regeneration was determined for each stratum.

RESULTS

Mortality Area

The percentage of area in the mixedwood cover type classified as light mortality ranged from 61 percent (104,500 acres) in New Hampshire to 83 percent (99,000 acres) in Vermont (Table 1). The percentage of area in the spruce-fir slope cover type classified as light mortality ranged from 41 percent (86,500 acres) in the Adirondack region to 100 percent (9,100 acres) in the Tug Hill region. The highest percentage of the mixedwood cover type classified as heavy mortality was in New Hampshire. The highest percentage of the spruce-fir slope cover type classified as heavy mortality was also in New Hampshire. Although high elevation balsam fir occurs in the Adirondack Region and in Vermont, none was found in the sample blocks for these regions. In New Hampshire, high elevation balsam fir, based on this survey, occurs on an estimated 14,200 acres. Of this amount, 96 percent was classified as heavy mortality.

Number of Dead Trees

The number of standing dead spruce and balsam fir per acre, based on the photo plots, ranged from 9.4 to 27.2, with the highest number in the heavy mortality class of the spruce-fir slope cover type, New Hampshire (Table 2).

Regeneration Stocking

Spruce regeneration stocking in most cases was higher in the heavy mortality class than in the light mortality class. Not considering Tug Hill, spruce regeneration stocking in the heavy mortality class, mixedwoods, ranged from 20 percent in Vermont to 75 percent in the Adirondacks. Spruce regeneration stocking in the heavy mortality class, spruce-fir slope, ranged from 36 percent in Vermont to 66 percent in the Adirondacks. When balsam fir and other commercial species are included, regeneration stocking in the heavy mortality classes of spruce-fir slope (not considering Tug Hill) ranged from 75 percent in Vermont to 86 percent in the Adirondacks. Spruce regeneration was lowest in Tug Hill, where the stocking of spruce was 17 percent or less.

* Walton, G. Personal communication. USDA Department of Agriculture, Forest Service, Northeastern Forest Experiment Station, Hamden, CT.

Table 1. Summary of the area of spruce-fir cover types by region and mortality class in the Cooperative Survey of Red Spruce and Balsam Fir Decline and Mortality, 1984 (based on photo stratification of sample blocks)

In thousands of acres ± percent of standard error of mean

Cover type and mortality class	Region			
	Adirondack	Tug Hill	Vermont	New Hampshire
Mixedwood				
Light	254.6 ±25	10.9 ± 38	99.0 ±24	104.5 ±24
Moderate	149.7 ±29	2.2 ± 72	17.0 ±39	21.3 ±22
Heavy	50.5 ±40	.2 ±103	3.9 ±86	45.9 ±73
Spruce-fir slope				
Light	86.5 ±24	9.1± 44	103.0 ±34	73.4 ±42
Moderate	65.4 ±27	.0± 0	17.2 ±34	25.6 ±28
Heavy	58.1 ±62	.0± 0	12.4 ±53	76.3 ±59
Spruce-fir swamp				
Light	118.9 ±24	.5± 97	9.2 ±76	1.1 ±76
Moderate	73.9 ±34	.1±103	8.0±100	.4 ±76
Heavy	31.6 ±34	.1± 66	5.9 ±74	.9 ±84

Table 2. Number of standing dead spruce and balsam fir trees (5.0+ inches dbh) per acre by cover type and mortality class in the Cooperative Survey of Red Spruce and Balsam Fir Decline and Mortality, 1984 (based on photo plot samples)

In number of trees per acre ± percent of standard error of mean

Region	Mixedwood			Spruce-fir slope		
	Light	Moderate	Heavy	Light	Moderate	Heavy
Adirondack	9.4± 6	11.4±4	13.0± 5	11.0±11	15.0±4	19.0±4
Tug Hill	13.2±12	14.7±5	23.7±29	12.0±14	.0±0	.0±0
Vermont	10.1± 6	10.5±4	15.8± 7	10.0± 4	15.0±6	16.5±6
New Hampshire	11.6± 7	13.9±2	19.5± 4	12.9± 7	16.5±3	27.2±4

DISCUSSION AND CONCLUSIONS

Mortality

In almost all cases, the proportion of cover types classified as light mortality was greater than that classified as heavy mortality. Nevertheless, intensive examination is needed of areas classified as having moderate or heavy mortality to determine which of the historically important mortality factors may be involved and to what extent the mortality may be considered normal. The differences in amount of mortality between regions also need to be explained. Areas with high percentages of trees dead need to be intensively examined for all factors that could have led to higher mortality.

Historical Records of Mortality

Historical records were searched for reports of mortality and causes in the survey regions. Several factors have been reported causing damage and mortality in one or more of the regions in recent years. At high elevations, the major factors have been wind, ice, snow, winter desiccation, drought, overmature stand conditions, and water level fluctuations. At low elevations, the major factors have been drought, winter desiccation, the spruce beetle (spruce over 10 inches dbh), root diseases, and water level fluctuations. Long-term changes in climate could be important in stand survival. Griggs (1946) suggested that changes observed in high elevation stands of red spruce and balsam fir in New Hampshire during the 1930's and 1940's were related to climatic change. Vulnerability of red spruce at lower elevations to climatic change was discussed by Siccama (1974) in relation to mortality observed from 1963 to 1965 at Camel's Hump in Vermont. Wright (1955) pointed out the limited genetic adaptability of red spruce and possible implications of change in climate. Others have discussed how human activities have affected forest quality (Westveld 1953, Cronon 1983).

Regeneration Stocking

In general, regeneration stocking was highest in the heavy mortality classes, indicating that as overstory trees die, the stands are regenerating.

REFERENCES

Allison, J.R., 1975, Forest pest conditions in the Northeast, 1974: U.S. Department of Agriculture, Forest Service, Northeastern Area, Upper Darby, PA.

Ciesla, W. H., 1984, Photo interpretation guidelines-cooperative survey of red spruce and balsam fir decline and mortality in New Hampshire, New York, and Vermont: U.S. Department of Agriculture, Forest Service, Methods Application Group, Ft. Collins, CO.

Cronon, W., 1983, Changes in the land: Hill and Wang, New York, N.Y.

Curry, J. R., Church, T. W., 1952, Observations on winter drying of conifers in the Adirondacks: Journal of Forestry, 50:114-116.

Ford, R. P., Murray, R.L., 1969, Aerial detection survey, White Mountain National Forest: U. S. Department of Agriculture, Forest Service, Northeastern Area, A-69-5, Amherst, MA.

Ford, R. P., 1966, Aerial detection survey on the White Mountain National Forest: U. S. Department of Agriculture, Forest Service, Northeastern Area, 66-NEZ-5, Amherst, MA.

Friedland, A. J., Gregory, R. A., Karenlampi, L., Johnson, A.H., 1984, Winter damage to foliage as a factor in red spruce decline: Canadian Journal of Forest Research, 14:963-965.

Griggs, R.F., 1946, The timberlines of North America and their interpretation: Ecology 27(4):275-189.

Hadfield, J.S., 1968, Evaluation of diseases of red spruce on the Chamberlin Hill Sale, Rochester Ranger District, Green Mountain National Forest: U. S. Department of Agriculture, Forest Service, Northeastern Area, A-68-8, Amherst, MA.

Hadfield, J.S., 1969a, Survey for insect and disease activity in the high elevation spruce and fir on the White Mountain National Forest: U. S. Department of Agriculture, Forest Service, Northeastern Area, A-69-8, Amherst, MA.

Hadfield, J. S., 1969b, Aerial detection survey, Green Mountain National Forest: U. S . Department of Agriculture, Forest Service Northeastern Area, A-69-7, Amherst, MA.

Hopkins, A.D., 1901, Insect enemies of the spruce in the northeast: Bulletin 29, U. S. Department of Agriculture, Bureau of Entomology, Washington, D.C.
Johnson, A. H., Siccama, T. G., 1983, Acid deposition and forest decline: Environmental Science and Technology, 17(7):194-305.
Kucera, D. R., Taylor, R. G., 1984, Spruce budworms situation in North America: Miscellaneous Publication 1441, U. S. Department of Agriculture, Forest Service and Environment Canada, Canadian Forestry Service.
Mook, P. V., Eno, H. G., 1956, Relation of heart rots to mortality of red spruce in the Green Mountain National Forest: Forest Research Note 59, U. S. Department of Agriculture, Forest Service, Northeastern Forest Experiment Station, Upper Darby, PA.
Siccama, T. G., 1974, Vegetation, soil and climate in the Green Mountains of Vermont: Ecological Monographs, 44:325-349.
Siccama, T. G., Bliss, M., Vogelmann, H. W., 1982, Decline of red spruce in the Green Mountains of Vermont: Bulletin of the Torrey Botanical Club, 109:162-168.
Snowden, P., 1972, Aerial and ground detection survey of the White Mountain National Forest: U. S. Department of Agriculture, Forest Service, Northeastern Area, P-72-8, Portsmouth, NH.
Sprugel, D. G., 1976, Dynamic structure of wave generated Abies balsamae forests in the northeastern United States: Journal of Ecology, 64:889-913.
U.S.Department of Agriculture, Forest Service, 1970, Forest insect and disease conditions in New England, New York, and New Jersey, 1970: U.S. Department of Agriculture, Forest Service, Northeastern Area, Portsmouth, NH.
U.S.Department of Agriculture, Forest Service, 1973, Aerial and ground detection survey of the White Mountain National Forest: U.S. Department of Agriculture, Forest Service, Northeastern Area, Portsmouth, NH.
U.S.Department of Agriculture, Forest Service, 1985, Forest insect and disease conditions in the Northeastern Area, 1984: U.S. Department of Agriculture, Forest Service, Northeastern Area, Forest Pest Management, Durham, NH.
Vermont Department of Forests and Parks, 1960, Biennial report 1959-1960: Environmental Conservation, Vermont Department of Forests and Parks, Montpelier, VT.
Vermont Department of Forests and Parks, 1962, Biennial report 1961-1962: Environmental Conservation, Vermont Department of Forests and Parks, Montperlier, VT.
Vermont Department of Forests and Parks, 1966, Biennial report 1965-1966: Environmental Conservation, Vermont Department of Forests and Parks, Montpelier, VT.
Vogelmann, H. W., 1982, Catastrophe on Camel's Hump: Natural History, 91:8-14.
Wear, J. F., Pope, R. B., Orr, P. W., 1966, Aerial photographic techniques for estimating damage by insects in western forests: U. S. Department of Agriculture, Forest Service, Pacific Northwest Forest and Range Experiment Station, Portland, OR.
Weiss, M. J., McCreery, L.R., Millers, I., O'Brien, J.T., Miller-Weeks, M. 1985. Cooperative survey of red spruce in New Hampshire, New York, and Vermont-1984 Interim Report, U. S. Department of Agriculture, Northeastern Area, Forest Pest Management, Durham, NH.
Westveld, M. 1953. Ecology and silviculture of the spruce-fir forests of Eastern North America: Journal of Forestry, 51:422-431.
Wright, J. W. 1955, Species crossability in spruce in relation to distribution and taxonomy: Forest Science 1(4):319-349.

MAPPING SOIL AND WATER RESOURCES

Moderator: E. Horvath

LANDSAT IMAGERY: A TOOL FOR MONITORING SNOWMELT AND PREDICTING RUNOFF PATTERNS FOR MOUNTAIN WATERSHEDS

Ronald W. Marrs
Department of Geology and Geophysics
University of Wyoming, Laramie, Wyoming 82071

ABSTRACT

Snowpack has been traditionally monitored by on-site recording stations supplemented by periodic sampling for water content. Snowcover estimates from Landsat imagery can be used to effectively augment these measurements and improve water forecasts. The method used is to interpret snowcover area from a sequence of images obtained during several snowmelt seasons. The snowcover measurements are plotted against measured streamflow summed over the melt season. As cumulative runoff increases, snowcover decreases each year. The empirical relationship derived from these data is evaluated and compared to similar curves for other snowmelt seasons to derive a composite relationship that represents the "typical" pattern of snowmelt vs. runoff for each watershed. This typical curve can then be used to forecast volume and timing of runoff during the spring snowmelt using the current Landsat imagery to assess the condition of the snowpack in a given watershed. Composite curves were compiled for those areas where adequate data were available for several seasons. Both the individual and composite curves are useful for predicting the expected runoff from the most recent satellite images. Dates for beginning, peak, and end of spring snowmelt can be predicted.

INTRODUCTION

Landsat provides a new perspective on many earth resources. Among these resources is the snowpack that accumulates in mountain watersheds each year. Landsat provides an excellent platform for periodically observing the snowpack so that its accumulation and depletion can be monitored. Researchers have used the capability provided by Landsat to monitor the size and conditions of snowpack and have developed methods for relating their measurements to the annual runoff that is released by the melting of the snow.

We have attempted to build on these efforts and to apply this new tool to water forecasting on a regional scale. Our test regions were all mountain watersheds of Wyoming in which snowpack contributes significantly to the annual runoff. The procedures for analysis were tailored to be as simple and economical as possible while maintaining both flexibility and accuracy. We believe that our results represent a first step in the development of an operational system for monitoring snowpack and estimating runoff across much of Wyoming.

BACKGROUND

Just one year after the launch of the first Landsat system, the first results of applications tests were published detailing the use of Landsat as a tool for periodically mapping snowcover. In October, 1973 Barnes and others reported:

> "...snow extent can be mapped from ERTS imagery in more detail than is depicted on aerial survey snow charts. For the areas tested, the agreement between the percentage snow cover as determined from ERTS data and from aerial survey snow charts is of the order of 5 percent for most cases. Moreover, it appears that although small details in the snowline can be mapped better from higher-resolution aircraft photographs, boundaries of the areas of significant snow cover can be mapped as accurately from the ERTS imagery as from the aircraft photography. Moreover, the costs involved in deriving snow maps from ERTS imagery appear to be very reasonable in comparison with existing data collection methods."

The work continued, and in 1975 a symposium was held (Rango, 1975 (ed.)) to discuss results of a number of studies in which the utility of satellite data for mapping snowcover and for predicting runoff had been demonstrated (Katibah, 1975; Leaf, 1975; Rango and Salomonson, 1975; and Thompson, 1975). Thompson (1975) and Rango and Salomonson (1975) demonstrated the application of snowcover mapping to runoff forecasting in Wyoming watersheds. Even so, the techniques were not immediately accepted as operational procedures, although the test studies indicated potential for efficient and cost-effective application.

Four years later, another conference was held to discuss "Operational Applications of Satellite Snowcover Observations" (Rango and Peterson, 1979 (eds.)). Several newly tested procedures for estimating runoff from satellite snowcover measurements were presented (Schumann and others, 1979, Shafer and Leaf, 1979; Moravec and Danielson, 1979; Brown and others, 1979; Hannaford and Hall, 1979; Martinec and Rango, 1979). Some workers used Landsat data together with data from meteorlogical satellites (Schumann and others, 1979), others correlated the snowcover data with more sophisticated models for runoff prediction (Dillard and Orwig, 1979; Moravec and Danielson, 1979; Hannaford and Hall, 1979). Rango and Martinec (1982) incorporated daily temperature data into a prediction model. Most studies emphasized the necessity for defining specially tailored snowcover/runoff relationships for each drainage area in order to accommodate variations in pattern due to topography, slope aspect, vegetation cover, elevation and other local conditions. Most workers reported that good results could be obtained using Landsat as a base for estimating snow cover.

METHODS

The methods used in our work were largely those pioneered by Rango and Salomonson (1975), Thompson (1976), and

Moravec and Danielson (1979). Every effort has been made to keep interpretation and analysis procedures as simple and cost-effective as possible without unduly jeopardizing accuracy of results. In this context, the methods used by early investigators have been applied with only minor modification and without embellishments that could produce slightly improved forecasts but at considerable cost in terms of manpower and data analysis time.

Bulk-processed Landsat image transparencies of spectral band 5 (red band) were selected for interpretation because they provided strong contrast between snow and vegetation even during the late spring snowmelt period. Single-band scenes were preferred to color-composite scenes because of greater availability and lower cost. Transparencies were used because they are more readily input to video analysis and provide better image detail than prints.

Streamflow data from selected gauge stations were obtained from the Wyoming Water Research Center in the form of computer listings of daily discharge values and monthly discharge totals (Table I). Drainage sub-areas were defined to provide for single-scene coverage of individual regions with consideration made for availability of appropriate streamflow data (Fig. 1). Discharge values for most sub-areas are represented by measurements from a single gauge station.

YELLOWSTONE RIVER AT YELLOWSTONE LAKE OUTLET Y.N.P. STATION NO. 061865.00
LATITUDE 44-34-03 LONGITUDE 110-22-48 SECTION 0 TOWNSHIP 0 ,RANGE 0 P.M.
ELEVATION 7727.77 FT DRAINAGE AREA 1006.00 SQ MI NONCONTRIBUTING 0.00 SQ MI BASIN 01030300
YELLOWSTONE PARK DATA FROM USGS (P)

MEAN DAILY FLOW IN CFS BY WATER YEAR 1977

DAY	OCT	NOV	DEC	JAN	FEB	MAR	APR	MAY	JUNE	JULY	AUG	SEPT
1	1180.00	747.00	516.00	300.00	290.00	300.00	444.00	575.00	1020.00	1940.00	1230.00	822.00
2	1170.00	742.00	512.00	290.00	290.00	310.00	444.00	561.00	1070.00	1930.09	1220.00	810.00
3	1170.00	736.00	503.00	280.00	290.00	310.00	444.00	625.00	1120.00	1910.00	1190.00	798.00
4	1140.00	731.00	494.00	280.00	300.00	310.00	440.00	650.00	1200.00	1900.00	1180.00	792.00
5	1140.00	720.00	494.00	280.00	300.00	310.00	440.00	660.00	1290.00	1890.00	1150.00	769.00
6	1100.00	714.00	480.00	280.00	300.00	310.00	440.00	665.00	1410.00	1870.00	1150.00	764.00
7	1090.00	709.00	480.00	280.00	300.00	310.00	440.00	676.00	1500.00	1830.00	1140.00	753.00
8	1070.00	697.00	470.00	280.00	300.00	310.00	432.00	681.00	1610.00	1790.00	1140.00	736.00
9	1040.00	692.00	460.00	280.00	300.00	310.00	432.00	681.00	1720.00	1770.00	1110.00	720.00
10	1040.00	681.00	450.00	280.00	310.00	310.00	432.00	714.00	1860.00	1740.00	1090.00	709.00
11	1030.00	676.00	420.00	280.00	310.00	310.00	432.00	731.00	1940.00	1700.00	1070.00	692.00
12	1020.00	655.00	410.00	290.00	310.00	310.00	432.00	747.00	2000.00	1670.00	1060.00	681.00
13	998.00	655.00	400.00	300.00	310.00	320.00	432.00	753.00	2010.00	1640.00	1040.00	676.00
14	984.00	640.00	400.00	300.00	310.00	320.00	436.00	775.00	2060.00	1610.00	1010.00	665.00
15	963.00	635.00	400.00	300.00	310.00	320.00	440.00	822.00	2060.00	1580.00	998.00	591.00
16	956.00	620.00	400.00	300.00	310.00	330.00	440.00	840.00	2080.00	1550.00	991.00	681.00
17	921.00	615.00	380.00	300.00	310.00	340.00	444.00	876.00	2070.00	1530.00	963.00	681.00
18	900.00	610.00	370.00	300.00	310.00	340.00	444.00	888.00	2070.00	1510.00	949.00	565.00
19	888.00	605.00	350.00	300.00	310.00	350.00	448.00	876.00	2060.00	1450.00	949.00	655.00
20	882.00	590.00	330.00	300.00	310.00	350.00	448.00	921.00	2130.00	1440.00	935.00	645.00
21	870.00	580.00	330.00	300.00	310.00	360.00	452.00	921.00	2110.00	1420.00	928.00	630.00
22	858.00	580.00	330.00	290.00	310.00	360.00	460.00	935.00	2120.00	1420.00	914.00	620.00
23	840.00	570.00	330.00	290.00	310.00	360.00	460.00	935.00	2110.00	1400.00	900.00	615.00
24	834.00	566.00	330.00	290.00	300.00	370.00	464.00	949.00	2090.00	1390.00	900.00	615.00
25	822.00	557.00	330.00	290.00	300.00	380.00	472.00	963.00	2080.00	1410.00	882.00	605.00
26	810.00	539.00	340.00	290.00	300.00	380.00	485.00	998.00	2080.00	1370.00	894.00	600.00
27	798.00	534.00	340.00	290.00	300.00	390.00	507.00	1030.00	2070.00	1360.00	888.00	590.00
28	780.00	530.00	330.00	290.00	300.00	390.00	521.00	1030.00	2040.00	1340.00	886.00	590.00
29	775.00	530.00	330.00	290.00		448.00	539.00	1030.00	2000.00	1330.00	876.00	600.00
30	769.00	525.00	310.00	290.00		456.00	552.00	1020.00	1960.00	1290.00	858.00	605.00
31	764.00		300.00	290.00		448.00		1010.00		1260.00	834.00	
TOTAL	29602.00	18981.00	12319.00	9000.00	8510.00	10722.00	13696.00	25538.00	54940.00	49250.00	31325.00	20465.00
MEAN	954.90	632.70	397.39	290.32	303.93	345.87	456.53	823.81	1831.33	1588.71	1010.49	682.17
AC-FT	58714.71	37648.26	24434.38	17851.23	16879.33	21266.77	27165.61	50653.88	108971.90	97685.95	62132.23	40591.73

TOTAL ANNUAL FLOW IN ACRE-FEET = 563995.98

INSTANTANEOUS PEAK IN CFS = **

** INDICATES MISSING DATA
* INDICATES COMPUTED FROM INCOMPLETE DATA
E INDICATES ESTIMATED VALUE

Total Annual Runoff (cfs) = 284,348.

Table 1. Daily discharge values (cfs) for area Y-1. Values were summed for periods preceeding each coverage cycle.

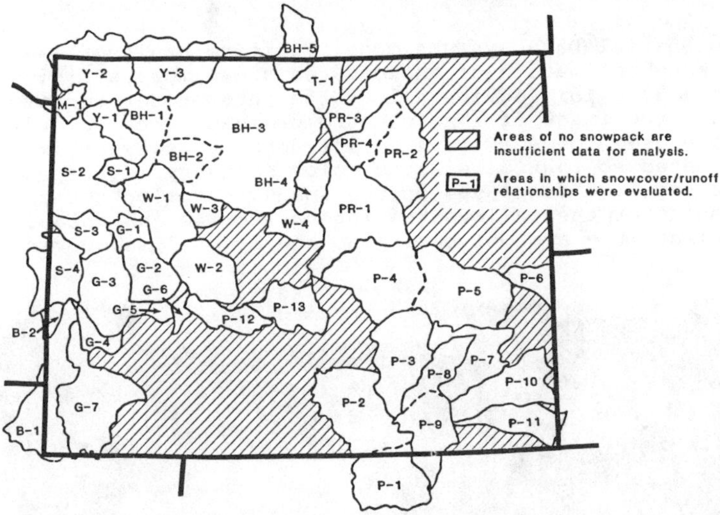

Figure 1. Index of areas analyzed for snowcover and runoff.

An image mask was constructed for each sub-area so that the region of interest could be isolated on each 1:1,000,000-scale Landsat scene (Fig. 2). Use of the image mask assures that the same area is evaluated on each successive Landsat scene. It also enables the interpreter to easily separate the sub-area to be evaluated from the surrounding areas with the electronic image analyzer.

Figure 2. Landsat image of the Yellowstone area partially snow-covered and masked for automatic analysis.

The Spatial Data Systems model 70 image analyzer was used to make estimation of snowcovered areas more efficient (Fig. 3). This device allows the interpreter to color slice the image of the selected drainage area according to image grey tones (Fig. 4). The density analyzer is adjusted so that a selected color boundary corresponds to the snowline displayed on the image. An electronic planimeter then allows the interpreter to instantaneously estimate the area of basin and the area of snowcover.

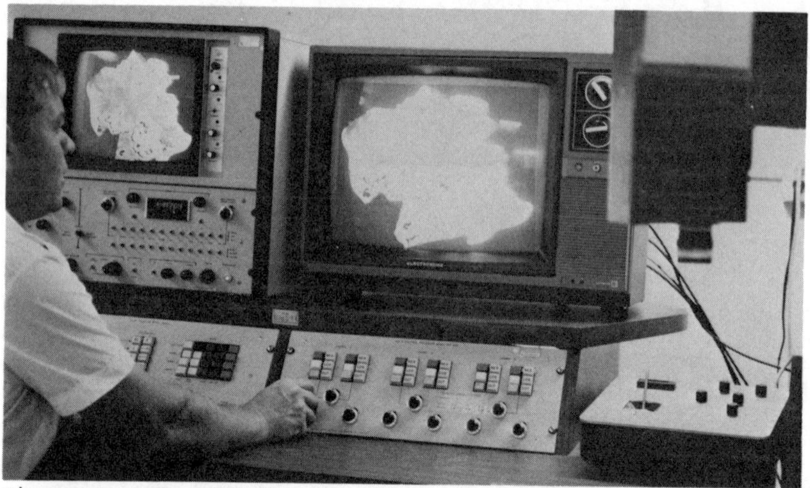

Figure 3. The Spatial Data Systems video image analyzer.

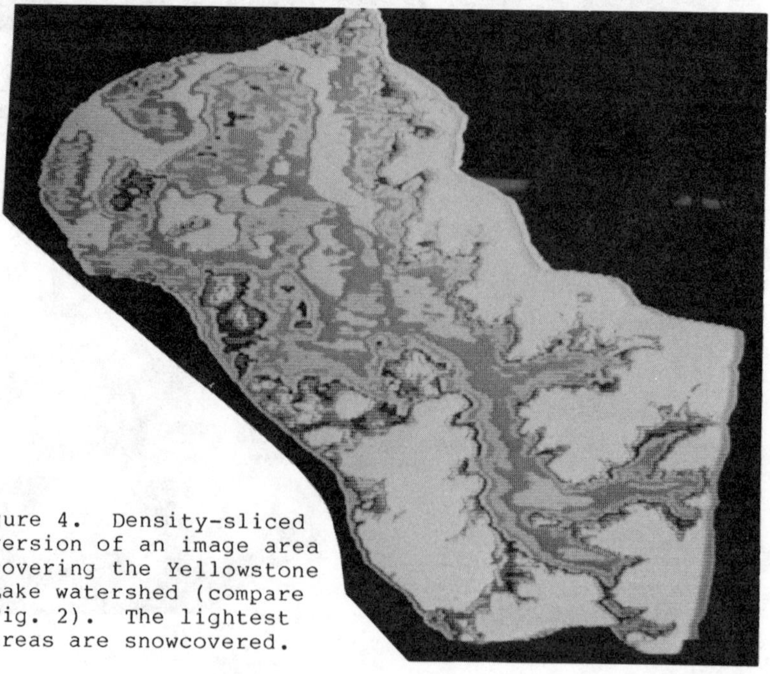

Figure 4. Density-sliced version of an image area covering the Yellowstone Lake watershed (compare Fig. 2). The lightest areas are snowcovered.

Daily discharge volumes measured at appropriate gauge stations (obtained from the digital data base of the Wyoming Water Research Center) are summed to produce cumulative runoff values for dates corresponding to the dates of the Landsat imagery. The cumulative runoff values are plotted against snowcover and against time to yield curves representing the snowmelt/runoff relationships of the drainage area (Fig. 5). Runoff versus time data are derived entirely from streamflow measurements and serve only as a guide to help define the snowcover/runoff relationship more accurately. Using the two curves together allows the interpreter to more accurately interpolate snowcover values between the widely spaced samples provided by Landsat images. Separate curves are generated for each season's data; then the curves are combined to produce a composite snowcover/runoff relationship for each drainage area. The composite curve is constructed by inversely weighting each season's data relative to its deviation from the long-term average runoff. As a result, the composite curve presents the snowcover/runoff relationship expected for an average year. The composite curve is presented along with curves representing individual seasons so that runoff predictions can be made using the composite curve as a standard and using the individual season curves to adjust for observed conditions indicating unusually heavy or unusually light snowpack or other abnormal conditions.

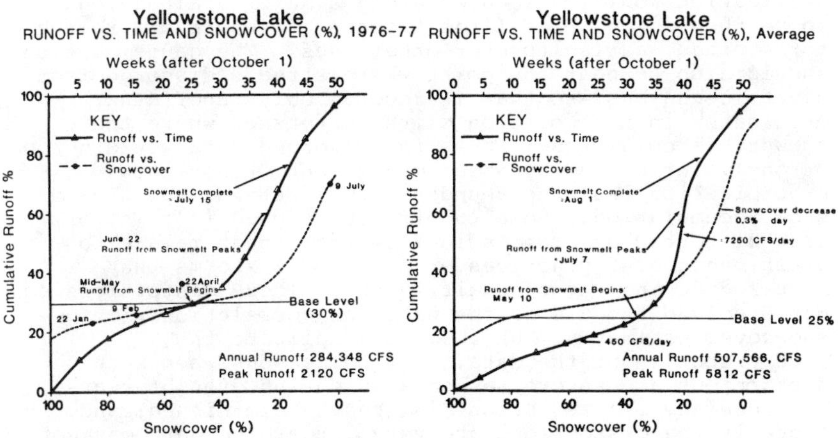

Figure 5. Snowcover and runoff curves for the Yellowstone Lake watershed. The average curve (right) defines the typical relationship snowcover and runoff in this area.

For each drainage area, a brief discussion is prepared detailing the quality of the data used in constructing the standard curves, critical values determined in the analysis, and other important considerations with regard to interpretation of graphs.

PROBLEMS AND PROGRESS

The procedure works well in most Wyoming watersheds and produces snowcover/runoff curves that show a consistent pattern and are readily interpretable. The typical curve shows that runoff is reasonably constant or diminishes slightly until the onset of spring snowmelt. As snowmelt progresses, the runoff increases steadily to a peak value and then wanes as the snowpack is depleted. For a given drainage the curve tends to shift laterally (to earlier dates) for seasons of light snowpack and toward later dates for seasons of heavy snowpack. Also, the runoff versus snowcover curve tends to steepen (percent snowcover diminishes rapidly) when snowpack is below normal. These typical patterns hold true for most areas; but considerable variability is apparent in comparing relationships from one drainage area to another, demonstrating the necessity for defining a separate set of reference curves for each area.

One problem that critically affects analysis in a few areas is that of availability of streamflow data. Gauge stations have not been installed on some drainages; and, for few others, the streamflow data are available only for a few years or only during certain parts of the year. If no streamflow data are available, the snowcover information can provide only a relative estimate of expected runoff. Also, few gauging stations are ideally located for measuring the actual water that results from snowmelt. Often, the gauge stations are below dams or along heavily used portions of the drainages where flows are regulated in one fashion or another. The ideal gauge station for this application would be located along a major drainage but above the region where large amounts of runoff are removed for storage, agriculture or other uses. The gauge station should also be near the point where streams disgorge from mountain canyons, so that it records only runoff that originates in that portion of the watershed where the snowpack accumulates. Thus, the recorded data would not be strongly influenced by water management or non-snow precipitation (such as thunderstorms) outside the area of snow accumulation. Some runoff curves used in this work were distorted by changes in flow rates that relate to water management practices or occasional storms that occurred during the snowmelt season. These distortions do not preclude the use of the data for correlation with snowcover estimates, but they do complicate the interpretation of the data. During seasons when such distortions are severe (especially during drought years), the accuracy of predictions is greatly diminished; and, in fact, if the distortions are very large, the data may not be adequate for defining a useful runoff/snowcover relationship.

A more severe problem is that of limited satellite coverage due to equipment failures or cloud cover. The 18-day cycle of Landsat satellites 1, 2, and 3 and the 16-day cycle of Landsats 4 and 5 are marginally adequate for runoff forecasts, even if every image acquired during the snowmelt season is cloud-free. Unfortunately, at least half of the imagery acquired during the snowmelt season in the Rocky

Mountain region is not useable because of cloud cover. Cloud cover is not a critical limitation in constructing snowcover/runoff reference curves because 14 years of relatively continuous operation of Landsat has produced one or more seasons of good, cloud-free coverage for most areas. These data can be used to define the reference relationships. But, if forecasts are to be made, the timeliness and availability of cloud-free images are critical, and the imagery should be available for each area every few days. No solution to this problem is possible with the present Landsat system, but several eventual solutions are possible:

1. The frequency of satellite imaging could be increased to provide greater opportunity to image on cloud-free days. This could be accomplished by using more than one satellite in alternating orbit cycles or by using a pointable satellite system (such as the planned SPOT system to be launched by the French in late 1985).
2. Landsat data might be supplemented by images from the geostationary weather satellites. The coverage provided by these systems has been used successfully for broad, regional snow surveys; and previous research suggests that it might be possible to effectively combine the data from the geosynchronous satellites with the less frequent, but higher resolution data from Landsat.
3. Cloud-penetrating radar systems could be used instead of the visible and near-IR bands of the current systems. Such a system would allow imaging at night and through cloud cover. Seasat and the Shuttle Imaging Radar experiments may be precursors to an operational radar system to be launched at some future time.

One final consideration is that snowcover measurements can not be used to forecast runoff until the snowmelt season is under way and the snowpack has begun to diminish. This precludes very early projections of expected runoff and necessitates continued use of snow course measurements or other methods to obtain early assessments of snowpack.

Forty-four drainage areas in Wyoming were identified in which winter snowpack contributes significantly to runoff. Data for all but three of these areas was adequate to attempt an evaluation of the relationship between snowcover and runoff. In the past two years images and streamflow data for several snowmelt seasons have been analyzed to determine the relationship between snowcover and expected runoff in each of the forty-one areas. Variations in runoff pattern have been noted for dry, normal, and wet seasons, so that estimates can be adjusted appropriately to accomodate variations in climate from year to year and month to month. We have undertaken preliminary tests of the predictive accuracy of the technique and have proposed a thorough test of both accuracy and applicability. In general, results are better in regions where runoff from snowmelt is a large proportion of the annual runoff; but other parameters (including cloud-cover, image availability, a quality of streamflow data) also affect the results.

CONCLUSION

The groundwork is now sufficiently complete and the accumulated data base is minimally adequate for operational application of this technique to water management in Wyoming. The techniques appear to be both effective and practical. Current and proposed tests will establish timeliness and reliability with which these new techniques can provide runoff estimates from satellite data. We are fast approaching the point where we must decide whether or not to implement the technique as a routine part of a statewide water management program. The commitment to use the Landsat estimation procedures in water management will involve data, equipment, and manpower; but the total cost should be very small relative to the potential gains to be realized through even the most modest improvement in runoff forecasts. Total cost for upgrading the data base (imagery and streamflow data), analyzing the data, and issuing forecasts of expected runoff each month during the snowmelt season should be on the order of $20,000-30,000 per year. The investment is certainly appropriate if even one mistake in water management can be averted due to improved forecasting capabilities.

REFERENCES

Barnes, J.C., Clinton, J.G., and Simmes, D.A., 1973, Use of satellite data for mapping snow cover in the western United States: in Anson, A., 1973, Symposiumon management and utilization of remote sensing data, American Society of Photogrammetry, Falls Church, VA, pp. 166-176

Brown, A.J., Hannaford, J.F. and Hall, R.L., 1979, Application of snowcovered area to runoff forecasting in selected basins of the Sierra Nevada, California: in Rango, A., and Peterson, R., 1979, Operational applications of satellite snowcover observations, NASA Conference Pub. 2116, pp. 185-200

Dillard, J.P. and Orwig, C.E., 1979, Use of satellite data in runoff forecasting in the heavily-forested, cloud-covered Pacific Northwest: in Rango, A., and Peterson, R., 1980, Operational applications of satellite snowcover observations, NASA Conference Pub., 2116, pp. 127-150

Hannaford, J.F., and Hall, R.L., 1979, Application of satellite imagery to hydrologic modeling snowmelt runoff in the southern Sierra Nevada: in Rango, A., and Peterson, R., 1979, Operational applications of satellite snowcover observations, NASA Conference Pub. 2116, pp. 201-222

Katibah, E.F., 1975, Operational use of Landsat imagery for the estimation of snow areal extent: in Rango, A., 1975, Operational applications of satellite snowcover observations, NASA SP391, pp. 129-142

Leaf, C.F., 1975, Applications of satellite snow cover in computerized short-term streamflow forecasting: in Rango, A., 1975, Operational applications of satellite snowcover observations, NASA SP-391, pp. 175-186

Martinec, J., and Rango, A., 1979, Discharge forecasts in mountain basins based on satellite snow cover mapping: in Rango, A., and Peterson, R., 1979, Operational applications of satellite snowcover observations, NASA Conference Pub. 2116, pp. 223-238

Moravec, G.F., and Danielson, J.A., 1979, A graphical method of stream runoff prediction from Landsat derived snowcover data for watersheds in the upper Rio Grande Basin of Colorado: in Rango, A., and Peterson, R., 1979, Operational applications of satellite snowcover observations, NASA Conference Pub. 2116, pp. 171-184

Rango, A., 1975, Operational applications of satellite snowcover observations: NASA SP-391, Washington, D.C., 430 p.

Rango, A., and Martinec, J., 1982, Snow accumulation derived from modified depletion curves of snow coverage: in Hydrological aspects of alpine and high mountain areas, IAHS Pub. no. 138, pp. 83-90

Rango, A., and Peterson, R., 1979, Operational applications of satellite snowcover observations: NASA Conference Pub. 2116, 301 p.

Rango, A., and Salomonson, V.V., 1975, Employment of satellite snowcover observations for improving seasonal runoff estimates: in Rango, A., 1975, Operational applications of satellite snowcover observations, NASA SP-391, pp. 157-174

Schumann, H.H., Kirdar, E. and Warskow, W.L., 1979, Satellite snowcover and runoff monitoring in central Arizona: in Rango, A., and Peterson, R., 1979, Operational applications of satellite snowcover observations, NASA Conference Pub. 2116, pp. 107-126

Shafer, B.A., and Leaf, C.F., 1979, Landsat derived snowcover as an input variable for snowmelt runoff forecasting in south central Colorado: in Rango, A., and Peterson, R., 1979, Operational applications of satellite snowcover observations, NASA Conference Pub. 2116, pp. 151-170

Thompson, A.G., 1975, Utilization of Landsat monitoring capabilities for snowcover depletion analysis: in Rango, A., 1975, Operational applications of satellite snowcover observations, NASA SP-391, pp. 113-128

MICROWAVE REMOTE SENSING OF SNOW COVER IN FORESTED AND NON-FORESTED AREAS

D. K. Hall, J. L. Foster, A. T. C. Chang
Hydrological Sciences Branch
NASA/Goddard Space Flight Center
Greenbelt, MD 20771

ABSTRACT

Determination of the variability of worldwide snow extent and depth is a problem which has important implications to global climate. The ability to input snow extent and depth information into General Circulation Models (GCMs) is currently being developed. Snow extent can be monitored using the 37 GHz passive microwave sensor on the Nimbus 6 and 7 satellites. Snow depth has been highly correlated with 37 GHz microwave brightness temperature (T_B) under dry snow conditions. However, there are two major factors that contribute to poor microwave T_B/snow depth relationships: dense forest cover and snow structure effects. Using Electrically Scanning Microwave Radiometer (ESMR) data and Scanning Multichannel Microwave Radiometer (SMMR) data, correlations between snow depth and microwave T_B have been found to be as high as -0.94 in homogeneous prairie areas, and as low as 0.06 in forested areas. Models have been developed to simulate the influence of forest cover and snow structure effects on the microwave T_B. The microwave emission from snow with large (depth hoar) crystals at the base of the snowpack has been modeled using a radiative transfer model. The formation of ice lenses and layers in a snowpack causes structural changes that are more difficult to model and thus the effects of ice lenses and layers on the microwave emission are largely unknown. Presence of forest cover results in higher T_Bs than would otherwise be expected according to modeling and observation.

INTRODUCTION AND PREVIOUS WORK

The extent and depth of worldwide snow cover is highly variable. The presence of even a thin layer of snow exerts a major influence on the local and regional energy balance because of its high albedo. In addition to extent, snow depth is important for climate and energy balance studies because deeper snow can take longer to melt than shallow snow and thus the highly reflective surface is present for a longer period of time.

In order to study the interactions between snow and climate, the capability to input snow extent and depth information into General Circulation Models (GCMs) is currently being developed. Such extent and depth data would also provide input to snowmelt runoff models and to snow energy balance models. The potential to measure snow depth using passive microwave sensors has been demonstrated under dry snow conditions through cloud cover (Rango et al., 1979; Burke et al., 1984), however, the quantitative determination of snow depth on a world-wide basis remains a difficult, but apparently solvable problem (Kunzi et al., 1982).

Chang et al. (1976) show that the depth of penetration of upwelling microwave radiation through snow is wavelength dependent and that, depending upon the properties of the snow, the emission can emanate from a depth that is 10 to 100 times the length of the wavelength. As liquid water content in snow increases, the ability of the microwaves to propagate through the snow decreases.

For temperatures generally encountered on Earth, the emitted intensity of radiation is expressed as brightness temperature (T_B) in Kelvins and follows the Rayleigh-Jeans approximation which shows that the radiance from a blackbody is proportional to its temperature (Chang et al., 1976).

It has been found that when using 37 GHz passive microwave data, an inverse relationship exists between snow depth and microwave brightness temperature (T_B) and correlations can be quite high especially in relatively flat, uniform areas comprised of low, homogeneous vegetation (Rango et al., 1979). In heavily forested areas, the microwave emission from forest cover tends to override the emissivity of the snow, and thus correlations between snow depth and microwave T_B tend to be low in these areas (Hall et al., 1982).

Data from the Nimbus 6 Electrically Scanning Microwave Radiometer (ESMR) and the Nimbus 7 Scanning Multichannel Microwave Radiometer (SMMR) are discussed. The Nimbus 6 ESMR (Wilheit, 1975) was launched in June 1975 and operated at a frequency of 37 GHz (0.81 cm wavelength), recording both horizontally and vertically polarized data at a resolution of approximately 30 km. The Nimbus 7 Scanning Multichannel Microwave Radiometer (SMMR) was launched in 1978, obtaining dual-polarized data at five wavelengths at a resolution of 30 km for the 37 GHz sensor (Gloersen and Barath, 1977).

A range of snow and ice features can be analyzed using different frequencies of the SMMR, but it is generally agreed that the 37 GHz data are most useful for studying internal snowpack properties and snow extent. In general, the horizontally polarized data are discussed in this paper because some evidence (Hall et al., 1982) suggests that the horizontally polarized data are preferable for analysis of snow depth.

Studies have been performed in various areas with varying vegetative and snow conditions and results differ widely (Foster et al., 1984). In this paper, two factors which complicate the microwave T_B/snow depth relationships will be discussed: vegetation and snow structure. The effect of vegetation on the microwave T_B is emphasized; there have been numerous studies in areas with vastly different types of vegetation. Studies have also been done in areas with quite different snow structures, but without ground verification, snow structure is difficult to identify and its effects on microwave emission are often unclear.

STUDY AREAS

Numerous snow covered areas in the United States and Canada having varied vegetative cover have been studied intensively in terms of their passive microwave response using ESMR and SMMR data. Some of these areas are 1) the Canadian high plains, 2) the Arctic Coastal Plain of Alaska, 3) Montana and North Dakota, 4) Illinois and Indiana, 5) Pennsylvania and Maryland, 6) northern Minnesota, 7) portions of Vermont, New Hampshire and Maine, and 8) the lower peninsula of Michigan.

The Canadian high plains are in southern Alberta and Saskatchewan and are covered by prairie grasses. The study area is located between 49° and 52°N latitude and 103° and 114°W longitude. The Arctic Coastal Plain of Alaska (north of 70°N, and between 145°W and 160°W) is underlain by permafrost and has low, shrubby tundra vegetation; there are no trees. The study area in the high plains of Montana and North Dakota is quite uniformly vegetated with prairie grasses.

The mid-western states of Illinois and Indiana and the eastern states of Maryland and Pennsylvania are characterized by mixed forest, pastureland and cropland.

The study area in northern Minnesota (44.5° - 49°N x 93.5° - 97°) is primarily comprised of spruce-fir, conifer bog and Great Lakes pine forests. The northeastern U.S. study area which covers portions of Vermont, New Hampshire and Maine, is forested with spruce hardwoods and spruce-fir forests, and the lower peninsula of Michigan has pine, fir, oak-hickory, elm, aspen and birch trees.

DISCUSSION OF RESULTS IN HOMOGENEOUS PRAIRIE AREAS

For the Canadian study area, correlations were obtained by determining snow depths (from snow course measurements) for 15 March 1976 and correlating the snow depths with the 37 GHz ESMR T_B data (Rango et al., 1979). For the Montana and North Dakota study areas, snow depths, measured at airports on 15 March 1976, were correlated with the Nimbus 6 ESMR data. For the Alaskan study area, the spatial variability in snow depth over the Arctic Coastal Plain was quite low and thus data covering a longer period of time were needed to obtain greater variability in snow depth in order to obtain meaningful correlations between snow depth and microwave T_B. Data from this area were analyzed for September through June, 1979-1982.

In the homogeneous prairie areas of the Canadian high plains the coefficients of correlation can be quite good, e.g. R = -0.93 as seen in Table 1. Also, in the Montana and North Dakota study area, the correlation seen in Table 1 is good where R = -0.94 (Foster et al., 1980). In addition to the fact that these areas have homogeneous vegetation, the fact that the correlations were run using only one day of data helped to reduce some of the variability in snow structure that may contribute to lowering the correlations when a time series of data is analyzed. Enough variability in snow depth was present over the Canadian and Montana and North Dakota study areas to allow meaningful regression relationships using only one day of data.

From Table 1, it can be seen that coefficients of correlation vary from -0.52 to -0.84 for the Arctic Coastal Plain of Alaska for the

Table 1. Microwave Responses (37 GHz Horizontally Polarized) in Homogeneous Prairie and Tundra Areas

Study Area	Vegetation	Study period	Range of Snow Depths	Coefficient of correlation	Significant at 0.05 Level	n	Source
North Dakota & Montana	prairie grasses	15 Mar 76	0-36	-0.94*	Yes	25	Foster et al. (1980)
Canadian high plains	prairie grasses	15 Mar 76	0-31	-0.93*	Yes	30	Rango et al. (1979)
Arctic Coastal Plain of Alaska	tundra	7 Oct 79 - 4 Jun 80	1-41	-0.79	Yes	41	This paper
		2 Oct 80 - 17 May 81	4-36	-0.52	Yes	39	This paper
		2 Oct 81 - 24 May 82	6-37	-0.84	Yes	40	This paper
		21 Sept 82 - 25 May 83	3-15	-0.63	Yes	42	This paper

*vertically polarized data used

4 years studied. Though this area is uniformly vegetated with low, shrubby vegetation, the correlation between microwave T_B and snow depth is not always strong. The variability of snow structure is undoubtedly responsible for some of the differences in T_B that cannot be explained by snow depth changes. This will be discussed in more detail later.

DISCUSSION OF RESULTS IN TRANSITIONAL AREAS

Snowfall occurs quite regularly in the mid-western United States, though a seasonal snowpack lasting more than 60 days is unusual. However, in Illinois and Indiana, snow was on the ground during most of January and February, 1979 and, during this period, air temperatures were considerably below normal. In general, using 37 GHz horizontally polarized SMMR data, the snow depth was significantly correlated with T_B as seen in Table 2. The snow depths and the coefficients of correlation for 5 nighttime satellite overpasses are shown in Table 2. The average of the 5 values shown is -0.76. In the winter, the croplands should be quite homogeneous and leaves should be absent from the deciduous trees allowing more upwelling emission through the snowpack to be sensed than if the area were fully forested with coniferous trees. Thus, this area is considered transitional.

An area in Pennsylvania and Maryland was analyzed for the 19 February 1979 SMMR overpass which occurred just after a major snowstorm. Though this area is forested, it has a mixture of cropland, pastureland and some urban areas and is also considered here to be transitional. The correlation between snow depth and microwave T_B was $R = -0.75$.

From these limited studies it can be seen that the correlations are poorer in these transitional areas than they are in the areas covered with prairie grasses. The variability of the Indiana and Illinois correlations is similar to the variability of the correlations found on the Arctic Coastal Plain of Alaska. It is presumed that snow structure effects are causing the variability in the correlations of microwave T_B and snow depth in Indiana and Illinois and in Alaska. In Indiana and Illinois, snow structure changes are probably the result of some melting and re-freezing within the snowpack. This will be discussed in more detail later.

DISCUSSION OF RESULTS IN FORESTED AREAS

Previous work has shown that surface cover will influence the passive microwave response to snow (Abrahms and Edgerton, 1977). Tiuri and Hallikainen (1981) used Nimbus 7 18 and 37 GHz data to analyze snowcovered areas in different types of terrain, e.g. farmland, bogland and forests. They concluded that, even with 35% shadowing due to trees, it was still possible to remotely sense snowcover through forests.

In the heavily forested lower peninsula of Michigan, Hall et al. (1982) found that the average 37 GHz horizontally polarized T_B was approximately 19°K lower in February 1979 than in February 1980.

Table 2. Microwave Responses (37 GHz Horizontally Polarized) in Transitional Areas Mixed Forest, Pastureland and Cropland

Study Area	Vegetation	Study period	Range of Snow Depths	Coefficient of correlation	Significant at 0.05 Level	n	Source
Illinois & Indiana	mixed forest & cropland	8 Jan 79	2-28	-0.80	Yes	28	Hall et al. (1984)
"	"	20 Jan 79	1-36	-0.87	Yes	28	Hall et al. (1984)
"	"	26 Jan 79	3-51	-0.62	Yes	28	Hall et al. (1984)
"	"	17 Jan 79	2-44	-0.86	Yes	28	Hall et al. (1984)
Pennsyl-vania & Maryland	mixed forest, pastureland & cropland	19 Feb 79	15-35	-0.75	Yes	62	Hall et al. (1984)

The average snow depth was 43.8 cm in February 1979 and 9.9 cm in February 1980. Even through dense forest cover, a markedly lower T_B was observed in the year with the deeper snow (1979). This lower T_B could not be explained by physical temperature differences between 1979 and 1980. However, the correlations between T_B and snow depth were low and positive (e.g. R = 0.25) and not significant at the 0.05 level (Table 3). It remains unclear why the correlations are so low for this area when such a large T_B drop occurs between the two years. One might expect a significant and negative correlation between snow depth and microwave T_B since the difference in T_B was so great between years. Other factors such as snow structure effects may be influencing the microwave emission.

A forested area in New England covering portions of Maine, Vermont and New Hampshire (43.5° - 45°N, 69.5° - 74°W) was also studied. For January through March 1979, a correlation was run for all available nighttime SMMR overpasses and a low coefficient of correlation (R = -0.11) was found (Table 3).

In the northern Minnesota forested study area, during the time period of January through March 1979, Nimbus 7 SMMR 37 GHz horizontally polarized data were correlated with snow depth and the correlation was again very low and not significant at the 0.05 level (R = 0.06) as seen in Table 3.

DISCUSSION OF THE EFFECTS OF SNOW STRUCTURE ON MICROWAVE EMISSION

In addition to the obvious and very important effect of forest cover on the microwave T_B/snow depth relationship, the effects of snow structure on the microwave T_B must be explained in order to determine the microwave response to snow depth. Snow structure is the result of local and regional energy balance processes including amount and timing of precipitation, meteorological and soil conditions and degree of melting and re-freezing within a snowpack caused by changes in air temperature.

Observation and modeling have indicated that, within a snowpack, variability in the thickness of the depth hoar layer can change the T_B. The depth hoar layer is a layer that forms at the base of a snowpack when there is a temperature gradient in the snowpack. Warmer temperatures at the base of the snowpack allow the formation of large crystals due to sublimation and redeposition of vapor from adjacent crystals. The large crystals found in the depth hoar layer are more effective scatterers of the microwave radiation and thus tend to lower the T_B and emissivity of the snowpack (Chang et al., 1976; Chang et al., 1982; Hall et al., in press). Depth hoar at the base of snowpacks is present in many areas of the world especially in very cold areas where the snow remains on the ground for a long period of time. The effect of depth hoar on microwave emission has been modeled using a radiative transfer model; results show a considerable lowering of the T_B resulting from the presence of depth hoar (Hall et al., in press).

The winter T_Bs on the Arctic Coastal Plain of Alaska should be consistently highly correlated with snow depth if homogeneity of terrain and absence of obscuring vegetation were the only factors that are significant to the microwave T_B. However the presence and

Table 3. Microwave Responses (37 GHz Horizontally Polarized) in Forested Study Areas

Study Area	Vegetation	Study Period	Range of Snow Depths	Coefficient of correlation	Significant at 0.05 Level	n	Source
Northern Minnesota	forested with marshes & lakes	4 Jan - 31 Mar 79	22-65	0.06	No	15	This paper
Maine, New Hampshire & Vermont	forested with crop- and pasture-land	4 Jan - 31 Mar 79	5-57	-0.11	No	16	This paper
Lower peninsula of Michigan	forested with crop- and pasture-land	1-28 Feb 80	5-28	0.25	No	34	Hall et (1982)

variability of the depth hoar layer will change the snow structure and thus the T_B because the thickness of the depth hoar layer and the crystal sizes within the depth hoar layer may increase as the winter progresses (Hall et al., in press). Correlations of snow depth and microwave T_B will be affected by these snow structure changes.

Other snow structures that are thought to influence the T_B/snow depth relationship are ice lenses and layers (Rango et al., 1979). Melting of even the top layer of a snowpack can substantially change the snowpack structure. Re-freezing as the water percolates down through the snowpack can result in the formation of ice lenses and layers.

The effect of ice lenses and layers on the microwave T_B is largely unknown but it is considered that these structure changes could alter the T_B/snow depth relationships in temperate areas where the air temperature fluctuates above and below the freezing point (Rango et al., 1979; Hall et al., 1984). This is undoubtedly true for the Illinois and Indiana and eastern study areas. These areas may have structural changes that result from meltwater percolation through the snowcover. This effect coupled with the mixed forest cover in these areas, is probably responsible for the poorer correlations which are found in temperate areas relative to those found in the homogeneous prairie areas. Snow structure changes may also be important in the forested areas.

CONCLUSION

Additional work is necessary in order to develop an improved understanding of the T_B/snow depth relationships in heavily forested areas, very cold areas that are prone to depth hoar development, and temperate areas in which the snowpacks are subjected to melting and re-freezing. The latter two conditions alter the snow structure, and the forests mask the microwave emission from snow.

Some work has been done that has increased our understanding of the complicating factors that affect the microwave scattering in snowpacks. Some snowpack structural chararcteristics that affect the microwave response have been identified and modeled using a radiative transfer model. Additional modeling coupled with actual snow measurements in different areas is necessary to improve our understanding of the T_B/snow depth relationship on a global scale.

REFERENCES

Abrahms, G. E. and A. T. Edgerton, 1977: Snow parameters from Nimbus 6 Electrically Scanned Microwave Radiometer. Final Report Aerojet Electro-Systems Co. Report No. 1932FR-1.

Burke, H. K., C. J. Bowley and J. C. Barnes, 1984: Determination of snowpack properties from satellite passive microwave measurements. Remote Sensing of Environment, V. 15, pp. 1-20.

Chang, A. T. C., P. Gloersen, T. Schmugge, T. T. Wilheit and H. J. Zwally, 1976: Microwave emission from snow and glacier ice. Journal of Glaciology, V. 16, pp. 23-39.

Chang, A. T. C., J. L. Foster, D. K. Hall, A. Rango and D. K. Hartline, 1982: Snow water equivalent estimation by microwave radiometry. Cold Regions Science and Technology, V. 5, pp. 259-267.

Foster, J. L., A. Rango, D. K. Hall, A. T. C. Chang, L. J. Allison and B. C. Diesen, 1980: Snowpack monitoring in North America and Eurasia using passive microwave satellite data. Remote Sensing of Environment, V. 10, pp. 285-298.

Foster, J. L., D. K. Hall, A. T. C. Chang and A. Rango, 1984: An overview of passive microwave snow research and results. Reviews of Geophysics and Space Physics, V. 22, pp. 195-208.

Gloersen, P. and F. T. Barath, 1977: A Scanning Multichannel Microwave Radiometer for Nimbus-G and Seasat-A. IEEE Journal of Oceanic Engineering, V. OE-2, pp. 172-178.

Hall, D. K., J. L. Foster and A. T. C. Chang, 1982: Measurement and modeling of microwave emission from forested snowfields in Michigan. Nordic Hydrology, V. 13, pp. 129-138.

Hall, D. K., J. L. Foster and A. T. C. Chang, 1984: Nimbus 7 SMMR polarization responses to snow depth in the mid-western U.S. Nordic Hydrology, V. 15, pp. 1-8.

Hall, D. K., A. T. C. Chang and J. L. Foster (in press): Detection of the depth hoar layer in the snowpack of the Arctic Coastal Plain of Alaska using satellite data.

Kunzi, K. F., S. Patil and H. Rott, 1982: Snow-cover parameters retrieved from Nimbus-7 Scanning Multichannel Microwave Radiometer (SMMR) data. IEEE Transactions of Geoscience and Remote Sensing, V. GE-20, pp. 452-467.

Rango, A., A. T. C. Chang, J. L. Foster, 1979: The utilization of spaceborne microwave radiometers for monitoring snowpack properties. Nordic Hydrology, V. 10, pp. 25-40.

Tiuri, M. and M. Hallikainen, 1981: Microwave emission characteristics of snow covered earth surfaces measured by the Nimbus-7 satellite. Proceedings of the 11th European Microwave Conference, Amsterdam, September 1981.

Wilheit, T. T., 1975: The Electrically Scanned Microwave Radiometer (ESMR) Experiment, J. E. Sissala (ed.), Nimbus-6 User's Guide, pp. 87-108.

PHOTOGRAMMETRIC INPUT TO A GEOGRAPHIC INFORMATION SYSTEM FOR MODELING SOIL EROSION

R. Welch and T.R. Jordan
Laboratory for Remote Sensing and Mapping Science
Department of Geography
University of Georgia
Athens, GA 30602

and

A. W. Thomas
Southern Piedmont Conservation Research Center
U. S. Department of Agriculture
Agricultural Research Service
Watkinsville, GA 30677

Photogrammetry is defined as the art and science of deriving accurate measurements from image data. Although photogrammetry has proved useful in a number of disciplines, it is only beginning to receive serious attention as a method for quantifying the amount of soil eroded from agricultural lands (Welch, Jordan and Thomas, 1984; Welch and Thomas, 1985).

Soil erosion, and in particular gully erosion, has been identified as a problem of major concern by the U.S. Department of Agriculture (USDA). Gullies produced along natural drainage paths in cultivated fields by storm runoff are periodically filled-in by normal tillage operations. These gullies are termed "ephemeral" and produce enormous amounts of sediment. However, the amount of soil lost from agricultural lands due to ephemeral gully erosion is difficult to measure. The Universal Soil Loss Equation (USLE) has been employed by the USDA to estimate sheet and rill erosion, but was never intended to include gully erosion. Consequently, efforts are presently underway to develop procedures that will lead to better estimates of total soil erosion.

A joint study team from the Laboratory for Remote Sensing and Mapping Science, University of Georgia, and the USDA Agricultural Research Service's Southern Piedmont Conservation Research Center, working in conjunction with the USDA Soil Conservation Service, have developed photogrammetric techniques for quantifying gully erosion using low altitude aerial photographs recorded over cultivated fields in North Georgia. It is anticipated that these studies will aid in the development of predictive methodologies for estimating the amount of soil lost from croplands in the Southeast.

The basic problem facing scientists seeking quantitative information on soil erosion is one of deriving extremely accurate elevation measurements at periodic intervals keyed to the cropping cycle. When it is realized that for average conditions a field will lose only a fraction of an inch of topsoil per year to erosion processes, it is evident that the measurement accuracy requirements for open field areas are extremely stringent and cannot be easily met by any current techniques. Gullies, however, which are developed along the paths of concentrated flow, may undergo erosion ranging from an inch or two to over one foot during the year and serve as transport corridors for soil eroded from the surrounding watershed. Consequently, vertical measurement accuracies of ±1/2 to ±1 inch and contours at intervals of 6 inches to 2 feet will permit the quantification of erosion and deposition along the gully. Specific measurement requirements will vary from site to site and may depend on slope, soil type, precipitation patterns and cropping practices typical of local conditions. The measurements, however, must be referenced to a stable datum to reflect the changes which occur with time.

Seven agricultural field test sites located near Athens, Georgia have been selected for detailed study. These sites range in size from 5 to 20 acres and are subjected to normal tillage operations for a winter wheat(December-May)/soybean(June-November) cropping cycle.

Stereo aerial photographs of 9 X 9 inch format are recorded with a photogrammetric camera (f = 6 inches) from altitudes of 600 to 900 feet above the terrain at four critical periods during the year determined by the agricultural practices. Depending on the flying height, a single stereomodel will cover an area of 7 to 17 acres and encompass an entire study site.

Prior to flying an aerial photographic mission, it is necessary to establish a reference datum based on a few ground control points. These ground control points must be accurately positioned (in X, Y and Z) at key locations by conventional ground survey techniques, and paneled so as to appear on the aerial photographs. Flight line markers are also established to guide the pilot and provide targets over which the camera operator will record the exposures. Centering exposures over the markers ensures that the paneled ground control points will be located at optimum positions within the photographs, thus permitting the stereoplotter operator to precisely scale and level the model. In order to meet the pre-defined accuracy requirements, the positioning of control points and markers must be carefully planned and executed.

The photogrammetric data reduction involves the selection of stereopairs which have an optimum distribution of control for scaling and leveling the model, plus additional points for conducting independent checks of the accuracy to which the X, Y, Z terrain coordinates are measured. All data reduction is carried out on a Wild B8 stereoplotter equipped with high intensity illumination and X, Y, Z tri-axis digitizing systems. This plotter has been carefully adjusted to optimize the accuracy to which Z-coordinates can be determined. Numerous operational checks have yielded root-mean-square error (RMSE) values of between ±0.5 and ±0.9 inch for the Z-coordinate (or approximately 1/12,000 of the flying height) and verified that contours are correct to 3 inches or better. These accuracies equal or exceed those which might be reasonably obtained by ground survey procedures.

In order to consistently achieve vertical accuracies of less than ±1 inch, special procedures for the orientation of the stereomodel have been developed. Measurements and plotting of contours are conducted by a skilled operator and detailed topographic maps at scales of 1:240 to 1:360 (1" = 20 ft. to 1" = 30 ft.) are compiled directly from the photographs. In addition, all measurements in the form of terrain coordinates are recorded on magnetic tape for subsequent generation of digital elevation models (DEMs) or other map products with the aid of computer-assisted mapping routines.

Because each map compiled by the photogrammetric procedures is referenced to the ground survey network, changes in the longitudinal or cross-sectional profiles of the gullies can be determined by comparing maps from two or more dates. By placing two maps in register and digitizing the areas between contours, the changes in volume at points along the gully as well as the total volume change between the two dates can be computed. These changes in volume may then be plotted as a function of position in the downslope direction to clearly depict those regions of the gully undergoing erosion and those at which material is being deposited.

Using these techniques, the change in volume of two representative ephemeral gullies was determined for the interval from March through December, 1984, a period of about 9 months (Table I). The gullies, located at separate sites, each occupied an area of about 0.1 acre and had drainage areas of approximately two acres. For Gully A, the host field was eroded by the first date of photograph (March 23) and was not planted until July 6. The tillage operation moved about 24 tons of soil into the gully area. By December 29, erosional processes had removed approximately 39 tons of soil, producing a net loss of about 16 tons over the 9 month study period. At Gully B, the net loss was much greater, totalling about 37 tons.

Table I - Erosion Data for Two Gullies

Gully A

Photography Date 1	Photography Date 2	Change in Volume (ft.^{3})	Change in Weight (tons)*
3/23/84	6/1/84	-36.7	-1.7
6/1/84	7/6/84	+539.7	+24.3
7/6/84	12/29/84	-859.4	-38.7
	NET CHANGE	(-356.4)	(-16.1)

Gully B

Photography Date 1	Photography Date 2	Change in Volume (ft.^{3})	Change in Weight (tons)*
3/23/84	7/6/84	+334.0	+15.0
7/6/84	12/31/84	-1144.8	-51.5
	NET CHANGE	(-810.8)	(-36.5)

* Based on an estimated soil weight of 90 lbs/foot^3

An alternative procedure which can be used to show changes caused by erosional processes with respect to time involves the construction of DEMs from photographs of the gully sites recorded at periodic intervals. Because a DEM is a set of elevation values (Z-coordinates) arranged in a regular grid referenced to a given coordinate system, the resulting matrix of X, Y, Z values defines the terrain surface. However, in order for the DEM to closely represent the terrain, a grid interval of 1 to 3 feet is required. If the critical area of a gully site is assumed to have dimensions of only about 30 X 300 feet, a matrix of 1000 to over 9000 points must be carefully measured by the photogrammetrist. While this may be possible for selected sites, the time required to construct a DEM containing thousands of points meeting stringent accuracy requirements is likely to prove prohibitive unless the measurement process can be automated.

To date, efforts have centered on developing photogrammetric techniques for use with very low altitude aerial photographs. Future research will be directed at reducing ground survey requirements, automating the data reduction procedures and achieving comparable accuracies with aerial photographs obtained from significantly higher altitudes, thus permitting intensive sampling schemes to be conducted over large areas at minimum cost. In addition, experiments are underway to integrate photogrammetric measurements of elevation, slope and gully parameters for the study sites with information on soil type, precipitation and cropping practices to create a data base which can be utilized with geographic information system (GIS) software. It is anticipated that a GIS approach will aid USDA personnel in developing models for soil erosion processes and provide a basis for recommending appropriate measures aimed at conserving the soil resource and improving crop productivity.

References

Welch, R., T. R. Jordan and A. W. Thomas, 1984, A Photogrammetric Technique for Measuring Soil Erosion, Journal of Soil and Water Conservation, Vol. 39, No. 3, pp. 191-194.

Welch, R. and A. W. Thomas, 1985, Aerial Photogrammetry: A Potential Technology for Monitoring Soil Erosion, Photomethods/Imaging Technology in Research and Development, May, pp. 4-7.

THE RELATIONSHIP BETWEEN SOILS DATA AND FOREST CLEARING AND FOREST REGROWTH TRENDS IN COSTA RICA

R. E. Pelletier and S. A. Sader
NASA/Earth Resources Laboratory
National Space Technology Laboratories
NSTL, Mississippi 39529

ABSTRACT

Trends indicate that rates of deforestation in tropical regions have greatly increased in recent years having a substantial impact on local resources as well as global processes. Techniques utilizing remotely sensed data can provide a means by which present and past forest conditions can be assessed, but future estimates of forest condition must also include input from a variety of biophysical factors which influence the propensity with which forested land may be subject to clearing and subsequent regrowth. A study was conducted to evaluate the relationships between soil type and trends in deforestation and the potential for land use abandonment and forest regrowth. The larger study, regarding deforestation trends, was conducted in the San Jose 1:20000 quadrangle area of Costa Rica, while a more intensive study focused on the Sarapiqui area and adjacent La Selva Biological Preserve due to the region's great diversity in climate and landforms in a relatively small area. A geographically referenced data base was created containing different taxonomic groupings of soils data, forested condition from five dates of aerial photography between 1940 and 1983, biophysical data from maps (e.g., slope), and land cover classifications from two dates of Landsat data over the intensive study area. Co-occurrence matrices demonstrated that certain soil types were quite likely to experience deforestation, other soil types would experience very little, and many showed only moderate correlations. This latter occurrence indicates that other biophysical factors may have a greater influence on the propensity towards deforestation than does soil type in those cases. Similar relationships could be drawn between soil type and the likelihood that an area would be abandoned and returned to forest regeneration.

INTRODUCTION

Background

Though there is some controversy about actual rates, trends indicate that deforestation in tropical regions has greatly increased over recent years (Myers, 1980; U.S. Dept. of State, 1980). Since forests and woodlands cover approximately 40 percent of tropical regions (Lanly, 1982), these high rates of deforestation can translate into significant changes in climatic processes and related biogeochemical cycling. Tropical forests account for at least 25 percent of global terrestrial carbon fixation. According to Dickinson (1981), the net global release of CO_2 due to tropical deforestation is approximately 1.0×10^{12} kg C annually, which is about 17 percent the CO_2 released by fossil fuels. Others have also investigated the present and potential influences of deforestation on the carbon cycle and organic matter storage depletion (Brown and Lugo, 1982) and atmospheric CO_2 from burning (Wong, 1978). Ehhalt and Schmidt (1978) have similarly noted the production of other "greenhouse" gases, namely CH_4, that can be attributed to deforestation. The energy balance of tropical regions and subsequently the globe is also being altered by changes in albedo from forest clearing (Potter, 1975; Dickinson, 1981). Likewise, the global water cycle, for which tropical regions contribute approximately 58 percent of the water vapor, could be drastically affected by deforestation (UNESCO/UNEP/FAO, 1978).

In addition to climatic consequences, soil properties and fertilizers are generally adversely affected by deforestation (Sanchez, 1981). Allen (1985) studied some of the short- and long-term effects. Burning often results in a short-lived increase in nutrient availability but most nutrient levels decrease significantly over the long-term. Additionally, soils derived from older parent material demonstrate substantially lower fertility levels than those derived from younger parent materials. While available phosphorous levels initially increase after forest clearing, phosphate adsorption is a major problem in the tropics (Fox and Searle, 1978). Erosion is another considerable problem for deforested regions in the tropics. The lowered fertility status of the soil may decrease crop density and vigor resulting in increased soil exposure. In addition many erosion problems are associated with cultivation of hilly regions in the humid tropics (Sheng, 1982).

Approach

A system for inventorying and monitoring forest change over time is greatly needed. A number of procedures can be employed to determine past forest areal coverage and clearing trends. Written records exist for many areas but their accuracy and degree of completion may be at a minimum. Techniques utlizing aerial photography and other forms of spatially determined remotely sensed data (e.g., satellite and aircraft data) can provide for a means by which present and past forest conditions can be assessed (Woodwell et al., 1983; Sader, 1980). From such change detection techniques, trends can be formulated for the time frames and geographic locations selected (Joyce et al., 1980).

However, future estimates of deforestation trends must include input from not only remotely sensed data sources but also input from the variety of biophysical factors which influence the propensity with which forested land may be subject to clearing and subsequent regrowth (Sader and Joyce, 1984). Such factors would include soils, geology, slope, elevation, geomorphology, life zone and climatic indices. Social, economic and political factors would also be important considerations in determining the potential for deforestation.

This paper will focus on biophysical factors and primarily discuss the relationships between soil type and trends in deforestation and the potential for land use abandonment and forest regrowth. The effect of clearing trends on critical erosion areas will also be examined. Social, economic and political factors will only be included as a means to describe some otherwise unexpected relationships but not in any more quantitatively analytical form. From this approach of developing relationships between these biophysical and other factors predictive models may be formulated to determine future trends in both areal extent and geogrpahic location.

Study Site Description

Costa Rica, as an example indicative of many areas in tropical and subtropical regions, is originally assumed to have been nearly all forested while only about 18 percent of the area is in primary forest today (Sader and Joyce, 1985). The specific broad-scale test site covers an area approximately 21,000 square kilometers corresponding to the San Jose 1:200,000 quadrangle map. Though the land mass considered is relatively small there is great diversity in geomorphology, soils, life-zone, slope and land use. However, unlike most other tropical/subtropical regions, Costa Rica also has better coverage of this diversity in the form of thematic maps at both large and small scales.

A more intensive study site is located in the Sarapiqui and adjacent La Selva Biological Preserve in the northeastern portion of the San Jose quadrangle. The Sarapiqui is a region located along the bottomland of the Rio Succio and has undergone large amounts of conversion from forest to agricultural crops (largely bannanas) and pasture land. The La Selva Biological Preserve is a 1400 hectare tropical wet forest dedicated to biological research which is owned and administered by the Organization for Tropical Studies. La Selva is located to the northwest of the Sarapiqui region and exhibits much steeper slopes and a variety of soil types not present in the bottomland. Together the two areas cover a large diversity of land forms in a relatively small area.

METHODS

Data Acquisition/Sources

Forest condition maps depicting location and extent of primary, dense (>80% crown closure) forest in Costa Rica at four reference dates were acquired from the Costa Rican National Planning Agency (Perez and Protti, 1978). The forest maps were derived through interpretation of aerial photos (1940, 1950, 1961) and image analysis of Landsat MSS data (1977). While these four forest condition maps cover the entire San Jose quad, additional Landsat MSS data from 1976 and 1984 were acquired for the intensive study site.

Several thematic maps of biophysical data, with special emphasis on soils, were also accumulated for the San Jose quad from the Institute Geografico National for incorporation into a multi-file data base. The original thirty-nine class soil data set (Perez, et al., 1979) was regrouped into six additional classifications based on U.S. soil taxonomy; order (7 classes); suborder (15 classes); great group (18 classes); subgroup (31 classes); clustered subgroup (25 classes); and landscape/climatic factors (e.g., parent material, landscape position, etc.) (7 classes). The other biophysical thematic maps included were slope gradient, geology, geomorphology and life zones.

Data Processing

All data was processed using the Earth Resources Laboratory Applications Software (ELAS) (Junkin, et al., 1980). All map data (forested condition, soils, slope, geology, geomorphology and life zones) were digitized and entered into a common georeferenced data base. In this form, individual files could be queried to generate tabular summaries of land area occupied by each class.

The Landsat MSS data from the intensive study area was likewise georeferenced and entered into the data base. Land cover classifications (including the following classes; forested, banannas, cropped or pastured, cleared and/or wet, and river or water bodies) were developed for both the 1976 and 1984 (January scenes) data sets. Upon completion of this procedure, both data sets were geographically overlaid to develop a change detection classification from 1976 to 1984 (forest-to-forest, nonforest-to-nonforest, forest-to-nonforest, and nonforest-to-forest).

Frequency of co-occurrence matrices were developed between each soil classification and the other digitized biophysical data files. For the intensive study site the soil classifications and slope data were similarly compared with the four change detection classes.

RESULTS AND DISCUSSION

Broad-Scale Study of the San Jose Quadrangle

A large variety of soils exist in Costa Rica due to the great diversity exhibited in its climate and landforms. For example, seven out of ten possible soil orders (of significant mapping size at a 1:200000 scale) can be found in the San Jose quad. Those present include Alfisols, Entisols, Histosols, Inceptisols, Mollisols, Ultisols, and Vertisols, of which the Inceptisols are the predominant order. According to a frequency of co-occurrence evaluation all the Alfisols had been cleared as early as 1940. By 1950 all of Vertisols had been cleared, and by 1961 all the Mollisols had been added to the "cleared" list as well. These soil orders are generally quite favorable for cropping or grazing purposes when climatic conditions are also suitable, which is the case here since these soil orders spatially correspond to favorable tropical dry and tropical moist life zones as well.

The other soil classes were similarly compared for frequency of co-occurrence with the nonforested classes from the four forest condition data sets and reclassified as having a high propensity for deforestation (>75% frequency of co-occurrence); moderate propensity (45-75%); and low propensity (<45%). Though the original (39 class) classification gave the most detailed account it showed few consistent and therefore decisive patterns for future modelling purposes based on the 1940 data set. The "clustered" subgroup classification gave similar results as the original classification while the subgroup level classification helped to significantly differentiate the propensities among the Inceptisols. The great group level illustrated the same general trend of correlation between clearing and soil type but demonstrated that most Inceptisols showed no strong propensity to deforestation. The landscape/climatic level showed a strong propensity for deforestation on fluventic, ustic and lithic classes with a moderate propensity on typic and highly organic soils.

The 1950-1977 data sets showed the same general trends when compared with the different soil classifications. Many more classes became 100 percent cleared. Some unexpected (considered less suitable) classes began to show a moderate or high propensity for clearing; typic Sulfaquents, several Histosols, lithic subgroups and aquic subgroups. However, all but the lithic subgroups tend to be wet for significant portions of the year and much of their clearing can be explained by increases in rice hectarage. The increased clearing of lithic subgroups, often the rocky soils of the mountainsides, can in part be explained by increased planting of coffee. On the other hand, those lithic subgroup soils planted for pasture are subject to severe erosion problems.

Soil types which demonstrated a low propensity for deforestation could generally be grouped as those soils inundated by water during some part of the year (with the exceptions as noted above which went into rice production), those young, rocky soils derived from volcanic parent material, and those soils subject to aluminum toxicity. These same trends were apparent no matter which classification scheme was evaluated. The landscape/climatic level demonstrated a strong correlation to unsuitability for forest clearing with aquic (wet) and andic (volcanic) classes, but the great group and subgroup levels offered the most consistent and relatively detailed levels for deforestation potential. Unexpected classes (those that should be suitable) with a low percentage of clearing included typic Paleudults and Dystrandepts. Upon further evaluation, both of these soil types were generally located on very steep slopes - more than 50 percent of their hectarage found on slope gradients greater than 30 percent. Though soil physical and chemical properties (however with possible phosphorous deficiencies) could have been suitable for agricultural use the steep slopes prevented their exploitation.

When the soil classifications were compared with the other biophysical factors for frequency of co-occurrence most soil types could be accounted for in three or less of the biophysical classes, often in just one class. For example, the fluvaquentic Hapludoll consisted of 97 percent of the geology class QAl derived from fluvial deposits. As with the correlations between soil classes and forested condition the great group and subgroup classifications offered the best means of correlation between soil type and the other biophysical classes. These two classifications offered the best procedure for generalizing relationships while still maintaining a needed amount of specificity. Understanding these relationships and the potential for co-occurrence should allow some interpretations to be made about a particular biophysical factor, should it be missing from the data base, if the others are present.

Intensive-Scale Study

For the intensive study site in the Sarapiqui and La Selva Biological Preserve a 50 meter spatial resolution data base, containing the same data planes as for the broad-scale study, was developed from the digitized data. In addition, two Landsat MSS scenes, (January, 1976; January, 1984) which cover the intensive study site, were added to the data base. Land cover classifications were developed from each MSS scene. Percentages of each landcover class and its percent change over the period 1976 to 1984 are found in Table 1. While this region

Table 1. Land Cover Classes from Landsat MSS Data for the La Selva and Sarapiqui Region

Class	Percent (%) Class Coverage 1976	Percent (%) Class Coverage 1984	Percent (%) Class Change
Forest	66.2	59.7	-6.5
Banannas	3.0	4.6	+1.6
Crop/Pasture	9.6	15.2	+5.6
Cleared/Wet	20.4	19.7	-0.7
River/H_2O Bodies	0.8	0.8	0.0

has a higher percent coverage of forest than the San Jose quad as a whole it has experienced an approximate 6.5 percent net clearing of forest over the six year period. The larger portion of the change in forested land took place in the bottomlands of the Sarapiqui. This is also the portion of the region in which all the bananna fields can be found.

A change detection classification was developed to indicate the type or direction of change over the 1976-1984 time frame (Figure 1). The La Selva Preserve occurs in the north central portion of the image and the Sarapiqui along the eastern half of the image. While there was a gross loss in forested land of approximately 12 percent over the six-

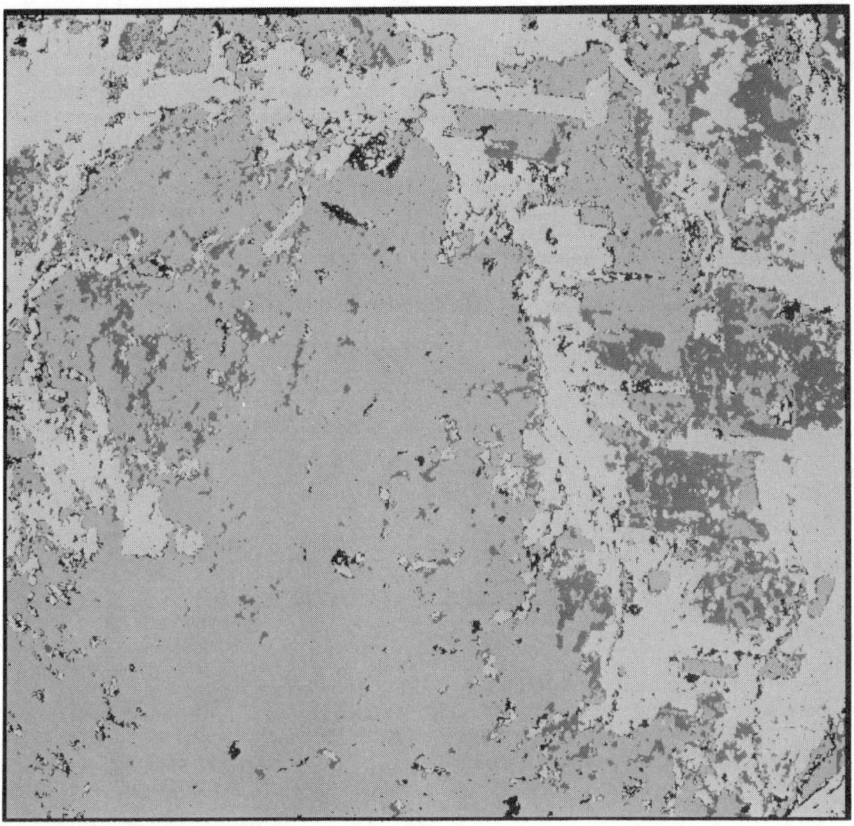

Figure 1. Landcover Change Classification for the La Selva and Sarapiqui Region

Forest ('76)-Forest ('84)
Nonforest ('76)-Nonforest ('84)
Forest ('76)-Nonforest ('84)
Nonforest ('76)-Forest ('84)

year period, more than five percent of previously nonforested land reverted back to a forested condition. Likewise, approximately 82 percent of the area experienced little change at all (Table 2). Most

Table 2. Landcover Change Classes for the La Selva and Sarapiqui Region

Class	Percent (%) Class Coverage
Forest ('76) - Forest ('84)	54.3
Nonforest ('76) - Nonforest ('84)	28.2
Forest ('76) - Nonforest ('84)	12.0
Nonforest ('76) - Forest ('84)	5.5

of the hectarage converted from a nonforested condition to forest occurred in the La Selva Biological Preserve where certain properties were acquired by the state from private owners and returned to their natural condition. Other areas reverted to forested cover occur along streams perhaps where riparian vegetation has returned after earlier clearing up to the streambank for crop or pasture land. Other areas generally fall along class boundaries and result either from gradual woodland spreading into cleared areas or from a slight misregistration of the MSS scenes.

The frequency of co-occurrence between slope classes and change detection classes already demonstrates the trend for less land to be cleared and the higher retention of forested land as slope gradient increases (Table 3). Approximately 75 percent of the very low (0-5%)

Table 3. Frequency of Co-occurrence Between Change Detection Classification and Slope Class

% Slope Class	% Change Detection Classes			
	F ('76)- F ('84)	NF ('76)- NF ('84)	F ('76)- NF ('84)	NF ('76)- F ('84)
0-5 (very low)	22	40	34	4
6-15 (low)	35	45	11	8
16-30 (moderate)	77	9	9	5
31-45 (high)	92	2	4	2
46-60 (very high)	95	1	1	3
60+ (extremely high)	95	0.5	2.5	2

F = Forest; NF = Non-forest

slopes are nonforested. Approximately 57 percent of low (6-15%) slopes are nonforested with the rest being retained in a forested condition or converted back to a forested status. At moderate (16-30%) slopes there is a drastic shift to over 80 percent forested, increasing still to over 94 percent forested at all greater slope levels. The correlation coefficient between percent hectarage forested and slope gradient is approximately 0.90.

Out of the original 39 classes only eight exist in the intensive site area, representing 3 orders; 5 suborders; 7 great groups; 7 subgroups; 7 clustered subgroups and 4 landscape/climatic factor groups. At the order level, Histosols showed a decisive pattern (>75% cleared) while the others were considerably mixed. At the suborder level, Orthents (~90% cleared) were differentiated from other Entisols. The great group and subgroup levels represented the same clustering of soils with the following significant patterns: Tropofibrists, Troporthents and Ustropepts (>75% cleared); but Humitropepts (>70% forested). The other great groups (Dystrandepts and Hydraquents) were relatively mixed in propensity for deforestation response. Despite the limited nature of the soils data set for the intensive site, some similar results are recognizable in comparing the deforestation propensity of soils between the broad-scale and intensive scale, such as high clearing on Histosols and a very mixed response with Inceptisols.

Additional differentiation in deforestation propensity may be necessary at the original level classification. For example, two different typic Dystrandepts (original classification) demonstrated very different responses to landcover change detection, while they were clustered together at all higher level classifications (subgroup, great group, etc.) and therefore imparted a "mixed" response (Table 4). Further evaluation of these two soils illustrates that typic Dystrandept-a,

Table 4. Frequency of Co-occurrence Between Change Detection Classification and Selected Soil Classes (Dystrandepts)

Soil Class	% Change Detection Classes			
	F ('76)- F ('84)	NF ('76)- NF ('84)	F ('76)- NF ('84)	NF ('76)- F ('84)
Subgroup Level typic Dystrandept	53	27	16	4
Original Level typic Dystrandept-a	96	1	1	2
Original Level typic Dystrandept-b	19	48	27	6

F = Forest; NF = Non-forest

mostly forested, occurred largely on high (steep) slopes, while typic Dystrandept-b, mostly cleared, was located much more often on very low slopes (Table 5). The likelihood of two such different classes being

Table 5. Frequency of Co-occurrence Between Slope Class and Selected Soil Classes (Dystrandepts)

Soil Class	% Slope Class					
	0-5	6-15	16-30	31-45	46-60	60+
Subgroup Level-typic Dystrandepts	39.7	18.5	7.7	25.6	3.3	5.2
Original Level-typic Dystrandept-a	0	6.6	16.5	57.8	7.5	11.6
Original Level-typic Dystrandept-b	71.3	27.9	0.8	0	0	0

grouped together was indicated by the bimodal increases at the subgroup level in the 0-5% and the 31-45% slope classes. In evaluating the subgroup level classification (or had the subgroup level classification been the most detailed available), incorporation of both slope class and change detection class would have further broken out this bimodal effect and indicated the potential presence of two distinct subclasses of typic Dystrandepts (Table 6). This relationship of

Table 6. Frequency of Co-occurrence Between Slope Class and Change Detection Class by Selected Soil Classes (Dystrandepts)

Soil Class	% Slope Class						Change Detection
Subgroup Level	0-5	6-15	16-30	31-45	46-60	60+	Class
typic Dystrandepts	8.4	4.7	6.7	24.8	3.3	4.8	F('76)-F('84)
typic Dystrandepts	16.5	10.6	0.4	0.1	0	0.1	NF('76)-NF('84)
typic Dystrandepts	13.3	1.7	0.2	0.2	0	0.2	F('76)-NF('84)
typic Dystrandepts	1.5	1.5	0.4	0.5	0	0.1	NF('76)-F('84)
typic Dystrandept	39.7	18.5	7.7	25.6	3.3	5.2	Cummulative Change

F = Forest; NF = Non-forest

soils compared with both slope and forest change is better than the relationship between soil subgroup and forest change class alone, but not quite as detailed as the relationship between original soil class and forest change (see Table 4).

CONCLUSIONS

The great group or subgroup levels offered the most consistent and relatively detailed classification for this study in determining deforestation potential on both broad and intensive scales. However, in some cases, observations demonstrate the need to often incorporate other biophysical information (e.g., slope) in addition to soils data and/or to use more detailed soils data to obtain better predictive methods. These observations support the contention of Stoner, et al. (1983) that a global soils data base, suitable for most purposes - large and small scale, could be developed at the subgroup level and improved to the family level by incorporation of such biophysical factors as topography and geology.

Alfisols, Mollisols and Vertisols quickly experienced deforestation for conversion to upland crops and pastures due to their favorable soil properties and location within suitable climatic zones. A significant percentage of Histosols also underwent forest clearing for use in rice production and some bananna plantations. Low slope gradient was an important consideration for most other soil types experiencing clearing. For those soil types remaining in forest or undergoing land use abandonment and forest regrowth, factors such as erosion, unsuitable soil properties and soil fertility problems were important. Steep slopes, poor soil structure, and lithic or aquic subgroups were the major physical impediments to forest clearing. Phosphorous retention problems, low pH and low base saturation were the major fertility problems. Socio-political factors also explain some unexpected patterns not indicated by biophysical factors. Future efforts will investigate a more quantitative approach in understanding the relative influence of each factor on deforestation trends through step-wise discriminant analysis and will develop modelling procedures to predict future trends (amount and location) based on incorporation of several biophysical factors.

REFERENCES

Allen, J. C., 1985, Soil Response to Forest Clearing in the United States and the Tropics: Geological and Biological Factors: Biotropica, Vol. 17(1), pp. 15-27.

Dickinson, R. E. 1981, Effects of Tropical Deforestation on Climate: In Blowing in the Wind: Deforestation and Long-Range Implications, Studies in Third World Societies. Publication #14, p. 411-441.

Ehhalt, D. H. and U. Schmidt, 1978, Sources and Sinks of Atmospheric Methane: Pure Appl. Geophys., Vol. 116, pp. 452-464.

Fox, R. L. and P. G. E. Searle, 1978, Phosphate Adsorption by Soils of the Tropics: In Diversity of Soils in the Tropics, Am. Soc. Agron. Sp. Pub. #34, pp. 97-119, ASA, Madison, WI.

Joyce, A. T., J. Ivey, and G. Burns, 1980, The Use of Landsat MSS Data for Detecting Land Use Changes in Forest Land: Proc. of the 14th Int. Symp. on Remote Sensing of the Environment. San Jose, Costa Rica, ERIM, Vol. 2, pp 979-988, Environ. Res. Inst. of Mich., Ann Arbor, MI.

Junkin, B. G., R. W. Pearson, B. R. Seyfarth, M. T. Kalcic, and M. H. Graham, 1980, ELAS: Earth Resources Laboratory Software - A Geobased Information System, NASA/NSTL/ERL Report #183, NSTL, MS 39529.

Lanly, Jean-Paul, 1982, Tropical Forest Resources, FAO Forestry Paper #30, FAO, Rome, Italy.

Myers, N., 1980, Conversion of Tropical Moist Forests, A Report for the Committee on Research Priorities in Tropical Biology of the National Res. Council, National Academy of Sciences, Washington, D.C.

Perez, S. and F. Protti, 1978, Comportamiento Del Sector Forestal Durante El Periodo 1950-1977, Oficina De Planificacion Sectorial Agropecuaria, Doc. - OPSA-15.

Perez, S., E. Ramirez, A. Alvardo, and E. G. Knox, 1979, Manual Descriptivo Del Associaciones De Subgroupos Desuelos De Costa Rica, Oficina De Planificacion Sectorial Agropecuaria, San Jose, Costa Rica.

Potter, G. L., H. Wellsaesser, M. MacCracken, and F. M. Luther, 1975, Possible Climatic Impact of Tropical Deforestation: Nature, Vol. 258, pp. 697-698.

Sader, S. A., 1980, Renewable Resource Analysis and Monitoring in Costa Rica: Proc. of 14th Int. Symp. on Remote Sensing of the Environment, San Jose, Costa Rica, ERIM, Vol. 2, pp 1251-1259, Ann Arbor, MI.

Sader, S. A. and A. T. Joyce, 1984, Relationship Between Forest Clearing and Biophysical Factors in Tropical Environments: Implications for the Design of a Forest Change Monitoring Approach, NASA/NSTL/ERL Report #230, NSTL, MS 39529.

Sader, S. A. and A. T. Joyce, 1985, Global Tropical Forest Monitoring: Proc. Int. Conference on Advanced Technology for Monitoring and Processing Global Environmental Information, London, England, September 9-13, 1985.

Sanchez, P. A., 1981, Soils of the Humid Subtropics: In Blowing in the Wind: Deforestation and Long-Range Implications. Studies in Third World Societies, Pub. #14, pp. 347-410.

Sheng, T. C., 1982, Erosion Problems Associated with Cultivation in Humid Tropical Hilly Regions: In Soil Erosion and Conservation in the Tropics, Am. Soc. of Agron. Special Pub. #43, pp. 27-39, ASA, Madison, WI.

Stoner, E. R., A. T. Joyce, and H. C. Hogg, 1983, A Potential Global Soils Data Base: Proc. IEEE-1983 Int. Geoscience and Remote Sensing Symp. (IGARSS), San Francisco, CA., Aug. 31-Sept. 2, 1985, Vol. 1, WP-1, pp. 8.1-8.7, Institute of Electrical and Electronic Engineers, New York.

UNESCO/UNEP/FAO, 1978, Tropical Forest Ecosystems, UNESCO, Paris, France.

U.S. Department of State, 1980, The World's Tropical Forests: A Policy, Strategy, and Program for the U.S., Report to the President by U.S. Interagency Task Force on Tropical Forests, U.S. Government Printing Office.

Wong, C. S., 1978, Atmospheric Input of Carbon Dioxide from Burning Wood: Science, Vol. 200, pp. 197-200.

Woodwell, G. M., J. E. Hobbie, R. A. Houghton, J. M. Melillo, B. J. Peterson, G. R. Shaver, T. A. Stone, B. Moore and A. B. Park, 1983, Deforestation Measured by Landsat: Steps Toward a Method, DOE/EV/10468-1, Dist. Category UC-11, U.S. Dept. of Energy, Wash. D.C.

FIRE FUELS MAPPING

Moderator: W. Bonner

VEGETATION AND FIRE FUEL MODELS MAPPING OF NORTH CASCADES NATIONAL PARK

Ralph R. Root, Susan C.F. Stitt, Maurice O. Nyquist,
Gary S. Waggoner, and James K. Agee
National Park Service, Geographic Information Systems Field Unit
P.O. Box 25287, Denver, Colorado 80225, and Cooperative Park
Studies Unit, University of Washington, Seattle, Washington 98195

ABSTRACT

A digital data base was created for North Cascades National Park and immediate surroundings using geographic information system (GIS) techniques. Landsat 2 and Landsat 3 multispectral scanner (MSS) data, Defense Mapping Agency (DMA) digital terrain models, and a manually digitized precipitation map were the data sources. A 22-class vegetation/landcover map was derived from the GIS which had an overall accuracy of 84% determined by field verification at 820 ground sites. The vegetation map was used to produce best-fit National Fire Danger Rating System (NFDRS) and Northern Forest Fire Laboratory (NFFL) fire fuel models. The NFFL models were verified by comparison with custom fuel models created via U.S. Forest Service BEHAVE modeling software from information collected at 425 ground sites. The North Cascades data base was determined to be a working management tool for making conservative fire behavior predictions and assisting in other resources management decisions. It can be updated or augmented with additional data layers at minimal cost. As a next step, park acquisition of appropriate software/hardware and staff training will place this tool in the hands of local park management, where its fullest potential can be realized as it is used on a day to day basis.

INTRODUCTION

Vegetation of the North Cascades National Park Service (NPS) complex and the surrounding region is described by a substantial body of literature. This information varies from generalized regional reports to detailed watershed specific documents. However, the only vegetation map covering the entire park complex is a 1936 U.S. Forest Service commercial forest type map. Although the map serves some purposes, it is out of date, and largely inadequate for contemporary resources management needs. The purpose of this project was to map the vegetation of the park, derive fire fuel models and incorporate topographic and other ancillary data into a digital data base for geographic information systems (GIS) analyses.

DESCRIPTION OF THE PROJECT AREA

The park complex, which consists of north and south park units, Ross Lake National Recreation Area, and Lake Chelan National Recreation Area, occupies 674,000 acres of northwestern Washington state (Figure 1). It is within the North Cascades physiographic province, described by Franklin and Dyrness (1973). Geologically, the area consists of folded, partially metamorphosed late Paleozoic and early Cretaceous age sediments that have been intruded by numerous Tertiary age granitic plutons. About 5 million years ago, during the Pliocene epoch, gradual uplift began, creating the present-day Cascade Range. During the

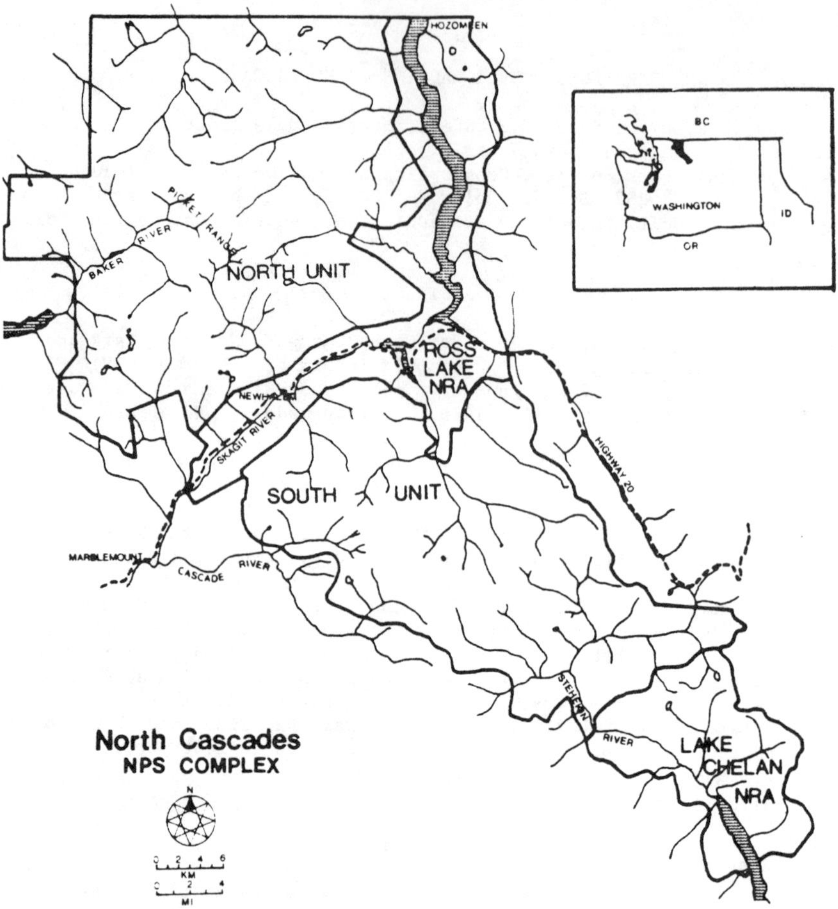

Figure 1. Location of the North Cascades National Park Service Complex in Washington.

Pleistocene epoch glacial till was deposited in major valleys of the area. Great variety in parent rock materials and widespread glaciation have helped to create a highly complex and intermixed pattern of soil types in the area. Maritime climate is predominant in the western portion of the project area, grading into a significantly drier climate in the eastern portion, due to the rain shadow effect of the Cascade Range. Vegetation in the park complex is mostly mesic temperate coniferous forest typical of the Northwest, one of the most densely forested areas in the United States. Franklin and Dyrness (1973) recognize three general vegetation zones in the western project area: the Tsuga heterophylla (Western hemlock) zone, the Abies amabilis (Pacific silver fir) zone, and the subalpine Tsuga mertensiana (Mountain hemlock) zone. The Pseudotsuga menziesii (Douglas-fir) and Abies lasiocarpa (Subalpine fir) zones are common in the drier eastern portion of the project area. Within the predominantly coniferous forests in all zones are numerous deciduous tree species, particularly along drainage bottoms. Avalanche chutes are typically shrub covered, and grass and herbaceous plants form meadows and tundra which are common in the subalpine and alpine areas.

METHODS

As an overall approach, Landsat multispectral scanner (MSS) data were used in conjunction with digital terrain data (elevation, slope, and aspect) and precipitation data to derive vegetation and fire fuel models. This approach was modeled after an earlier project at Olympic National Park (Cibula and Nyquist, in press) in which the same concept was initially applied with success.

Two summer Landsat scenes were required to cover the project area. A July 20, 1979 scene obtained from Landsat 2 covers most of the area, except for a small portion of Lake Chelan National Recreation Area in the southeast corner of the project area. A June 27, 1978 Landsat 3 scene was used to provide data for this area.

The 1979 MSS data set was first processed through Earth Resources Laboratory Applications Software (ELAS) module SRCH, which is an automated "unsupervised" technique for deriving homogeneous training statistics by systematically moving a 3 X 3 pixel window throughout the raw multi-channel data set. After collection of training statistics, 38 spectral signatures were derived. These statistics were input into MAXL, an ELAS module which is a maximum likelihood classifier. The resulting 38-class output product was viewed using a color table generated by ELAS module CIRT which assigns colors as they would appear in a color infrared photograph, by computing appropriate color values from the mean radiance values in the green, red, and infrared channels. Control points were then picked for this data set and ELAS modules PMGC and PMGE were used to resample the data to 50 meter cells and fit them to the Universal Transverse Mercator (UTM) grid projection. The resulting RMS error was +/- 59 meters or about one pixel.

Digital topographic information which was derived from 1:250,000 scale topographic data was obtained from the Defense Mapping Agency (DMA). ELAS modules TOP1 through TOP5 were used to reformat, rotate, compute mapping coefficients, and resample the data to the UTM grid. The ELAS module TOP6 was used to derive elevation (100 foot intervals), aspect (nine classes including flat), and slope (15 class intervals), all of which were formatted in a 50 meter cell size.

Precipitation zone isohyets and park boundary lines were mapped. These data were manually digitized using modules in the Systems Applications Group Information System (SAGIS) software package. The resulting vector files were converted to ELAS raster (cell) data files through ELAS module SAGE.

A five-layer geographic information system (GIS) data base was created by registering the Landsat 2 MSS classified data, elevation, slope, aspect, and precipitation files according to the UTM references in each file. A single 6-channel data file was used to store the 5-layer data base. The extra channel was used for storing products produced by subsequent analyses of the data base.

At this stage the entire GIS data base had been created except for Landsat MSS classification of the southeastern corner of the park complex. Past experience had shown that a technique called "signature extension" might appropriately be used in this situation. Signature extension is simply the application of a set of training statistics from one scene to an adjoining scene. In this case, the statistics were developed from the July, 1979 scene, which was from Landsat 2, and were to be applied to the June, 1978 scene, which was from Landsat 3.

Channel 7 in the Landsat 3 data had a dynamic range of 0-63 compared to 0-127 for the same channel in Landsat 2 data. The Landsat 3 data was "normalized" to the Landsat 2 data by doubling the channel 7 raw data values in the 1978 scene. Then the SRCH statistics and MAXL parameters generated from the July, 1979 MSS data were applied to the altered June, 1978 MSS data. The false color infrared color table used for the July, 1979 data was applied to the resulting June, 1978 classification. Areas where the data sets overlapped were visually inspected one class at a time for inconsistencies. The only detectable difference was the areal extent of snowpack. This indicated that the signature extension technique was valid, and that both MSS scenes were producing comparable classifications of vegetative land cover.

Control points for the June, 1978 scene were picked. ELAS modules OCGN and OGEO were used to perform the georeferencing. These are different modules than those used on the 1979 scene due to differences in Landsat data formatting (Goddard versus EROS). The RMS error was again +/- 59 meters, or about one pixel.

The portion of the June, 1978, MSS classified scene required to fill in the southeast corner of the park complex was merged with the July, 1979 data set using ELAS module JCOP, producing one continuous data file of consistent multispectral signatures. The fit of the two data sets was excellent, both geographically and spectrally. The only discernable evidence of using two data sets is the difference in extent of snowpack along the line joining the two sets.

The 38 spectral classes commonly identified on both Landsat scenes were then manually lumped into nine broad cover type classes; dark conifer, light conifer, dark deciduous/mixed, light deciduous/mixed, herbaceous, bare ground/rock, snow/ice, water, and shadow. To produce a vegetation map at the community type level required for adequate resource management applications, additional environmental information was needed. Extensive field sampling and analysis was designed and carried out to determine predictive direct gradient models based on the layers of data input into the GIS data base. A total of 425 field plots were located in or adjacent to the park complex for sampling the environments of a wide variety of the five broad Landsat vegetation classes. Sampling points were subjectively established to assure representative inclusion of the broad Landsat classes with their various associated topographic positions (elevation, slope, and aspect). A detailed description of sampling procedures is described in Agee et al (1985).

Data from the field sites were then examined using multivariate analysis (Gauch, 1982) to relate community variation to environmental gradients that could be represented by layers of information in the data base, and to classify all 425 sites in a coherent and consistent manner. An ordination technique called detrended correspondence analysis (Hill and Gauch, 1980) was employed, since it provided the most objective means of defining important environmental gradients in the data set. Description of the technique and how it was applied is discussed at length in Agee et al (1985). The results of the analysis were direct ecological models that predict plant community type from general Landsat classes, elevation, slope, aspect, and average annual precipitation.

Figure 2 illustrates the process of developing the vegetation and fuel type maps from the GIS data base. The five-layered data base, covering the entire project area, was used to produce a vegetation community type map with 22 vegetation classes, water, shadow, inert materials (non-vegetated areas such as rock, talus, bare soil, etc.) and snow/ice. ELAS module DBAS was utilized for extracting each ecological model, which resulted in one of 22 vegetation classes. DBAS utilizes a pseudo-BASIC language for developing the appropriate algorithm for extraction of the pertinent ecological model information from each data base layer. Limitations of computer memory and disk storage required the GIS data base and DBAS programs to be broken down into multiple parts, making it necessary to do 57 different computer runs to produce the vegetation map. Once completed, the map was plotted in color at scales of 1:100,000 and 1:24,000 to aid in field checking. A total of 820 plots was sampled to check for final classification accuracy of the plant community types.

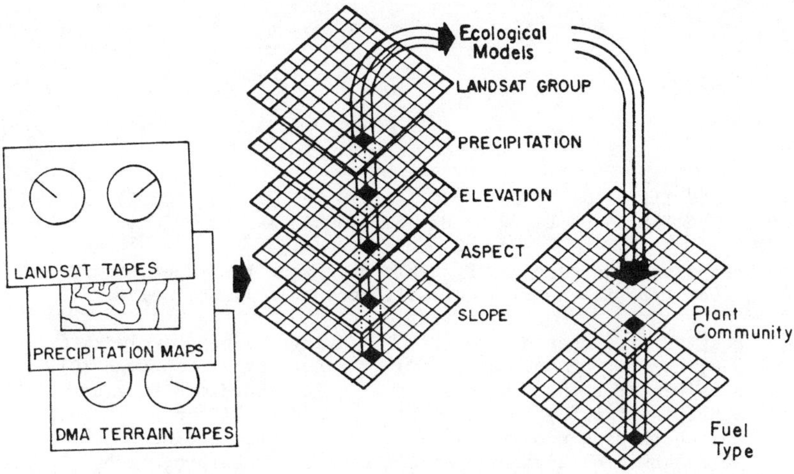

Figure 2. Conceptual scheme of a geographic information system.

Four classes were eliminated because they covered so little area they could not be adequately field checked. One of the four had inadequate environmental definition of its boundaries. In all cases the eliminated classes were absorbed into a more common type which was ecologically similar. Final classification accuracy for the remaining 18 vegetation classes averaged 84.3%. It was observed that classification errors between open and closed canopies of the same type were relatively frequent. If these are ignored, the overall average accuracy of classification is 88.2%. Results of accuracy evaluation for each of the vegetation classes and a description of all mapping classes are shown in Table 1.

Using data collected from 425 field sites, the vegetation community type classes were combined in two different ways to produce the forest fire fuel model type maps. The first is the National Fire Danger Rating System (NFDRS) (Deeming et al, 1977), with a total of 11 classes, and the second is the Northern Forest Fire Laboratory (NFFL) model (Albini, 1976) with a total of 8 classes. Table 2 shows how the vegetation classes were combined to produce each of these fuel models. The field data showed considerable variation in fuel model types within individual vegetation types. The most common NFDRS and NFFL models within each plant community type were selected by comparison with the field data as the "best-fit" models. Where two models had about equal

Table 1. Contingency table for accuracy of final vegetation classification of North Cascades NPS complex

	*1	2	3	4	5	6	7	8	9	10	11	12	13	14	15	16	17	18
*1	59	12								1		1					1	
2	4	91		3														
3	1		46	2			3		1								1	
4	1	2	5	48														
5	1				25	1	1											
6						11		1										
7			13	1			37	1										
8		2	1	6		1	1	26		1								
9			3		1				13		1							
10		1	3	11		1		2	2	69	1	3						
11		2								1	22	3						
12	1	5								4	2	63						
13	2	1											4					
14												1		9	2			
15									1	1				1	52		1	
16																16		
17																1	85	
18																	3	14
Total	69	116	71	71	26	14	42	30	17	77	26	71	4	10	54	17	91	14
% OK	85	78	65	67	96	79	88	87	76	90	84	89	100	90	96	94	93	100

* Numbers along the edge of the table represent the vegetation/landcover type in the final classification as listed below:

1. Douglas fir, open canopy (O)
2. Douglas fir, closed canopy (C)
3. Subalpine fir (O)
4. Subalpine fir (C)
5. Whitebark pine/subalpine larch (O)
6. Whitebark pine/subalpine larch (C)
7. Mountain hemlock (O)
8. Mountain hemlock (C)
9. Pacific silver fir (O)
10. Pacific silver fir (C)
11. Western hemlock (O)
12. Western hemlock (C)
13. Ponderosa pine (C)
14. Hardwood forest
15. Tall shrub
16. Lowland herb
17. Subalpine herb
18. Heather meadow
19. Water
20. Shadow
21. Inert
22. Snow/ice

Table 2. Development of NFDRS and NFFL fuel models from vegetation community types.

NFDRS models	vegetation type class number from Table 1
C	12
H	1,2,3,17
L	9,10
O	8
R	7
H/G	4,5,6,13
H/F	16
H/Q	15,18
F/Q	14
S/L	11
Non-Fuels	19,20,21,22

NFFL models	vegetation type class number from Table 1
8	1,2,3,4,5,6
9	7
4/5	8
1	9,10
1/5	11
8/1	12,15
8/5	13,14,16,17,18
Non-fuels	19,20,21,22

coverage within a community type both were included to best represent the variation in that type. These two maps are the first comprehensive fire fuel models for the park complex, and are useful for making general predictions of surface fire behavior. Such models are generalized however, and even more accurate fire behavior predictions can be made using site-specific fuels information. Such information was collected throughout the project area at 425 sampling locations. To further substantiate the validity of applying standard fuel models, site-specific fuels data were used for each community type to develop custom fuel models for comparison with fire behavior as predicted by the standard NFFL models. Using bulk density and fuel height obeservations from each field site, fuel models were developed for each of the 18 vegetation types mapped by using U.S. Forest Service BEHAVE fuel modeling and testing software (Burgan and Rothermel 1984). These custom models were then run through the fuel model testing portion of BEHAVE (TSTMDL) to produce fire behavior outputs over a range of environmental conditions. Although this range does not include all conceivable environmental conditions that could exist at the time of a fire, the TSTMDL outputs permitted some rough comparisons to be made between the field data and the standardized NFFL models. Results of the comparisons suggest that the best-fit NFFL models are sufficiently accurate to predict fire behavior in the fuel types at the North Cascades NPS complex. Where differences occurred in comparisons, the NFFL models showed higher burning characteristics, particularly in the grass, herbaceous, shrub, and hardwood cover types. Use of the NFFL fuel models under most environmental conditions would therefore tend to overestimate fire activity, which puts it on the conservative side as a management tool.

CONCLUSIONS

The GIS data base for the North Cascades NPS complex, with its eight layers of information (general Landsat classification, elevation, slope, aspect, precipitation, plant community types, NFDRS fuel models, and NFFL fuel models) is a working tool for management. An example of its practical use was a recent DBAS analysis using the existing data layers to predict locations for potential nesting sites for peregrine falcon. In the future, new layers of information can be added, and existing layers can be updated, increasing the power of the data base and keeping it current. Future updates of the vegetation map can be made by either manually outlining and digitizing changes such as recent forest fires, timber cutting adjacent to the park complex, and other landuse changes affecting surface cover, or by creating a new classification from a more recent Landsat scene and applying the existing vegetation type models in the same fashion as was done in the current project. Terrain and climate data are long-term in nature, and would not be expected to change as rapidly as vegetation.

Even though the North Cascades NPS complex now has a useable, working data base, which is potentially a very powerful resources management tool, its maximum utility will only be realized if the GIS can be used by the park staff in a timely manner. Two major factors, training of park staff in the use of the system, and timely access to processing software/hardware, display, and printing/plotting equipment are necessary to make the most effective use of this technology. The National Park Service Geographic Information Systems Field Unit will be working with parks such as North Cascades not only to create data bases, but also to provide training in their use, and to aid the park in selection and acquisition of appropriate hardware and software for local in-park use on a day-to-day basis.

REFERENCES

Agee, J.K., and S.G. Pickford. 1985. Vegetation and fuel mapping of North Cascades National Park Service complex. Final Report. NPS Contract CX-9000-3-E029. NPS Cooperative Park Studies Unit, College of Forest Resources, Univ. of Washington, Seattle, Washington 98195

Albini, F.A. 1976. Estimating wildfire behavior and effects. USDA For. Serv. Gen. Tech. Rpt. INT-30. 92pp.

Burgan, R.E. and R.C. Rothermel. 1984. BEHAVE: Fire behavior prediction and fuel modeling system - FUEL subsystem. USDA For. Serv. Gen. Tech. Rpt. INT-167.

Cibula, W.B. and M.O. Nyquist. (In press). Use of topographic and climatological models to improve Landsat MSS classification for Olympic National Park, Washington. Photogram. Eng. and Remote Sensing.

Deeming, J.W., R.E. Burgan, and L.D. Cohen. 1977. The National Fire-Danger Rating System - 1978. USDA For. Serv. Gen. Tech. Rpt. INT-39.

Franklin, J.F., and C.T. Dyrness. 1973. Natural vegetation of Oregon and Washington. USDA For. Serv. Gen. Tech. Rpt. PNW-8.

Gauch, H.G. 1982. Multivariate analysis in community ecology. Cambridge Univ. Press. New York. 298 pp.

Hill, M.O., and H.G. Gauch. 1980. Detrended correspondence analysis, an improved ordination technique. Vegetation 42:47-58.

OPERATIONAL FIRE FUELS MAPPING WITH NOAA-AVHRR DATA

Randy A. McKinley and Edward P. Chine
Technicolor Government Services
Bureau of Land Management Operations
Bldg. 50, D-442T Denver Federal Center
Denver, CO 80225

Lee F. Werth
Bureau of Land Management
Division of Advanced Data Technology
Branch of Remote Sensing
Bldg. 50, D-442 Denver Federal Center
Denver, CO 80225

ABSTRACT

The Bureau of Land Management is currently developing a geographic fire fuels data base for the eleven western contiguous states, excluding Washington. This project is in support of the Boise Interagency Fire Center's (BIFC) effort to develop the Initial Attack Management System (IAMS). An operational technique has been developed which utilizes NOAA-AVHRR remotely sensed data to produce information suitable for broad scale fire fuel classes. This technique utilizes processing steps which include raw data evaluation and manipulation, multi-scene registration, and multi-date data base development. To transform base data into useful vegetation information, digital image processing procedures including ratioing (ie. normalized difference calculation), principle components analysis, clustering, and classification are employed. This paper expands strictly upon the above technical procedures and alerts prospective users to inherent limitations of similar data sets.

INTRODUCTION

The Bureau of Land Management (BLM) is currently developing a digital fire fuels data base for the eleven western contiguous states excluding Washington. This project is in support of the Boise Interagency Fire Center's (BIFC) effort to develop the Initial Attack Management System (IAMS). IAMS is a real time geo-based system designed to incorporate lightning strike, fire fuels (based on vegetation), terrain, and weather information to assist in the fire dispatch decision making process (IAMS Program Manager, 1981). Digital data base themes, registered to a 30 arc second grid (cell size of aprx. 700 by 925 meters, varying with latitude), include fire fuels, roads, elevation, slope, and aspect. The sources of this digital information are National Oceanic and Atmospheric Administration Advanced Very High Resolution Radiometer (NOAA-AVHRR) spectral data for vegetation, United States Geologic Survey (USGS) 1:2,000,000 Digital Line Graph (DLG) data for roads, and NOAA 30 Arc Second Point Elevation data for elevation, slope, and aspect information. The methodology utilized by the BLM to transform NOAA-AVHRR data into fire fuels information is described in this paper.

BACKGROUND

A number of data sources and methodologies have shown promise for broad scale vegetation mapping for large geographic areas (Gatlin et al. 1983, Gervin et al. 1983, Horvath et al. 1982, Sader et al. 1981). Faced with the task of providing fire fuels information for IAMS, the BLM evaluated a variety of data sources. The manual interpretation of single date Landsat false color composites and the digital processing of multi-date NOAA-AVHRR data were determined to have the most potential for providing the broad scale fire fuels information required. After an extensive pilot project designed to compare these two methodologies (Miller et al. 1983), the digital processing of multi-date AVHRR data was selected. This decision was based on the lower cost per acre, decreased time requirement, superior update/revision potential, and other factors estimated for the AVHRR methodology. Research conducted by Miller et al. (1982 and 1983) provided a guideline for digitally analyzing AVHRR data. Although that methodology was adopted, specific refinements to it have evolved with time and experience. A more complete discussion of the data source and project methodology selection process can be obtained in Werth et al. (1984), and Miller et al. (1982 and 1983).

METHODOLOGY

AVHRR Data Limitations and Concerns

Potential complications, relating to radiometric considerations, scanner geometry, etc. are present when working with AVHRR multi-date imagery. Data sets can typically be made up of data collected in the morning and/or afternoon, data ranging from nadir to approximately 56 degrees off nadir (Figures 1 & 2), data from four unique satellites, and data collected with varying atmospheric conditions. Recent interest in the use of AVHRR data for various applications, has brought increased attention to many of these concerns. The current literature often addresses these problems, and at times corrections or adjustments necessary for the use of AVHRR data have been proposed (Holben and Fraser 1984, Kimes et al. 1984). However, the operational environment surrounding the BLM fire fuels mapping project requires that many image processing decisions still be made with less than optimum quantitative date or pertinent research on which to base those decisions. Questions such as; What is the maximum usable portion of the AVHRR image swath?; What corrections should be employed for dates of imagery with extreme variation in sun elevation and azimuth?; are often left to qualitative assessment. These questions are definitely worthy of research, however, operational constraints limit the time and funds available to provide answers.

Figure 1. September 5, 1983 NOAA-7 near infrared image which illustrates the full across track coverage for the AVHRR scanner. Note Great Salt Lake at center right and California Central Valley at center left.

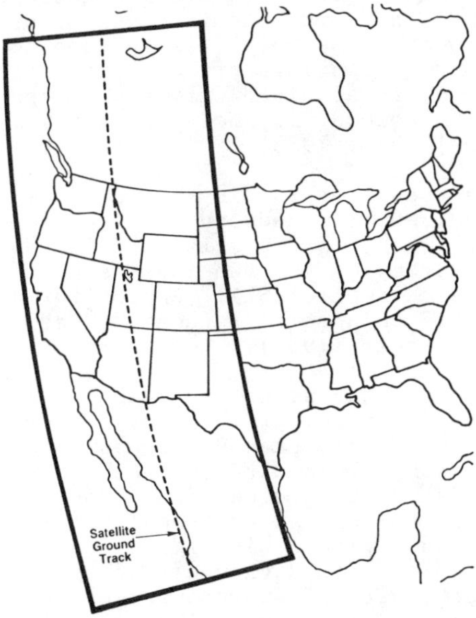

Figure 2. Example of NOAA-7 ground track and swath width as recorded at Redwood City. (Miller et al. 1983)

Data Acquisition

NOAA-AVHRR data is broadcast from satellites in two modes, real-time (direct readout data) and recorded (on-board recording for later playback or broadcast). The NOAA National Environmental Satellite Service (NESS) operates two data acquisition stations, one in Wallops Island, Virginia and one in Gilmore Creek, Alaska, to receive both recorded and direct readout data (Kidwell 1981). From these acquisition stations data are transmitted via DOMSAT (a communications satellite link) to the World Weather Building in Suitland, Maryland. A data receipt-only station has also been established in Lannion, France that transmits data to Suitland. Further, a NOAA/NESS ground station in Redwood City, California receives data for the North American west coast only. These data are not permanently archived, however, they are maintained in Redwood City for 90-120 days (Kidwell 1981).

Three types of AVHRR data are currently being used for investigative purposes; High Resolution Picture Transmission (HRPT) data, Local Area Coverage (LAC) data, and Global Area Coverage (GAC) data. HRPT and LAC data have one (1) kilometer resolution. GAC data have four (4) kilometer resolution (Daghir personal communication 1985).
Five distinct spectral channels are available with afternoon data, while only four channels are available with morning data. Channels 1 and 2 from morning and afternoon data are used for fire fuels mapping purposes. These channels approximate Landsat channels 4 - 7.
HRPT data is direct readout data only, and consists of 1024 (10 bit) quantitizing levels for all bands (Kidwell 1981). It is acquired at all of the above mentioned stations. HRPT data is LAC data in as much as it consists of local area coverage (ie. west coast only), as opposed to global area coverage. LAC data consists of both direct readout and recorded data and is acquired only at Wallops Island and Gilmore Creek. Redwood City does not acquire LAC data (Daghir personal comm. 1985).

All AVHRR data used thus far in the BLM fire fuels mapping effort was acquired from Redwood City. This data is "field station format" with 8 bit precision. The 8 bit precision is achieved by dropping the two least significant bits from the 10 bit data (Kidwell 1981). Analog format 10 bit data is also available on special request at Redwood City (Daghir personal comm. 1985).

Approximately one AVHRR scene per week has been collected in the BLM archive since April, 1983. To date, the data archive consists of more than 90 scenes. NOAA personnel attempt to furnish scenes with correct geographic coverage (from central Canada to Southern Baja and from west of the California coast to central Colorado), minimum cloud and snow coverage, and overall high image quality. However, approximately 40% of all data received is unusable due to excessive cloud coverage, snow coverage, or problems associated with errors in receiving satellite data or writing data tapes (i.e. bad lines, skips of several hundred lines, missing bands). Less obvious data quality problems such as slight band to band misregistration and one line or column data shifts, also eliminate otherwise acceptable scenes from consideration or necessitate extensive image correction activity. A large portion of every AVHRR full scene is considered unusable due to image distortion resulting from excessive distance from nadir. Utilizing data where scan angle does not exceed 10-15 degrees would be optimal. In practice, however, it has been necessary to use data with a scan angle of 25 - 30 degrees, in order to achieve adequate geographic coverage. The quality of this data appears to be acceptable and provides reasonable classification results.

Data Organization

The BLM fire fuel mapping project is organized on a state by state basis. Mapping is completed for one state and then moves to the next. A priority listing of states has been established. However, AVHRR sensor characteristics (i.e. usable swath width less than the total western U.S.), common regional cloud and snow coverage phenomena, and overall organizational considerations require the division of the data base building process into multi-state units (i.e. California and Nevada). Experience has shown that the probability of acquiring even one usable AVHRR scene, from approximately 14 scenes per week, is minimal. For the northern states snow and cloud coverage in the early spring has made data acquisition especially difficult. For example, this concern becomes important in an operational sense when trying to acquire data during the early spring green-up of annual grasses. Data bases, for which raw imagery is organized, typically range from 1-4 states in size. This allows for the collection of the best possible multi-date imagery set for a specific geographic area. For the state of Idaho it proved to be difficult to acquire early spring imagery which did not possess extreme cloud or snow coverage in the adjoining states of Oregon and Montana. In contrast, sufficient AVHRR data of a high quality should be available to develop multi-date coverage for the states of Colorado, Utah, Arizona and New Mexico as one unit, dependent upon distance from nadir considerations.

Data Base Development

Selection of specific dates of imagery for a project area is determined by plant phenology. Field personnel attempt to target specific windows of time most likely to encompass the peak green period of vegetation types which significantly contribute to available fire fuels. In theory, the spectral separability of important fuel types should be increased when analyzing imagery possessing maximum contrast for given cover types. For example, annual grasses tend to have an early spring peak green period during which they are more easily separated from woody vegetation which is still dormant. Upon senescence the annual grass cover type is easily confused with low density shrubs. In practice, limited imagery availability has made the acquisition of even one scene per month during the Spring, Summer, and Fall seasons somewhat difficult. For the states of California and Nevada eight scenes covering an eight month time span were utilized in the image classification process. For the state of Idaho five acceptable scenes covering a time span of 4 months were utilized.

Digital Processing

Registration. Image processing in support of the fire fuels mapping project is performed utilizing an *Electromagnetic Systems Laboratory (ESL) Interactive Digital Image Manipulation System (IDIMS). The digital processing sequence is illustrated in Figure 3. All AVHRR data selected for a specific project data set, receives extensive visual examination for data inconsistencies (i.e. band shifts, bad lines, etc.) prior to registration to a 30 second longitude/latitude grid. Registration of numerous scenes in relatively short periods of time, roughly one scene per 3 hours, is achieved utilizing software which automates the usually laborious process of ground control point selection. This capability is partly due to the availability of Navy ephemeris data for the NOAA satellites. This data facilitates geometric correction based on assumed satellite position. Experience has shown that the automatic registration process is reliable, although some scene to scene shifting (1 to 4 pixels) is often required to achieve a best fit. A traditional registration approach utilizing manually selected control points could provide equivalent or improved accuracy. However, this would require greatly increased analyst time and computer resources.

Normalized Difference Calculation. Five to eight, 1500 line by 1500 sample, multi-date AVHRR two band images (visible and Near IR), are generally available for each fire fuel mapping project. In order to locate areas having high proportions of herbaceous vegetation (corresponding to flash fuels) a Normalized Difference Greenness Index (ND) is calculated for each of the AVHRR images. The actual technique used is similar to that described by Deering and Haas (1980). Using AVHRR data, the calculation is as follows:

$$ND = \frac{\text{Channel 2 (IR)} - \text{Channel 1 (VIS)}}{\text{Channel 2 (IR)} + \text{Channel 1 (VIS)}}$$

The difference between infrared and red reflectance (the numerator) has been shown to be related to the amount of herbaceous vegetation in the scene (Deering and Haas, 1980). Therefore, large values represent areas containing high proportions of standing green biomass. The addition of IR plus red in the denominator allows the index to be compared from date to date and site to site by reducing the effect of daily variance of incoming radiation caused by atmospheric conditions and landscape position (Miller et al., 1983). Each resulting ND image is then merged with the visible and IR channels of the "best" scene (the one scene thought to be most representative and the one that would be chosen if only one date could be analyzed). The

---

* Use of trade and company names is for descriptive purposes only and does not imply endorsement by the Bureau of Land Management.

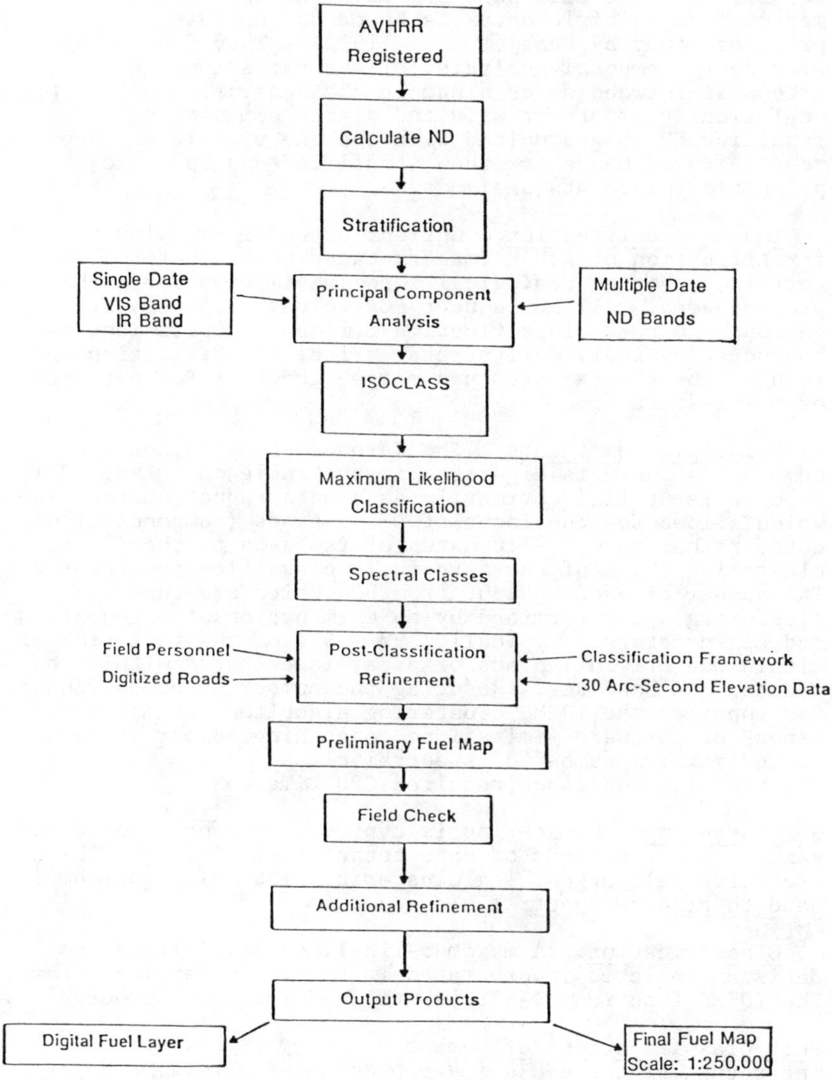

Figure 3. Digital analysis flowchart for developing a fire fuels classification.

visible and IR bands are utilized primarily to retain spectral information that the ND calculations tend to obscure. For example, non-vegetated cover types (i.e. water and snow) have distinct spectral responses in the IR and visible bands, while the ND value calculated for both will be similar. The rationale for using ND (greenness) information from multiple dates is based on the results of a previous study by Haas et al., (1982). They found that with multi temporal analysis, slight variations in greenness between dates aided in the separation of ecologically important wildland plant communities. The resulting ND images united with the one visible and one near infrared bands are then stratified and input to principle components analysis.

Image Stratification. Current procedures employ stratification of AVHRR imagery based upon landforms, terrain, or other ecological zone information for each project area. All subsequent PCA, clustering, classification, and post classification refinement operations are conducted by individual strata. Final classification results (by strata) are then merged together for each state or multi-state area.

Principle Components. The Karhunen-Loeve principle components analysis algorithm is available on IDIMS. This process is utilized primarily as a data reduction technique which allows for the input of 4 - 5 bands (components) of data, rather than 7 - 10 bands of raw data to the clustering phase of the fire fuels classification process. The number of bands output from PCA which are input to clustering are determined by an examination of eigenvalues and eigenvectors. Typically, 92 - 97% of the variation within the original bands of data, is captured within the reduced PCA band set. Reducing the number of bands required for input to the IDIMS clustering algorithm minimize the impact of software limitations concerning number of bands in and maximum number of spectral classes out. This process also decreases required CPU time/cost.

Clustering. Clustering is typically performed on a 100% sample of 4 - 6 bands of data output from PCA. The IDIMS Iterative Self-Organizing Clustering (ISOCLS) algorithm is used to produce spectral classes.

Classification. A maximum-likelihood classification decision logic is incorporated in the fuels mapping effort. The IDIMS function CLASFY is utilized for this purpose.

Post Classification Refinement

The result of the AVHRR image processing effort, a spectral map, is then subjected to post-classification refinement. A fire fuels classification is produced by assigning fire fuel classes to each spectral class. This is accomplished through the efforts of Service Center and field personnel. Personnel from a specific BLM district office are brought to the Service Center in Denver to utilize the IDIMS interactive color display. To assist them in locating fuel classes on the display, various information such as 1:250,000 quadrangle boundaries and road networks are embedded in the classified data. Ancillary information, which assists in making resource class assignments and is routinely utilized in the post-classification refinement process, includes slope, aspect, elevation, and resource area, district, and state boundaries.

SUMMARY

The BLM fire fuels mapping effort is based upon the analysis of NOAA-AVHRR spectral data. A methodology incorporating standard image processing and Geographic Information System (GIS) data manipulation techniques has been developed. A number of limitations or concerns not commonly associated with the processing of Landsat data must be addressed when utilizing NOAA-AVHRR data. Based upon BLM experience, NOAA-AVHRR data is capable of providing information suitable for broad scale fire fuels mapping over large geographic areas.

REFERENCES

Deering, D. W., and Haas, R. H., 1980. Using Landsat Digital Data for Estimating Green Biomass: NASA Technical Memorandum 80727, Goddard Space Flight Center, Greenbelt, Maryland, 21 p.

Gatlin, J. A., R. J. Sullivan, and C. J. Tucker, 1983, Monitoring Global Vegetation using NOAA -7 AVHRR data. NASA-Goddard Space Flight Center, Greenbelt, MD. 5 pp.

Gervin, J. C., A. G. Kerber, R. G. Witt, Y. C. Lu, and R. Sekhon, 1983, Comparison of Level I Land Cover Classification Accuracy for MSS and AVHRR data, in proc. Seventeenth International Symposium on Remote Sensing of Environment, Ann Arbor, MI. p. 154-164.

Haas, R. H., Newcomer, J. A., and Horvath, E. H., 1982. A Multitemporal Approach for Classifying and Mapping Rangeland Vegetation in U.S. Geological Survey Open-File Report 80-000.

Holben, B., R. S. Fraser, 1984, Red and Near-Infrared Sensor Responsive to Off-Nadir Viewing, Int. Journal Remote Sensing, Vol. 5, No. 1 p. 145-160.

Horvath, N. C., T. I. Gray, and D. G. McCrary, 1982, Advanced Very High Resolution Radiometer (AVHRR) Data Evaluation for Use in Monitoring Vegetation, Volume 1, Channels 1 & 2. AgRISTARS Report EW-L2-04303, 80 pp.

IAMS Program Manager, 1981, Review of Initial Attack Management System: unpublished memo, IAMS 9124 (400), on file at Bureau of Land Management, Boise Interagency Fire Center, Boise, ID.

Kidwell, D. B., 1981, NOAA Polar Orbitor Data Users Guide, Department of Commerce, National Climate Center, Satellite Data Services Division, Washington, D.C. 168 pp.

Kimes, D. S., B. N. Holben, C. J. Tucker, and W. W. Newcomb, 1984, Optimal Directional View Angles for Remote Sensing Missions, Int. Journal Remote Sensing, Vol. 5, No. 6, p. 887-908.

Miller, W. A., Draeger, W. C., and Chine, E. P., 1982. Applications of AVHRR and CZCS Data for Wildlife Fuels Assessment, USGS, Earth Resources Observation Systems (EROS) Data Center Internal Report to BLM-Denver, EROS Data Center, Sioux Falls, SD, 30 pp.

Miller, W. A., E. P. Chine, and S. M. Howard, 1983, Evaluation of AVHRR Data to Develop Fire Fuels Information as an Input to IAMS, USGS, Earth Resources Observation Systems (EROS) Data Center Internal Report to BLM-Denver, EROS Data Center, Sioux Falls, SD, 77 pp.

Sader, Steven A., David S. Linden, and Mark McGuire, 1981, Fuels Mapping from Landsat Imagery and Digital Terrain Data Aid Fire Suppression Decisions, in Proc. of the ACSM-ASP Fall Technical Meeting, 1982, 7 pp.

Werth, L., Randy McKinley, and Richard Franklin, 1984, Use of NOAA Satellite Data for BLM Wildfire Fuels Mapping in California and Nevada, in Proc. NOAA Conf. on Pathways and Future Directions for Environ. Data and Information Users, Denver, CO. 11 pp.

COMPARISON OF FIRE FUEL MAPS PRODUCED USING MSS AND AVHRR DATA*

Wayne A. Miller** and David C. Johnston***
Technique Development and Applications Branch
EROS Data Center
Sioux Falls, South Dakota 57198

ABSTRACT

Fire fuel information, in support of the Bureau of Land Management's (BLM) national fire program, has been obtained through the manual interpretation of Landsat multi-spectral scanner images and digital image analysis of Advanced Very High Resolution Radiometer (AVHRR) data. The BLM, in cooperation with the Earth Resources Observation Systems Data Center, determined that the accuracy (approximately 90 percent overall) was similar for deriving fire fuel information for Malheur County in eastern Oregon using either approach, and for an area the size of Malheur County (6.4 million acres), the costs were about the same (0.19 cents per acre). But the cost per acre was substantially lower (0.04 cents) where digital analysis of AVHRR data were used to derive fire fuel information for a 42-million-acre area in eastern Oregon. Based on these results, the BLM is using digital analysis of AVHRR data to support its operational fire fuel mapping program.

INTRODUCTION

The Bureau of Land Management (BLM) has a national fire program that uses an Initial Attack Management System (IAMS) located at the Boise Interagency Fire Center (BIFC) in Idaho which provides real-time information to fire management officers. IAMS is a geographic information system that uses data from three different sources registered to a 1-minute latitude/longitude grid. The three data sources are: (1) an automatic lightning detection system, (2) a remote automatic weather station, and (3) a digital fire fuel data base. The fire fuel data base contains elevation, slope, aspect, roads, and fire fuel information. The fire fuel information has been obtained through: (1) manual interpretation of Landsat multispectral scanner (MSS) data (Sader and others, 1981) and (2) digital image analysis of National Oceanic and

---

* Publication authorized by the Director, U.S. Geological Survey, on July 8, 1985.

** U.S. Geological Survey.

*** Technicolor Government Services, Inc. Work performed under U.S. Geological Survey contract no. 14-08-0001-20129.

Atmospheric Administration's (NOAA) Advanced Very High Resolution Radiometer (AVHRR) data (Miller and others, 1983). To determine which method was more appropriate to BLM's national fire program, BLM initiated a study with the USGS National Mapping Division's Earth Resources Observation Systems (EROS) Data Center to determine the cost and accuracy of using each method to develop fire fuel information in support of IAMS. Malheur County in eastern Oregon (fig. 1) was selected as the study site because fire fuel information had been developed for this area using both methods.

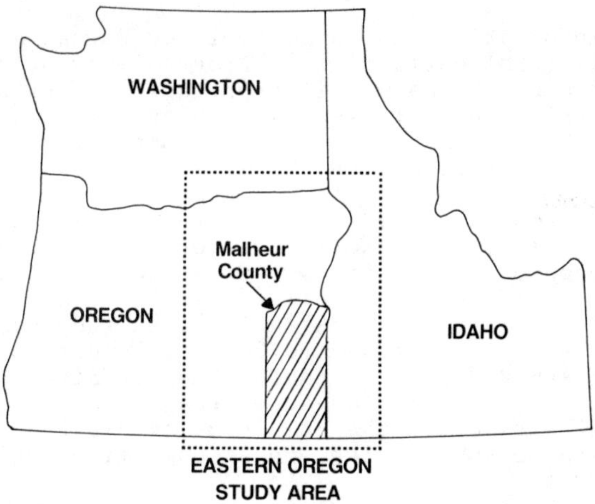

Figure 1. Map showing the location of the eastern Oregon study area and Malheur County.

The AVHRR and MSS sensors are both multispectral scanners. The AVHRR is carried by the NOAA series of meteorological polar-orbiting satellite (Kidwell, 1983) and the MSS is carried by Landsat, which is also a polar-orbiting satellite. The main difference between these two sensors is ground resolution. For the AVHRR it is 1.1 km at nadir, and for the MSS it is 80m. These two sensors are similar in that channels 1 and 2 of the AVHRR have spectral response ranges similar to MSS bands 5, 6 and 7 (fig. 2).

MSS APPROACH

A fire fuel map was produced by Sader and others (1981) for Malheur County using six different Landsat scenes. These scenes, collectively, were a combination of different seasons and different years. The 1:250,000-scale color prints of these scenes were mosaicked together and photointerpreted to identify spectrally different areas. To determine the fire fuel type, Sader and others (1981) visited each area in the field and assigned it a fire fuel type according to the classification framework

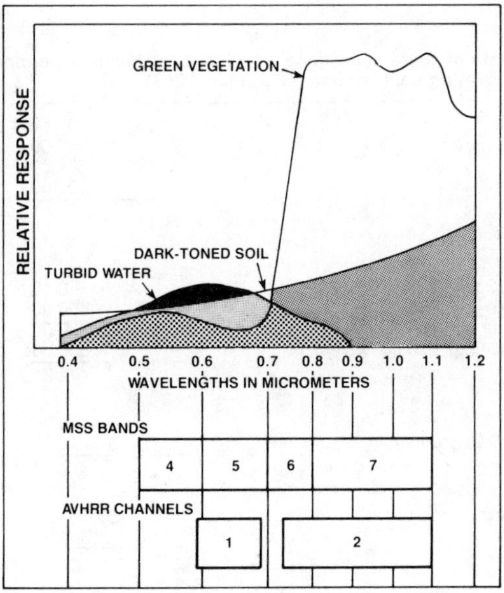

Figure 2. Generalized reflectance curves for vegetation, soil, and water, and the spectral response ranges of NOAA AVHRR channels 1 and 2 and Landsat MSS bands 4, 5, 6, and 7.

in table 1. This framework is based on the National Fire Danger Rating System (NFDRS) fuel model descriptions (Deeming and others, 1978) and was developed for this area by BLM field personnel. After field checking, each photo-interpreted area was digitized and registered to a 100-m Universal Transverse Mercator (UTM) digital data base already containing terrain information, roads, and administrative boundaries. Since the IAMS program requires data to be registered to a 1-minute latitude/

Table 1.—Classification framework used for mapping fire fuels in Malheur County, Oregon using Landsat data (Sadar and others, 1981).

Land Cover/ Fuel Type	NFDRS Fire Fuel Model	Land Cover/ Fuel Type	NFDRS Fire Fuel Model
Annual Grass	A	Desert Shrub	X
Perennial Grass	L	Conifer	U
Sagebrush	T	Agriculture/Riparian	X
Juniper	F	Water/Barren	X

longitude (approximately 1-km) grid, the 100-m data base was degraded to this resolution by re-registering it using nearest neighbor resampling techniques. The total cost of producing the fire fuel information for this 6.4-million-acre area using this approach was approximately $12,000 (table 2).

Table 2.—Estimate of costs to produce a fuel map for Malheur County by manually interpreting Landsat images (Sadar, 1983).

Travel		
6/14-6/19	$1,020.00	
6/14-6/19	569.00	
6/14-6/17	500.00	
		$ 2,089.00
Computer Usage		$ 3,000.00
Data Acquisition		
USGS Maps @ $3.25 (-30%) x 4	$ 9.10	
Landsat Images @ $300 x 6	1,800.00	
Color Composite @ $50 x 6	300.00	
DMA Terrain Data @ $75 (1° x 1°) x 8	600.00	
		$ 2,709.10
Final Products		
Applicon @ $8.50 x10	$ 85.00	
Lamination @ $.01/sq. inch	$ 72.00	
		$ 157.00
Labor		
Direct 354 hours	$3,694.00	
Overhead 10%	369.00	
		$ 4,063,00
		$12,018.10

Cost per acre = $12,018.10/6.4 million acres = .0091/acre or 2/10 cents/acre.

AVHRR APPROACH

A fire fuel map was produced by digitally analyzing AVHRR data for a 42-million-acre area in eastern Oregon (fig. 1) that contained Malheur County (Miller and others, 1983). This required creating a digital data base (fig. 3) that was registered to a 1-km UTM grid.

Using multi-temporal analysis and unsupervised classification procedures, four dates of AVHRR data were classified into 45 different spectral classes for the entire study area. A classification framework (table 3) developed by BLM field personnel was used to aggregate these 45 spectral classes into 11 land cover types associated with the NFDRS fire fuel models. Spectral classes associated with more than one fuel type were split into additional classes using elevation data (Miller and Shasby, 1982) and the knowledge of field personnel as to the distribution of these cover types in this area. This resulted in the original 45 spectral classes becoming 85 spectral classes. Each of the 85 spectral classes was assigned to one of the 11 land cover/fuel types and a preliminary fuel map was produced for field checking. Based upon the fieldwork, discrepancies in the preliminary fuel map were removed, which resulted in the 85 spectral classes becoming 138 spectral classes. To produce the fire fuel layer required by the IAMS program, these final 138 spectral classes were aggregated into 11 land cover/fuel types. Since 1-km is approximately the same scale as the 1-minute latitude/longitude grid used by the IAMS program, the resolution of the final fuel layer

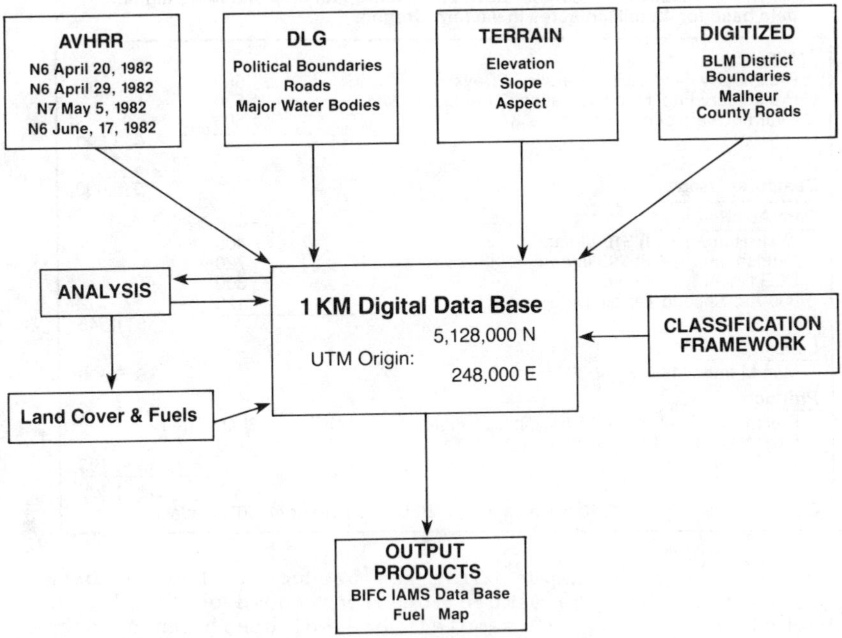

Figure 3. A digital data base developed for a 42-million-acre area in eastern Oregon.

Table 3.—Classification framework used for mapping fire fuels in eastern Oregon by digitally analyzing AVHRR data .

Land Cover/ Fuel Type	NFDRS Fire Fuel Model	Land Cover/ Fuel Type	NFDRS Fire Fuel Model
Grass	A	Mountain Shrub	L
Grass/Shrub	L	Dwarf Mountain Shrub	X
Shrub/Grass	T	Agriculture/Riparian	X
Low Shrub	X	Barren	X
Mixed Conifer	H	Water	X
Juniper/Shrub	L		

produced using this method was not affected. The total cost of developing the fire fuel information for this 42-million-acre area using this approach was approximately $16,000 (table 4).

ACCURACY ASSESSMENT

Each fuel map was sampled independently using a stratified random sampling design (Scheaffer and others, 1979), where each fuel class was considered a stratum. Simple random samples were selected in each stratum, and a file containing the line/sample (x/y, latitude/longitude)

Table 4.—Breakdown of costs for developing a fire fuel map and IAMS digital data base for 42 million acres in eastern Oregon.

Travel		
Preliminary Field Visit (1 Person, 3 Days)	$ 560	
Preliminary Fuel Map Evaluation (2 People, 5 Days)	1,530	
Aircraft Overflights	1,500	
		$ 3,590
Computer Usage		$ 5,000
Data Acquisition		
AVHRR Images (4) $100/Image	$ 400	
Landsat Images (8) $90/Image	720	
DLG Data (3) $100/File	300	
300-Arc-Second Terrain Data	125	
		$ 1,545
Labor		
Man Months (4) @ $1,400/Month		$ 5,600
Products		
Preliminary Fuel Map (3) $15/Applicon Plot	$ 45	
Final Map (10) $15/Applicon Plot	150	
		$ 195
		$15,930

Cost per acre = $15,930/42 million acres = $0.00038/acre or 4/100 cents/acre.

location for each sample point was produced. Due to cost constraints, only 165 sample locations could be field checked. Therefore, 10 samples were selected within each stratum (table 5) except the agriculture/riparian, water,

Table 5.—The number of samples allocated by strata for the 11-class AVHRR and 8-class Landsat MSS derived fuel maps.

AVHRR Strata No.	Land Cover/ Fuel Description	n	MSS Strata No.	Land Cover/ Fuel Description	n
1	Grass	10	1	Annual Grass	10
2	Grass/Shrub	10	2	Perennial Grass	10
3	Shrub/Grass	10	3	Sagebrush	10
4	Low Shrub	10	4	Juniper	10
5	Conifer	10	5	Desert Shrub	10
6	Juniper/Shrub	10	6	Conifer	10
7	Mountain Shrub	10	7	Agriculture/Riparian	5
8	Dwarf Mountain Shrub	10	8	Water/Barren	5
9	Agriculture/Riparian	5			
10	Barren	5			
11	Water	5			

and barren strata. For these strata only five samples were selected from each because these strata were considered less important from a fire fuel perspective. A total of 70 and 95 samples were selected from the MSS and AVHRR derived fuel maps, respectively. Using a Calcomp flatbed plotter, the locations of all 165 1-km-square samples were plotted onto a mylar overlay, along with the roads and county boundary, at a scale of 1:250,000 (fig. 4).

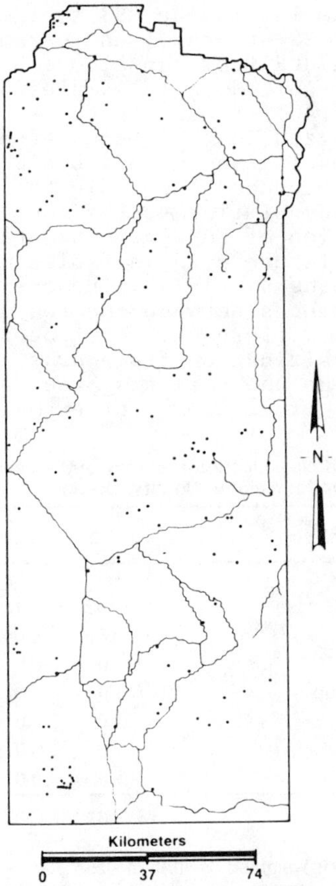

Figure 4. Location of roads and the distribution of 165 sample points within Malheur County, Oregon.

Each sample point was manually transferred from the mylar overlay to a mosaic of 1:250,000-scale topographic maps. This mosaic, along with the sample point locations and a line printer map of each fuel map, was used in a helicopter to locate each sample point in the field. The correctness of each fuel map was recorded based upon the classification framework used to produce that fuel map. This resulted in two separate determinations of correctness for each of the 165 sample points. One was based on the classification framework used to digitally analyze AVHRR data, and one was based on the classification framework used in the manual interpretation of the Landsat MSS color prints. The data on the field forms were digitally encoded using an electronic field recorder and transferred to a computer analysis system (Newcomer and Szajgin, 1982) for statistical analysis. Since the correctness for each of the 165 sample points had been determined for each fuel map regardless of which map it was selected from, all samples were pooled for analysis. For example, in the MSS fuel map 10 samples were allocated to MSS stratum 1 and 6 samples that were allocated to the AVHRR fuel map also corresponded to MSS stratum 1. This resulted in 16 samples being used to determine the accuracy of MSS stratum 1. All strata in each map were analyzed in this manner (tables 6 and 7). The statistical results shown in table 6 and 7 indicate that the fuel map produced using digital analysis of AVHRR data had an accuracy similar to the fuel map derived by manual interpretation of Landsat MSS data. The statistical nomenclature used in tables 6 and 7 is defined in Schaeffer and others, (1979). However, there were some interesting differences between the two different fuel maps. Some of the differences were the result of using two different classification frameworks, but others were due to the data type and analysis procedures used to derive the specific fuel map. For example, in the AVHRR

Table 6.—Estimate of classification accuracy of the Landsat MSS derived land cover/fuel map for Malheur County, Oregon.

MSS Strata	Land Cover/ Fuel Type	$\hat{P}$	S.E.	n	N
1	Annual Grass	1.0	0.0	16	1348
2	Perennial Grass	.82	.12	11	1434
3	Sagebrush	.87	.04	62	19562
4	Juniper	1.0	0.0	33	1243
5	Desert Shrub	1.0	0.0	11	243
6	Conifer	1.0	0.0	11	94
7	Agriculture/Riparian	1.0	0.0	11	1519
8	Water/Barren	1.0	0.0	10	245
Overall Accuracy*		.87	.03	165	25688

*This overall estimate is weighted by stratum size.

Table 7.—Estimate of classification accuracy of the AVHRR derived land cover/fuel map for Malheur County, Oregon.

AVHRR Strata	Land Cover/ Fuel Type	$\hat{P}$	S.E.	n	N
1	Grass	.77	.12	13	1794
2	Grass/Shrub	.96	.04	28	8790
3	Shrub/Grass	.93	.05	30	9916
4	Low Shrub	.79	.11	14	1739
5	Conifer	.90	.07	20	319
6	Juniper/Shrub	.80	.11	15	1095
7	Mountain Shrub	1.0	0.0	11	332
8	Dwarf Mountain Shrub	1.0	0.0	10	11
9	Agriculture/Riparian	1.0	0.0	10	1342
10	Barren	1.0	0.0	9	164
11	Water	1.0	0.0	5	186
Overall Accuracy*		.92	.02	165	25688

*This overall estimate is weighted by stratum size.

fuel map some of the treatment areas that had been reseeded to grass and an old burned area had been omitted. This is possibly due to not using data acquired early enough in the growing season to catch the greenup of these grassy areas. Likewise, in the Landsat MSS-derived fuel map two areas in the Trout Creek mountains were mapped as juniper and should have been mapped as mountain shrub. This is probably due to mis-labeling of these areas during the analysis portion of the project.

The pooling of all samples for the analysis is essentially a post-stratification procedure which inflates the variances of the accuracy estimates (Cochran, 1977). The inflation factors were very minor--four percent for the MSS map and six percent for the AVHRR map. Therefore, they were not incorporated into the variance estimates.

Time and cost constraints precluded field checking more than 165 sample locations. This resulted in rather small sample sizes across all strata, which resulted in several strata on both maps having an estimated 100 percent accuracy. Had it been possible to collect more samples, a more realistic estimate of the accuracy might have been obtained for these strata. However, since all samples were selected randomly, it is assumed that an increase in sample size would have only slightly affected the estimates shown in tables 6 and 7.

COSTS

It cost $16,000, or 0.04 cents per acre, to develop a fuel map using digital analysis of AVHRR data for a 42-million-acre study site in eastern Oregon, which included Malheur County (Table 4). In contrast, it cost $12,000, or 0.19 cents per acre, to derive a fuel map by manually interpreting Landsat MSS data for Malheur County, Oregon which covers 6.4 million acres (table 2). If the AVHRR mapping project had been restricted to only Malheur County, the project costs were estimated to be approximately the same as those for the Landsat MSS mapping project, due to the fixed costs of doing digital analysis of satellite data regardless of the size of the area. However, digital analysis of AVHRR data becomes more cost effective as the size of area to be mapped increases.

CONCLUSION

Based on a comparison of the cost and accuracy for deriving fire fuel information for Malheur County in eastern Oregon using both MSS and AVHRR data, digital analysis of multi-date AVHRR data was found to be a viable alternative to deriving fire fuel information using Landsat MSS data and manual interpretation techniques. As a result, the BLM is presently using digital analysis of multi-date AVHRR data to derive fire fuel information for all 180 million acres of the land under its jurisdiction in the western United States. This information will be used in support of its national fire program.

REFERENCES

Cochran, W. G., 1977, Sampling techniques, 3rd edition: John Wiley and Sons, New York, New York, 428 p.

Deeming, J. E., Burgan, R. E., and Cohen, J. D., 1978, The National Fire Danger Rating System -- 1978: U.S. Department of Agriculture General Technical Report INT-319, 63 p.

Kidwell, K. B., 1983, NOAA polar orbiter data (TIROS-N, NOAA-6, NOAA-7, and NOAA-8) users guide: National Oceanic and Atmospheric Administration, Satellite Data Services Division, World Weather Building, Washington, D.C., 100 p.

Miller, W. A., Howard, S. M., and Chine, E. P., 1983, Evaluation of AVHRR data to develop fire fuels information as an input to IAMS: Unpubl. report, EROS Data Center, Sioux Falls, South Dakota, 77 p.

Miller, W. A., Shasby, M. B., 1982, Refining Landsat classification results using digital terrain data: Journal of Applied Photographic Engineering, v. 8, no. 1, p. 35-40.

Newcomer, J. A. and Szajgin, J., 1982, Evaluation of an automated field data entry system: in Proceedings of the In-Place Resource Inventories National Workshop, 1982, Society of American Foresters, Bethesda, Maryland, p. 1084-1087.

Sader, S. A., 1983, Estimation of 1981 Vale manual fuels mapping costs: Unpubl. memo to the record, Technicolor Government Services, Inc., Denver Federal Center, Denver, Colorado, June 13, 1983, 1 p.

Sader, S. A., Linden, D. S., and McGuire, M., 1981, Fuels mapping from Landsat imagery and digital terrain data aid fire suppression decisions: in Proceedings of the ACSM-ASP Fall Technical Meeting, 1982, American Society of Photogrammetry, Falls Church, Virginia, p. 92-99.

Scheaffer, R. L., Mendenhall, W., and Ott, L., 1979, Elementary survey sampling, 2nd edition: Duxbury Press, North Scituate, Massachusetts, 278 p.

COMPUTERIZATION OF FIRE DISPATCH UTILIZING SATELLITE IMAGERY "OKANOGAN PROJECT"

Philip W. Gum, Fire and Aviation Staff
U. S. Forest Service
Okanogan National Forest, Okanogan, WA 98840

ABSTRACT

On the Okanogan National Forest, traditional data bases and decision processes were incapable of fully supporting fire management direction developed as part of the national forest planning process. Therefore, new natural resource data bases were acquired utilizing landsat multispectral imagery, digitized terrain data, and ancilary information. Data bases are to a .6 acre resolution, tree specie specific, tree size ranges, forest crown closures, grass and brush component, slope, aspect and elevation. Accuracy of the data at a 4.4 acre resoltuion is: (1) vegetation classification 88%, and (2) crown closure 85%. These data bases are a key basic component of a fully integrated real time computerized fire dispatch system being programmed on a DG MV 8000 and based on Geoloc (latitude and longitude) mapping system. Additional data bases such as current and potential timber productivity, land ownership, etc., are currently being acquired and installed. Value of data bases for other than fire purposes was recognized in the acquisition process and are realized in the Forest's operational G.I.S.

BACKGROUND

The 1976 National Forest Management Act directed the Forest Service to develop integrated land management plans for the national forests. As part of the planning process, fire protection and use strategies were developed for the various land management objectives. These fire strategies were refined into specific fire management direction for each of the major land management allocations such as wilderness, general forest, developed recreation, deer winter range, etc.

As an example, fire management direction for key deer winter range is:

Fire Suppression Strategy

a) Control all wildfires at 30 acres or less from May 1 to November 1.
b) Contain all wildfires of fire intensity Level 2 or less at 100 acres or less from February 1 to May 1 in fire tolerant stands.
c) Control fires at fire intensity Level 3 or above.
d) Protect bitterbrush (May-November) crown cover (thermal and hiding), habitat improvements and old growth stands.
e) Minimize loss of snags, whips, and large diameter logs (+20" diameter) during "mop-up operations.

Fire Intensity Levels reflect flame length as a measure of fire behavior:

I 0-2' II 2-4' III 4-6' IV 6-8' V 8-12' VI 12'+

Cost efficiency was a basic issue in developing the various fire management directions. Our approach to this as well as natural resources management concerns was to base the fire plan on fire effects.

Therefore, specific knowledge of affected vegetation, fire behavior parameters and resource values at risk become important to initial attack decisions on the 1.7 million acre Okanogan National Forest and to a much lesser extent approximately 200,000 acres of reciprocated cooperative protected lands.

Additionally, if the fire escapes initial attack, Forest Service policy requires the use of suppression strategies which minimize total cost with total cost being defined as suppression cost plus resource loss plus rehabilitation cost, minus resource benefit. Therefore, the natural resource information needed to make the initial attack decision is also important in making suppression strategies decisions on fires which have escaped initial attack.

Our analysis indicated, that if we could fully implement the fire management direction in the Forest Plan, an average of $340,000 could be saved annually in pre-suppression, suppression, resource damages (loses) and rehabilitation costs, when compared to the average annual fire expenditures for 1970 through 1979. Therefore, there was substantial incentive to develop the data bases and process which would allow us to fully implement our fire management direction.

COMPUTERIZED FIRE DISPATCH

Once we had completed fire planning, the question or challenge became "how do we implement?" It was obvious to us, traditional data bases and decision processes were unacceptable for implementing the new fire management direction with its requirements for site specific fire effects information for dispatching and initial attack decisions. Therefore, we decided to develop a computerized fire management strategy system which could manipulate data bases, do computations and display desired information within established 2 minute initial attack decision time frames. The sole initial input by the duty fire dispatcher being incident latitude and longitude. The computer then displays the type of information previously determined to be important in deciding appropriate response Level and initial fire suppression strategy. Additional, the system was to be designed to handle logistic questions and to allow for alternate strategy selection.

A basic schematic of the entire computerized fire dispatch system is presented in Figure 1. As you will note, many of the different parts have already been developed by other individuals and therefore our effort is really a matter of bringing them all together into a fully integrated real time decision making process.

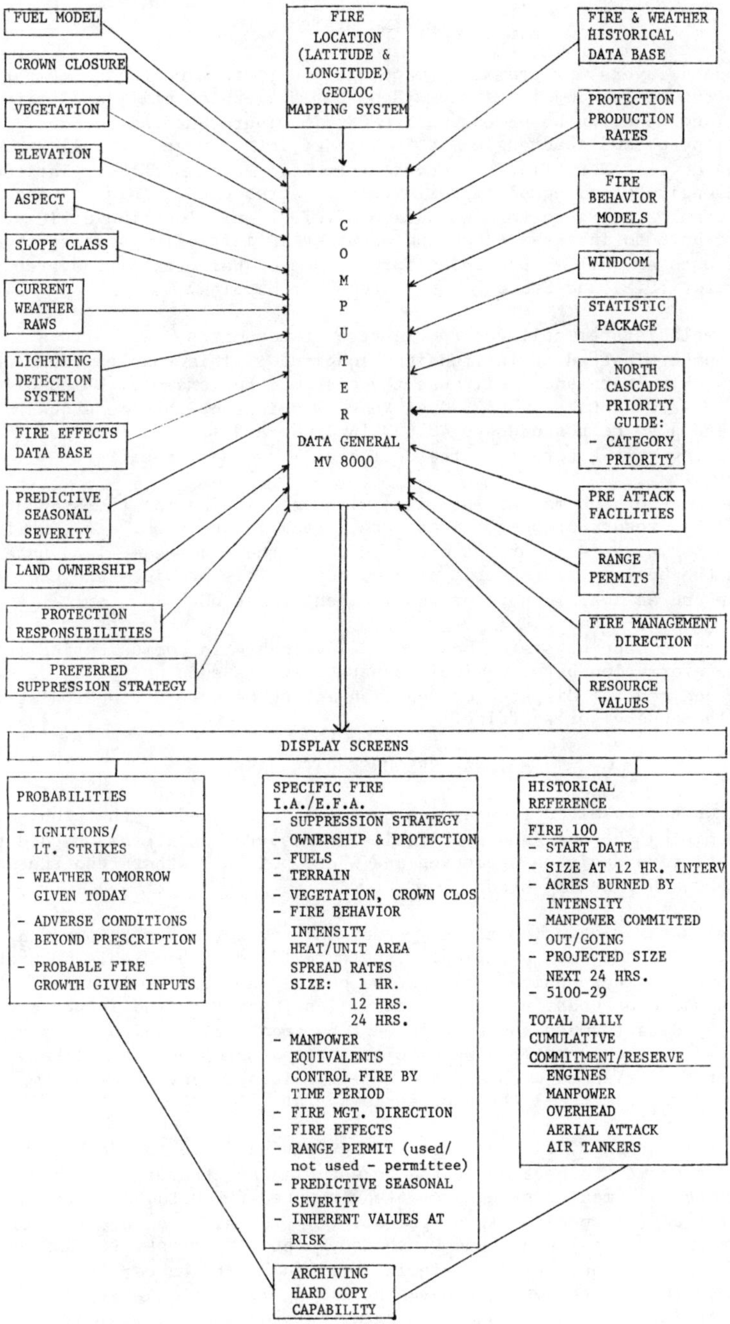

Inputs to the Okanogan National Forest Computerized Fire Dispatch System

MAPPING SYSTEM

All data layers, programming, mapping, etc., are tied to and compatible with the Geoloc mapping system (Renslow 1979) which utilizes latitude and longitude, ortho photos, and USGS 7.5 minute quadrangle maps. Each 7.5 minute USGS quadrangle map is gridded into 400 geoloc cells which at 54 degrees North latitude constitute 22.5 seconds of both latitude and longitude and equal approximately 82 acres each. This mapping system enables users to take advantage of electronic location device such as Loran C to increase speed and accuracy in initially locating fire incidents as well as providing for the use of Universal Transverse Mercator (UTM) and State Plane systems for mapping.

The Geoloc system utilizes two letters, two numbers, two letters and one number instead of latitude and longitude. This enables field units to quickly learn and easily use the system. For example, DE13GA6 locates West Oval Lake on USGS Oval Peak, Washington 7.5 minute quadrangle map and also corresponds to 481630 latitude and 1202723 longitude utilizing the SE corner of the Geoloc cell as the identifier.

Both sides of the map are utilized with one side having the standard contours, roads, property lines, pre-attack information, etc., while the other side is a current ortho photo of the exact same land base with the Geoloc grid superimposed on both. This provides the user both a map and an aerial photo of the incident location.

The Geoloc mapping system has proved invaluable in communicating incident information between widely separated units such as aerial detection and central dispatch or decision making between a ranger district and the supervisor's office.

NATURAL RESOURCE DATA BASES

Once we had resolved the mapping question, we turned to the problem of determining the type and specificity of resource data bases needed to support fire decision processes and where possible other resource management and planning needs.

Early in the discussion stage, we set some criteria for resource data acquisition and data updating. They were (1) must be highly accurate, (2) must be inexpensive, (3) must be quickly and easily acquired, (4) maintenance or updating must be easy, fast, accurate and inexpensive, and (5) data formats and storage must be in a total manipulative mode. Obviously, these criteria were relative to other processes we knew of or were experienced with and are a little idealistic. However, we believe we've met all these criteria so far.

The utilization of fire effects as the driving force for developing fire management direction in the Forest Plan and therefore in any real time decision making process created the need for data bases which identified (1) specific tree species, (2) tree size ranges, (3) forest crown closure, (4) grass and brush component, (5) slope, (6) aspect, and (7) elevation. After reviewing "in hand" data bases, it was decided to acquire new data bases utilizing landsat imagery, digitized terrain information and ancilary data and processes.

Although vegetation by species, diameter class, crown closure, and terrain by slope, aspect and elevation were originally mapped and data base integrity maintained at a .6 acre resolution, we required the information to be summarized by Geoloc cell (82 acres) for use in our fire management computerized dispatch system. The Geoloc cells become the geographical data base size for collecting and storing the necessary data to accurately, within model limitations, calculate fire behavior and therefore fire effects of any given fire.

We further made the decision to contract the acquisition of these data bases. ERDAS, Inc., of Atlanta, Georgia, was the successful bidder at a delivered cost for all contract items of $.06 per acre. Dr. Glen Klock, Wenatchee, Washington, accomplished the necessary field work (selection of training sites, ecological modeling, etc.) for ERDAS, Inc. It should be noted the contract actually required mapping of approximately 2,960,000 acres but the cost per acre was calculated using the Okanogan National Forest land base of 1.7 MM acres.

Specific contract requirements were:

I. Conifers

A. Species

1. Ponderosa pine, Douglas-fir, subalpine fir, western larch, Engelmann spruce, lodgepole pine, Western red cedar, western hemlock, white bark pine, etc.

a. Classification by species

(1) Separate ponderosa pine, Douglas-fir, subalpine fir, western red cedar, western hemlock, and western larch when more than 25% of the stand. (Could include one or more species within each stand.)

(a) 25-40% of the stand
(b) 40-60% of the stand
(c) 60% or more of the stand

b. Stratification by size:

(1) 0-4" DBH (2) 4-14" DBH (3) 14" DBH or larger

2. Lodgepole pine

a. Separate from species listed in A1, when more than 25% of stand.

b. Stratification by type of stand.

(1) Old growth decadent stands-characterized by large amounts of downfall with little or no reproduction.

(2) Healthy, vigorous stands-characterized by limited downfall, reproduction in open or fringe areas, the presence of other species in the understory and no obvious signs of stunted growth.

(3) Seedlings/saplings-characterized by large amounts of reproduction; generally vigorous in appearance.

(4) "Dog-hair" stands - severely stunted stands,

often decadent, with an obvious lack of vigor.

B. Crown Closure (for all species)
(1) 0 (2) 1-30% (3) 31-70% (4) 71-100%

II. Aspen and Other Deciduous Hardwoods

III. Other

A. Grasslands
1. Annuals (includes cheatgrass)
2. Bunchgrasses
3. Other

B. Brushlands
1. Species
a. Bitterbrush
b. Sagebrush
c. Willow
d. Other (list by predominate species)

C. Crown closure
(1) 0 (2) 1-30% (3) 31-70% (4) 71-100%

D. Rock Covered Areas

E. Snow Covered Areas (glaciers)

F. All Water Bodies and Streams

G. Topographic Features

1. Elevation - summarize for each geoloc grid cell
2. Aspect - general aspect (N, NE, NW, W, S, SW, SE, E) to be summarized for each geoloc grid cell
3. Slope - average slope to be summarized for each geoloc grid cell as follows: 0-23%, 24-40%, 41-55%, 56-75%, 75+%.

Contract accuracy requirements for each species, crown closure, and cover type defined in the contract required the data to meet or exceed an aggregate 80% correct classification (PCC) with the confidence interval at 85%.

Primary Sample Unit (PSU) - Geoloc grid cell (approximately 82 acres).

Aggregate accuracies within each PSU for species, crown closure, and cover type will meet or exceed 80% correct classification (PCC) with the confidence interval at 85%.

Secondary Sample Unit (SUU) - (4.4 acres).

Site specific (i.e., 4.4 ac.) accuracy standard for classification/stratification of ponderosa pine, Douglas-fir, subalpine fir, and western larch stands with more than 25% pure type within each SSU will meet or exceed 80% correct classification (PCC) with the confidence interval at 85%.

Site specific (i.e., 4.4 ac.) accuracy standard for crown closure for each species within each SSU will meet or exceed 65% correct classification (PCC) with the confidence interval at 85%.

The contract was originally designed into two phases with the first phase constituting a test of the concept and involving only two 7.5

minute quadrangles. Actual accuracy achieved during Phase 1, according to our review was:

Primary Sampling Unit (82 acres)		Secondary Sampling Unit (4.4 acres)	
Tiffany Quad		Tiffany Quad	
Species Classification	81%	Species Classification	82.8%
Crown Closure	90%	Crown Closure	90.3%
Loup Loup Quad		Loup Loup Quad	
Species Classification	88%	Species Classification	80.2%
Crown Closure	85%	Crown Closure	84.8%

With the achievement of these accuracies, we gave the contractor the go ahead for the second phase of the project which was for similar mapping of the remaining 92 quadrangles.

As of this date, the contractor has completed the additional quadrangles with an overall accuracy at the SSU resolution (4.4 acres) of (1) Vegetation Classification 88%, (2) Crown Closure 85%.

The technical aspects for Phase I of completing this classification at these accuracy Levels were presented at the International Symposium on Inventorying Forest and Other Vegetation of the High Latitude and Altitude Regions (Klock 1984) and again here for Phase II by Dr. G. Klock.

You will note we have not tried to determine fire behavior fuel models from data derived from landsat imagery. We had previously accomplished this task utilizing traditional processes (aerial photographs, field checking, personal knowledge, ecological associations, etc.) at a resolution and accuracy which will meet our present modeling and decision needs.

However, we believe our data bases could be used to map fire behavior fuel models particularly those models where there is little tree canopy. Burgan and Shesby's (1984) work, near Missoula, Montana, indicated the feasibility of this approach to fuel modeling at an accuracy Level of 73%. A review of our vegetative and crown closure data bases, because of the species specificity and .6 acre resolution, provides for relative easy mapping of the brush/grass fuel models. We have not to date reviewed the timber models.

Before I get into the use of these data bases in the fire management decision process, I'd like to mention our effort to arrive at inherent values at risk. Before we could deal with this issue we realized we needed to determine for forested sites (1) current timber productivity, and (2) potential timber productivity in cubic feet per year. Again, we decided to contract the acquisition of these data bases as well as some support programming. G.O. Klock and Assoc., Wenatchee, Wash., was the successful bidder on this 20 USGS 7.5 minute quadrangle test effort.

The contract requires (1) current productivity be estimated utilizing species and condition of the stand presently occupying a specific site, and (2) potential productivity for the same stand be estimated by identifying the most productive species (plant association) given the site physical characteristics. Productivity will be estimated in cubic feet per year and will be applied to all national forest land, within the designated quadrangles, that are/were at least 10% tree covered before management practices were applied.

Plant association type classes to be used are those listed in publication R6-Ecol-132A-1983, <u>Forest Plant Associations of the Okanogan</u>

National Forest, by Dr. Clinton K. Williams, Area Ecologist and Terry L. Lillybridge, Association Area Ecologist.

This approach to measuring and mapping productivity was conceived because of the data being acquired with the vegetation contract (species, size class, crown closure, slope, aspect, elevation at a .6 acre resolution) and publication of the recently completed Forested Plant Associations of the Okanogan National Forest. The thought was to use both basic information bases to generate current and potential productivity data bases at a 10 acre resolution and with following cubic feet/acre/year ranges: 0-20, 21-49, 50-65, 66-84, 85-105, and 105+.

Preliminary results are very encouraging and if success continues, we will be able to aggressively pursue the inherent value at risk question which is so important to both planning and operational decisions.

OTHER MANAGEMENT APPLICATIONS

The importance and value of, when possible, developing resource data bases which can meet more than one management need cannot be over emphasized. This need and importance was recognized very early and resulted in some basic decisions such as summarizing data bases by 82 acre cells for initial attack decisions but for natural resource management uses insisting the data base integrity be maintained at the .6 acre resolution and at the greatest specificity possible.

This early decision has proved to be solid. Data bases primarily generated by and for fire management purposes have become exceedingly valuable to other natural resource disciplines. The relationship of fire's efforts to other resource areas and users is best depicted by the following schematic:

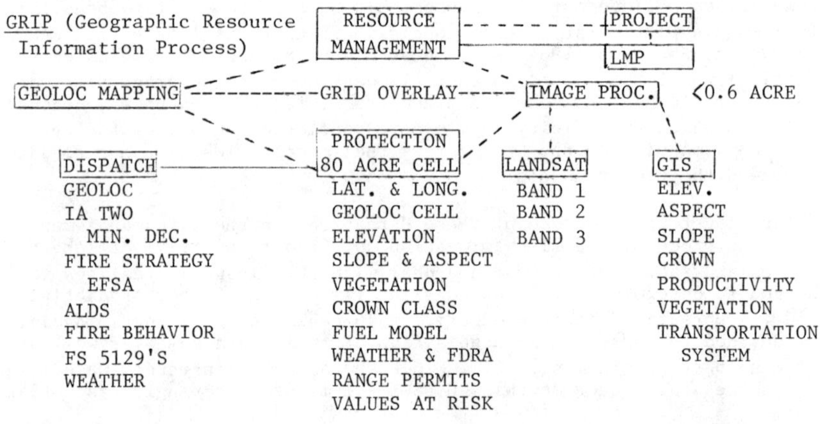

A quick summary of experienced and estimated cost to completion are as follows:

Item	Fire Cost	G.I.S.	Total Cost
Experienced:			
Geolock Mapping (lat, long) 94 quads	$30,000		
Vegetative Mapping 94 quads	110,000	*	
Productivity Mapping 20 quads	37,255	*	
Image Processing equipment/software		69,450	
Fire and Wind Behavior Modeling	8,000		
System Analyst (salary, travel)	15,000	5,000	

Item	Fire Cost	G.I.S.	Total Cost
Support Costs (extra travel, etc.)	2,000	1,000	
	$202,255	75,450	$277,705
Estimated to completion:			
Productivity Mapping 74 quads	50,000	*	
Riparian Areas Layer 94 quads		13,000	
Transportation Layer 94 quads	1,000	500	
Land Ownership Layer 94 quads	1,000	500	
System Analyst (salary, travel)	40,000	5,000	
	92,000	19,000	$111,000
GRAND TOTAL			$388,000

*Items acquired for computerized dispatch that also would need to be completed for a G.I.S., but maybe not to the same specificity.

Grand total cost completes computerization of dispatch and all the custom software plus completed 11 data layers for the Forest G.I.S. as well as the G.I.S. software and hardware.

INITIAL ATTACK DECISION OUTPUTS

Computerization of Fire Dispatch is not complete. As of this writing we have many of the data bases either acquired or in the final stages of completion. Most of the programming is completed or using the Geoloc mapping system with conversion to latitude and longitude, incident bearing and distance, closes aerial attack force, etc. (being converted to national program as of July 1985).

Much of the initial attack programming is completed or in the final stages of completion. We have not spent much time on the alternative strategies portion and do not expect to have that completed until September 1986 or later. We do have automatic lightning detection system and probabilities of a fire start gives a lightning strike up and running. So as you look back at Figure 1, you can see that in the two years we've been at this effort, we've completed about 1/2 to 2/3 of project and have also been able to acquire and equip with data bases a Forest G.I.S.

Two examples of present computer outputs which we currently use in our Central Dispatch follow:

Example #1 CG91BD9 - 82 Acre Geoloc Cell

Screen 1

```
Geoloc Cell ID:  CG01BD9      NFDRS Model:  C      NFFL Model:  02
Latitude:        0485831      NFDRS Model:         NFFL Model:
Longitude:       1185230      FDRA:         2
Land Ownership:  Okanogan NF  Prot Resp:    Okanogan NF
Recip Prot:      No Recip Prot
IA Response:     Air
               Category (((2 CONTAINMENT or CONTROL)))
Mgt Area:  6008      Key Deer Winter Range
```

- o CONTROL wildfire 30 acres/less May 1 to November 1
- o CONTAIN wildfire FIL 2/less - 100 acres/less from February 1 to May 1 in fire tolerant stands
- o CONTROL fires of FIL 3/above
- o Protect bitterbrush (May-November) crown cover (thermal and hiding), habitat improvements and old growth stands

- o Minimize loss of snags, whips, large DBH logs (+20 inch diameter) during MOP-UP operations

Range Permit Allotment Name: CEDAR
Active Dates - Start: 060185 End: 093085

Screen 2

Geoloc Cell ID: CG01BD9 Latitude: 0485831 Longitude: 1185230
Elevation: 3100 Aspect: North Slope: 0-23%

CROWN CLOSURE		VEGETATION			
Acres	Class	Acres	Species	Size	% of Stand
71.660	70-100%	17.297	Douglas-fir	Mature 14+	25-40
11.120	30-70%		Western Larch	Mature 14+	25-40
.000			DF/WL	Poles 4-14	25-40
.000		12.355	Douglas-fir	Poles 4-14	25-40
			Lodgepole Pine	Poles 4-14	25-40
			Western Larch	Poles 4-14	25-40
		8.649	Ponderosa Pine	Poles 4-14	25-40
		8.031	Ponderosa Pine	Poles 4-14	25-40
			PP/DF/WL	Poles 4-14	25-40
		6.178	Douglas-fir	Poles 4-14	25-40
			SF/WL/LP	Poles 4-14	25-40
		5.560	Douglas-fir	Poles 4-14	25-40
			Western Larch	Poles 4-14	25-40

82.780 Ac. Crown 58.07 Acres in Vegetation sample

Example #2 CG10AC3 - 82 Acre Geoloc Cell

Screen 1

Geoloc Cell ID: CG10AC3 NFDRS Model: C NFFL Model: 02
Latitude: 0485123 NFDRS Model: NFFL Model:
Longitude: 1185100 FDRA: 2
Land Ownership: Okanogan NF Prot Resp: Okanogan NF
Recip Prot: DNR
IA Response: Air
Category (((2 CONTAINMENT or CONTROL)))

Mgt Area: 3011 General Forest
- o CONTROL fires at 10 acres/less
- o Exception 10 acre objective for fire tolerant stands FIL 2/lower will be CONTAINMENT
- o Limited destructive burned acreage in stands under intensive management

Range Permit Allotment Name:
Active Dates - Start: End:

Screen 2

Geoloc Cell ID: CG10AC3 Latitude: 0485123 Longitude: 1185100
Elevation: 3400 Aspect: Southwest Slope: 0-23%

CROWN CLOSURE		VEGETATION			
Acres	Class	Acres	Species	Size	% of Stand
63.630	0%	50.657	Dry Meadow		
6.795	30-70%	29.035	Ponderosa Pine	Poles 4-14	60-100
6.178	0-30%	.618	Ponderosa Pine	Poles 4-14	25-40
3.707	70-100%	.618	PP/DF/WL	Mat/poles	25-40

80.310 Ac. Crown 80.31 Acres in Vegetation sample

I would feel remiss in my responsibilities and obligations if before I closed I did not share with you the organization which is working on the entire project. Please be advised some of these folks do not directly work for me, but are "borrowed" as their regular duties permit. You will also note the reliance on contracted expertise and staff energy.

Computerization of Dispatch

H. Richter*	D. Finch	S. Watts*	R. Hayward*	R.Whited*
System Analyst	Geoloc Mapping	Data Bases Sys. Design	Computer Specialist	Vegetative & Product. Mapping COR

J. Russell	M. Bennett	B. Moody		C. Williams
Forest Plan Fire	Predictive Seasonal Fire	Appropriate Response Plan	Suppression Implementation	Fire Effects Date Base

ERDAS, Inc., Atlanta, GA - Prime Contractor
Vegetative Terrain Mapping

Dr. Stew Pickford - Contr.
Windcom/Behave Adaptions

G.O. Klock & Assoc., Wenatchee, WA - Prime Contractor
Productivity Mapping

*P.Gum - Fire Staff - Star Wars - Pot Stirrer *Core Group

SUMMARY

Information from automatic lighting detection system (ALDS), fire weather utilizing remote automated fire weather stations (RAWS), wildland fuel models, land management objectives, current and potential productivity mapping, inherent value at risk assessments, priority guides, fire weather seasonal predictive guides, slope, aspect, elevation, crown closure, and vegetation by species and size class are all combined by interactive computer programs utilizing DF MV8000 to display to decision makers desired information. Information is to be displayed for both the 2 minute Initial Attack decision as well as alternative fire strategy analysis.

LITERATURE CITED

1. Burgan, Robert E. and Shesby, Mark B., 1984. Mapping Broad-Area Fire Potential from Digital Fuel, Terrain, and Weather Data, Journal of Forest, Vol. 82, No. 4
2. Consentino, M.J., C.E. Woodcock, and J. Franklin. 1981. U.S. Forest Service Vegetative Fuels Research - Final Report. Geography Remote Sensing Unit, Univ. of California, Santa Barbara
3. Williams, C.K., and T. R. Lillybridge. 1983. Forest Plant Associations of the Okanogan National Forest. Pac. Northwest Rep., USDA For. Serv., R6-Ecol-132b-1983. 140 pp.
4. Klock, G.O., L.E. Jordan III, and P.Gum. 1984. A forest vegetation inventory with Landsat imagery for use with fire management strategies in Washington State. International Symposium on Inventorying Forest and Other Vegetation of the High Latitude and High Altitude Regions, Fairbanks, AK, July 23-26. p. 182-187.
5. Renslow, M.S. 1979. Geo-referencing/locator grid and map numbering system. (unpublished) Pacific Southwest Region, USDA Forest Service, 24 pp.

THE USE OF WILDLAND FIRE FUEL MAPS PRODUCED WITH NOAA AVHRR SCANNER DATA

Lee F. Werth
Bureau of Land Management
Branch of Remote Sensing
Building 50, Denver Federal Center
P.O. Box 25047
Denver, Colorado 80225-0047

Randy A. McKinley and Edward P. Chine
Technicolor Government Services
Building 50, Denver Federal Center
P.O. Box 25047
Denver, Colorado 80225-0047

ABSTRACT

In the near future, Bureau of Land Management (BLM) fire managers will feed information about weather, lightning strikes, fire fuels, and terrain into a fire prediction model for dispatch decisions on wildlands. The BLM system is called Initial Attack Management System (IAMS). The Denver Service Center (DSC) in conjunction with Boise Interagency Fire Center (BIFC) and the Washington Office is working with BLM Districts to map fire fuels and produce terrain data layers for the IAMS data base that will be accessed from the BIFC facility by local micro-computers. This paper will discuss and illustrate the capabilities and limitations of wildland fire fuel maps produced with NOAA AVHRR scanner data. By 1988, fire fuel maps will be generated for BLM Districts within 10 western States.

INTRODUCTION

Wildland fires in the ten western States and Alaska blacken about 914,000 acres of Bureau of Land Management (BLM) lands on the average each year. About 72% of these fires are caused by lightning strikes. In the past, BLM fire managers would dispatch crews and equipment to lightning caused fires based on aerial observation, lookout tower sighting, or from reports by individuals or other agencies. Sometimes, personnel and equipment would be sent on a wild goose chase because the fire location was inaccurate or the fire was already out by the time they got there. The dispatch decision for deployment was based on knowledge of the area, believed fire spread risk, and just plain gut feeling. Usually, fire personnel who had been on the district for a long time and really knew local conditions made very sound dispatch decisions. However, shrinking federal budgets have put a premium on the availability of personnel and equipment for fire fighting. The BLM has, therefore, initiated a new system called Initial Attack Management System (IAMS) in an attempt to reduce costs and improve efficiency.

BACKGROUND

IAMS is a computerized data base system headquartered at Boise Interagency Fire Center (BIFC), Boise, Idaho that can be accessed by local office microcomputers or Remote Interactive Grid Systems (RIGS). IAMS is a real time geo-based system that determines if a dispatch is necessary based on location of lightning strike or series of strikes, fire fuels, terrain, and weather data. The geographical location of a lightning strike is detected by the automatic lightning detection system (ALDS) which consists of direction finders and a position analyzer. The direction finder stations are positioned throughout the Western U.S. so at least two stations can record a negative cloud-to-ground strike and plot its location within 1 mile. The risk and subsequent fire spread rate can be evaluated by fire fuels (which are changed into NFFL fuel models) and terrain at that site and the weather conditions recorded by Remote Automatic Weather Stations (RAWS). RAWS stations record wind speed and direction, temperature, precipitation, humidity, and fuel moisture. The data are telemetered to the GOES satellite which in turn transmits them to a receiving antennae at BIFC. The fuel models, terrain, and weather data are input into the BEHAVE prediction model which outputs rate of spread, flame length, fireline intensity, crews needed, containment prediction, heat per unit area, maximum spotting distance, size prediction, and whether dispatch is necessary. (Figure 1)

The BLM Service Center (SC) in Denver, Colorado, has been charged with the responsibility of mapping fire fuels and terrain for the ten western States by 1988. Initially, fire fuels were mapped on the Craig, Colorado and Lewistown, Montana Districts by manual interpretation of Landsat color composites enlarged to 1:500,000 scale. Later a pilot study was done in the Arizona Strip, Burns, and Vale, Oregon Districts utilizing digital processing of NOAA Advanced Very High Resolution Radiometer (AVHRR) scanner data. The results of the pilot study showed that AVHRR mapping was of comparable accuracy to manually interpreted Landsat data and was considerably cheaper than Landsat (.04 cents vs 0.2 cents/acre) (Miller et al. 1983). The BLM fire community made the decision to use NOAA AVHRR because of the reduced costs, repetitive daily coverage (greater chance of getting cloud-free coverage) and the 1.1 km resolution of AVHRR is similar to the IAMS data base. AVHRR data also can be used for rapid update of fire fuel changes.

FIRE FUEL CLASSES

Fire fuel classes are an aggregation of vegetation types that have similar fire behavior. The coarse resolution of the NOAA satellite (300 acres or greater) does not permit mapping of detailed vegetation types that are needed for range, wildlife and forestry management purposes. For example, the sagebrush-grass class may include the following vegetation types: big sagebrush (*Artemisia tridendata*), rabbitbrush (*Chrysothammus spp*), buckwheat

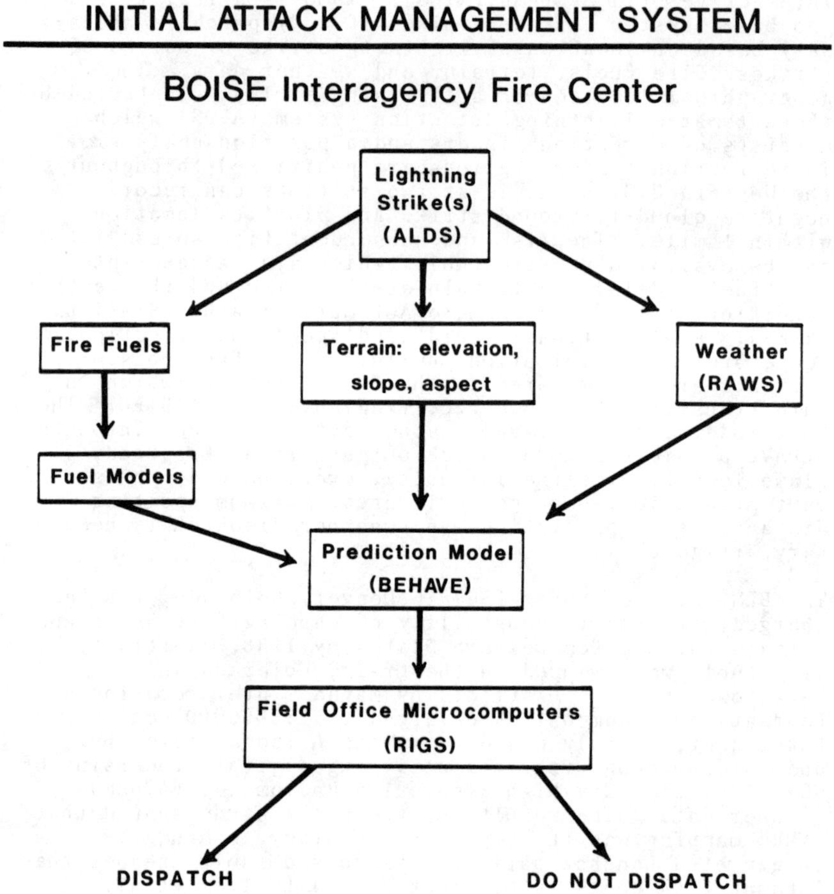

Figure 1 - IAMS Flow Chart: From Lightning Strike to Dispatch

(*Eriogonum spp*), needle and thread (*Stipa comata*), greasewood (*Sarcobatus vermiculatus*), fourwing saltbush (*Atriplex canescens*), scrub oak (*Quercus gambelii* and *Quercus turbinella*), cliffrose (*Cowania mexicana*), snakeweed (*Gutierrezia sarothrae*), shadscale (*Atriplex confertifolia*), gelleta (*Hilaria jamesii*), squirreltail (*Sitanion spp*), blue grama (*Bouteloua gracilis*), horse-brush (*Tetradymia spp*), mountain snowberry (*Symphoricarpos oreophilus*), serviceberry (*Amelanchier alnifolia*), desert peach (*Prunus andersonii*), ephedra (*Ephedra spp*), blue-bunch wheatgrass (*Agropyron spicatum*), cheatgrass (*Bromus tectorum*), Great Basin wildrye (*Elymus cinereus*), Idaho fescue (*Festuca idahoensis*), Sandberg's bluegrass (*Poa secunda*), Indian ricegrass (*Oryzopsis hymenoides*), Columbia needlegrass (*Stipa columbiana*), lupine (*Lupinus spp*), penstemon (*Penstemon spp*), arrowleaf balsam root (*Balsamorhiza sagittata*), green needle grass (*Stipa viridula*), and Western wheatgrass (*Agropyron smithii*).

Table 1 shows the 31 fuel classes that have been identified for the ten western States and Alaska. The fuel classes represent the major fuels identified in the states that have been already mapped or where mapping is in progress. New fuel classes can be added in the future if they are important and if they can be mapped by the combination of NOAA AVHRR scanner and terrain data. The fuel classes in turn have to be aggregated into the 13 National Fire Fuels Laboratory (NFFL) fuel models or a custom model prepared before they can be input into the BEHAVE prediction model.

TABLE 1 - STANDARD BLM FIRE FUEL CLASSES FOR IAMS

1.	Grassland	16.	Mixed Conifer - Sparse
2.	Salt Desert Shrub	17.	Mixed Conifer - Dense
3.	Mixed Desert Shrub	18.	Aspen/Shrub
4.	Blackbrush	19.	Subalpine
5.	Chaparral	20.	Agriculture/Wetland
6.	Sagebrush/Grassland	21.	Barren
7.	Sagebrush	22.	Barren/Shrub
8.	Low Shrub	23.	Water
9.	Sagebrush/Pinon-Juniper	24.	Urban
10.	Mountain Shrub	25.	Black Spruce/Feather Moss
11.	Dwarf Mountain Shrub	26.	Tussock
12.	Foothill Woodland	27.	White Spruce
13.	Pinon-Juniper	28.	Hardwood/Mixed Forest
14.	Mixed Conifer	29.	Nonburnable
15.	Lowland Conifer	30.	Forest Brush
		31.	Juniper

Fire fuels are being mapped to a 30-second cell size which means a uniform cell size of approximately 166 acres (38° N Latitude) when registered to 1:250,000 USGS map base. However, the digital tapes that are delivered to BIFC will have fuels data degraded to match the IAMS data base of 1-minute cells or approximately 664 acres. The fuel class for each 1-minute cell was determined by evaluating the corresponding four 30-second cells and selecting the most important fuel class (by each state) in terms of fire behavior (rate of spread). For example, if grassland occurred in any of the four 30-second cells, the entire 1-minute cell was called grassland. It is possible in the future that the IAMS data base cell size may be changed to 30 seconds to permit greater detail in fuel modeling.

OUTPUT PRODUCTS

1. Types of Outputs:
In addition to the digital tapes, BLM districts can request hardcopy *Versatec plot maps that show fuel classes with black and white symbols. The Versatec plots have corner registration marks on them so they can be overlaid with 1:250,000 base maps. The fire community will use the maps in fire planning for initial attack (placement of men and equipment, risk values, cost/acre, and budget), short and long term fire plans, and prescribed fire plans. For example, the fuels maps will help in prescription plans for calculating the number of acres to burn, cost, size of burn, and whether the burn will meet resource requirements such as range improvements.

Although the NOAA AVHRR reflectance data was the primary tool for producing fire fuel maps, elevation was used to separate those fuel classes that were spectrally confused. Fuel classes that could not be classified with the combination of NOAA and terrain data were force-classified within *IDIMS (Interactive Digital Image Manipulative System) or digitized corrections (after district review) were imbedded into the classification. This happened on the California Desert District where annual grass that was misclassified (or may not have been present during satellite overpass) had to be hand digitized from corrected maps. We would like to emphasize that the key to producing good maps that the field offices have confidence in is the direct participation of field office fire personnel at the workshops and during field checking. The field personnel's perception of overall classification accuracy after the workshops varied from 70-80%. Field verification of this level of accuracy has yet to be accomplished. The real test is going to come when field offices start to access fire fuels information at the local level via the microcomputers.

2. Field Utility of Outputs:
Field personnel's perception of the utility of fire fuels maps varies. Most see it as a chance to have fire fuels information for all ownerships within a district and a Bureauwide standardization of initial attack information. In the past, vegetation maps were only available for BLM lands and the quality and detail of the maps varied. Some fire personnel have expressed their belief that the AVHRR

*Any use of trade names is for identification purposes only and does not imply endorsement by the U.S. Bureau of Land Management.

maps are more accurate than some of the old resource planning maps. Others have stated that even the 30-second cell size maps are too coarse and should have more detail. They contend that annuals should be separated from perennial grasses and the conifer types should be separated rather than lumped into mixed conifer because they have different understories and hence fire spread rates.

The 5 districts in California and Nevada which represent about 116,480,000 acres were mapped for a cost of about .1 cents/acre during 1984. This cost includes aircraft support, output products, contract cost, field trips, workshops, and BLM personnel. Operational cost has proven to be somewhat higher than the Burns/Vale pilot project (.1 cents/acre vs .04 cents/acre).

Budget and time restrictions have made it necessary for field office personnel to perform the field checking of the preliminary fuel maps without much SC assistance. Districts are generally emphasizing checking those areas that have a history of high fire occurrence and are on BLM lands. Thus, non-BLM lands within a district boundary could contain some fire fuel misclassifications. Presently, we are mapping fire fuels in Idaho and soon will move on to Oregon, Utah, and Colorado for fiscal year 1986.

SUMMARY

The NOAA AVHRR scanner appears capable of broad scale fire fuel mapping over large areas of the Western U.S. This level of mapping may not satisfy all fire management requirements.

If, after IAMS implementation, field offices identify critical areas where AVHRR produced maps are not meeting their requirements or more detail is needed in priority areas, we will consider using other remote sensing data sources such as National High Altitude Photography (NHAP). We may also digitize existing vegetation maps where available. The extent to which we use the other sources, of course, will depend on available funds and time.

REFERENCE

Miller, Wayne A., E.P. Chine, and S.M. Howard, 1983. Evaluation of AVHRR Data to Develop Fire Fuels Information as an Input to IAMS. Final Report, USGS, Earth Resources Observation Systems (EROS) Data Center Internal Report to BLM-Denver. EROS Data Center, Sioux Falls, South Dakota, 77 pp.

MAPPING AND MONITORING WILDLIFE HABITAT

Moderator: B. Schrumpf
R. Murray

LANDSAT-FACILITATED VEGETATION CLASSIFICATION OF THE KENAI NATIONAL WILDLIFE REFUGE AND ADJACENT AREAS, ALASKA

S.S. Talbot
U.S. Fish and Wildlife Service
1011 East Tudor Road
Anchorage, AK 99503 USA

M.B. Shasby
Technical Government Services
U.S. Geological Survey Building
4230 University Drive
Anchorage, AK 99508 USA

T.N. Bailey
Kenai National Wildlife Refuge
P.O. Box 2139
Soldotna, AK 99669 USA

ABSTRACT

A Landsat-based vegetation map was prepared for Kenai National Wildlife Refuge and adjacent lands, 2 million and 2.5 million acres respectively. The refuge lies within the middle boreal subzone of south central Alaska. Seven major classes and sixteen subclasses were recognized: forest (closed needleleaf, needleleaf woodland, mixed); deciduous scrub (lowland and montane, subalpine); dwarf scrub (dwarf shrub tundra, lichen tundra, dwarf shrub and lichen tundra, dwarf shrub peatland, string bog/wetlands); herbaceous (graminoid meadows and marshes); scarcely vegetated areas; water (clear, moderately turbid, highly turbid); and glaciers.

The methodology employed a cluster-block technique. Sample areas were described based on a combination of helicopter-ground survey, aerial photo interpretation, and digital Landsat data. Major steps in the Landsat analysis involved: preprocessing (geometric correction), spectral class labeling of sample areas, derivation of statistical parameters for spectral classes, preliminary classification of the entire study area using a maximum-likelihood algorithm, and final classification through ancillary information such as digital elevation data.

The vegetation map (scale 1:250,000) was a pioneering effort since there were no intermediate-scale maps of the area. Representative of distinctive regional patterns, the map was suitable for use in comprehensive conservation planning and wildlife management.

INTRODUCTION

The Alaska National Interest Land Act of 1980 required the U.S. Fish and Wildlife Service to identify and describe wildlife habitats as part of developing refuge comprehensive conservation plans for nine new national wildlife refuges in Alaska. Vegetation maps were necessary to examine these habitats since habitat is essentially a vegetational phenomenon (Egler 1977).

Kenai National Wildlife Refuge (NWR) was selected as the first refuge for analysis. During the literature review it was apparent a paucity of vegetation data suitable for conservation or management plans existed.

Published vegetation maps that included the refuge such as Kuchler (1966; scale 1:7,500,000) were too broad to convey much information appropriate for planning. It was therefore suggested new 1:250,000 scale vegetation maps be prepared, and a Landsat-derived map was selected as the most practical approach. The reasons included: 1) short completion time 2) remoteness and large refuge size 3) ease of data base registration to the Landsat-facilitated map.

This paper: 1) presents a Landsat-derived classification and 2) describes the variation in the physiognomy and composition of the vegetation in relation to broad ecological factors. The classification is a first approximation preparing the way for more detailed quantitative methods.

GEOGRAPHICAL SETTING

The study area is Kenai NWR (2 million acres) and adjacent lands (2.5 million acres) that encompass most of the Kenai Peninsula. The refuge lies within two diverse land resource areas: Cook Inlet-Susitna Lowland and South Central Alaska Mountains (Reiger et al. 1979).

Soil development occurs on a variety of surficial deposits: slope, moraine, glacio-lacustrine, glacio-fluvial, and modern floodplain (Karlstrom et al. 1964). Three soil orders are represented--Spodosols, Histosols, and Inceptisols. The most widespread soil subgroups are Typic Cryorthods, Sphagnic Borofibrists, and Dystric Cryandepts (Reiger et al. 1979). The Kenai Mountains are classed as rough mountainous land.

The study area is situated within the middle boreal subzone (Hamet-Ahti 1976). The climate of the eastern half of the Kenai Peninsula is maritime while the western section is transitional between maritime and continental (Joint Federal-State Land Use Planning Commission for Alaska 1973). The average annual temperature and total precipitation for Anchorage (20 miles north of the peninsula) are 35.8 F and 14.7 in, and for Homer (on the southern peninsula), 36.5 F and 23.0 in, respectively.

Vegetation maps of Alaska published by Kuchler (1966; 1:7,500,000) and the Joint Federal-State Land Use Planning Commission for Alaska 1973; scale 1:2,500,000) are relatively similar. Classes recognized by Kuchler (1966) are: 1) hemlock-spruce forest (Tsuga-Picea) 2) spruce-birch forest (Picea-Betula) 3) black spruce forest (Picea) 4) muskeg (Eriophorum-Sphagnum-Betula) 5) cottonsedge tundra (Eriophorum) and 6) dryas meadows and barren (Dryas-Carex-Betula).

METHODS

The vegetation map was developed in a stepwise procedure as follows:

Inventory

Two Landsat scenes (#22043-20390 and -20393; 26 August 1980) covering the entire study area were acquired prior to initiating the field investigation. Twelve sampling areas (9x9 mi) or training blocks were selected from the scene within the refuge using the cluster block technique of Flemming (1975). Criteria for block placement involved visually interpreting the Landsat image and locating the blocks in areas which represented the totality of spectral variation within the refuge. Other criteria included differences in landforms, soils, surficial geology, and spatial distribution. Color infrared aerial photographs (scale 1:60,000) were obtained for each training block. The photographs were visually interpreted using stereo triplets and delineated into spectrally homogeneous units or polygons.

Following a fixed-wing aerial reconnaissance of the study area, a combined helicopter-ground survey was conducted. In this survey vegetation description was accomplished at two levels: general and detailed. At the general level, the physiognomy and dominant woody species were recorded for each polygon. All polygons were classified using a combination of the UNESCO (1967) and Ellenberg and Mueller-Dombois (1967) systems. Sampling intensity ranged from approximately ten to twenty polygons per training block. At the detailed level, quantitative descriptions were recorded of the floristic composition, structure, and major site features of 30 stands selected to represent the spectrum of major environmental and vegetational variation.

To be acceptable for sampling a stand had to be homogeneous and representative of the community from which it was sampled. Stand descriptions were from single plots employing a 30x30 ft quadrat for non-forest vegetation and a 70x70 ft quadrat for forest vegetation. Cover values were estimated for each species using nine cover-abundance classes (Westhoff and Maarel 1973).

Preprocessing

The 1980 Landsat scenes were obtained in computer compatible format. All digital processing was conducted at the U.S. Geological Survey's EROS Field Office,

Anchorage on the Interactive Digital Image Manipulation System (IDIMS, ESL Incorporated 1981).

Geometric correction of the Landsat scenes to a Universal Transverse Mercator (UTM) projection was performed to allow for registration of other related data sets (e.g., terrain information, refuge and watershed boundaries, Landsat winter data and fire history maps) to the classified image. For scene registration, three points were located on each USGS topographic map (scale 1:63,360) and corresponding points on the scene by viewing the image on a color display. The 75 points covering each scene were used to define two second-order, least-squares polynomial transformations relating UTM northing and easting values on the map to row and column position for each of the Landsat scenes. The coefficients in the polynomial equations were used to correct the scenes to a UTM projection with individual pixels $50m^2$. Examination of mean residual errors associated with the second-order transformation equation indicated a registration accuracy of +1 pixel. Following registration, the two scenes were mosaiced together to produce one data set which provided complete coverage for the study area. All subsequent processing addressed the registered scenes as a single data set.

Development of Training Classes and Associated Statistics

A cluster block approach (Flemming 1975) was used to develop training classes. The 12 blocks visited in the field were grouped into three sets of four blocks each. A clustering algorithm (ISOCLASS) was used to define discrete groups of pixels on the basis of their reflectance in the four Landsat spectral bands. Each of the four block data sets were clustered independently to produce three discrete statistical files. It was arbitrarily specified that 50 discrete clusters be produced for each data set. The algorithm classified pixels of similar reflectance values, maximized statistical distance between classes of dissimilar pixels, and provided statistical estimates (mean vector and convarience matrix) for the individual classes. Thus, from clustering the Kenai study blocks, three statistical files with 50 classes each were generated.

Generating three sets of statistics provided three independent estimates of the spectral characteristics of the scenes being classified. Following statistical tests for cluster separability, the three files were merged into one final statistical file. Mean spectral reflectance values for clusters which were redundant or where significant cluster overlap occurred were either combined into one cluster or one of the redundant files was deleted. After completing the pooled file, analysis of the field data and aerial photographs was begun.

Spectral Class Labeling

The process of assigning spectral classes to individual land cover classes was accomplished by using field reference data, true color, and high altitude color infrared aerial photography. The individual study blocks that had been used in the development of the spectral classes were displayed on the IDIMS system. Each pixel had been classified into one of the 48 classes present in the final pooled statistics file by using a maximum-likelihood classification algorithm. Individual classes of displayed pixels were evaluated for their relationship to the field data and aerial photographs and notes were made on their apparent composition in terms of the vegetation types which they were found to contain. Problem classes were noted for later refinement in the post-classification phases.

Several classes were found to be comprised of more than one land cover category. Most frequently these cases occurred where variations in slope and aspect caused one vegetation type to have inconsistent reflectance properties based on topographic position and/or different cover types to have similiar reflectance properties due to the influence of terrain.

Another factor serving to greatly alter the reflectance characteristics of any one vegetation type was the amount of water present in association with the vegetation. Wet herbaceous meadows, wet graminoid peatlands and string bogs were consistently confused with open and closed conifer forests which have low reflectance in the infrared. A solution not involving the use of terrain data was required to deal with the problem in post classification refinement

Classification

The training class statistical file developed within the study blocks was applied to the entire two-scene mosaic using a maximum-likelihood classification algorithm. The algorithm compared each pixel's reflectance value to the mean and covariance matrix values obtained for each training class and assigned it to a class with maximum probability. Thus, every pixel in the two scene study area was assigned to a spectral class developed and labeled in the previous phases resulting in a preliminary classification.

Post Classification Refinement

A visual evaluation of spectral class integrity of the preliminary map was performed to see how classes developed within the study blocks correlated with classes outside the study blocks. The most significant problems related to those spectral classes containing more than one vegetation class due to the influence of terrain and/or water on the spectral reflectance values. To reduce these problems, two ancillary data types--Digital Elevation Model (DEM) and winter Landsat MSS data--were registered and intergrated into the data base.

DEM data, produced by digitizing elevation contours from USGS topographical maps (scale 1:250,000), was registered to the same 50 meter UTM grid as the Landsat MSS data. Algorithms in the IDIMS system were subsequently used to derive slope and aspect information from the elevation data to provide a total of three terrain-related sets to complement the Landsat data. The relationship of vegetation types to terrain variables was documented by field observation. Aerial photo interpretation was then used to correct classification errors such as: 1) slope and aspect to distinguish mountain shadows from the water classes 2) elevation data to separate lowland cover classes from subalpine and alpine cover types 3) elevation and aspect data to label spectral classes on north and northwest aspects which were shadowed due to their position on the terrain.

Winter Landsat MSS data was used to overcome problems relating to the relative amounts of water found in association with particular vegetation types. Because there is less reflectance by water in the infrared, wetland communities which commonly have open standing water associated with them tend to have much lower mean reflectance values when the vegetation and water reflectance is averaged over the pixel. This resulted in wetland communities such as string bogs having a spectral signature almost identical to coniferous forest. To resolve this problem, winter Landsat MSS data (Band 7) was registered and analyzed. Because low growing wetland communities are normally frozen and covered with an even layer of snow in winter, their reflectance is much greater than taller communities such as conifer forests. Accordingly, a density slice was performed on the winter data and a mask was generated which contained all areas having a brightness value greater than what was determined to be associated with areas of forest or scrub. This mask was then applied to the classified summer data to stratify out the misclassified wetland communites which were being classified as forest.

RESULTS

The vegetation map documents the spatial dimensions and recurrent patterns of the vegetation throughout the Kenai study area as seven vegetation classes and sixteen subclasses.* A representative portion (Kenai Quadrangle; scale 1:785,000) is shown in Figure 1. Types recognized are:

* The map is available in color for the entire study area (scale 1:250,000) from the second author at the cost of reproduction.

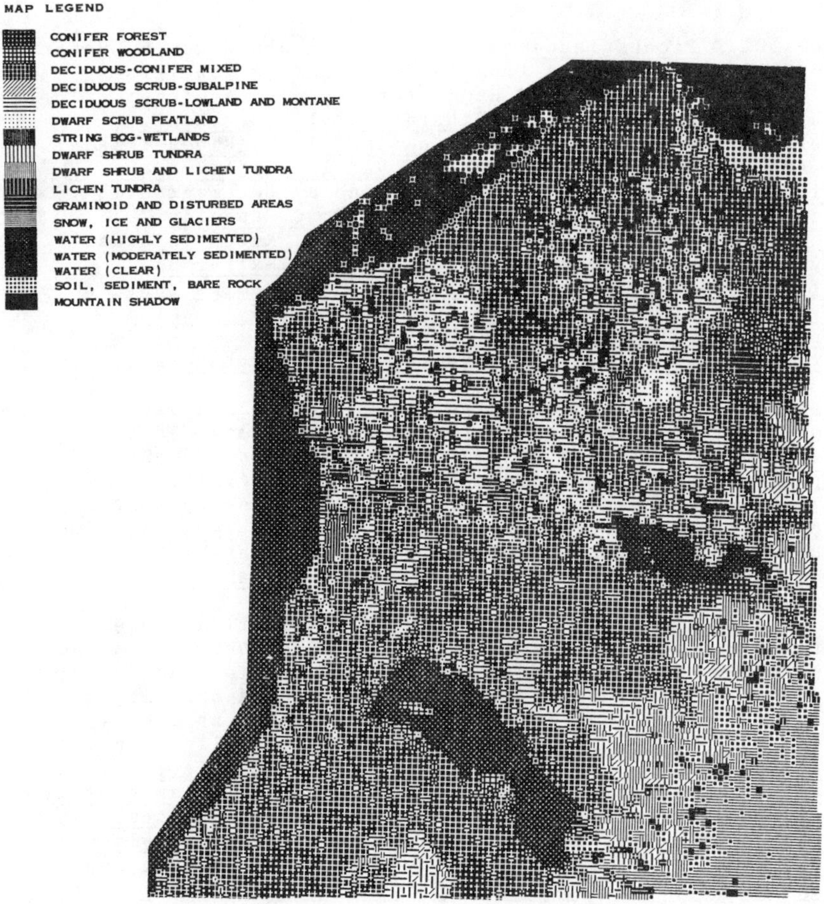

Figure 1. Vegetation map of the Kenai Quadrangle (scale 1:785,000).

Forest

Forests are formed of tree species 16 ft tall. Included within the concept of forest is secondary tree growth temporarily less than 16 ft in height, i.e., intermediate successional stages. The forest class is the most widespread vegetation type covering 37% of the refuge. The major forest subclasses are: closed needeleaf forest, needleleaf woodland, and mixed forest.

Closed needleleaf forest. This subclass is dominated by forests of *Picea glauca*, *P. sitchensis*, and *Tsuga mertensiana*. In closed forests tree cover ranges from

60-100%. *Picea glauca* forests peak in abundance on moist to well drained sites from the lowlands to subalpine slopes. Characteristic understory species include: shrubs, *Rosa acicularis*, *Viburnum edule*, *Menziesia ferruginea*, *Salix* spp. *Oplopanax horridus*, *Ribes triste*; dwarf shrubs, *Linnaea borealis*, *Vaccinum vitis-idaea*, *V. uliginosum*, *Empetrum nigrum*; graminoid, *Calamagrostis canadensis*; forbs, *Cornus canadensis*, *C. suecica*, *Geocaulon lividum*, *Trientalis europaea*, *Rubus pedatus*, *Gymnocarpium dryopteris*, *Streptopus amplexifolius*, *Dryopteris austriaca*, *Actaea rubra*, *Lycopodium* spp., *Pyrola* spp., and *Equisetum* spp.; mosses, *Hylocomium splendens*, *Pleurozium schreberi*, *Ptilium crista-castrensis*; and lichen, *Peltigera aphthosa*.

Dense *Picea sitchensis* forests occur on well drained coastal montane areas chiefly in the southern Kenai. Forest and krummholz of *Tsuga mertensiana* are found on upper montane and subalpine sites, respectively. Floristic understory composition is similiar to that of *Picea glauca* forests with the exception of *Tsuga* krummholz characterized by the presence of *Cassiope stelleriana*, *Luetkea pectinata*, and *Spiraea beauverdiana*

Needleleaf woodlands (10-60% tree cover) are open stands of trees at least 16 ft tall with their crowns not touching. They are dominated by *Picea mariana* and occur on moderately to poorly drained mineral and organic soils of the lowlands and lower montane. Typical species are: dwarf shrubs, *Betula glandulosa*, *Vaccinium uliginosum*, *V. vitis-idaea*, *Empetrum nigrum*, *Ledum groenlandicum*, *L. decumbens*, *Andromeda polifolia*; graminoids, *Calamagrostis canadensis*; forbs, *Geocaulon lividum*, *Rubus chamaemorus*, *Pedicularis* spp;, mosses, *Pleurozium schreberi*, *Hylocomium splendens*, *Polytrichum strictum*, *Sphagnum fuscum*; and lichens, *Nephroma arcticum*, *Cladina rangiferina*, *C. mitis x arbuscula*, *Cladonia* spp., *Peltigera aphthosa*, and *Peltigera* spp. Woodlands of *Picea glauca* and *P. sitchensis* with an understory of *Alnus crispa* also occur in the subalpine.

Mixed Forest. This subclass is comprised of mixed deciduous broadleaf (*Betula papyrifera* and occasionally *Populus tremuloides*) and evergreen needleleaf coniferous (*Picea glauca*) forests. They occur on moderately to well drained sites in the lowlands and montane. Understory composition is similiar to closed needleleaf forest.

Deciduous Scrub

This vegetation class is predominantly composed of woody plants ranging between 1.5 and 16 ft in height that shed their foilage simultaneously in connection with winter. Within the class concept are included multiple-stemmed woody plants such as *Salix* and *Alnus* as well as immature stands of single-stemmed tree species like *Betula papyrifera*.

Subalpine deciduous scrub is found above timberline as *Alnus* or *Salix* thickets. *Alnus crispa* scrub predominates on steep, well-drained sites. Characteristic associates are the shrubs, *Sorbus sitchensis*, *Sambucus racemosa*, *Ribes triste*; graminoid, *Calamagrostis canadensis*, and forbs, *Trientalis europaea*, *Streptopus amplexifolius*, *Dryopteris austriaca*, and the mosses, *Plagiothecium* spp. and *Brachythecium* spp.

Salix thickets usually occur on relatively gentle slopes. Typical species are the shrubs, *Salix barclayi*, *S. planifolia* ssp. *puchra*, *S. glauca*, *Betula glandulosa*; dwarf shrub, *Empetrum nigrum*; graminoids, *Calamagrostis canadensis*, *Luzula parviflora*; forbs, *Epilobium angustifolium*; *Aconitum delphinifolium*, *Rubus arcticus*, *Veratrum eschscholtzii*, *Equisetum* spp.; and mosses, *Climacium dendroides*, *Mnium magnifolium*.

Lowland and montane deciduous scrub occurs along water courses, poorly drained sites, areas regenerating from wildfires, avalanche sites, and on sandy, wind-exposed sites on the northernmost Kenai Peninsula. Principal shrubs are of the genera *Alnus* and *Salix* and are usually underlain by *Calamagrostis canadensis*. Areas regenerating from fire are dominated by immature *Betula papyrifera* and *Populus tremuloides*.

Dwarf Scrub and Related Types

This vegetation class is composed of slow growing dwarf shrubs less than 1.5 ft composed chiefly of Ericaceae and Empetraceae. There are an abundance of mosses and lichens growing admidst the dwarf shrubs. As used below the term "tundra" refers to low growing dwarf scrub communities above the limit continuous forests in the mountains.

Lichen tundra. These alpine communities grow on relatively bare ridges and mountaintops often snow free in winter. The rich lichen cover is represented by *Cladina alpestris*, *Thamnolia subuliformis*, *Sphaerophorus globosus*, *Alectoria ochroleuca*, *A. nigricans*, and *Cetraria* spp. Creeping dwarf shrubs form a significant component, *Dryas octopetala*, *Loiseleuria procumbens*, *Arctostaphylos alpina*, *Empetrum nigrum*, *Vaccinium vitis-idaea*, *V. uliginosum*, and *Diapensia lapponica*. Characteristic forbs include *Arenaria macrocarpa*, *Campanula lasiocarpa*, and *Oxytropis nigrescens*.

Dwarf shrub tundra. Elevationally, this alpine subclass occurs below lichen tundra. It is dominated by erect and prostrate dwarf shrubs such as *Betula glandulosa*, *Empetrum nigrum*, *Ledum decumbens*, *Vaccinium uliginosum*, and *V. vitis-idea* Mosses and lichens are abundant in the understory: *Pleurozium schreberi*,

Hylocomium splendens, *Ptilidium ciliare*, *Peltigera aphthosa*, *Lobaria pulmonaria*, *Cladina* spp., *Thamnolia subuliformis*, and *Cetraria* spp.

Dwarf scrub and lichen tundra. This subclass forms a mozaic of the two tundra types discussed above.

String bog/wetlands. On water impermeable material with flat or nearly imperceptible slopes, a type of organic terrain develops termed string bog. These are open bogs either devoid of trees or with a few trees (*Picea mariana*) of low vigor. The bogs have a patterned microrelief of raised strings covered by dwarf schrubs--*Betula glandulosa*, *Vaccinium uliginosum*, *Potetilla fruticosa*, and *Andromeda polifolia*--and rich in sphagnum peat mosses. Wetter depressions, flarks, occur between the strings and are usually dominated by graminoids, *Eriophorum russeolum*, *E. angustifolium*, *Carex livida*, *C. rotundata* and *C. pluriflora*.

Dwarf shrub peatland is formed mainly by sphagnum or other mosses that cover the land surface as organic terrain. Soils are poorly to very poorly drained. Dwarf shrubs such as *Ledum decumbens*, *Betula glandulosa*, *Vaccinium uliginosum*, *Andromeda polifolia*, *Myrica gale*, *Empetrum nigrum* are concentrated on the relatively drier portions of the peatland or are loosely scattered. Graminoids like *Carex aquatilis*, *C. limosa*, *C. livida*, *C. chordorrhiza*, *C. rotundata*, *C. plurifora*, *Scirpus caespitosus*, *Eriophorum angustifolium*, and forbs, *Menyanthes trifoliata* and *Potentilla palustris* may dominate locally. Slow growing trees, *Picea mariana*, or shrubs, *Salix* spp., are often found and grow as isolated individuals, in groups, or in woodlands that are marginal to the dwarf shrub peatlands.

Herbaceous Vegetation

Herbaceous plants--graminoids and forbs--are without significant woody tissue. They die back to the ground surface each year.

Graminoid marsh and meadows. This subclass includes sedge marshes (*Carex rostrata*); cultural grasslands such as hayfields (*Phleum commutatum*) and lawns (*Poa*); lowland and subalpine meadows (*Calamagrostis canadensis*); and areas regenerating from wildfires.

Scarcely Vegetated Areas

Plants are scattered or absent in this class and bare mineral substrate or rock determines the overall appearance of the landscape. Major examples and their characteristic species are tidal mud flats (*Triglochin maritima*, *Puccinellia grandis*, *Scirpus paludosus*, *Carex lyngbyei*, *C. ramenskii*), coastal sandy cliffs and beaches (*Elymus arenarius*, *Artemisia tilesii*), and talus (*Umbilicaria*, *Grimmia maritima*, *Andreaea rupestris*).

Water

The water class includes lakes, rivers, and marine waters. Subclasses are distinguished based on water clarity from clear to highly turbid.

Clear water. Contains little particulate matter.

Moderately turbid water. Water with some visible sediment.

Highly turbid water. Considerable particulate matter and appears opaque or milky.

Glaciers

This class identifies glaciers and late-melt snowbeds.

DISCUSSION

There are no intermediate-scale vegetation maps published for the Kenai Peninsula. This vegetation map is a pioneering effort and the first attempt to map the vegetation in a quantitative and systematic manner. Information on vegetation composition extends knowledge of the North America boreal zone.

A Landsat study of Tetlin NWR (Talbot et al. 1984) from the northern boreal subzone of interior Alaska distinguished seven major vegetation classes and nineteen subclasses. In conjunction with the present study these indicate what is possible with Landsat at intermediate mapping scales in different areas of Alaska.

The use of ancillary data such as elevation facilitated our ability to extend the vegetation classification and assisted in establishing ecologically meaningful units. Winter MSS data was helpful in reducing misclassification between treeless wetlands and coniferous forests. Accuracy of future Landsat studies in the boreal forest may also be enhanced by applying the mask procedure to the winter MSS data.

One of the major reasons for selecting Landsat-facilitated mapping was the relatively short time period that was available for study completion. The final map was produced within four months meeting our objectives. As a baseline vegetation study for the comprehensive planning effort the map was successfully used to derive wildlife-habitat relationships (Bailey 1984).

ACKNOWLEDGMENTS

We are grateful to the following taxonomists who identified or confirmed voucher specimens: bryophytes (W.C. Steere), *Sphagnum* (R.E. Andrus), Gramineae (W.W. Mitchell), *Salix* (G.W. Argus), Cyperaceae (J.H. Hudson).

LITERATURE CITED

BAILEY, T.N. 1984. Terrestrial habitats and wildlife species. Technical Supplement _in_ U.S. Fish and Wildlife Service, Kenai National Wildlife Refuge Comprehensive Conservation Plan. Anchorage, Alaska. 73 p.

EGLER, F.E. 1977. The nature of vegetation. Aton Forest, Norwalk, Connecticut. 527 p.

ELLENBERG, H. and D. MUELLER-DOMBOIS. 1967. Tentative physiognomic-ecological classification and mapping. Ber. geobot. Inst. ETH Stifg Rubel, Zurich. 37:21-46.

ESL INCORPORATED. 1981 IDIMS Functional Guide. Tech. Manual ESL-TM705. vol, I 716 P. Vol II. 319 p. ESL Incorporated, Sunnyvale, California.

FLEMMING, M.D. 1975. Computer aided analysis of Landsat-1 MSS data: a comparison of three approaches including a modified clustering approach. Purdue Univ. Lab. Applications of Remote Sensing, Lars Inf. Note 072475. 9 p.

HAMET-AHTI, L. 1976. Biotic subdivisions of the boreal zone. Geobotanicheksoe Kartografirovanie 1976:51-58 (In Russian).

JOINT FEDERAL-STATE LAND USE PLANNING COMMISSION FOR ALASKA. 1973. Major ecosystems of Alaska. U.S. Geological Survey, Fairbanks, Alaska.

KARLSTROM, T.N.V., H.W.COULTER, A.T. FERNALD, J.R. WILLIAMS, D.M. HOPKINS, T.L. PEWE, H. DREWES, E.H. MULLER, AND W.H. CONDON. 1964. Surficial geology of Alaska. U.S. Geol. Surv. Misc. Geol. Invest. Map I-357.

KUCHLER, A.W. 1966. Potential natural vegetation of Alaska. Page 92 _in_ the National Atlas of the United States of America. U.S. Geological Survey, Washingtn, D.C. 1970.

REIGER, S.,D.D. SCHOEPHORSTER, and C.E. FURBUSH. 1979. Exploratory soil survey of Alaska. USDA, Soil Conservation Service, Anchorage, Alaska. 213 p. and maps.

TALBOT, S.S., C.J. MARKON, and M.B. SHASBY. 1984. Landsat-facilitated vegetation classification of Tetlin National Wildlife Refuge, Alaska. Pages 143-151 _in_ V.J. LaBau and C.L. Kerr, eds. Inventorying forest and other vegetation of the high latitude and high altitude regions. Proc. Int. Symp. Soc. American Foresters Reg. Tech. Conf. 23-26 July 1984, Fairbank, Alaska

UNESCO. 1973. International classification and mapping of vegetation. United Nations Educational, Scientific and Cultural Organization, Paris, France. 93 p.

WESTHOFF, V. and E. VAN DER MAAREL. 1973. The Braun-Blanquet approach. Pages 619-726 *in* R.H. Whittaker, ed. Ordination and classification of communities. Handbook of vegetation science, Pt. 5. Dr. W. Junk B.V. Publ., The Hague.

ELK HABITAT EVALUATION USING DISTANCE-MAPPED LANDSAT DATA

RJay Murray
ERSAL/Milne Computer Center
Oregon State University
Corvallis, Oregon, 97331

Donavin A. Leckenby
Oregon Department of Fish and Wildlife
Forestry and Range Sciences Laboratory
USDA Pacific NW Forest and Range Experiment Station
La Grande, Oregon, 97850

ABSTRACT

Landsat Multispectral Scanner (MSS) data that has been transformed into distance-from-edge classes is a basic information source used in an operational system developed for Rocky Mountain elk habitat evaluation. Using this computer-based system, field biologists delineate areal blocks of MSS data, make spectral class-to-resource class assignments, distance-transform the data, and print maps and summaries. Output data from computer-assisted processing is used to calculate a Habitat Effectiveness Index (HEI) for alternate management plans, especially for different cutting patterns within timber harvest areas.

HEI's based on distance-to-edge classes have been used in operational projects to implement actions that have maintained or enhanced cover-forage ratios, stand size, and stand spacing. An example based on proposed timber sales will illustrate the computer-assisted procedures. Visual presentation of distance-transformed data can be used to emphasize balanced habitat components, and information presented in this format can be assimilated quickly by the expert and by the lay person. Obtaining accurate minimum distances is not a trivial computational task and some techniques used for distance mapping are discussed.

INTRODUCTION

Landsat MSS digital data have been used to map, inventory, and monitor components of elk habitat (Bright and Alexander 1983; Isaacson et al. 1982; Leckenby et al. 1985). These studies have summarized vegetation-associated habitat components discernible when using Landsat digital MSS data, and the information about vegetative ground cover was based on the spectral and radiometric responses measured by the MSS sensors. Processing of MSS data to obtain information about the spatial properties of elk habitat components has been reported by Murray (1984), and results given here extend this previous work.

Descriptions of elk habitat components for the Blue Mountains of Oregon and the importance of a balance among components have been detailed by Thomas et al. (1979). A consensus was that the cover-to-forage ratio for Blue

Mountain elk habitat should be 40-to-60. Direct field observation of elk has verified the relative use and importance of the primary habitat components: thermal cover, hiding cover, and forage; and the preference for sites at the interface between habitat components (Leckenby and Adams 1981; Leckenby 1984). In particular, 80 percent of observed elk use of thermal cover was within 100 meters of a forage edge and 80 percent of forage used was within 400 meters of thermal cover. As a result, computing the distance-to-edge for each Landsat ground resolution element in two habitat components, forage and combined hiding-thermal cover, has been investigated.

The two component case is simple, yet it is an improvement over current practices. Cover and forage tabulations with components discounted by areas greater than the preferred distance from an interface are an immediate result of distance-to-edge processing. Additional factors are needed to rate the value of a specific geographic or management unit as habitat. For example, road density can be an important factor, and the definition of cover can vary (Lyon 1984). The results presented here will only emphasize computing distance-to-edge values from Landsat classification data. However, of the data needed to compute habitat effectiveness, distance information probably is the most difficult to obtain and tabulate by the usual methods available to habitat managers.

METHODS

The term pixel (<u>pic</u>ture <u>el</u>ement) refers to a symbol in an image representation and to the surface area represented by a symbol; a ground pixel approximates a 200 (E-W) feet by 250 (N-S) feet surface area. <u>Distance transformation</u> and <u>distance mapping</u> are used synomymously to describe the process of computing distance-to-edge values.

Preprocessing

Methods for the computer-assisted classification of Landsat MSS data and for the assignment of spectral classes to ground (resource) classes are well-known. Here, the assumption is that a satisfactory assignment of MSS data into resource classes has been completed prior to any distance transformation. The PIXSYS software used in this study has been developed at Oregon State University for use on a large mainframe computer. Additional details about PIXSYS-based classifications are given by McGraw and Tueller (1983) and by Isaacson et al. (1982); also, methods for relating ground-based resource classes to spectral classes are described in these articles. Preprocessing and distance mapping programs are accessed <u>via</u> remote terminals equipped with printers. MSS data are stored on magnetic tape, and a user extracts a smaller data block from these files. Usually the block selected is equivalent to an area covered by 1 to 4 U.S. Geological Survey (USGS) 7.5 minute quadrangles (<u>ca</u>. 30,000 to 120,000 acres).

MSS-based data used as input to the distance transformation program and the transformed output are in scanline-column format. File structure is analogous to a line printer mapping of classified data with each record corresponding to a scanline. Classified data and transformed data are scaled 1:24,000 and are registered to the Universal Transverse Mercator (UTM) grid on 7.5 minute quadrangles. Line printer maps can be generated quickly (Murray and Alexander 1983), and a user can thereby associate the printed output directly to the stored data file contents and to aerial photographs. Areal delineations are done by specifying the start and stop column coordinate pairs that delimit within-area scanline segments. For small areas, pixel coordinates can be read manually from the line printer maps, but for large areas computer-assisted digitizing is needed.

MSS data clustered into classes with similiar spectral properties still must be assigned to ground-based resource classes. Starting at the spectral class level, a resource manager can make optimal selections for a specific geographic area, and once a satisfactory spectral-to-resource class grouping is available, a distance mapping can be done. Subsequently, a PIXSYS module is available to change ground class assignments within a selected sub-area, e.g., a proposed clearcut to forage. Thus, as measured by changes in distance-to-edge categories, the effect of different management actions on habitat components can be simulated.

Distance-transformation

The distance transformation is based on an algorithm given by Danielsson (1980). The basic technique is to start, using Danielsson's term, at an "object pixel" and then to use a iterative, one-step propagation in all directions. To meet the "all directions" requirement, a pixel image representation is first processed in a top-to-bottom, left-to-right, right-to-left raster scan and then by a bottom-to-top, left-to-right, right-to-left scan. At each of the six comparisons, the row and column increments from the nearest object pixels are computed. However, there are several paths to different object pixels, and the minimum distance must be selected. A straightforward minimum distance is the Euclidean distance computed by converting the row and column increments into distances, e.g., meters, and then using the resulting values as the sides of a right triangle; the smallest resulting diagonal is the minimum Euclidean distance. An example of the steps in the distance transformation algorithm is presented in Figure 1.

RESULTS AND DISCUSSION

Results from distance-to-edge calculations are being used in operational projects. However, only one example based on an actual project will be used to illustrate the procedures used and to show the results obtained with distance-transformed data. In Figure 2, the La Grande, Oregon, municipal watershed is delineated on 1:130,000 scale photography, and the inset shows the location within Oregon. Timber harvesting has been proposed in the 21,360 acre watershed where there are an estimated 102 million board feet of salvable lodgepole pine infested with

a)

O	X	X	X	O
X	X	X	X	X
X	X	X	X	X
X	X	O	X	X

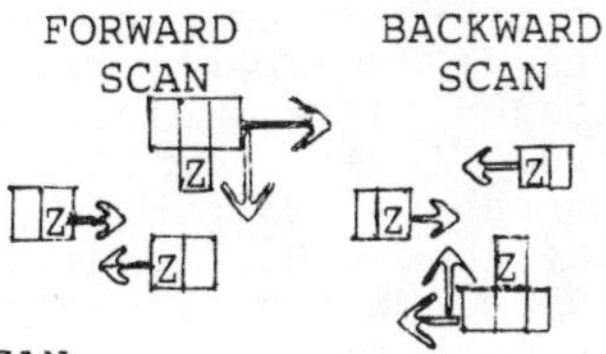

AFTER FORWARD SCAN

b)

ROW, COLUMN STEPS					MIN. DIST., METERS				
(0,0)	(999)	(999)	(999)	(0,0)	0	999	999	999	0
(1,0)	(1,1)	(1,2)	(1,1)	(1,0)	76	97	144	97	76
(2,0)	(2,1)	(2,2)	(2,1)	(2,0)	152	164	195	164	152
(0,2)	(0,1)	(0,0)	(0,1)	(0,2)	122	61	0	61	122

AFTER BACKWARD SCAN

c)

ROW, COLUMN STEPS					MIN. DIST., METERS				
(0,0)	(0,1)	(0,2)	(0,1)	(0,0)	0	61	122	61	0
(1,0)	(1,1)	(1,2)	(1,1)	(1,0)	76	97	144	97	76
(1,2)	(1,1)	(1,0)	(1,1)	(1,2)	144	97	76	97	144
(0,2)	(0,1)	(0,0)	(0,1)	(0,2)	122	61	0	61	122

Figure 1. A minimum distance transformation. In a), O's are source pixels and X's are non-source pixels assigned the value 999. Row and column steps are 76 and 61 meters respectively, and in b) and c) the paired values in parentheses are the (row, column) steps to the nearest source pixel. The six scanning operators are shown in a) where the Z's are replaced with the (row, column) pair corresponding to the minimum distance value.

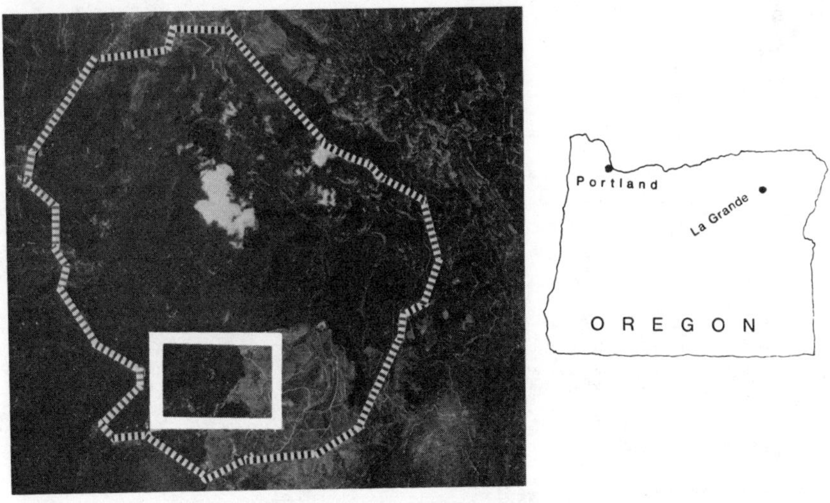

Figure 2. La Grande watershed outlined on 1:130,000 scale photography. The rectangle encloses an area shown in subsequent figures.

mountain pine beetle. This condition occurs on about 7,265 acres (29 percent) of the watershed. For this essentially unlogged area, nine action alternatives, including no harvest, were developed in response to multiple resource constraints and public issues. Using 1:24,000 scale MSS data, the U.S. Forest Service biologist for the La Grande Ranger District manually aligned sale boundaries for the nine alternatives. Subject to the constraints for each option, the alignments were used to maintain optimal size and spacing for habitat components. Habitat effectiveness was then based on the resulting cover-forage ratios. (More details about the use of Landsat-based data in the La Grande watershed project are given by Leckenby et al. [1985].)

To illustrate the results that could have been obtained with distance-transformed data, part of the work done manually in the previous study has been repeated. A rectangular block of classified data enclosing the La Grande watershed was selected from a magnetic tape file corresponding to half of a Landsat scene. The watershed boundary was converted into scanline segments, and the included area was extracted (Figure 3). Then, the 51

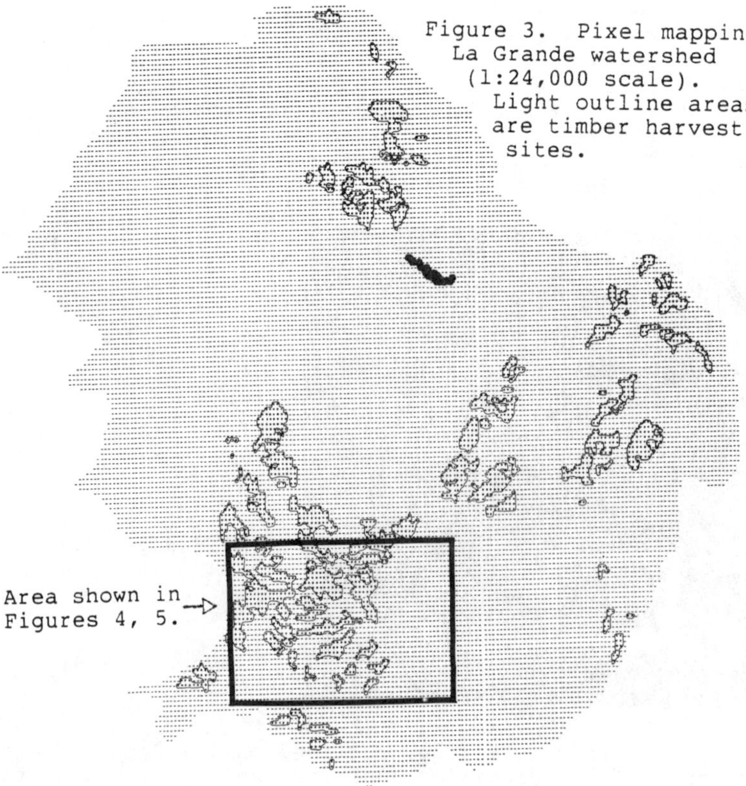

Figure 3. Pixel mapping of La Grande watershed (1:24,000 scale). Light outline areas are timber harvest sites.

spectral classes occurring within the watershed were assigned to only two ground classes: cover and forage (Figure 4). Next, this tailor-made grouping of cover and forage pixels was distance-mapped into four distance-to-edge categories (183 meters = 600 feet): cover 0-183 meters, cover >183 meters, forage 0-183 meters, forage >183 meters. The results of these operations are illustrated in Figure 5-a.

Several timber sales within the watershed have been proposed for the next few years. Site boundaries for the first three sales were converted into corresponding Landsat pixel representations and are shown in Figure 3. Of the nine alternatives, the size and configurations (selected by the above manual interpretation) of these three sites are the most satisfactory for maintaining or enhancing elk habitat. The three sale sites were converted successively to the forage category, the distance transform was applied after each change, and the accumulative effect on edge classes were tabulated (Table 1). In the pre-harvest condition, the area is 69 percent cover and therefore is forage-limited as elk habitat when compared to ideal 40-60 cover-forage ratios. Implicit in the 40-60 ratio is that the size and spacing of the stands comprising the habitat components are optimal. From the distance-mapped pre-harvest components, 14 percent of the cover and 17 percent of the forage are greater than 183 meters from an edge, and for the post-harvest components the corresponding values are 11 and 8 percent. Therefore, the stand sizes are not optimal and each component should be discounted by those acres more than 183 meters from an edge (Figure 5-b).

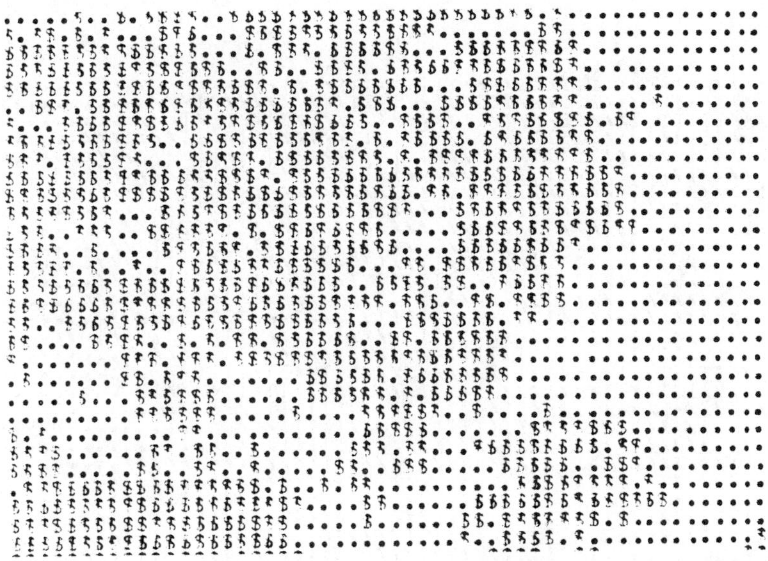

Figure 4. Cover($) and forage(.) map. See Figure 2 for location within watershed. A symbol represents 1.15 acres.

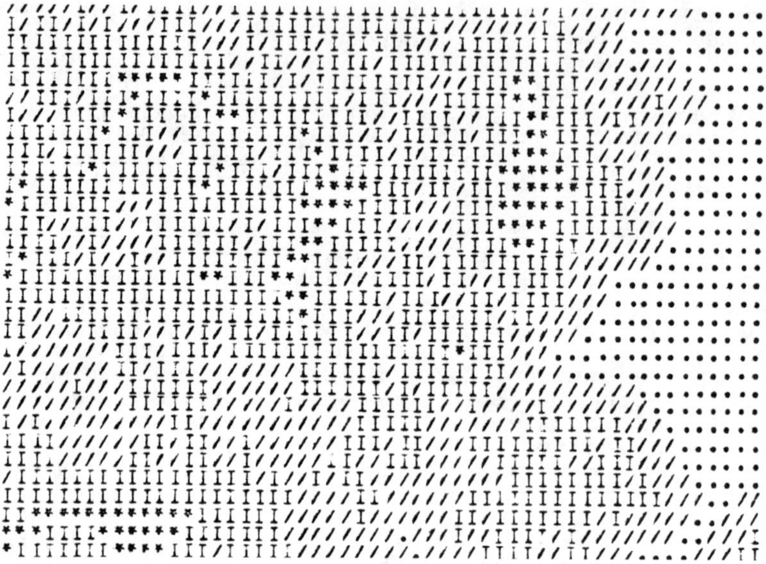

Figures 5-a (above) and 5-b (below). Distance-mapping of cover-forage components (<u>cf</u>. Figure 4). Symbol key is: cover 0-183m(I), cover >183m(*), forage 0-183m(/), and forage >183m(.). Figure 5-a is before harvest and Figure 5-b is simulated post-harvest.

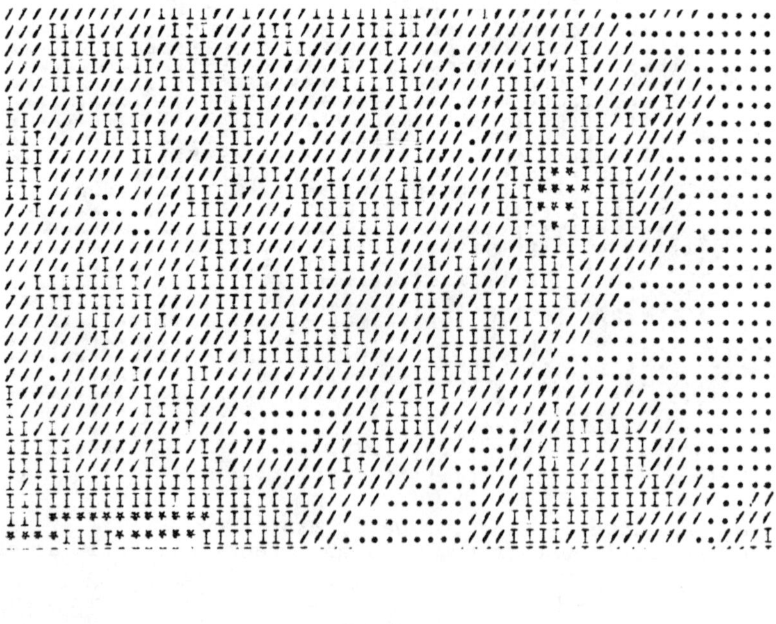

Table 1. Distance-to-edge categories in the La Grande, Oregon, municipal watershed.

Distance* category	Acres in categories$: Before	After harvest: Site 1	Sites 1+2	Sites 1+2+3	Change, percent
Cover, 0-183	11,755	11,606 (-149)	11,004 (-750)	10,758 (-997)	-8.5
Cover, >183	2,920	2,835 (-58)	2,604 (-316)	2,329 (-591)	-20.2
Forage, 0-183	5,095	5,294 (199)	5,996 (901)	6,489 (1394)	+27.4
Forage, >183	1,548	1,583 (36)	1,713 (166)	1,742 (195)	+12.6
Other	42	42	42	42	0.0
		Site size, acres**			
	21,360	234	1,066 [832]	1,588 [523]	NA

$ The values in parentheses are the acre changes from the before-harvest values.
* Distance values are in meters.
** The values in brackets are the acres in site 2 and in site 3 respectively.

One method accounting for the differences between edge classes is a weighted sum of distance classes (Thomas, et al. 1985). Based on the relative use by elk, a reasonable scheme is to assign both cover and forage areas within 183 meters of an edge a weight of 1.0, cover >183 a weight of 0.25, and forage >183 a weight of 0.5. Multiplying the percent of the area in each class by the weighting factor and then summing the results leads to a measure of habitat effectiveness which will be called Habitat Effectiveness Spacing (HEsp). For the pre-harvest data in Table 1, the HEsp calculation is:

$$0.87 = (0.55*1.0)+(0.14*0.25)+(0.24*1.0)+(0.07*0.5).$$

If the three sites are harvested as proposed, the amount of cover is 50 percent, and although the cover-forage ratio is closer to the 40-60 goal, the overall site is still forage limited. However, analyzing the distance classes leads to a different result: the after-harvest HEsp is 0.87, the same as the pre-harvest value. Although the amount of forage has increased and cover >183 meters has decreased, both favorable for improved elk habitat, the improvements are balanced by a loss of cover in the 0-183 meters class and a gain in forage >183 meters. However, the goal of maintaining an acceptable level of habitat effectiveness was met by the manual analysis. Using distance-transformed data is less labor-intensive and provides additional information.

The HEsp weights given above are not definitive indicators of habitat effectiveness and are used solely to illustrate the information obtained from distance-mapped data. Nevertheless, the format and the weights used to calculate the HEsp are reasonable and provide a starting point for improved models. The HEsp illustrated above does not account for all deviations from the optimal size and spacing, but it does alleviate some problems of habitat ratings based on cover-forage ratios.

A benefit from distance-mapping is its use as a display medium for showing the status of habitat components and for the results of different management plans. Distance-coded information is beneficial for displaying the simulated results from proposed actions: a before-after comparison. This data format has been readily accepted in public meetings and interpretation of the results easily grasped by lay persons.

CONCLUSIONS

Inclusion of spatial information as distance-to-edge values corrects several defects inherent in elk habitat effectiveness measures based on cover-forage ratios, and also provides the opportunity to formulate Habitat Effectiveness Indices that can be adjusted to meet localized conditions. The availability of large blocks of computer-classified MSS data and of computer procedures provides an opportunity for resource managers to obtain more information about existing habitat components and to model actions affecting cover and forage balance. Not all of the knowledge needed for the management of habitat components can be extracted from distance-transformed data, but it does provide information otherwise difficult to obtain about stand size and spacing.

ACKNOWLEDGEMENTS

Information about the La Grande watershed was supplied by Sylvan (Tom) Thomas, U.S. Forest Service, La Grande Ranger District, and is based on his and Greg Green's heroic, by-hand analysis of distance-to-edge classes. Manuscript entry was done by student programmer Dean Huston. The study was supported by NASA Grant NGL-39-002-053, the Oregon State University Computer Center, and the Oregon Department of Fish and Wildlife.

REFERENCES

Bright, L.R.; Alexander, C.J. 1983, Monitoring Rocky Mountain Elk Habitat Using Satellite Data: Proc. Conf. Renewable Resource Inventories for Monitoring Changes and Trends, 14-19 August 1983, Corvallis, OR., 113-115.

Danielsson, Per-Erik. 1980, Euclidean Distance Mapping: Computer Graphics and Image Processing, Vol. 14, 227-248.

Isaacson, D.L.; Leckenby, D.A.; Alexander, C.J. 1982, The Use of Large-Scale Aerial Photography for Interpreting Landsat Digital Data in an Elk Habitat-Analysis Project: J. App. Photo. Eng., Vol. 8, 1(February), 51-57.

Leckenby, D.A.; Adams, A.W. 1981, Eastern Oregon Cover Study: Information Report Series, Wildlife 81-1, Oregon Dept. of Fish and Wildlife, Portland, OR., 68pp.

Leckenby, D.A. 1984, Elk Use and Availability of Cover and Forage Habitat Components in the Blue Mountains, Northeast Oregon, 1976-1982: Wildlife Research Report Number 14, Oregon Dept. of Fish and Wildlife, Portland, OR., 40pp.

Leckenby, D.A.; Isaacson, D.L.; Thomas, S.R. 1985, Landsat Application to Elk Habitat Management in Northeast Oregon: Wildl. Soc. Bull., Vol. 13, 2(Summer 1985), 130-134.

Lyon, L.J. 1984, Field Tests of Elk/Timber Coordination Guidlines, Research Paper INT-325, Intermtn. Forest and Range Exp. Sta., Ogden, UT., 10pp.

McGraw, J.F.; Tueller, P.T. 1983, Landsat Computer-Aided Analysis Techniques for Vegetation Mapping: J. Range Manag., Vol. 36, 5(September), 627-631.

Murray, R. 1984, Segmentation of Computer-Classified Landsat Data Into Spatially-Connected Regions of Elk Habitat Components: Pecora IX Symp., 2-4 October 1984, Souix Falls, SD., 208-212.

Murray, R.; Alexander, C.J. 1983, Practical Aspects of Producing Thematic Line Printer Maps from Landsat Data: Fall Meeting Am. Soc. Photogram.-Am. Cong. Map. and Surv., 19-23 September 1983, Salt Lake City, UT, 358-366.

Thomas, J.W.; Black, H., Jr.; Scherzinger, R.J.; Pederson, R.J. 1979, Chap. 8: Deer and Elk: In Wildlife Habitat in Managed Forests--The Blue Mountains of Oregon and Washington, J.W. Thomas, Tech. Ed., U.S. Dept. Agric. Handbook 553, U.S. Govt. Print. Off., Washington, D.C., 104-127.

Thomas, J.W.; Leckenby, D.A.; Pederson, R.J.; Henjum, M. 1985, Unpubl. manuscript, Region 6, U.S. Forest Service, titled: Habitat Effectiveness Index for Elk on Blue Mountain Winter Range, May, 1985, 32pp.

STREAM/RIPARIAN AREA INVENTORY AND MONITORING USING LARGE SCALE COLOR INFRARED AIRPHOTOS

Paul Cuplin and Fred Batson
Bureau of Land Management
Building 50, Denver Federal Center
P.O. Box 25047
Denver, Colorado 80225-0047

ABSTRACT

There is an increasing level of interest in inventory and monitoring of critical stream/riparian environments. The Bureau of Land Management has investigated the use of remote sensing procedures to increase the efficiency of traditional ground-based methodologies. Ground data collection combined with photo interpretation of large scale color infrared airphotos (1:2,000) in 9- x 9-inch format provide inventory and monitoring capability for riparian areas on arid land stream/riparian areas. Variables that can be photo interpreted from large scale color infrared airphotos and used to detect change in riparian areas are stream width, flood plain width, stream channel stability, stream bank stability, stream shade, ground cover of trees, shrubs, herbaceous vegetation, bare soil, riparian area and width, density and structure of large trees and shrubs.

INTRODUCTION

Stream/riparian area values are especially important because they are a critical source of biological diversity. Stream/riparian areas are popular recreation areas, excellent sources of rock and aggregate, preferred grazing areas, productive timber sites, road construction sites, and scenic areas. The majority of this country's wildlife either depends on riparian areas or uses them proportionally more than any other habitat type. Riparian areas are fragile and comprise an extremely small percentage of BLM administered public lands. Many have been destroyed or degraded. This degradation is influencing water quality and quantity; flood frequency and severity; pollution; commercial, recreational, and subsistence fisheries; area aesthetics; and a wide range of fish and wildlife, including many endangered, threatened, and sensitive species.

Color infrared photographs are especially valuable for vegetation analysis. Color tones, along with shape, size, pattern, shadow, and texture, are used to identify individual tree and shrub species in riparian areas.

Types of riparian vegetation information that can be collected on the ground are vegetation type and subtype; width of riparian vegetation zone; species composition of shrubs, trees, grasses, and forbs, bare soil; plant density by species; condition class; reproduction; trend; structure; and potential.

Photo interpretation of large-scale airphotos combined with ground truthing produces a smaller list of riparian vegetation information that can be collected. They are as follows: vegetation type and subtype; width of riparian vegetation zone; riparian area acreage; structure; percent ground cover of trees, shrubs, and herbaceous vegetation; percent bare soil; and density of large shrubs and trees.

Stream variables that can be measured from the airphotos are stream width, stream channel stability, streambank stability, floodplain width, and stream shade.

Subtle changes in riparian vegetation are difficult to detect on airphotos. Conversely, catastrophic changes due to flooding can be easily monitored and the amount of riparian area lost can be readily calculated.

APPLICATION OF LARGE-SCALE AIRPHOTOS FOR INVENTORY AND MONITORING

Ground sampling or familiarity with the riparian vegetation photographed is essential for accurate photo interpretation. Vegetation transects and tree, shrub circular plots are transferable in general terms to large-scale airphotos. Trees, and some shrubs, can be readily identified by species, grasses and forbs generally cannot be identified by species.

Variables that may be photo interpreted from 1:2,000 scale color infrared airphotos combined with on-the-ground data collection are as follows:

Variable	May Be Photo Interpreted or Calculated
Vegetation Type and Subtype	Yes
Riparian Area Width and Acreage	Yes
Plant Species Composition	
Trees	Yes
Shrubs	Yes, some shrubs such as willow and baccharis
Grasses	No
Forbs	No

Variable	May be Photo Interpreted or Calculated
Ground Cover	
% Trees	Yes
% Shrubs	Yes
% Herbaceous Vegetation	Yes
% Bare Soil	Yes
Density	Yes, large trees and some shrubs such as cottonwood and willow
Reproduction	Yes, young trees and shrubs but not seedlings
Condition Class	No
Trend	Yes, as related to change in the amount of ground cover and bare soil
Potential	No
Structure	Yes, height of trees and shrubs
Streambank Shade	Yes
Stream Width	Yes
Flood Plain Width	Yes
Streambank Stability	Yes
Streambed Silt	Yes
Stream Channel Stability	Yes

BASELINE

Original large-scale airphotos of a riparian area provide an overview of existing conditions in terms of the readily interpreted variables. Subsequent airphotos over time provide the means of detecting change in these variables. Subtle change due to the lack of or above normal precipitation may not be evident as compared to the catastrophic change caused by a 100-year flood.

The most easily detected change in a riparian area would be reduction in foliar cover and an increase in bare soil. This change may be obvious upon inspection of the baseline (original) photo compared with the monitoring photo taken 5 to 10 years later. The cause of the change may be answered only by on-the-ground inspection; thus the photos offer the opportunity to monitor change, but not the cause.

GROUND DATA COLLECTION

Select ground data sites that are accessible and representative of the riparian vegetation and stream conditions. One site per 1- to 3-mile stream segment may be sufficient for airphoto interpretation. If there are significant differences in a stream segment, additional data collection sites should be established.

ON-SITE DATA COLLECTION

- Take color print 35-mm photographs of the stream/ riparian area (upstream, across stream, and downstream) at the site where the airphoto target is placed.
- Collect stream inventory of 1/10 mile of stream segment.
- Determine the dominant and subdominant herbaceous vegetation, shrubs, and trees, and the percent bare soil.
- Take additional field notes that may assist in photo interpretation.

ACQUISITION OF LARGE-SCALE, LARGE-FORMAT (9- x 9-INCH) AERIAL PHOTOGRAPHS

The 9- x 9-inch format is easy to work with, and most photo interpretation can be accomplished with a minimum of equipment. Interpretation can be accomplished via transparencies on a light table, with contact paper prints or with enlarged prints. Although stereo analysis can be very helpful in some situations (vegetation heights), riparian inventory and monitoring can be greatly enhanced simply with additional magnification. The 9- x 9-inch photos are identified by date, agency, photo scale, state symbols, roll number, and exposure number. Photos can be easily filed, and extra prints can be ordered as needed for field use.

A 9- x 9-inch format photo covers approximately 52 acres, or 1,500 feet on a side, at a scale of 1:2,000. At this scale, riparian zones on first, second, and third order streams are easily photographed with a good margin of the adjacent upland.

Assistance on airphoto acquisition should be sought from a remote sensing specialist. The publication "Aerial Photography Specifications," June 1983, BLM, is available from the Branch of Remote Sensing (D-442), Denver Service Center, Bureau of Land Management. This publication provides detailed specifications for acquiring aerial photographs via contract.

Provide the photographer with a USGS quad map that shows the beginning and end of each stream/riparian area to be photographed. Delineate flight lines on the map to indicate to the photographer the direction of airplane travel and the midpoint and width of each flight line according to the photo scale that you have requested.

ACQUISITION OF LARGE-SCALE AIRPHOTO WITH SMALL-FORMAT CAMERAS

The previously described use of 9- x 9-inch format aerial photography is preferred for remote sensing riparian areas. The larger format yields more coverage per exposure and is more easily handled and interpreted.

Procedures have been developed for field personnel to acquire 35-mm aerial photos of small riparian areas. These procedures are not suited for large stream/riparian areas unless the photographer is very experienced and the topography is conducive to aerial photography.

The procedure involves the use of a hand-held 35-mm camera and a small bubble-type helicopter to acquire the photography (Meyer, Batson, and Whitmer [1982]).

CONCLUSIONS

The use of large-scale color infrared airphotos for inventory and monitoring riparian areas is best applied at a scale of 1:2,000 in 9- x 9-inch format.

Airphotos taken from a hand-held 35-mm camera from a small bubble type helicopter have application to small projects. However, if several miles of stream/riparian area are to be photographed, it can best be done by an airphoto contractor with a 9- x 9-inch mapping camera.

Variables that can be photointerpreted from large-scale color infrared airphotos with ground truth and that can be used to detect change in the stream/riparian area are stream width; flood plain width; stream channel stability; stream bank stability; stream shade; ground cover of trees, shrubs, and herbaceous vegetation; bare soil; riparian area and width; and density and structure of large trees and shrubs.

REFERENCES

Colwell, Robert N. ed., 1983, Manual of Remote Sensing: American Society of Photogrammetry, Vol. 1, 1232pp., 210 Little Falls Street, Falls Church, VA 22046.

Colwell, Robert N. ed., 1983, Manual of Remote Sensing: American Society of Photogrammetry, Vol. II, 1232pp., 210 Little Falls Street, Falls Church, VA 22046.

Cuplin, P., 1978, Remote Sensing Streams, Proceedings of the Int. Symp. on Remote Sensing of Observation and Inventory of Earth Resources and the Endangered Environment, July 2-8, 1978: Freiburg Federal Republic of Germany, Vol. III, 2391pp.

Meyer, M. and P. Grumstrup, 1978 Operational Manual for the Montana 35mm Aerial Photography System, 2nd Revision: Univ. of Minn. Remote Sensing Laboratory Rep. 78-1, 22p., St. Paul, MN.

Meyer, M., F. Batson and D. Whitmer, 1982, Helicopter-borne 35mm Aerial Photography Applications to Range and Riparian Studies: Remote Sensing Laboratory, Univ. of Minn., 80pp., St. Paul, MN.

Meyer, M. D. Meisner, W. Johnson and O. Lindstrom, 1983, Small aircraft 35mm Photography and Video Imagery Applications to Developing Country Surveys: Remote Sensing Laboratory, Univ. of Minn., 11pp., St. Paul, MN.

U.S. Dept. of Agriculture, Forest Service, 1978, Forester's Guide to Aerial Photo Interpretation: U.S. Forest Agriculture Handbook 308, 41pp.

U.S. Dept. of Interior, Bureau of Land Management, 1983, Interpretation of Aerial Photographers, 38pp., BLM Denver Service Center, Denver, CO.

U.S. Dept. of Interior, Bureau of Land Management, 1983, Aerial Photography Specifications, 15pp., BLM Denver Service Center, Denver, CO.

U.S. Dept. of Interior, 1983, Interior Dept. Aircraft Policy and Guidelines: Departmental Manual 2530 Office of Aircraft Services, U.S. Dept. of Interior, OAS, 25pp.

EMERGING TECHNOLOGIES FOR THE ASSESSMENT AND MONITORING OF WILDLIFE HABITAT IN DEVELOPING COUNTRIES[1]

B. Dean Treadwell[2], Peter L. Warren[3], and Stephen Berwick[4]

ABSTRACT

Remote sensing has a proven applicability for providing a wide variety of information required for wildlife habitat assessment and monitoring. Successful implementation of the technology in many developing countries can be limited by various constraints such as the timely acquisition of permits and other logistical requirements; the availability and/or cost of suitable aircraft; the requisite refrigeration of color infrared film; capability in sampling and interpretation and delays in processing films. Monitoring programs require that all aspects of the technology, including the equipment, the means for processing and analysis and the trained counterparts remain in-country and operational. Three technologies appear to have the potential for mitigating these constraints and providing pertinent wildlife habitat information. Small format (35 mm) aerial photography offers a versatile, reliable, easy to operate and economical system which can provide either complete coverage of moderate sized areas, or sample coverage for regional applications or monitoring programs. Dual camera configurations can be adapted to most aircraft to provide multi-scale or multi-emulsion photography. New developments in solid state video cameras and shuttering capabilities have resulted in low-cost, rugged, self-contained systems which can record high resolution data in several spectral bands including the near- and mid-infrared. The instant availability of the overflight record, as well as microcomputer interface capabilities, provide an entirely new dimension for data acquisition and analysis. The ultralight aircraft is an exceptionally economical platform for various remote sensing systems and

---

[1]Paper prepared for the 10th Pecora Symposium: Remote Sensing in Forest and Range Resource Management. Fort Collins, CO. August 20-22, 1985.

[2]Wildlife and Remote Sensing Consultant. P.O. Box 93, Moundsville, WV 26041.

[3]Research Assistant. Arizona Remote Sensing Center, Office of Arid Lands Studies, College of Agriculture, University of Arizona, Tucson, AZ 85719.

[4]International Institute for Environment and Development (IIED), 1717 Massachusetts Ave. NW, Washington, DC 20036.

appears to be ideally suited for wildlife habitat studies in developing countries because they are easy to fly and very transportable. The low cost of operation enables frequent flights when needed.

INTRODUCTION

Wildlife management programs in many developing countries are sometimes considered a luxury. In spite of altruistic, or even ecologically sound conservation objectives, the exigencies of simply providing food for an ever-growing human population amidst a decreasing natural resource base (and periodic natural disasters amplified by land mis-use) dictate that the vast majority of development assistance funds be directed towards the production of food, infrastructure, energy, or other national priorities.

Whereas expenditures for agricultural programs are in the billions of dollars, budgets for wildlife projects are often five or six orders of magnitude smaller. The average budget for the 267 projects approved for funding in 1984 by WWF/IUCN was slightly less than $26,000. With the cost of a single Land Rover in Africa being as much as $20,000.00, or airplanes renting for $200 to $500 per hour, the funding limitations are obvious.

Thus it is incumbent upon the wildlife biologist to identify technological packages which are financially feasible and can provide the data required. Remote sensing has a proven capability for providing a wide variety of useful information for wildlife habitat assessment and monitoring (Anderson et al. 1980; Carneggie et al. 1980; Best 1982; Conant et al. 1983). The extremely diversified array of available remote sensing systems (e.g., small to large format cameras; panchromatic, natural color and color infrared films; satellite imagery; visual interpretation or digital classification, etc.) requires that the user select the most appropriate tool for the task at hand. Beyond the technological package, it is equally important that the user recognize the constraints of working in developing countries (Meyer et al. 1983). These constraints impose an additional set of criteria for selecting the most advantageous remote sensing system.

This paper summarizes those kinds of wildlife habitat information which can be provided by remote sensing. Next, many of the constraints frequently encountered in developing countries which can limit the utility of remote sensing systems are enumerated. With this background, three technologies are described which can provide the necessary data for wildlife habitat assessment and monitoring, and can circumvent many of the constraints. These operational and emerging technologies are Small Format Aerial Photography, Solid State Video Cameras, and Ultralight Aircraft.

WILDLIFE HABITAT INFORMATION OBTAINABLE BY REMOTE SENSING

The following kinds of information relevant to wildlife habitat assessment can be obtained by remote sensing: (1) definition of habitat types based on landform, substrate, and vegetation communities; (2) the spatial distribution or juxtaposition of these habitat types; (3) indices of interspersion, or measurements of proximity to other habitats; (4) areal measurements of habitat types; (5) varying degrees of floristic information, depending on the scale of photography and diagnostic characteristics of the plant species; (6) certain aspects of forage quality and biomass, especially the identification of early green-up areas; (7) the seasonal location of water supplies; (8) the distribution and frequency of wildlife; (9) recognition of special habitat features.

Regional information needs, such as the distribution of riparian forests, or the encroachment of agricultural lands, are perhaps the easiest to accomodate. More specific data, however, such as the current distribution of water points, or the locations of available forage after a particularly dry year are more useful for actual wildlife management. Both types of data requirements can be addressed by remote sensing; it is merely a function of the scale of photography and how current the photography is. Certain information needs, such as the activity areas of poachers as evidenced by trails through tall grass, require virtually instant availability of the overflight record.

Radio-telemetry is not generally included in reviews of remote sensing. Data on animal movements can, however, be used to identify preferred habitat types, daily activity areas, maximum distances travelled to water, and seasonal migration patterns -- all complementary information to the habitat parameters enumerated above. One of the emerging technologies described in this paper, the ultralight aircraft, appears to be especially well suited for frequent radio-tracking due to its minimal cost of operation. Even if a conventional aircraft was available, the operations cost alone for a Cessna 172 in Senegal is $75.00 per hour, compared to $10 to $15 for an ultralight. Dhungel (pers. comm., 1985) calculated that airborne radio-tracking would have increased the sample size by 440 percent (compared to surveys by elephant-back and Land Rover) during a project in the Royal Chitwan National Park, Nepal, simply because every collared animal could have been located during each aerial mission. Accordingly, if both radio-tracking and habitat mapping are envisioned, ultralights offer phenomenal potential.

Poaching activity can be as much of a habitat modification factor as wildfires (many of which are set by poachers), distribution of surface water or forest destruction. In Niokolo-Koba National Park (Senegal), years of regular poaching have driven elephants to the interior of the Park, away from the Gambia River which delineates the southern boundary. Many of the remaining elephants will no longer travel to the river, even during the dry season, and this occasionally results in the death of an animal from lack of water and nutritionally related problems as water points disappear from the interior of the Park during particularly dry years. This is another instance when regular airborne capability can generate valuable habitat information via remote sensing. An ultralight used by a wildlife husbandry project in Burkina Faso (West Africa) was able to direct ground patrols along poaching trails which were undetected by experienced personnel. The same overflights were also used to delineate unauthorized burns. Simultaneous photographic records, had the equipment been available, would have further assisted vegetation mapping efforts. The ultralight missions were so effective in controlling illegal hunting, that a substantial increase in poaching was noted after the aircraft was damaged on the ground during a storm (Lungren, pers. comm. 1984).

Visual observations fall well within the definition of remote sensing. Thus, combined aerial missions with trained observers and photographic or video equipment constitute an ideal technique for obtaining wildlife habitat information (Dunford et al. 1981; Norton-Griffiths et al. 1982; and Milligan 1982). This combined visual/photographic technique has been perfected by researchers in East Africa, and is commonly referred to as the Systematic Reconnaissance Flight (SRF). Originally designed for studies of population size, density, distribution and movements of large wildlife and livestock, the SRF was modified in recent years to collect both point data (numbers of animals, houses, etc.) and cover data (percentage of ground covered by various types of topography, soil, natural vegetation and cultural land use patterns, including agricultural crops).

It is well known that many wildlife habitats exist under precarious conditions in developing countries. There seems to be ever expanding human land use activities associated with rapid population growth. Monitoring activities, whether to provide data for wildlife research or to measure the encroachment of access roads, human settlements and associated uses of the land, must be an integral aspect of wildlife management programs. Large scale aerial photography is unsurpassed for monitoring changes in wildlife habitat. Sites for sample coverage can be specified to record the impacts of anticipated events, or randomly distributed to provide statistically testable data. Multiscale photography (Treadwell and Bursink, 1981) or precise navigational records can be used to locate sample plots for repetitive coverage. Large scale photos provide sufficient detail to document these changes (even for the lay audience) and to serve as a permanent record. Video systems employing ultralight aircraft may offer new advantages beyond the aerial record: on-site footage can be combined with the aerial images to produce inexpensive documentation and monitoring for decisionmaking and enforcement. Ultralight aircraft can facilitate monitoring activities by providing an inexpensive platform for repetitive overflights.

CONSTRAINTS TO APPLIED REMOTE SENSING IN DEVELOPING COUNTRIES

Although the remote sensing technology is available to address all of the wildlife habitat assessment and monitoring information requirements discussed in the previous section, successful implementation in many developing countries can be so limited by various constraints as to render the technology essentially useless. Some of these limitations can be mitigated by careful advanced planning and a little common sense, such as buying extra camera batteries or trying out a new system prior to taking it overseas. It is difficult and time-consuming to resolve technical problems via telexed communications to purchasing agents. Other constraints are much more real, such as the cost of airplane rentals, the lack of film processing capabilities and the logistical problems of acquiring aviation fuel in remote localities. Some of the problems enumerated below can perhaps be circumvented by technological innovations reviewed here.

1. Acquisition of Permits

Permit acquisition can be a lengthy process. Many developing countries prohibit aerial photography for national security reasons. Although special dispensations are granted, these generally require an official or special project status. If an aircraft is to be imported (possibly including ultralights) license and registration numbers may well be requested for both aircraft and designated pilot. In addition to aviation and aerial photography permits, a host of other official documents are often necessary, including visas and travel permits; national park entry and work permits; duty waivers and custom clearances. The latter can be especially aggravating if technical equipment such as computers, video or camera systems, tranquilizing and radio-tracking equipment are withheld for months. Specialized communication systems sometimes require radio frequency assignments, which must be obtained prior to purchase. Adequate lead time for permit acquisition should be provided.

2. Availability of Suitable Aircraft

Often suitable airplanes with high wings, slow cruising speeds and/or removable windows simply are not available. An aircraft with a camera bay would be a rare find. Even if external camera mounts are used, fabrication of special brackets will be necessary. In addition to commercial aviation firms represented in-country, private pilots, flying clubs, other international projects, and missionary groups are sources of airplanes. In Dakar, Senegal, Treadwell spent well over a man-month trying to locate a suitable plane and arrange an overflight. This was never successfully accomplished and several aspects of the work plan (e.g., aerial reconnaissance missions) were deleted from the project.

3. Cost of Rented Aircraft

In Francophone West Africa, rental prices can range from $200 to $500 per hour; costs in Southeast Asia can run as high as $800 (Dhungel, pers. comm. 1985). Thus, for frequent aerial missions or large area coverage the costs are prohibitive unless included in the original budget.

4. Scheduling Aerial Missions

Studies of wildlife and plant communities are extremely sensitive to seasonal changes. Missing the desired (or critical) season can have serious implications for the remainder of the project. Constraint #1 summarized the limitations pertaining to permit acquisition, but other factors including the availability of the aircraft or fuel supplies can cause cancellation of missions. Some aspects of wildlife investigations (e.g., monitoring high water periods in riparian habitats or poaching control surveys) require virtually on-call, instant availability of the aircraft.

5. Logistical Support for Overflights

Supplying aviation fuel for remote sites has proven to be a major constraint (both in availability and cost) in some areas. Distribution of landing strips can be another limiting factor. Very thorough logistical planning can partially mitigate these constraints.

6. Large Format Aerial Photography

Standard aerial mapping photography (9 x 9 inch) is available virtually worldwide, although it may be only black and white, of moderate scale (ca. 1:50,000), and somewhat dated. New coverage with this format is very expensive because these cameras are not generally available in-country and commercial aerial survey companies must be hired. Although there are circumstances when large format photos are appropriate, most wildlife habitat studies can use small format (35 to 70 mm.).

7. Preservation of Color Infrared Films

CIR film which is extremely useful for vegetation and wildlife habitat mapping is quite sensitive to high ambient temperatures and must generally be kept refrigerated in tropical countries. Not only must portable refrigerators be used for optimal results, but care must be taken when shipping the film for processing.

8. Delays in Processing

Aerial film is not processed in most developing countries. This is almost invariably true for 70 mm format, long film rolls (250 exposure), CIR and Kodachrome films. Black and white and Ektrachrome films can sometimes be processed at either commercial or project photolabs in-country. One can expect to wait at least a month to receive processed film, even if travellers are used as couriers. Situations which require "instant" availability of the overflight record cannot generally be accomodated.

9. Level of Host-Country Expertise
Counterpart training is included in most development assistance programs, and the majority of trainees become quite proficient in their fields. An unexpected, but not infrequent occurrence is that an individual trained is not always re-assigned to the project, or is assigned a managerial rather than operational position. Proficiency in operating a remote sensing system does not usually extend to trouble-shooting and repairing system failures. Simple, reliable equipment with detailed users manuals should be selected.

10. Monitoring Activities
Programs based on remote sensing systems which are designed to detect changes over a period of time require that all aspects of the technology, including the equipment, supplies and upkeep, processing and analysis, as well as trained personnel, remain in-country and operational. Many of the sophisticated systems used for resource assessment involve expensive equipment which is never installed in-country (enhanced Landsat imagery, for example), or requires prohibitive recurrent costs for operation (e.g., large format, aerial mapping cameras, airplane support and processing). Since most monitoring activities involve selected sub-sample areas, the use of small format systems can provide the necessary coverage.

35 MM AERIAL PHOTOGRAPHY

35 mm aerial photography has been used for natural resource analysis for over ten years (Meyer 1982). The advantages of this format include relatively lost cost of equipment and film, familiarity of operation, portability, simplicity of film processing, and flexibility of photo interpretation from either print or transparency.

In addition, these compact and lightweight systems can be used with window or wing-strut mounts, thus eliminating the need for an aircraft with a camera bay. In other words, the 35 mm camera system can mitigate the previously identified constraints pertaining to aircraft suitability, large format aerial photography, delays in film processing, level of counterpart expertise, and certain monitoring program needs.

However, in spite of these advantages, 35 mm aerial photography is not yet widely used by wildlife managers and other natural resource specialists. Perhaps one of the most difficult obstacles to overcome is what appears to be a common, though not often explicitly stated perception that the 35 mm format is not the "professional" choice for aerial photography. Many investigators first exposed to 35 mm photography question whether it has comparable resolution as larger format photography. The answer is clearly "no" when comparing 35 mm prints with prints from larger format cameras at the same scale and taken from the same altitude. The 35 mm prints are not as sharp (Clegg and Scherz, 1975). However, the important question is not whether they have comparable resolution, but whether they have comparable information content.

Recent work in southern Arizona has evaluated the utility of 35 mm large-scale aerial photography for quantitative sampling of such habitat characteristics as total vegetation cover, vegetation structure and growth form and changes in vegetation cover over a period of several years. The results of these studies (in press) are summarized below to indicate appropriate scales of photography, comparisons with larger format cameras, and ability to measure vegetation change. Two motor-driven cameras with 250 frame film cartridges were mounted side-by-side to produce photographs of two scales. Kodak Ektachrome 5017 was used. This film can be developed by most color photo labs, or can be developed using relatively simple portable equipment. Photointerpretation of vegetation cover was performed using both rear projection of the original transparencies on ground glass and photo prints. More detailed information on 35 mm camera systems and analytical methods appropriate for developing country situations can be found in Meyer, et al. (1983).

Selection of Photo Scale

The two-camera system was flown over representative stands of semiarid woodland and mixed grass/shrub habitat at three elevations to produce six photo scales ranging from 1:200 to 1:3,000. Vegetation cover on photos of all scales was measured using 100 point dot grids. A comparison of photo results with ground measurements made within the area of photographic coverage showed that both total cover and tree cover from the ground surveys significantly correlated with measurements made on photographs of all scales. At scales smaller than 1:1,000 however, photomeasurements of grass and shrub cover deviated significantly from the ground measurements. Thus, photo scales of approximately 1:1,000 or larger are a satisfactory substitute for ground surveys for quantifying vegetation cover of shrub, tree and grass in these semi-arid habitats.

35 mm Versus Larger Formats

Another photo interpretation trial was performed in which 35 mm and 70 mm cameras were fired simultaneously and the resulting photographs printed at the same final scale. Interpretation of the matched photo pairs at two scales, 1:600 and 1:1,000, using 100 point dot grids showed that 35 mm photos were at least as accurate as 70 mm photos for measuring vegetation cover, at a lower per-frame cost (details in press).

Photo measurements in this study were made by four different interpreters, and there was a substantial amount of variation between the interpretations. This observation led to the conclusion that in any comparison of photographic measurements of vegetation parameters, inconsistency among interpreters can contribute a larger amount of error than the format of photography. It is recommended that in future comparisons of photo measurements (e.g., between dates or between locations), all measurements should be made by the same personnel to ensure consistency. This is a constraint when personnel are ephemeral as described earlier.

Measuring Vegetation Change

Monitoring habitat change is another aspect of wildlife habitat assessment that is critically important. Analysis of change using aerial photography requires repetitive coverage, and the low cost of 35 mm photos is especially appealing.

Large-scale photography of approximately 1:1,200 was used to evaluate the accuracy of vegetation change estimates made in arid desertscrub over a period of nine years (in press). Large-scale photography was acquired over a series of permanent study plots in 1975 and again in 1984. Ground measurements of vegetation cover were made during the same season that the photos were taken and photo measurements were correlated with ground measurements. During the intervening nine years, livestock grazing was halted in the study area and the severely degraded vegetation was allowed to recover for five years before the second sampling period.

This analysis showed that measurements of cover change made on the 35 mm photos were highly correlated ($r = 0.936$) with actual increases in vegetation cover. One additional observation was that cover measurements made on 9-inch contact prints (1975 photography) were virtually identical in accuracy to those made on 35 mm prints enlarged to the same scale (1984 photography) with correlation coefficients of (respectively) 0.972 and 0.976 for total cover, and 0.891 and 0.858 for shrub cover.

Considering the numerous advantages of 35 mm over other camera systems, such photography is a clear choice for wildlife biologists working in remote locations with limited facilities.

VIDEO CAMERA SYSTEMS

New developments in video camera technology have resulted in a complete and economical remote sensing system well worth examining for developing country applications. Solid state technology is replacing tube-type cameras, and the ability to shutter video now permits low altitude, large scale imagery with remarkable resolution. Another improvement from solid state technology is the elimination of geometric distortion. Flash digitizer/frame store devices are available to convert the analogue video tape to a digital image format for analysis employing off-the-shelf GIS software.

Description of the Video System

The following components comprise the basic video recording and visual analysis system. Descriptions and salient specifications are provided, rather than trade names, because there is considerable latitude in available products. Prices are included as guidelines rather than gospel.

- o Shuttered, solid state, single chip, color video camera ($3500).
- o Latest generation (4 or 5 head), 1/2 inch format, portable VHS or Beta video cassette recorder (VCR). The multiple heads provide vastly improved still framing, which is critical for on-screen interpretation ($900).
- o Small, portable color TV monitor ($300).
- o Rechargable, sealed (e.g., gel-type), 20 amp-hour battery ($50).
- o Battery charger ($50).
- o Miscellaneous cables and connectors ($25).

The basic equipment package can be acquired for less than $5,000. The approximate weight is 30-35 pounds. A camera mount will be necessary for aerial applications. Many overseas situations will also require a 220 v. transformer, power surge protector, additional batteries, and a small generator for recharging in remote localities.

It should be noted that this basic system records in natural color. A color infrared (CIR) version is not yet available in solid state video. Xybion electronic systems (San Diego, CA) markets a multispectral, shuttered solid state video camera, but this system requires computer processing to generate a color image. The entire package (camera, recorder, and computer image processor) is rather expensive (ca. $25,000). The USDA-ARS has produced false color composite video imagery using 3 spectrally filtered b/w tube-type video cameras and a color additive viewer (Everitt and Nixon, 1985). Meisner and Lindstrom (1985) have modified a 3-tube video camera to record CIR, which is currently being marketed for $10,000. This camera does not have the shutter capability to allow low altitude, large scale images. A 3-chip, solid state camera could be modified to produce CIR, but would cost about $15,000 or $17,000 with shutter capability. Beyond the basic system for natural color imagery, or one of the modified systems for CIR, 2 other components can be added.

- o Video printers (as low as $400) can be used to produce b/w hard-copy photos. Color hard copy is available, but at a substantially increased cost.
- o Digital Analysis Capabilities. There are numerous video digitizers on the market which can convert video images to an electronic format fully compatible with image processing systems, including those designed for micro-computers.

Once digitized into the system, video images can be treated as the familiar Landsat digital data. This includes image classification capabilities and Geographic Information System (GIS) analysis. A complete, basic analysis package (including the micro-computer) can be purchased for $10-12,000.

It is absolutely essential that which ever video system is selected, is assembled and thoroughly tested prior to shipment overseas.

Advantages of Video Camera Systems

The primary advantage of the basic video system is the ability to generate high resolution, instantaneous natural color imagery. This immediate availability of the overflight record offers a previously unknown dimension for time sensitive data requirements. Figure 1 is a b/w image (original in color) acquired at 1,000 feet AGL, and illustrates the excellent resolution capabilities of the basic system previously described.

Figure 1. Vertical video image of riparian wildlife habitat in Alaska acquired by low flying aircraft travelling at 140 mph. Although significant information content was lost in this reproduction, the ground resolution of the original shuttered (1/500 sec.) color video image was approximately 4 inches. One video tape ($6.00) is capable of recording 432,000 such images (video fields) at a cost of less than 0.002 cents per image. (Image courtesy W.H. Anderson, EROS Data Center)

Even though the cost of the basic video system is 2 - 3 times greater than for a 35 mm camera system (assuming a dual camera configuration), it is considered to be competitively affordable. By including the costs of the video cassette or film rolls in the analysis, the operational economy becomes apparent. A standard video tape costs less than the purchase and processing price of a 36 exposure, 35 mm film roll. Moreover, the tape can record 2 hours of imagery, compared to 1 or 2 minutes for the film: a savings factor of 60 to 120 times (Meisner and Lindstrom, 1985). For a modest increase in price (ca. $7,000.00), color infrared imagery can be produced without the usual problems of refrigeration and processing delays. THe spectral quality of the CIR imagery is comparable to that of photographic emissions (Everitt and Nixon 1985; Meisner and Lindstrom 1985). The additional cost for a CIR system must be evaluated according to the information needs of the project. If CIR is necessary, and particularly if turn-around time is important, video may well prove to be the most economical system. The loss of an entire photographic CIR overflight record from heat damage or postal problems is a definite possibility in developing countries.

Video systems are reliable and no more complicated to operate than small format cameras. Accordingly, they may well offer solutions to Constraints No. 9 and 10 (level of counterpart expertise, and continued operational capabilities for monitoring programs).

Two other advantages previously mentioned can be reiterated. The first is the potential for a complete digital analysis capability, in-country, for an additional $10,000. The second is the utility of the video system for complementary activities such as project documentaries, training courses and public relations.

More detailed information on video systems, operations and analysis methods, and advantages/disadvantages can be found in Meisner and Lindstrom (1985).

ULTRALIGHT AIRCRAFT

The legal definition of ultralight aircraft is based on Federal Aviation Authority (FAA) Regulation Part 103. This regulation describes a type of aircraft designated for pleasure use only and imposes specific limitations on maximum speed, weight, fuel tank capacity and other features. Within the confines of the FAA regulation, operators need not be licensed pilots. This unique distinction, coupled with the modest cost of these aircraft (ca. $4,000 to $19,000), have afforded ultralights a special popularity.

Commercial or other non-pleasure use of these aircraft, however, voids the applicability of FAA Reg. Part 103, and an entirely different set of regulations requires the licensing of both the plane and pilot. Although these new regulations, as well as State and Federal laws affecting pilot requirements for official business may effectively preclude the utility of ultralights for stateside applications, this does not seem to be the case in many developing countries. There are at least 3 countries in West Africa (Ivory Coast, Burkina Faso and The Gambia) where ultralight aircraft have been used for both personal and project related overflights. It is recommended that national regulations be thoroughly investigated prior to project commitment to ultralight aircraft.

Description of Ultralight Aircraft

There are a great diversity of ultralight aircraft available. While the purpose of this paper is neither to evaluate the products of this dynamic market, nor to endorse a specific aircraft, a general description should be of benefit to the uninformed reader. The following specifications represent a "composite ultralight" based on the characteristics of the American Aerolight Corp. "Eagle", the "Condor III+2", and the "Falcon XP."

- Weight, empty -- 155 to 500 pounds.
- Maximum payload (including pilot) -- 220 to 425 pounds.
- Size -- Wing span 32 feet; disassembled package measures 1 x 5 x 16 feet for the Condor III+2.
- Time required for assembly -- Approximately 1 hour with 2 people.
- Engine type -- 2 stroke, 20 to 50 horsepower.
- Take-off and landing roll -- 100 to 250 feet, but this distance will increase under tropical conditions.
- Stall speed -- 20 to 30 miles per hour and cruising speed of about 60 mph.
- Fuel tank capacity/flying time -- 2 to 15 gallons; 1.5 to 3.5 hours.
- Operation cost -- $5 per hour (1982 prices) for the Eagle; more for the larger aircraft.
- Purchase price -- $5,000 to $9,000 U.S.

There are several other features of these aircraft which augment their utility and safety. Both the Condor III+2 and Falcon XP are 2 seat aircraft, which provides either for additional payload capacity, or for a second observer/photographer. The Falcon XP has a completely enclosed cockpit which should be advantageous for the use of sensitive electronic equipment. Many of these aircraft can be equipped with pontoon landing gear. Finally, most ultralights are now available with ballistically deployed parachutes designed to bring both the passengers and aircraft safely to ground.

Advantages of Ultralights

The most obvious advantage of ultralights for remote sensing applications is as an inexpensive and highly transportable platform for aerial sensing devices or observations. Furthermore, virtually anybody can learn to fly these aircraft with 10 to 20 hours of airborne time (training costs were reported by Dhungel, pers. comm. 1985, to be $35.00 per hour.) It is highly recommended, however, that ultralight operators undertake the equivalent training of a private licensed pilot (40 hours in-flight) in order to be fully prepared for all exigencies.

The full-time availability of these aircraft to a project, the ability to rapidly transport them to various sites and prepare small landing fields, and the minimal cost of operation, should provide enormous benefits when time-sensitive data are required. In other words, the previously identified constraints pertaining to the availability of suitable aircraft, the exorbitant cost for rentals, adequate logistical support, and timely deployment can be mitigated or essentially eliminated.

The inherent limitations of ultralights for maximum payloads and cruising distances may well be a blessing in many developing countries. These limited capabilities may be much more acceptable when applying for permits in countries where overflights are restricted due to national security.

Limitations of Ultralights

It will be necessary to fully ascertain all laws, regulations and permit requirements applicable to the country and area of intended use. This task can be exceptionally time-demanding and should be vigorously pursued at the earliest opportunity.

Heavy turbulence, low cloud bases, storm conditions and strong wind currents can be especially limiting for ultralights. Substantial crosswinds have also been cited as a limiting factor, although this is less true for the newer generation aircraft. Since crosswinds are most hazardous during take-off and landing maneuvers, this situation can be largely alleviated if airfields are designed to allow multi-directional approaches. Familiarity with weather and climate evaluation and forcasting is as important to an ultralight operator as a sailor.

Spare parts and quality maintenance can be expected to limit the utility of ultralights in developing countries unless precautionary measures are taken. Advanced planning should include a complete supply of routinely needed spare parts and service tools, as well as specific agreements with the manufacturer for supplying additional materials as required. Such agreements should also include communication networks, copies of duty waivers, freight forwarding and receiving agents, etc. If feasible, an extra motor and other major parts should be acquired during the original purchase. If the operator can not perform the necessary maintenance, arrangements with a qualified airport mechanic should be made for routine inspections and servicing. Fuel quality can also be expected to be a problem, although this can be largely alleviated with a proper filtering system and additives. In this respect, the small amounts of fuel required by ultralights is very advantageous. In some areas fine dust can be a significant element of the atmosphere, and the effects and solutions for this problem should be discussed with the manufacturer.

The limited travel distance and flying time of these aircraft should be evaluated according to the intended use. With sufficient landing fields and coordinated ground support, this limitation can be effectively mitigated.

Experience in Burkina Faso has shown that proper hanger facilities are necessary. Long exposure to ultraviolet sunlight can deteriorate parts of the plane, and the strong winds associated with tropical storms can easily damage these lightweight aircraft on the ground (C. Lungren, pers. comm. 1984).

Two other problems have been recognized during the use of ultralights for aerial photography: aircraft "drift" and the effects of low altitude turbulence on vertical aerial photography (Vooren and Offermans, 1985). Drift, or the variance of the longitudinal axis of the aircraft from the true flightline caused by wind current can be considerable for ultralights. To resolve this problem, a "drift-indication-tube" was constructed so the pilot could measure the degree of drift, and correct the camera orientation immediately prior to exposing the film. Large scale, stereoscopic photographs exposed from an ultralight platform for a silvicultural study in the Ivory Coast were found not to be photogrammetrically accurate because low altitude turbulence frequently tilted the camera axis from the true vertical orientation. This resulted in sizable measurement errors derived from the 1:1000 to 1:2000 scale photographs. The authors recommended establishing a network of ground control points to resolve this problem, because a gyroscopically stabilized camera mount was too expensive and too heavy for the ultralight (Vooren and Offermans, 1985). Anderson (pers. comm. 1985) reports that it is now possible to purchase a 4-pound stabilizing gyroscope for approximately $2000.00. Most applications of large to medium scale, small format photographs acquired from an ultralight platform would not require such precise photogrammetric fidelity, and the user should decide if a gyroscopic mount is indeed necessary. A simple gimbel mount used on boats may be adequate to deal with this problem.

Some Example Applications of Ultralights for Wildlife Habitat Analyses in Developing Countries

The principle use of the ultralight will be to substitute for other conventional airplanes. In general, ultralights are best suited for localized applications (i.e., 100 mile roundtrips). If adequate airstrips are available (the tarmac rural roads in many developing countries being ideal), the ultralight could be transported via vehicle to distant locations to acquire subsample coverage for regional multi-stage sampling programs. Large scale photostrips, for example, concurrent with ground truth activities (keeping in mind that a vehicle is also in the vicinity for aircraft support) would constitute an ideal system for updating and/or extrapolating habitat interpretations derived from older aerial mapping photo coverage or satelite imagery. Similarly, ultralights can be used economically for seasonal or annual monitoring. These aircraft may even permit access to areas closed to vehicular traffic by flood waters during the rainy season.

In summary, the real advantage of ultralight aircraft is its independence -- when an overflight is required, the aircraft is available. Moreover, the cost of operation permits frequent use even with limited budgets. A private sector game ranching project in Burkina Faso, which has extremely modest financial support, was able to conduct 11 aerial missions during a 3-month period. In addition to the cost and convenience, the ultralight offers the possibility of achieving the often desired "ideal" remote sensing scenario: the pilot, photographer, photo-interpreter and field ecologist can now be the same person.

CONCLUSIONS

The adaptation and application of the methods described above are the tip of an appropriate technology iceberg limited only by the imagination and capabilities of the researcher. However, these novel and adaptive uses of technology require conscious and conscientious technology transfer lest they be disappointing or even dangerous. There is no free lunch here. For example, these methods imply capabilities in (among others):

- o photointerpretation;
- o coarse-level meteorological forcasting/evaluation;
- o navigational and recording skills for location specificity and aircraft log maintenance.

However, eager and capable counterparts should easily acquire the necessary skills. A rigid code of conduct and use of such project-associated and specialized equipment as ultralight aircraft must be maintained at the field site. This may assist local counterparts in weaker positions of their organizations by providing procedures to address unauthorized use. This should be developed in anticipation of need.

In spite of such caveats and potential problems, the need for good quality wildlife and wildland inventory and research in the developing world is so acute that decent work can be enormously important. A rule of thumb is that a single wildlife biologist in a developing country can have more than a thousand times the effect and influence that a thousand biologists have in the United States. There is an enormously satisfying magnifier effect to leverage the little work which is done.

Other advantages of the three technologies (more accurately, technology packages) not explicitly mentioned in this article include:

- o increased access to study sites which may be very difficult to approach;
- o complementarity of products useful in other related studies, i.e. of rangelands;
- o coordination with telemetry reception.

In fact, line of sight proximity is often necessary in heavy jungle and ultralights may afford the only feasible way of obtaining triangulated fixes of radio-tagged animals. A host of such ancillary but important contributions with such technology render these methods very cost effective to the project as a whole and provide additional reasons for their consideration.

ACKNOWLEDGMENTS

The authors wish to express their appreciation to the following people for important contributions to this paper. William H. Anderson graciously provided pre-publication information on the low-cost, shuttered, solid state video camera and Figure 1. Readers interested in further information are advised to contact him through the Applications Branch, USGS EROS Data Center, Sioux Falls, SD 57198. He also provided salient information on ultralights. Sanat K. Dhungel, Chief Warden, Royal Chitwan National Park, Narayani Anchal, Nepal, contributed an original analysis of the use of ultralights for radio-tracking and other wildlife related applications. Clark G. Lungren, Nazinga Project Director, Republique de Burkina Faso, West Africa, reported on the experience and utility of ultralights for a variety of wildlife husbandry applications. Paul Toomey of Sheffield, England confirmed some of the ultralight applications based on his experience.

Special thanks are due to Charles K. Paul, U.S. Agency for International Development, Washington, DC for reviewing this manuscript, and to the International Institute for Environment & Development (IIED), Washington, DC for support.

REFERENCES

Anderson, W.H., W.A. Wentz and B.D. Treadwell 1980, A Guide to Remote Sensing Information for Wildliife Biologists, In: The Wildlife Management Techniques Manual, The Wildlife Society, Washington, DC

Best, R.G. 1982, Handbook of Remote Sensing in Fish and Wildlife Management, South Dakota State University, Brookings

Carneggie, D.M., D.O. Ohlen and L.R. Pettinger 1980, A Selected Bibliography: Remote Sensing Applications in Wildlife Management, U.S. Geological Survey, Sioux Falls, SD

Clegg, R.H. and J.P. Scherz 1975, A Comparison of 9-inch, 70 mm, and 35 mm Cameras, Photogram. Eng. and Remote Sensing, 41:1487-1500

Conant, F., P. Rogers, M. Baumgardner, C. McKell, R. Dasmann and P. Reining, eds. 1983, Resource Inventory and Baseline Study Methods for Developing Countries, American Association for the Advancement of Science, Publication 83-3, Washington, DC

Dunford, C., D.A. Mouat, M. Norton-Griffiths an D.M. Slaymaker 1981, Remote Sensing for Village Planning in Africa, Science, Vol. 213, No. 4510

Everitt, J.H. and P.R. Nixon 1985, False Color Video Imagery: A Potential Remote Sensing Tool for Range Management, Photogram. Eng. and Remote Sensing, Vol. 51, No. 6, pp. 675-679

Fish, E.B. and E.L. Smith 1973, The Use of Remote Sensing for Vegetation Inventories in a Desert Shrub Community, Progressive Agriculture in Arizona, 25:3-5

Heintz, T.W., J.K. Lewis and S.S. Waller 1979, Low-level Aerial Photography as a Management and Research Tool for Range Inventory, Journal of Range Mangement, 32:247-249

Meisner, D.E. and O.M. Lindstrom 1985, Design and Operation of a Color Infrared Aerial Video System, Photogram. Eng. and Remote Sensing, Vol. 51, No. 5, pp. 555-560

Meyer, M.P. 1982, Place of Small-format Aerial Photography in Resource Surveys, Journal of Forestry, 80:15-17

Meyer, M.P., D.E. Meisner, N.L. Johnson and O.M. Linstrom 1983, Small Aircraft 35 mm Photography and Video Imagery Applications to Developing Country Resource Surveys, Proceed. National Conf. on Resource Mgmt. Applications: Energy and Environment, Vol. 1, San Francisco, CA.

Milligan, K. 1982, Assessing Human, Animal and Environmental Conditions by Low Level Aerial Survey: The ILCA Experience in the Sahel, Paper Prepared for the USAID Sahel Resource Inventory and Mapping Coordination Workshop, Bamako, Mali

Norton-Griffiths, M., T. Hart and M. Parton 1982, Sample Surveys from Light Aircraft Combining Visual Observation and Very Large Scale Photography, Remote Sensing Newsletter 82-2, University of Arizona, Tucson

Treadwell, B.D. and J. Buursink 1981, The Mali Land Use Project; A Multiple Resource Inventory in West Africa, In: Arid Land Resource Inventories: Developing Cost Efficient Methods, USDA Forest Service Gen. Tech. Rep. WO-28, pp. 159-164.

Vooren, A.P. and D.M.J. Offermans 1985, An Ultralight Aircraft for Low-Cost, Large-Scale Stereoscopic Aerial Photographs, Biotropica 17(1), pp. 84-88

WWF/IUCN (World Wildlife Fund/International Union for Conservation of Nature) 1984, List of Approved Projects, WWF/IUCN, 1196 Gland, Switzerland

MAPPING THE GREEN LEAF AREA INDEX OF RANGELAND WITH AIRBORNE MULTISPECTRAL SCANNER DATA

Paul J. Curran and H. Dawn Williamson
Department of Geography, University of Sheffield,
Western Bank, Sheffield S10 2TN, U.K.

ABSTRACT

Airborne multispectral scanner (MSS) data, collected in June 1984 were used to estimate green leaf area index (GLAI) for 60 km^2 of rangeland.

The methodology involved: (i) radiometric and atmospheric correction, (ii) production of a vegetation index image, (iii) calculation of a calibration relationship between a vegetation index and GLAI, (iv) production of an image of estimated GLAI by inversion of the calibration relationship in (iii) and (v) accuracy assessment. The initial accuracy of GLAI estimation was 17-40% at the 95% confidence level, for a six class classification. By refining the methodology the accuracy of GLAI estimation increased to 60-82% at the 95% confidence level, for a five class classification.

To test the methodology the experiment was repeated for the same study area using data collected in August 1984. The initial accuracy of GLAI estimation was 5-40% at the 95% confidence level for a six class classification. When the refined methodology was employed the accuracy of GLAI estimation increased to 58-93% at the 95% confidence level. These levels of accuracy were not significantly different, again at the 95% confidence level, to those obtained for the June 1984 survey.

INTRODUCTION

A suitable methodology for the estimation of vegetation characteristics from remotely sensed data would be first, to model the relationship between remotely sensed data and characteristics of a vegetation canopy; second, to invert this model to estimate characteristics of a vegetation canopy and third, to assess the accuracy of these estimates (Curran, 1983). Such a methodology, or a simplification of it, has been used for the remote sensing of vegetation amount in three similar types of environment. These three environments are cereal agriculture (eg. Hatfield *et al*., 1985), semi-natural vegetation (eg. Hardisky *et al*., 1984) and rangeland (eg. Wardley and Curran, 1984).

The aim of this paper is to refine this methodology and evaluate the effect that these refinements have upon the accuracy with which airborne multispectral scanner (MSS) data can be used to estimate the green leaf area index (GLAI is the one sided area of green leaves per unit area of ground) of rangeland. For further details refer to Curran and Williamson (1986a).

STUDY AREA

The study area is located in Derbyshire, U.K. (Figure 1). It covers 60 km^2 of limestone plateau at an altitude of around 300 metres. The rangeland which covers over 80% of the study area is ecologically very diverse comprising both unimproved and improved agricultural pasture (Anderson and Shimwell, 1981).

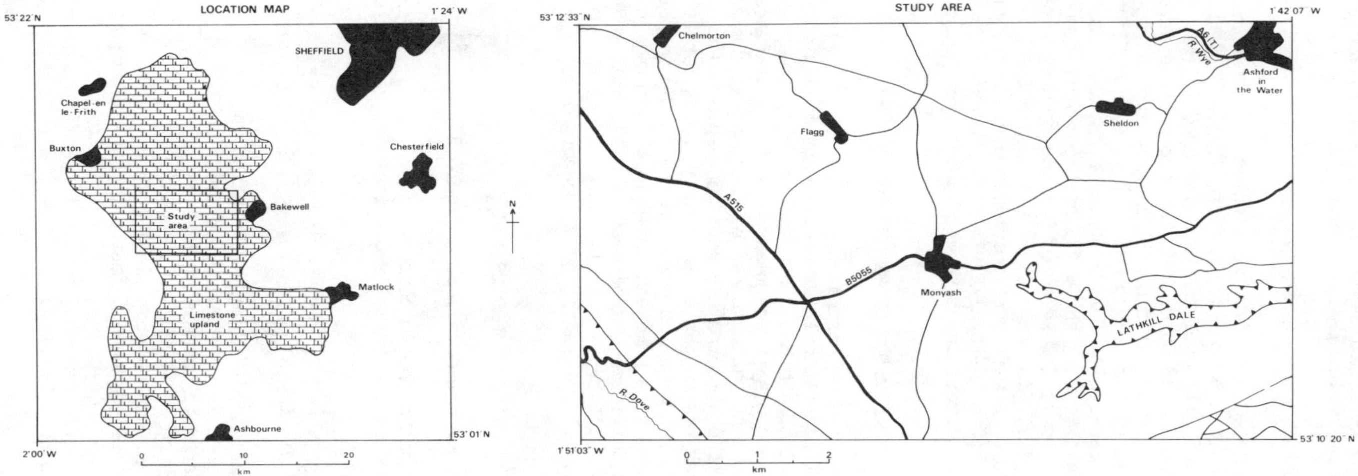

Figure 1 The study area.

very diverse comprising both unimproved and improved agricultural pasture (Anderson and Shimwell, 1981).

COLLECTION OF REMOTELY SENSED DATA

A twelve-band Daedalus AADS 1268 ATM multispectral scanner was employed for the survey, although only the red (0.605-0.69 µm) and near infrared (0.76-1.05 µm) wavebands are discussed in this paper. The overflights occurred on 18 June 1984 when the Sun was at a zenith angle of ≈ 44° and an azimuth angle of ≈ 111°, the sky had up to $^2/_8$ cloud cover and visibility was very good. Ten west/east flight-lines were recorded with a scan-angle of ± 38.5° and a spatial resolution of approximately 5 metres. Four of these flight-lines were found to have cloud and cloud shadow cover in excess of 50% and so they were not used for further analysis. The central ± 20° of data from the remaining flight-lines were extracted for radiometric correction using the formulae supplied by the Natural Environment Research Council (NERC) and atmospheric correction using the band-offset method (Billingsley, 1984).

COLLECTION OF GROUND DATA

Two vegetation samples, each covering an area of 400 cm² were harvested from each of 222 randomly sampled fields during the ten days prior to the overflight of the MSS. For each sample, green and brown leaf area index (GLAI and BLAI) were measured using a leaf area meter and canopy moisture content was measured by thermogravimetry. As a result of cloud cover the airborne MSS data corresponding to the majority of these sites was unusable, leaving a ground data set of only 74. A further 24 random sites were visited on the day of the MSS overflight. At each site a vegetation sample covering an area of 400 cm² was harvested for later measurement of GLAI, BLAI and canopy moisture content. In addition, the reflectance of red (0.6-0.7 µm) and near infrared (0.8-1.1 µm) radiation was recorded using a band-pass radiometer.

DATA PROCESSING

The methodological stages are detailed in Figure 2.

(i) Radiometric and atmospheric correction.
(ii) Produce an image of the normalised-difference vegetation index (Formulae).

$$ND = \frac{(L_{MSS7} + L_{MSS8})\ \rho TIR - (L_{MSS4} + L_{MSS5})\ \rho TR}{(L_{MSS7} + L_{MSS8})\ \rho TIR + (L_{MSS4} + L_{MSS5})\ \rho TR}$$

Where:
- ND = Normalised-difference vegetation index
- L_{MSS4} = Radiance in waveband 0.605-0.625 µm
- L_{MSS5} = Radiance in waveband 0.63-0.69 µm
- L_{MSS7} = Radiance in waveband 0.76-0.90 µm
- L_{MSS8} = Radiance in waveband 0.91-1.05 µm
- ρTIR = Mean reflectance of four calibration targets in waveband 0.6-0.7 µm
- ρTR = Mean reflectance of four calibration targets in waveband 0.8-1.1 µm

(iii) Derive an empirical model by calculating a calibration relationship between ND and GLAI on a training set of data.
(iv) Produce an image of estimated GLAI by inversion of the calibration relationship in (iii), (Figure 3).
(v) Calculate the accuracy of the GLAI image using the remaining ground data.

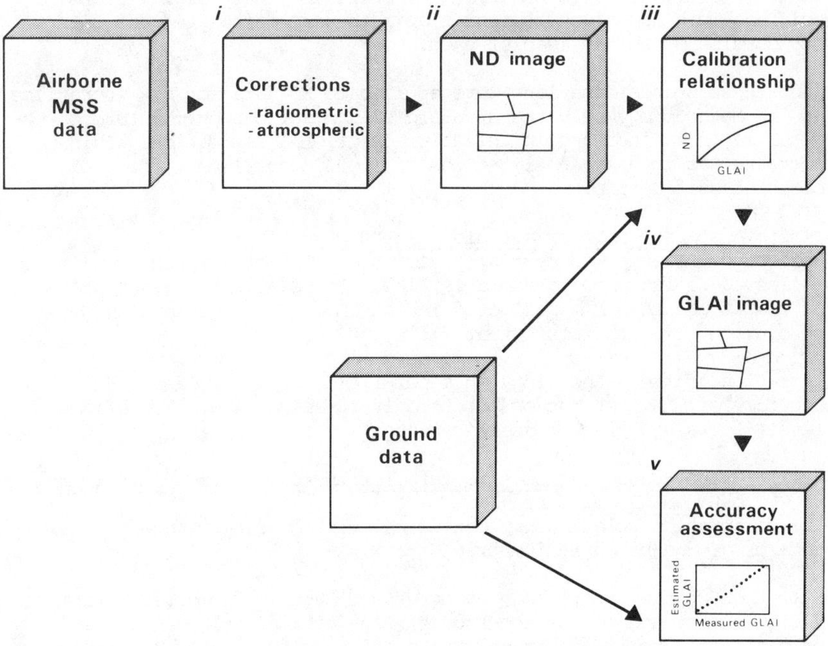

Figure 2 The methodology.

When this methodology was applied the accuracy of a GLAI classification of six classes, each 0.5 GLAI in size was 17-40% at the 95% confidence level, which is similar to that obtained for the same study area in the previous year (Curran and Williamson, 1985).

REFINING THE METHODOLOGY

Six possible refinements to the methodology were considered.

(i) Suppression of environmental influences on the MSS data.
(ii) A spatial unit of a field instead of a pixel.
(iii) Ground-based radiometric data for the calibration relationship between ND and GLAI.
(iv) Modified least-squares regression for the calibration relationship between ND and GLAI.

(v) Allowance for the asymptotic relationship between ND and GLAI.
(vi) Allowance for error in measured GLAI.

It would be impractical to evaluate the effect on accuracy of each refinement on every other refinement. Consequently the effect on accuracy of each refinement was assessed in relation to the original methodology.

(i) Suppression of environmental influences of the MSS data

The relationships between ND and the six variables of canopy moisture content, BLAI, slope to solar angle, slope to sensor angle, slope to solar azimuth and slope to sensor azimuth were calculated and used to standardise ND to the mean of each.

When these standardisations were applied to ND data for six iterations of the methodology only the standardisation for canopy moisture resulted in an increase in the accuracy of GLAI estimation. This increase was to only 18-41% at the 95% confidence level, for a six class classification.

(ii) A spatial unit of a field instead of a pixel

Reflected radiation is spatially variable. Therefore, the mean reflected radiation of a field is likely to be a better descriptor of that field than the reflected radiation of the individual ground resolution elements within that field.

Iteration of the methodology on a per-field basis increased the accuracy of GLAI estimation but to only 18-41% at the 95% confidence level, for a six class classification.

(iii) Ground-based radiometric data for the calibration relationship between ND and GLAI

The calibration relationship can be calculated using remotely sensed data or ground-based radiometric data.

Iteration of the methodology using ground-based radiometric data made little difference to the accuracy of GLAI estimation, changing it only slightly to 19-43% at the 95% confidence level, for a six class classification.

(iv) Modified least-squares regression for the calibration relationship between ND and GLAI

The assumptions that are inherent in a least-squares regression ie. no error in GLAI and X on Y being equivalent to Y on X, make it a statistically unsuitable technique for the derivation of a calibration relationship. The alternatives of reduced-major-axis, Wald-Bartlett and modified least-squares (Curran and Hay, 1985) all allow for error in ND and GLAI and define a suitable mid-line X on Y and Y on X.

When using these alternative techniques for three iterations of the methodology only modified least squares increased the accuracy of GLAI estimation, and then to only 20-44% at the 95% confidence level for a six class classification.

(v) Allowance for the asymptotic relationship between ND and GLAI

The relationship between ND and GLAI is known to be asymptotic (Curran, 1983).

Iteration of the methodology after grouping the GLAI values above the asymptote increases the accuracy of GLAI estimation to 23-45% at the 95% confidence level, for a five class classification.

(vi) Allowance for error in measured GLAI

The small errors in the methodology and remotely sensed data need to be separated from the large errors in the ground data (Curran and Hay, 1985). This can be achieved by adding the error in the measurement of GLAI to the class boundaries of the ground data used for the accuracy assessment of a GLAI classification. When the accuracy of the methodology was recalculated with this allowance for error in the ground data the accuracy of GLAI estimation increased to 57-81% at the 95% confidence level, for a six class classification.

THE METHODOLOGY REFINED

Error in the measurement of GLAI has been shown to have most effect on the accuracy with which MSS data can be used to estimate GLAI.

The procedures of environmental suppression and per-field processing were very time consuming and made little difference to the accuracy with which GLAI could be estimated. Changes in the way in which the calibration relationship could be calculated were less onerous and increased slightly the accuracy with which GLAI could be estimated. Together the effects of these changes were evaluated by a final iteration of the methodology in which the effects of canopy moisture variations were suppressed, the data were processed on a per-field basis, the calibration relationship was calculated using ground-based radiometric data and modified least-squares up to the asymptote of the vegetation index and error in the ground data was allowed for. The accuracy of GLAI estimation increased to 60-82% at the 95% confidence level, for a five class classification. A GLAI image produced using the refined methodology, albeit on a per-pixel basis, is shown for illustrative purposes in Figure 3.

THE METHODOLOGY TESTED

To determine if these levels of accuracy could be repeated the entire experiment was rerun on a second data set. This data set was collected for the same, but enlarged, study area reported previously, ground data were gathered on 22-25th August 1984 and MSS data were gathered on 25th August 1984 (Curran and Williamson, 1986b). These data were first processed using the methodology in figure 2. This was done without suppression of environmental influences, on a per-pixel basis with a calibration relationship derived from airborne MSS data and a least-squares regression and with no allowance for error in the measurement of GLAI. The accuracy with which GLAI could be estimated was 5-40% at the 95% confidence level, for a six class classification. This level of accuracy is not significantly different, again at the 95% confidence level, to that reported for the June experiment.

Iteration of the methodology after suppression of the effect of canopy moisture variations, processing on a per-field basis, calculating the calibration relationship using ground-based radiometric data and modified least-squares regression up to the asymptote of the vegetation index and allowing for the large error in the measurement of GLAI resulted in an increase in the accuracy of GLAI estimation to 58-93% at the 95% confidence level, for a five class classification. This level of accuracy is not significantly different, again at the 95%

confidence level, to that reported in the June experiment (Curran and Williamson, 1986b).

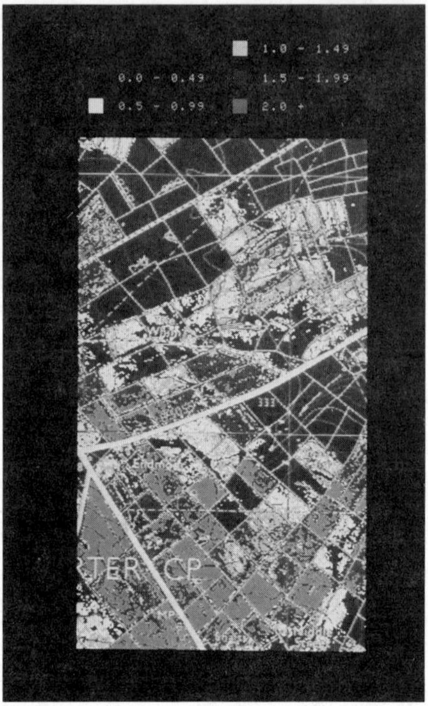

Figure 3 A five class GLAI classification of the area around Monyash (Figure 1) produced using the refined methodology on a per-pixel basis (original in colour).
The map overlay is used as an aid to location; it is reproduced from the 1979 Ordnance Survey, 1:25,000 scale topographic map with the permission of the Controller of Her Majesty's Stationary Office, Crown Copyright Reserved.

CONCLUSIONS

(i) Airborne MSS data and a methodology based upon model/inversion/accuracy assessment were used successfully to estimate the GLAI of rangeland.

(ii) Suppression of environmental influences on the remotely sensed data, processing on a per-field basis and the use of a calibration relationship derived from ground-based radiometric data made little difference to the accuracy with which MSS data could be used to estimate GLAI.

(iii) Calculation of a calibration relationship using modified least-squares regression up to the asymptote of the vegetation index resulted in a modest increase in the accuracy with which MSS data could be used to estimate GLAI.

(iv) Allowance for error in the measurement of GLAI resulted in a large increase in the accuracy with which MSS data could be used to estimate GLAI.

(v) By making all six refinements to the methodology the accuracy with which MSS data could be used to estimate GLAI increased. This increase was from 17-40% at the 95% confidence level, to 60-82% at the 95% confidence level, for a six and five class classification respectively.

(vi) To test this methodology the entire experiment was rerun using a second data set. By making all six refinements to the methodology the accuracy with which MSS data could be used to estimate GLAI increased. This increase was from 5-40% at the 95% confidence level, to 58-93% at the 95% confidence level for a six and five class classification respectively.

ACKNOWLEDGEMENTS

The authors wish to acknowledge the financial support of NERC under research grant GR3/5096 to PJC and their sponsorship of airborne multispectral scanner flights as part of their MSS-84a and MSS-84b campaigns. Thanks are also due to Mike Pedley for help with image processing, Dr. Alan Hay for statistical advice, Mrs. Angela Morrison (NERC) for writing the images to film and 'the team', Giles Foody, Charles Pattie, Tim Hobson, Susan Mason, Mike Pedley, David Talmage and Neil Wardley for assistance with the collection of ground data.

REFERENCES

Anderson, P. and D. Shimwell 1981, Wild Flowers and Other Plants of the Peak District. An Ecological Study, Moorland, Ashbourne.

Billingsley, F.C. 1984, Remote sensing for monitoring vegetation: an emphasis on satellites, In Woodwell, G.H. (ed.) The Role of Terrestrial Vegetation in the Global Carbon Cycle: Measurement by Remote Sensing, John Wiley and Sons: 161-80.

Curran, P.J. 1983, Multispectral remote sensing for the estimation of green leaf area index, Phil. Trans. Royal Soc. Lond., A, 309: 257-70.

Curran, P.J. and A.M. Hay 1985, The importance of measurement error for certain procedures in remote sensing at optical wavelengths, Photogramm. Engng. Remote Sensing, 51, (in press).

Curran, P.J. and H.D. Williamson 1985, The accuracy of ground data used in remote sensing investigations, Int. J. Remote Sensing, 6, (in press).

Curran, P.J. and H.D. Williamson 1986a, MSS data to estimate GLAI, I. Refining the methodology, Int. J. Remote Sensing, (submitted).

Curran, P.J. and H.D. Williamson 1986b, MSS data to estimate GLAI, II. Testing the methodology, Int. J. Remote Sensing, (submitted).

Hardisky, M.A., F.C. Daiber, C.T. Roman and V. Klemas 1984, Remote sensing of biomass and annual net primary productivity of a salt marsh. Remote Sensing Environ., 16: 91-106.

Hatfield, J.L., E.T. Kanemasu, G. Asrar, R.D. Jackson, P.J. Pinter, R.J. Reginato and S.B. Idso 1985, Leaf area estimates from spectral measurements over various planting dates of wheat. Int. J. Remote Sensing, 6: 167-75.

Wardley, N.W. and P.J. Curran 1984, The estimation of green leaf area index from remotely sensed airborne multispectral scanner data. Int. J. Remote Sensing, 5: 671-80.

SPECIAL POSTER SESSION

Chairperson: Richard Myhre

THE FOREST FIRE ADVANCED SYSTEM TECHNOLOGY SYSTEM CONFIGURATION

J. David Nichols
James R. Huning
Jet Propulsion Laboratory
California Institute of Technology
4800 Oak Grove Drive
Pasadena, CA 91109

John R. Warren
United States Department of Agriculture
Forest Service
Boise Interagency Fire Center
3905 Vista Avenue
Boise, ID 83705

Forest fire suppression and management is a complex operation involving trained professionals in many disciplines, a variety of equipment, extensive communications, and assorted means of acquiring fire information. The tactical nature of forest fire suppression and management requires timely, accurate, and reliable information on fire condition, behavior, and location. Vast amounts of timber, watershed resources, wildlife habitat, and human property are put in danger if fire information is not delivered to the fire suppression management in a timely manner. A joint project between the United States Department of Agriculture-Forest Service and NASA's Jet Propulsion Laboratory was initiated in 1984 to design a system to provide this information in a timely manner to the fire Incident Command Post (ICP). The primary objective of the Forest Fire Advanced System Technology (FFAST) task is to prepare a conceptual design for a forest fire detection and mapping system, based upon technology that will be available in the early 1990s.

Alternative FFAST System Configurations

After consultation with Forest Service regional personnel, several alternative FFAST system configurations have been developed in light of future Forest Service information needs, and what current and emerging technologies might be available in the 1990s. The components of the system configurations were derived from identified operations that will meet the functional requirements of an advanced forest fire detection and mapping system. The alternatives are intended as baseline configurations (constructed in a modular format) from which a detailed design will be made. The alternative system configurations are:

Option A: Thermal infrared line scanner, geo-referencing utilizing a Global Positioning Satellite (GPS) system, data processing on-board the aircraft, mobile satellite communications, data display at the ICP,

Option B: Thermal infrared linear array detector with dual bands (3-5 and 8-12 Microns), geo-referencing utilizing a GPS system, data processing on-board the aircraft, mobile satellite communications, and advanced data display at the ICP,

Option C: Single band (10-12 Microns) thermal infrared linear array, geo-referencing with a GPS system, automated data processing on-board the aircraft, mobile satellite communications, and advanced display at the ICP,

Option D: Dual band linear array, audio transmission of GPS and uncorrected infrared digitized data to a central ground facility for automated processing, re-transmission of processed data to the ICP for advanced data display.

Candidate FFAST system configuration alternatives were evaluated on factors considered to be of major importance in future systems. The present Forest Service infrared system, the Fire Logistics Airborne Mapping Equipment (FLAME), was used as a basis of comparison for the alternative system configurations.

The amount of time required to deliver fire information to fire management personnel is of prime importance if a new system is to be cost-efficient. The FFAST alternative configurations, all of which feature the delivery of fire data to the ICP in less than thirty minutes, have at least a three hour advantage over the present FLAME system, but have other attributes which make them either more or less attractive to the Forest Service.

The Preferred System Configuration

The preferred system alternative configuration (Option B) will meet or exceed the users' functional requirements for an advanced forest fire detection and mapping system. The modular design of the preferred system provides flexibility by allowing the incorporation of both present and developing technologies at a later date. The system will rely on thermal infrared linear array detector technology, coupled with a Global Positioning Satellite system and on-board signal processing capabilities, to produce fire data that can be geo-referenced to a common map base. The reduced size of the on-board system will allow for a smaller aircraft platform to be used. The processed data will contain fire perimeter location, hot spot detection, and fire intensity information. The processed fire data will be telemetered via mobile satellite communications to the ICP for display in a final map format for use by fire suppression personnel.

The preferred FFAST system configuration addresses the dual goals of meeting the stated needs for fire management while being realistic in terms of the technology that is likely to be available by the 1990s. The technology will make it possible to acquire the requisite information and to integrate the various mapping and detection technologies into a high resolution, user-friendly system which would perform fire detection and mapping on a real-time basis.

LANDSAT-A TOOL FOR MAPPING FUEL TYPES IN THE BOREAL FOREST OF MANITOBA: A PILOT STUDY

R. Dixon
W. Shipley
A. Briggs

Department of Natural Resources
Manitoba Remote Sensing Centre
Winnipeg, Manitoba R3H 0W4

INTRODUCTION

Forest lands are a major contributor to the economic and social development of Manitoba. Demand for forest resources are expanding, and therefore the need for intensive forest management is increasing.

Fire management is an essential component in the administration of forest land. Wildfire managers require specific information to accurately predict fire behavior. One important factor affecting fire spread and intensity is the type of vegetation or fuel type. Consequently, maps which display the location, areal extent and type of fuel complexes are helpful in assessing probable wildfire behavior.

The objective of this study was to demonstrate the feasibility of Landsat MSS digital data for fuel type mapping in the Boreal Forest of Manitoba. The study area was located in northwestern Manitoba between latitude 54 and 55 degrees north, and between longitude 100 and 102 degrees west. The Boreal Forest region of the area includes three forest sections as described by Rowe (1972).

METHODS

Landsat Digital Image Correction System (DICS) scenes were chosen for the study. This geometrically corrected format was easily registered to National Topographic Series (NTS) map sheets and other reference data.

Landsat digital data can be beneficial to the fire manager in predicting fire behavior when used in conjunction with the Canadian Forest Fire Behavior Prediction (FBP) System (Alexander, Lawson, Stocks and Van Wagner, 1984).

Fire Management and Communications Section required eight fuel types to be mapped. Landsat digital data were used to identify land cover/land use types which were then matched, to corresponding FBP System fuel type identifiers. In addition to fuel types, rock outcrops and water bodies which represent natural barriers to fire spread, were also identified and classified for mapping.

Digital image analysis was conducted at the Manitoba Remote Sensing Centre with the Dipix Systems Ltd. ARIES II

(Applied Resources Image Exploitation System). A supervised classification method was employed and training areas were selected with the aid of 1:15,840 panchromatic aerial photography, forest inventory and topographic maps.

RESULTS

Confusion matrices were generated to determine classification results. The Lowland Forest section had an overall classification accuracy of 70 percent. The majority of the misclassification was attributed to difficulties in separating jack pine (*Pinus banksiana*) from other softwoods. On the Northern Forest and Nelson River Forest sections, the softwoods were combined resulting in an overall accuracy of 80 percent for the area. An Applicon inkjet plotter was used to produce colour thematic maps of the themes generated by the classification. The maps display the location and distribution of the land cover/land use types and their corresponding (FBP) fuel type identifiers.

CONCLUSION

The project demonstrated the usefulness of Landsat MSS digital data for fuel type mapping in Manitoba. The Manitoba Department of Natural Resources, Fire Management and Communications Section integrated the study information into their fire management program for the 1984 fire season and have expanded forest fuel mapping to other areas of the province.

ACKNOWLEDGEMENTS

The authors wish to thank Bryan S. Lee and Martin E. Alexander, Northern Forest Research Centre, Canadian Forestry Service, Edmonton for their direction on the fuel type classification system used.

REFERENCES

Alexander, M.E., B. O. Lawson, B. J. Stocks and C. E. Van Wagner, (listed in alphabetical order) 1984. *User Guide to the Canadian Forest Fire Behavior Prediction System: Rate of Spread Relationships - Interim Ed.*, Environ. Can. For. Serv. Fire Danger Group, Unpubl. Rep., 73 p.

Rowe, J. S. 1972. *Forest Regions of Canada*, Dept. of Fisheries and Environment, Canadian Forestry Serivce Publ. 1300. pp. 172.

APPLICATIONS OF LANDSAT IMAGERY FOR FIRE FUELS MAPPING PROJECTS COVERING LARGE GEOGRAPHIC AREAS

Julie Michaelson Willis
Bureau of Land Management
Anchorage, Alaska 99513

INTRODUCTION

The Alaska Fire Service (AFS), Bureau of Land Management (BLM), is building a data base and management system for the Alaska Initial Attack Management System (AIAMS). Fire fuels is one data layer for this geographical information system. BLM, Branch of Photogrammetry is responsible for mapping fire fuels for 250 million acres in Alaska. The project area shows extensive climatic, geomorphic, and vegetative diversity including coastal and taiga forestland, arctic and alpine tundra. In most of Alaska, vegetative variation is linked closely with fire history. Little or no pre-existing vegetative reference data are available.

MAPPING DATA AND METHODOLOGY

Landsat imagery is a useful in mapping generalized fuel classes over large, diverse areas of Alaska. Fuels interpretations are made using three types of Landsat imagery: 1) black and white winter data, 2) standard false color composite summer data and 3) custom computer enhancements generated to emphasize fire fuels types (Miller, 1984). Manual interpretation techniques are used to extract preliminary fire fuel types on each image. Images are then compared to provide additional fuel type detail. This comparative analysis expands the interpretative capability of the mapper (Table 1).

Field reconnaissance is conducted using fixed wing over flights. Data in ecologically representative areas are collected on color infrared photographs. Topographic data are overlayed on each image as it is examined to incorporate topography into the ecological interpretation. Aerial photographs are used as ancillary data with other resource data and to spot check finished fuels maps.

CONCLUSION

Manual interpretations using Landsat data can be enhanced and map category accuracy increased using a variety of imagery and comparative analysis. This approach is particularly useful in conducting mapping projects over large, diverse geographic areas.

ACKNOWLEDGMENTS

The author wishes to thank Dr. Page Spencer for her valuable suggestions and support in the design of this poster session.

REFERENCES

Mallot, N. 1984, Alaska Fuels Fact Sheet Summary, 5390 Immediate Fire Behavior State of Alaska, Dept. of Natural Resources, Div. of Forestry, 3 pages.

Miller, M. 1984, Landsat Enhancement Procedures and Key Central Yukon Planning Area, 10 pages, BLM, Fairbanks, Alaska.

Table 1. Use of Landsat Imagery in Fire Fuel Interpretation. Comparison of black and white winter (B/W winter), false color composite (FCC summer), and custom computer enhancements (CCE) are used in fuel type line placement and refinement of the interpretation. Black and white winter imagery is used to make initial distinction between forested and non-forested lands. Fuel model codes (Mallot, 1984) are those associated with map categories/symbols.

MAPPING SYMBOL	LANDCOVER CLASS	IMAGERY USED	ASSOCIATED FUEL MODEL
2	Black spruce forest	FCC Summer	9/5
	Mixed forest coniferous dominate	CCE	
2	Graminoid tussocks	CCE	1/3
3	Riparian white spruce	B/W winter	10
4	Deciduous forest	FCC summer	9
	Mixed forest deciduous dominate	CCE	
	Shrubland tall and low	FCC summer	
	Recent burns	FCC summer	
5	Alpine tundra dwarf shrubs	CCE	None
	Barren	FCC summer	
	Ice	FCC summer	
6	Water-Lakes	FCC summer	None

THEMATIC MAPPER ANALYSIS OF CONIFEROUS FOREST STRUCTURE AND COMPOSITION

Janet Franklin
Department of Geography
University of California
Santa Barbara, CA 93106

INTRODUCTION

The spectral characteristics of aircraft-acquired Thematic Mapper Simulator (TMS) data were examined for their relationship to species composition of coniferous forest stands, and forest structure, as measured by basal area and foliage biomass, in the Mendocino National Forest (MNF), California.

In montane regions of California the dominant high elevation forest type (above 1800 m) is Red Fir, while lower elevations (750-1800 m) are characterized by a Mixed Conifer type. In previous studies using Landsat Multispectral Scanner (MSS) data, Red Fir and Mixed Conifer types could not be discriminated using spectral data alone (Strahler et al. 1983). In this study, TMS data, with its increased spectral and spatial resolution, was examined to see if these types could be discriminated. The characteristics of the TMS spectral data in the forested area were also examined using Principal Components Analysis, and the relationship of canopy reflectance to stand structure was established by regression analyses.

METHODS

The TMS data was acquire with a Daedelus Airborne Thematic Mapper flown on a U-2 aircraft. The reflectance vector for each sample plot was composed of the average reflectance numbers in each band for a five by five window around the plot location for each spectral band. Diameter at breast height (DBH) and species of every tree greater than 2.5 cm DBH were recorded in a circular plot of area 100 $\underline{m}^2$. The height of the largest tree was measured, as an index of maximum canopy height for the plot. Basal area was calculated for each species and all species in the plot, and ranged from 1-195 $\underline{m}^2/\underline{ha}$ (mean, 57 $\underline{m}^2/\underline{ha}$, standard deviation 34 $\underline{m}^2/\underline{ha}$). Conifer foliage biomass was calculated for each plot from basal area.

The reflectance vector for each stand was treated as a set of training statistics, and spectral separability was measured using a divergence index. Basal area and foliage biomass were treated as the dependent variable in simple regressions with plot reflectance in each spectral band, and with the Band 3/Band 4 greenness index. Principal components of all the bands were used in a stepwise multiple regression, with basal area as the dependent variable. Finally, several of the regression analyses were repeated for pooled plot data.

RESULTS

Measurement of spectral divergence showed no consistent separability between Red Fir stands and white fir-dominated Mixed Conifer stands, however, within the high density, large timber category red fir stands have a much lower near-infrared reflectance (.76-.90 microns) than the other types. The regression analyses of single band reflectances showed the red channel (Band 3, .63-.69 microns) to be the strongest predictor of basal area (R^2=.29; all R^2 values significant at least at the .01 level) or foliage biomass (R^2=.30), almost as good as the first five principal components (R^2=.30), and much better than the 4/3 ratio (R^2=.15). The grouped plot data showed a much better relationship between reflectance and stand characteristics (R^2=.64 for basal area versus Band 3 reflectance). Foliage biomass shows a slightly but consistently stronger relationship than basal area to red reflectance. This can be seen particularly well in the grouped data (R^2=.67 versus R^2=.64).

Principal components analysis revealed that the first principal component (63% of the total variance) had high positive weightings in all bands, and can be interpreted as a scene brightness measure. The results of the stepwise regression indicate that scene brightness is the composite spectral characteristic most strongly related to forest structure.

CONCLUSIONS

In this study, any potential for species discrimination can only be seen in dense forest stands, where crown closure approaches 100%. In this case, red fir stands appear to be spectrally distinct in TM Band 4. In regression analyses, conifer basal area and foliage biomass was inversely related to visible reflectance (particularly in the red band) and scene brightness, curvilinearly related to the near-infrared/red ratio, non-linearly related to near-infrared reflectance, and showed no no clear relationship to mid-infrared and thermal response. Greenness, as measured in the third principle component, or by the 4/3 ratio, is not very strongly related to vegetation amount in conifer forest. The greatly improved regression fit for pooled plot data indicates that the TM spectral data bears a relationship to forest amount when the forest is stratified by type, but due to the inherent spatial heterogeneity ofthe forest vegetation, TM reflectance measures are not good site-specific predictors of forest vegetation amount.

ACKNOWLEDGEMENTS

This research was sponsored in part by a grant from the California Space Institute, Donald Disraeli And Michael Barbour, co-investigators.

REFERENCES

Strahler, A. H., C. E. Woodcock, T. L. Logan, J. Franklin, H. Bowlin and J. Levitan, 1983, Automated forest classification and inventory of the Eldorado National Forest, USDA Forest Service, San Francisco, CA 38 p.

THE USE OF LOW-RESOLUTION SATELLITE IMAGE DATA FOR RANGELAND MONITORING

by Thomas M. Holm
U.S. Geological Survey
EROS Data Center
Sioux Falls, South Dakota 57198
(605) 594-6114

William C. Draeger
U.S. Geological Survey
EROS Data Center
Sioux Falls, South Dakota 57198

Patrick T. Williams
Papago Agency, Land Operations Branch
Bureau of Indian Affairs

Robert F. Buzzard
Papago Agency, Land Operations Branch
Bureau of Indian Affairs

ABSTRACT

The U.S. Geological Survey's Earth Resources Observation Systems (EROS) Data Center and the Bureau of Indian Affairs (BIA) jointly evaluated the use of Advanced Very High Resolution Radiometer (AVHRR) images for the calculation of normalized difference data as a green biomass indicator in an arid rangeland environment. The normalized difference data were compared to corresponding field-collected herbage data, and were used in a geographic information system to provide range management information.

The AVHRR images, with a nominal ground resolution of 1.1 kilometer, were acquired daily by the NOAA-6, -8 and -9 satellites. Data for the Papago Indian Reservation in south-central Arizona, acquired on representative dates during the spring and summer growing season of 1984 and the spring growing season of 1985, were selected for analysis and registered to a geo-referenced data base. AVHRR channels 1 (visible, 0.58-0.68 um) and 2 (reflected infrared, 0.725-1.1 um) were used to calculate vegetation greenness using a normalized difference algorithm: (channel 2 - channel 1)/(channel 1 + channel 2). Regression analysis was used to evaluate the relationship of normalized difference and herbage data for 25 ground plots.

The regression equation was significant at the 95-percent confidence level for all growing seasons. The regression model provided the basis for grouping normalized difference values into seven greenness categories, from high to low. These categories illustrated seasonal spatial green-up patterns for the entire reservation. These spatial greenness data, when analyzed in a geographic information system containing data on grazing districts, road networks, and water sources, provided the BIA Papago Agency Range conservationists with rangeland planning and management information.

SENSITIVITY OF VEGETATION INDICES TO VARIATIONS IN SOIL COLOR

Anthony Morse

Idaho Department of Water Resources
Boise, Idaho

Eugene L. Maxwell

Solar Energy Research Institute
Golden, Colorado

SUMMARY

Researchers have used vegetation indices (VI's) computed from Landsat MSS bands for many different applications. Among the simplest of the VI's are the IR/RED ratio (R), the normalized difference, IR-RED/IR+RED (ND), and the square root of the normalized difference (RND).

Problems related to variations in soil color are common to computer-assisted classification of Landsat data, especially where vegetation is sparse. These problems are potentially confronted also by the user of VI's. However, theoretical reasons and empirical work indicate that variations in soil color should be reduced or eliminated by some of the transformations used to calculate VI's.

An experiment was designed to test the hypothesis that VI's are relatively less sensitive to variations in soil color than are raw MSS data. The above-named VI's were computed with both MSS7 and MSS 6, for a total of six VI's. These were tested on twelve non-vegetated areas, which included light and dark sand, agricultural fields, salt flats, and rock waste in an open-pit copper mine. Although sand, salt, and rock are not soils, their inclusion provided a large range of "soil" colors, similar to what could be found under the sparse vegetation of western rangeland.

The experiment compared the F-values from a one-way analysis of variance. F-values were calculated individually for each MSS band and each VI. The results show that all the VI's tested were much less sensitive than the MSS bands to variations in soil color, and those VI's computed with MSS6 were much less sensitive than those VI's computed with MSS7.

The VI's and corresponding F-values are as follows. The lower the F-value, the lower the sensitivity to variations in soil color.

VI	F-VALUE
R6	64
RND6	70
ND6	74
R7	174
ND7	199
RND7	201
MSS7	1263
MSS6	1384
MSS5	1542
MSS4	1691

DETERMINATION OF THE APPROPRIATE REGRESSION MODELS FOR THE PREDICTION OF RANGELAND BIOMASS FROM LANDSAT VEGETATION INDICES

Anthony Morse

Idaho Department of Water Resources
Boise, Idaho

Eugene L. Maxwell

Solar Energy Research Institute
Golden, Colorado

SUMMARY

Reports of regression analyses between vegetation indices and various canopy variables such as biomass have appeared frequently in remote sensing literature. These reports have generally described the fitting of data with a linear regression model. However, tests for the appropriateness of the linear model for the data being analyzed have not been included. This research tests the appropriateness of the linear regression model for rangeland biomass and six common vegetation indices.

The linear model was tested on Landsat MSS data of 23 rangeland plots having standing, green biomass weights between 0 and approximately 4000 kg./ha. The Landsat data span two years and four separate scenes. The vegetation indices (VI's) chosen for evaluation were the ratio (IR/RED), normalized difference (IR-RED/IR+RED), and the square root of the normalized difference. Using both MSS 7 and MSS 6 in these transformations resulted in a total of six VI's.

Both qualitative and quantitative tests of the data show that a linear regression model is appropriate for only one VI, and that a quadratic model is needed for the other VI's. Visual inspection of the regression plots and of the residual plots shows a definite, albeit undramatic, curvilinear trend to all the regression plots except that of MSS7/MSS5. This qualitative analysis is confirmed by a chi-square test for the significance of additional polynomial terms in the regression equations.

PREDICTING IRON AND ORGANIC CARBON CONTENT IN SOILS, SEDIMENTS, AND MINE SAMPLES USING SPECTRAL REFLECTION MEASUREMENTS.

H. Schreier
Department of Soil Science
University of British Columbia
Vancouver, B.C. V6T 2A2 Canada

SUMMARY

With the expanded spectral range of new satellites it is imperative that more precise laboratory assessments of spectral reflection of soil and surficial materials be made. This is of particular interest if we want to predict chemical properties in a quantitative manner from reflection measurements. Three components which have a profound influence on surface reflection of soils were examined in the present study: organic matter content, total iron concentration, and sample provenance or overall genetic composition. Previous work by Stoner *et al.* (1981) and Schreier (1977) has shown that organic carbon is often correlated with reflection. In the present study we also found a significant relationship between reflection and total iron content. However, these relationships are complex and do not apply to all soil, rock, and sediment samples and simple linear and curvilinear regressions do not suffice in making routine quantitative predictions.

The difficulties of using regressions to predict organic carbon and iron from reflection in a routine manner are best exemplified in Figures 1 and 2. In the case of organic carbon a good exponential relationship is possible between % reflection and % organic carbon if the samples contain organic carbon values which are greater than 2.0%. In the present example the regression accounted for 85% of the total variance. Once we include soil and mine tailing samples which had undergone little soil development and which contained less than 2% organic carbon the reflection/carbon relationship was no longer significant. Similarly an excellent relationship between reflection and total iron content was obtained in Figure 2 if we use mine tailing samples which originated from oxidized environments, and which contained virtually no organic carbon. 31 samples from 14 different mine sites were used and the curvilinear regression accounted for 87% of the total variance. Again the relationship no longer held once soil samples were included in the regression. In the above equations Landsat band 4 reflections, measured with an Exotech-100 instrument in the laboratory, provided the best results.

Based on these results it is clearly evident that routine quantitative predictions of organic carbon and total iron cannot readily be made from reflection measurements alone. The reflection at a specified wavelength will only provide a partial picture of the chemical content and the overall provenance or genetic composition of the sample should also be examined. This can be done by analysing the spectral reflection curve over the entire wavelength range. Multivariate statistical techniques such as average distance cluster analysis can be used to classify the different reflection curves of the samples, and the methods proved to be very useful in separating samples according to origin and selective chemical content.

A cluster analysis using Landsat bands 4, 5, 6, and 7 and band ratios 4/5, 5/6, and 6/7 of 31 mine tailing samples produced five unique types of spectral reflection curves and the mine samples were separated according to provenance and total iron content. A similar analysis was carried out on 44 soil samples with less than 2% organic carbon and between 3 and 8% total iron. Four unique types of spectral curves were obtained and the grouping provided an excellent separation according to sample provenance. Finally a cluster analysis of the spectral characteristics of a mixed set of 162 soil, sediment and mine samples was carried out and in this case 8 types of characteristic reflection curves were identified and the samples were separated according to organic carbon content, total iron content, and sample provenance. It is anticipated that with the speciation of total Fe into Fe^{2+} and Fe^{3+} and by using more narrow band reflection values this analysis can be further refined.

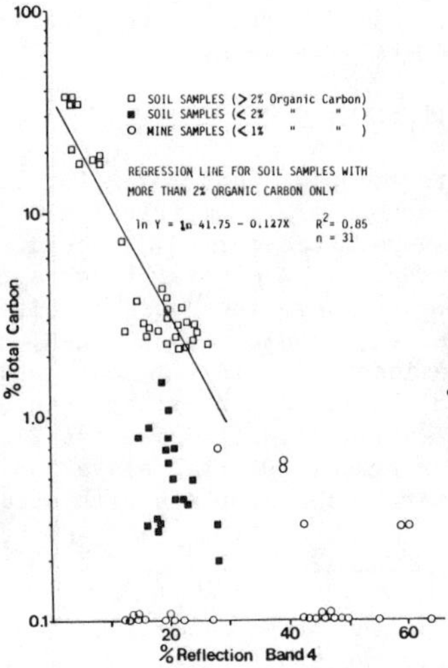

Figure 1: Relationship between organic carbon content and spectral reflection.

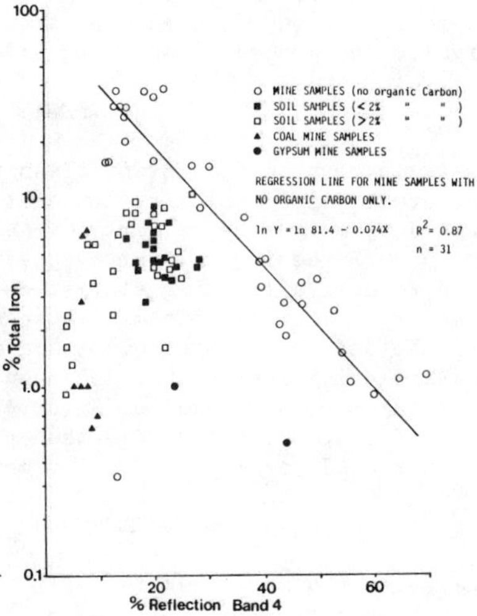

Figure 2: Relationship between total iron content and spectral reflection.

REFERENCES

Schreier, H. 1977. Quantitative predictions of chemical soil conditions from multispectral airborne, ground, and laboratory measurements. Proc. 4th Canadian Symposium on Remote Sensing. pp. 106-112.

Stoner, E.R. and M.F. Baumgardner. 1981. Characteristic variations in reflectance of surface soils. Soil Sci. Soc. Am. Proc. Vol. 45: 1161-1165.

MULTIPHASE SAMPLE OF VEGETATION AND RELATED RESOURCES USING LANDSAT, AERIAL PHOTO, AND GROUND SURVEY DATA

Kenneth C. Winterberger and Frederic R. Larson
Pacific Northwest Research Station
Forestry Sciences Lab
201 E. 9th. Avenue, Suite 303
Anchorage, Alaska 99501

ABSTRACT

The Alaska Integrated Resource Inventory System was developed as a cost effective method of inventorying renewable resources over vast remote areas. The sample population consists of 8.1-ha plots on colocated grids that employ the Universal Transverse Mercator grid system. The phases include Landsat data analysis at 5-km intervals, small-scale (1:60,000), color infrared photo interpretation at 10-km intervals, large-scale (1:3,000), color infrared photo interpertation at 20-km intervals, and ground subsampling at 40-km intervals.

LANDSAT PHASE

Landsat can provide consistent classification of land cover for large remote areas fairly quickly. Landsat scenes are geometrically corrected to match the Universal Transverse Mercator (UTM) coordinate system on U.S. Geological Survey quadrangles. A cluster-block approach is used to derive training statistics, which are then used to classify the imagery. Training blocks are selected, clustered, and then, using aerial photos, the cluster classes are identified in terms of land cover--including coniferous, deciduous, or mixed forest; herbaceous vegetation; barren ground; or water. Maximum likelihood classification of the entire Landsat image is then performed. Digital elevation models are used to help sort out vegetation and problems with shadows.

EXTRACTING LANDSAT PLOT DATA

A study area boundary mask is applied to the final classification to eliminate that part of the image outside the study unit. The resulting image is then subsampled to extract the 8.1-ha sample plots from the land cover classification data with sample centers on the 5-km UTM grid interval. To sample a circular 8.1-ha area, an image mask approximating this area is developed and applied to the output image. Proportions of areas in each cover class are then determined. This constitutes the first phase of the sample design.

SMALL-SCALE PHOTOGRAPHY

A mylar overlay depicting an 8.1-ha circular plot at an average photo scale of 1:60,000 is used as a photo aid to delineate the areas centered on a 10-km UTM interval. This overlay also includes 19 equidistant points, each representing 5.26 percent of the plot, to aid in estimating the percentage of cover by type over the plot area. Additional measurement of distance to water, cultural influence, mature trees, hiding cover, and edge are made for wildlife interpretations. Estimates of land use are noted. This constitutes the second phase of the sample design.

LARGE-SCALE PHOTOGRAPHY

A scale of 1:3,000 or 1:4,500 is used for photos in the third phase of the sample. This photography is limited in coverage to a stereo triplet of 25 percent of the small-scale photo plots, or those on a 20-km UTM grid interval. The plot area is delineated by land cover types and each cover type polygon is digitized to determine areas. Each polygon in the 8.1-ha plot is classified by land cover or vegetation found; in addition, broadleaf and needleleaf cover, crown diameters, and tree heights are measured, and estimates of gross timber volume made. Height and cover of the understory are measured, the number of nest or den trees noted, and land class and tree stand size class determined.

GROUND SAMPLE PHASE

Ground plots are established in the fourth phase on a 40-km UTM grid interval. Cover classifications are made by areas in different vegetation classes, and detailed measurements are made of trees, shrubs, herbaceous vegetation, downed wood, soils, and wildlife habitat.

RESULTS

Results are fairly consistent regarding cover type throughout all phases. The percentage of coniferous vegetation increases as resolution or ability to "see" small black spruce crowns in herbaceous and mixed coniferous/deciduous areas increases. Missing data in the two photo phases would probably be classified as barren ground as missing data occurred in high-elevation cloud-covered plots where this cover type is found (table 1). Preliminary correlation analysis indicates that sampling efficiency can be improved by using four phases; however, cost analyses are yet to be made.

Table 1--Percent of land cover by classes as determined in each of the 4 phases in the 34.5-million-acre Tanana River Basin

	Landsat Data	Small-scale Photos	Large-scale Photos	Ground Sample
	- - - - - - -	-Percent of area-	- - - - - - -	
Coniferous	23	26	37	38
Deciduous	33	39	34	28
Mixed forest	16	14	7	11
Herbaceous	10	4	5	6
Barren ground	16	11	13	15
Water	2	2	3	2
Missing Data	0	4	1	0
Total	100	100	100	100
Sample size	5,535	1,343	331	88

SMALL FORMAT AERIAL PHOTOGRAPHY FOR MONITORING SOUTHERN PINE PLANTATIONS

Harlan R. Stoin
University of Arkansas
Fayetteville, AR 72701

Small format camera systems are useful for detailed vegetation assessment of small areas. With the use of rental aircraft they offer a means of obtaining economic, repetitive coverage of an area at varying scales including very large scales, which are especially useful for sampling, analyzing, and measuring small trees and ground vegetation in greater detail than is possible with conventional medium to small scale aerial photos.

A small format aerial camera system using motorized 35 mm cameras mounted on a small helicopter has been designed, built, and tested for its utility in monitoring intensively managed southern pine plantations. Two cameras at the ends of a boom provide stereo coverage at very large scales from a fixed air base of 4.5 meters. If desired a third camera with a wider angle lens can be added to provide for flight tracking and locating the small plots photographed with the twin cameras. Provision for precise leveling of the cameras has been made.

The fixed air base of the cameras allows for direct scaling of the stereo pairs of photographs. Regressions of photographic scales determined from the fixed air base with those determined from ground have been made to calibrate the system.

A modified Meyer mount flown on a Cessna aircraft is also used where direct scaling of the photographs is not needed or for wider area coverage at smaller scales.

Large scale photos taken with the small format camera provide the capability of detecting many vegetation and site features of southern pine plantations even at very young ages. Accurate measurements of vegetational features can be made.

AERIAL PHOTO STAND VOLUME TABLES FOR SOUTHEAST ALASKA

Theodore S. Setzer and Bert R. Mead
Forestry Sciences Lab
201 E. 9th. Ave., Suite 303
Anchorage, Alaska 99501

Good timber volume estimates are a historically and continuing important objective of many forest inventories. Reliable aerial photo volume estimates can be used to augment ground measurements, thus improving volume sampling error. Localized aerial photo volume tables can be expected to provide stand volume estimates similar to ground estimates. For this reason, the Alaska Integrated Resource Inventory System (AIRIS) needs localized aerial tables. Improved volume figures would also be a boon to supplying Alaska's input to national and local timber assessments. With this in mind, we developed aerial photo stand volume tables specifically for southeast Alaska.

THE DATA SET

Lack of funding for a separate aerial volume table study necessitated using data already existing in Forest Inventory and Analysis (FIA) files. The data set evaluated for this study was collected in the 1970s from five southeast Alaska inventory units. The original photo plots were systematically distributed over 1:15,840 black and white contact prints. Data used for volume table construction came from 962 ground measurement locations which were selected randomly from 66,731 photo points. All points were located on timberland and a number of forest types, as well as a wide range of stand conditions, were represented. Seedling/sapling stand-size plots were eliminated from the data set because residual trees in these cutover stands produced widely varying volumes. Additional plots were deleted because of missing aerial photos or data elements such as crown closure and tree height, leaving a total of 890 plots randomly located on timberland throughout southeast.

The 5 tallest ground-measured tree heights of both softwood and hardwood groups were averaged and used as the basis for our predictive equations. Previous studies by Gingrich and Meyer, and Haack have shown better correlation with the 3-5 tallest tree heights than with average stand height.

Crown closure was measured for a circular acre in the same forest type as the photo point. Closure was determined by comparing with crown density scales graduated in 10 percent classes and interpolated to the nearest 5 percent.

Gross cubic-foot volume was compiled from ground data using accepted cubic-foot local volume tables.

Aerial photo volume tables were produced for each major softwood type. It is often difficult to distinguish forest types on aerial photography and, therefore, the table shown was developed also. It is for all softwood types combined.

ANALYSIS

Data were first described and checked using BMDP statistical software. The literature was reviewed to see which variables and transformations had been used successfully in other stand volume tables. Regressions were then run using height, crown closure, and various transformations as independent variables. The most promising independent variables for species groups were used to run the final weighted regression equations which were used to produce table values. Mean values for each height-crown closure class were used and weights corresponded to number of observations in each class. An R^2 of .881 was obtained for the final general model data.

APPLICABILITY OF RESULTS

The southeast Alaska inventory area includes approximately 20.7 million acres. An estimate of the forestland area is 11.1 million acres. The 4 most important forest types (considering both area and volume) are western hemlock, mountain hemlock, western hemlock-Sitka spruce, and Sitka spruce. AIRIS plots will be established and measured at about 60 ground locations, not all of which are on timberland.

Individual interpreter biases, when making photo height measurements, may cause additional systematic error when using these tables. The tables are also dependent on the accuracy and confidence limits of the original local volume tables used to determine ground plot volumes.

Table 4.--Aerial photo volume table for all softwood types combined, southeast Alaska

(Gross cubic feet per acre)1/

Height2/	Crown Closure (percent)3/									
	5	15	25	35	45	55	65	75	85	95
Feet										
40	921	1375	1828	2281	2734	3188	3641	4094	4548	5001
50	1589	2042	2490	2949	3402	3855	4309	4762	5212	5668
60	2256	2710	3136	3616	4070	4523	4976	5429	5883	6336
70	2924	3377	3631	4284	4737	5190	5644	6097	6550	7004
80	3592	4045	4498	4951	5405	5858	6311	6765	7218	7671
90	4259	4712	5166	5619	6072	6526	6979	7432	7885	8339
100	4927	5380	5833	6287	6740	7193	7646	8100	8553	9006
110	5594	6048	6501	6954	7407	7861	8314	8767	9220	9674
120	6262	6715	7168	7622	8075	8528	8910	9435	9888	10340
130	6929	7383	7836	8289	8742	9196	9649	10100	10560	11010
140	7597	8050	8503	8957	9410	9863	10320	10770	11220	11680
150	8264	8718	9171	9624	10080	10530	10980	11440	11890	12340
160	8932	9385	9839	10290	10750	11200	11650	12100	12560	13010
170	9600	10050	10510	10960	11410	11870	12320	12770	13230	13680
180	10270	10720	11170	11630	12080	12530	12990	13440	13890	14350

1/ Gross volume per acre (inside bark, from one-foot stump to 4.0 inch top) for all live trees at least 5.0 inches d.b.h. Table volumes obtained from weighted regression equation: V = -1975.5437 + 66.7557 H + 45.3280 C. Standard error of estimate around mean volume = 10.5%

2/ Average height, H, in feet of 5 tallest softwoods.

3/ Crown closure, C, of all overstory from 10% crown density scales.

4/ Data obtained from 890 ground plots in Southeast Alaska. Boxes show cells which had actual observations used in equation development. Other cells were extrapolated using the equation.

FEASIBILITY STUDY USING ULTRALIGHT AND OTHER SMALL AIRCRAFT TO OBTAIN LARGE SCALE PHOTOGRAPHY FOR PHOTOGRAMMETRIC OPERATIONS

Wallace A. Crisco
Bureau of Land Management
Denver Federal Center, Bldg. 50
Denver, Colorado 80225

INTRODUCTION

This project was conducted to evaluate a system using large scale aerial photography, existing resource photography, image transfer techniques and devices, and semi-analytical photogrammetric procedures to determine geodetic coordinates of selected U.S. Public Land Survey System (PLSS) corners. These coordinates can then be used to mathematically compute coordinates for large numbers of other PLSS (township or section) corners. The utility of such a technique to develop geodetic positions could be applied nationwide but this experiment deals with public lands only.

SPECIFIC GOALS

Evaluate the costs, time, and practicability of obtaining large scale aerial photo coverage of paneled control stations, and township or section corners, utilizing platforms such as ultralight or conventional aircraft such as helicopters and small fixed wing (Cessna or equivalent) airplanes.

PROCEDURAL STEPS

1) Obtain existing photo coverage, 2) contract ultralight firm to obtain large scale photo coverage, 3) panel selected points, 4) field survey control, 5) transfer panel positions to small scale imagery, 6) photogrammetrically develop coordinate positions using semi-analytical procedures.

PROJECT DESCRIPTION

This project covered two and one-half square miles of mountainous terrain and required nine new field control points (eight were panels and one was photo identified) and one existing geodetic control station. Since the objective was to develop horizontal coordinate positions for the 10 points simulating PLSS corners, vertical data were of no concern. Field Survey Control Operations were accomplished by BLM personnel of the Denver Service Center, Branch of Survey and Mapping Development (D-416).

Acquisition of large scale aerial photography was unsuccessful with ultralight aircraft but was obtained utilizing a small fixed wing airplane. The nature of this experimental project did not require the use of helicopters.

Photogrammetric operations consisted of transferring the paneled and photo identified points from the large-scale photos (1:6800) to the small scale photos (1:26,400) and performing the analytical triangulation phases to develop coordinate positions for the selected points. This phase was completed by the Service Center's photogrammetry staff.

TEST RESULTS (FEET)

POINT I.D.	Surveyed Coordinates X	Surveyed Coordinates Y	Semi-Analytical X	Semi-Analytical Y	Difference X	Difference Y
100	499,543.21	93,724.77	Control			
101 (1)	500,166.05	95,557.92	500,186.37	95,509.82	+20.3	-48.1
102 (Green)	500,000.00	100,000.00	Control			
103	499,593.26	102,648.36	499;576.88	102,650.06	-16.4	+1.7
104	498,193.79	106,201.70	Control			
105	501,368.29	106,280.34	501,350.23	106,293.74	-18.1	+13.4
106	503,396.60	106,298.83	Control			
107 (2)	503,337.73	104,267.33	503,400.77	104,312.99	-63.0	+45.8
108	502,988.26	98,140.38	502,977.87	98,135.68	-10.4	-4.7
110	502,836.37	94,390.34	Control			

(1) Point 101 was placed in an area that did not have enough recognizable features to accurately transfer it from the large scale to small scale photos. (Care should be taken to assure that points have enough features to enable accurate transfer.)

(2) Point 107 was placed in an area that had been excavated since the small scale photos were taken. (Care must be taken to locate points on undisturbed terrain.)

NOTE: Project consisted of three (3) stereo models. Points 100, 110 and 102 (Green) controlled the southernmost model and points 104 and 106 controlled the northernmost model.

CONCLUSIONS

1) It is not feasible or practical to use ultralight aircraft for this type of work. They are very unstable during minimal turbulence thus preventing acquisition of the desired photo coverage. It was hoped (because these aircraft are easily transported to a field site) that this type of platform could enable us to get quick low cost large scale photo coverage. From our experience, the opposite was true because they are so very sensitive to any wind and other weather disturbances.

2) It is feasible and practical to rent a light aircraft such as a Cessna 206 to fly large scale coverage. The hourly rate is usually less than $100 for pilot and plane and most rural airports have someone willing to fly such coverage. The widespread availability of such platforms is very important to expedite project completion on a regional basis.

3) It is feasible to utilize the photogrammetric technique for developing PLSS coordinates with significant accuracy improvement over digitizing maps. The "Test Results" are from semi-analytical procedures demonstrating that points can be transferred with accuracy less than 20 feet. Most of that 20 foot error is probably caused by transfer problems since the semi-analytical procedure is capable of accuracies to 5 feet or less (utilizing 1:26,400 photography). Therefore, it is doubtful that fully analytical aerial triangulation would significantly increase the accuracy. The cost involved in acquiring this improved accuracy may be prohibitive on a nationwide basis. However, it is feasible and cost effective for special projects that require higher order accuracy.

EVALUATING BIOFUEL AMOUNTS AND SPATIAL PLACEMENT USING GIS AND REMOTE SENSING TECHNIQUES

By

Hal N. Anderson and Bruce L. Kessler
Idaho Image Analysis Facility
Idaho Department of Water Resources
Boise, Idaho 83720

ABSTRACT

Biofuel users need to locate and quantify local supplies for feasibility analysis and planning. Computer assisted techniques that tabulate and display spatially referenced biofuel information were proposed and developed in this project.

The McCall, Idaho 1:24,000 topographic quadrangle was chosen as the study area. A raster based geographic information system (GIS) was used to process multiple layers of data: forest management plans, forest stand and ownership boundaries, digital line graphs (DLG), a classified 1980 Landsat MSS image, and a mathematical model to predict logging slash volumes. From this, feasibility areas based on access and predicted slash volumes were assessed by category (high to low) and plotted as separate overlays (scale 1:24,000). For each feasibility area, summaries of acreage and expected slash volumes were tabulated. These data were combined with five-year logging plans to produce an expected volume of logging residue within five years. The overlays and the tabular data produced from the computer manipulation of these spatial data files provide information necessary to evaluate areas of potential biofuel supplies.

INTRODUCTION

Preliminary estimates indicate the amount of biofuels, including forest, agricultural, and municipal residues is significant. The location, magnitude and composition of biofuels has not been fully documented in Idaho. This study developed a framework for biofuel quantification and analysis.

The primary objective of this project was to outline the steps necessary for analysis of the expected logging residues using remote sensing and spatial data handling techniques. The developed methodology provides potential producers of bioenergy an efficient and cost effective tool for analysis of previously incompatible data types. Recommendations were included in the final report to help guide future investigators doing similar analyses.

METHODS

A 1980 Landsat scene covering the study area was classified to discriminate forested from non-forested land.

These data were combined with a mathematical model that predicts logging slash volumes, the 1:24,000 digital line graph (DLG) of McCall, Idaho, and digitized line files of forest logging, forest stand and ownership boundaries. The data were analyzed using a geographic information system (GIS), and the results were displayed in a map format that indicate classes of forest residue amount (Fig. 1).

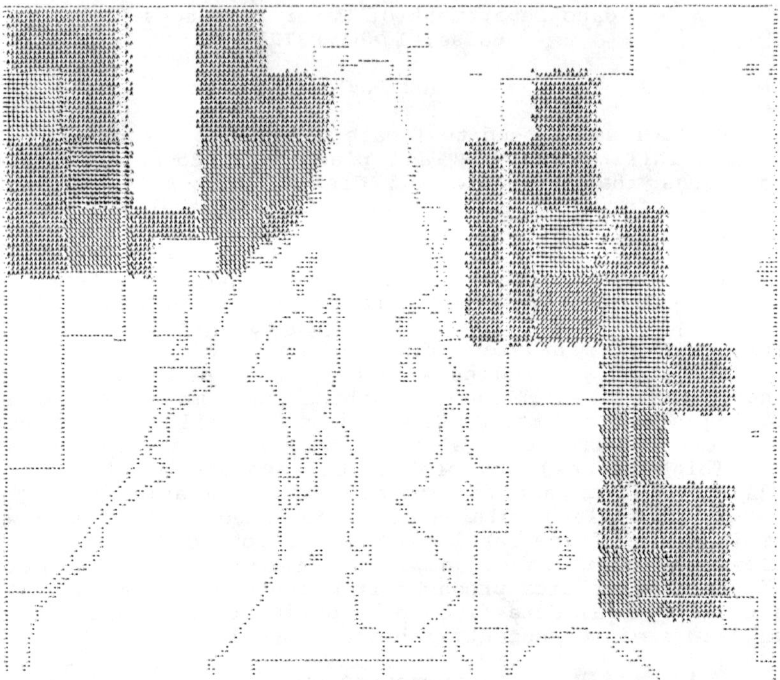

Fig. 1. A map overlay that indicates classes of predicted residue volume. Blank = not applicable, ^ = low volume, \ = med. low vol., + = med. vol., / = med. high vol., * = high vol, - = ownership and lake boundaries.

SUMMARY

In this study, techniques were developed and documented for combining remotely sensed, map, and tabular data into computer-readable data sets necessary for studying potential forest residue amounts and availability. These data can be used as mapped images or as summary satistics by bioenergy producers to assist in their feasibility assessment and planning activities. Due to the success of this study, we recommended in our final report to the funding organization that some refinements and a follow-up study be made. The results of that second study will provide bioenergy producers with verified methods for biofuels analysis.

A MICROCOMPUTER GEOGRAPHIC INFORMATION SYSTEM WORKSTATION

Don O. Hunter
United States Fish and Wildlife Service
Western Energy and Land Use Team
Fort Collins, CO 80526

Geographic information system (GIS) technology and remote sensing technology are essential tools for natural resource management. Remotely sensed data are increasingly being used as data layers in multi-resource GIS data bases. The efficacy of these tools is being increased by a new generation of more powerful and less expensive microcomputers.

The display consists of what is currently a "typical" GIS workstation. A self-running program shows remotely sensed data and thematic resource data in several analysis scenarios. Associated map products complement the display program.

The workstation consists of a Data General 10SP microcomputer configured with two 38-megabyte hard discs, a floppy disc drive, a cassette drive, and or 1.0 megabyte internal memory. The graphics terminal is a Visual 500 and the copier is a Data South 180. The GIS software used in the display program is the Map Overlay and Statistical system (MOSS) developed by the United States Fish and Wildlife Service, Western Energy and Land Use Team. MOSS is currently being used on microcomputers by a number of federal land management agencies including the Bureau of Land Management, Bureau of Indian Affairs, and the U.S. Forest Service.

Mention of trade names or commercial products does not constitute endorsement or recommendation for use by the Division of Biological Services, Fish and Wildlife Service, U.S. Department of Interior.

A MICROCOMPUTER PROGRAM FOR AERIAL PHOTOGRAPHIC MISSION PLANNING AND COST ESTIMATION

Charles W. Dull
USDA Forest Service
3620 Interstate 85, N. E., Suite 2103
Doraville, GA 30340

A software package which resides on the Apple II series minicomputers has been developed to assist those involved in aerial photographic mission planning. It is designed to assist not only those actually involved in aerial photographic acquisition, but also project managers. It allows for the comprehensive planning of time and materials, flight planning, and provides all the specifications required to design aerial photographic mapping operations and photographic sampling.

A new degree of flexibility is provided to Project Managers to quickly and comprehensively determine variations in time and cost of an aerial photographic project by developing a computer data file from which the inputs may be easily reviewed and changed. Utilization of the microcomputer will more effectively determine the greatest benefit of personnel, equipment and materials for resource data acquisition from aerial photography.

Quite often aerial photographic missions are requested upon which the project planner may not have complete details or he may have to tailor the mission to suit the contractor's wishes based on a wide variety of variables such as film types, number of prints required, and variations in scale. This software package not only allows for a comprehensive and quick turn-around for mission planning but may also serve as a data base to store aircraft and camera statistics, personnel and equipment costs, as well as film, film processing and printing costs.

A permanent record of all data may be stored on disk and may be reviewed on the computer monitor. Printed reports may also be generated in three different formats: 1) a project cost estimate which provides a one-page summary of cost and materials; 2) a project preparation report which provides a more detailed report on time and materials required for mission accomplishment; and 3) a comprehensive mission planning report which gives a complete record of all specifications involved in the project.

The software also allows the project manager to determine expected length of time to accomplish photographic acquisition activities based on historical records of climatic patterns of the target area. It provides an estimate on time to completion based on user inputs, climatic data, as well as the minimum accomplishment period, if the project can be completed in consecutive clear weather days. This package also allows the utilization of all focal length lenses for cameras with 9 x 9-inch, 70mm and 35mm formats. Reports stores on a disk may be easily recalled and changes of previous inputs allow for immediate turn around of a new time and cost projection. All of the basic computations required for aerial mapping are provided.

E. LINE TRACER: AN ECONOMICAL DIGITIZING SYSTEM

R. R. "Sky" Chamard, Consultant
E. Coyote Enterprises, Inc.
Far West Office:
P.O. Box 10761
Eugene, OR 97440

ABSTRACT

Digitizing of map data can be economically accomplished using the E. Line Tracer System. This paper describes in general terms the hardware-software of the system, how it works and gives some production accomplishments and costs.

THE HARDWARE SYSTEM

E. Line Tracer is a system utilizing all of the strengths of the U.S. Forest Service's Line Trace system. It is modified with proprietary software to make it less expensive, more operator friendly. All application software operations are in FORTH and the real strength of the system.

E. Line Tracer Hardware is configured as in Diagram 1. The Master Work Station (MWS) consists of a Data Capture Station (DCS), a Data Vectorizing Station (DVS), and a Data Output Station (DOS). Up to eight additional DVS can be added to the Master Work Station as per Diagram 2.

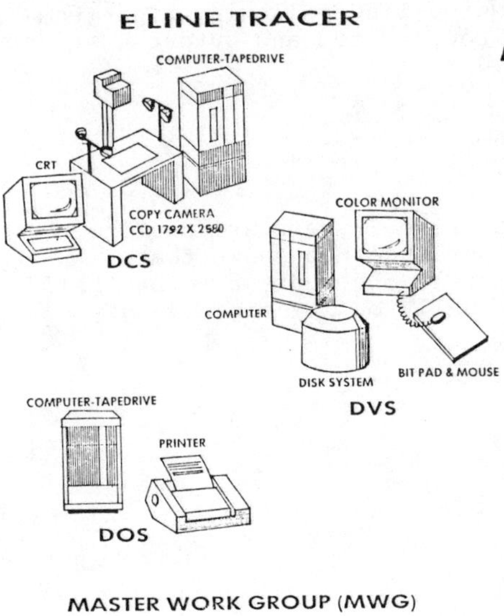

Diagram 1

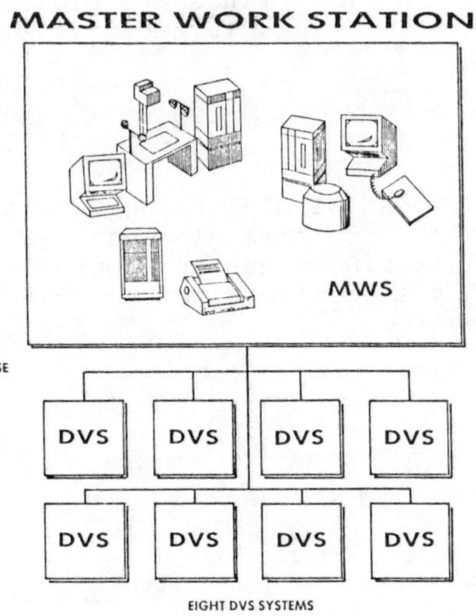

Diagram 2

The DCS consists of a 1792 X 2580 CCD camera, tapedrive, disk storage and operator terminal. The DVS consists of a 19" color monitor, 11/73 computer, disk drive and software. The DOS consists of a standard 1600 BPI tape output and an H.P. printer and necessary software to format output to various user systems.

- Project initialization and description
- Project registration and point collection
- Raster data edit tools and thinning capabilities
- Polygon lines - digitizing, tagging and edit
- Contour lines - digitizing, tagging and edit
- Hydrographic lines - digitizing, tagging and edit
- Joining of blocks within map area
- Joining with adjoining maps areas
- Data compression
- Zoom
- Report writer
- HELP

PROCESS

The process of converting map graphic data into digital is simple.

Step 1 - Prepare data, assure map is clean, etc.

Step 2 - Rasterize, using CED camera convert map data to 1792 X 2580 pixels.

Step 3 - Digitize-Vectorize. Pixel data is displayed as line data. Using line following algorithm lines are digitized, edited and output to selected medium.

COST

The cost of digitizing with E, Line Tracers are directly related to time elements. Experience has shown that digitizing times are approximately one quarter to one-half of the time required to do manually on X-Y digitizing tablets.

Forest Pest Management On The Nicolet National Forest Using The Automated Geographic Information System.

Bruce W. Morse, Autometric, Inc., Geographic Information System, 2629 Redwing Rd. #340, Ft. Collins, Co 80526.
William B. White, U.S. Forest Service, Methods Application Group, Forest Pest Management, Ft. Collins, CO 80524.

Management of forest pests is an important activity necessary to prevent unacceptable tree damage or loss. Components of forest pest management vary from region to region and from pest to pest; but in all cases , the knowledge of the location of the damage is vital. Locating past damage sites can be determined with aerial sketch mapping surveys, while locating future damage sites can be predicted using risk-rating methods. In both situations, a geographic information system (GIS) can assist in locating damag sites.

A cooperative project between the U.S. Forest Service's Methods Application Group and the University of Minnesota's Department of Entomology has been udertaken to use a GIS to assist the management of forest pests. The GIS being used is the Automated Geographic Information System (AUTOGIS), commonly refered to as MOSS. AUTOGIS, which was developed by Autometric, Inc., while under contract for the U.S. Fish and Wildlife Service, is a powerful and complete GIS with both raster and vector data processing capabilityies. AUTOGIS was recently installed on the 650,000 acre Nicolet National Forest, located in northern Wisconsin, for use in their forest planning. The following maps were digitized as data themes: administration units, land ownership, roads, water features, soil types, timber stands, elevation, recreation sites, genetically superior trees, fish and wildlife themes, and forest pest themes. Manipulation of these maps were performed on a 16-bit Data General microcomputer with 15 megabytes of storage capacity and 768 kilobytes of internal memory (RAM).

Delineating the boundary of past pest damage using aerial sketch mapping can be inaccurate. AUTOGIS can improve the accuracy of the survey by identifying and locating only those stands judged to be susceptible to the particular pest. For instance, with the use of AUTOGIS a flight-line map was produced and susceptible spruce and fir stands were color coded. This map assisted the observer in locating and accurately recording damage by the spruce budworm only in those susceptible types.

Risk-rating models are very helpful in identifying sites vulnerable to future damge. Risk-rating models exist for some economic important pests on the Nicolet National Forest sush as the Saratoga spittlebug and white trunk rot of aspen. With these models, the data themes can be manipulated and overlaid by AUTOGIS to identify and locate areas vulnerable to damage. A map of these hazard areas can be used by the forest manage to avoid potential pest problems.

It is anticipated that geographic information systems will greatly aid in facilitating pest management within the management of the overall forest.

AERIAL PHOTOGRAPHY FOR POST-MINE VEGETATION INVENTORY

Robert V. Dams
INTERA Technologies Ltd.
#1200, 510-5th Street S.W.
Calgary, Alberta T2P 3S2

Richard E. Francis
USFS, Rocky Mountain Forest and Range Experiment Station
2205 Columbia S.E.
Albuquerque, New Mexico 87106

The objective of this study was to evaluate the utility of large scale 70 mm color and color infrared aerial photography for plant species identification and ground cover estimation on revegetated coal-mined lands.

Large (1:600) and medium (1:10 000) scale 70 mm color and color infrared (CIR) aerial photography was acquired in July 1980 of reclaimed coal-mined lands near Steamboat Springs, Colorado. A concurrent field survey of vegetation cover was also carried out.

A plant species identification key and point method for estimating ground cover were developed for the 1:600 scale photography. Six volunteer photo interpreters were tested for identification of ground marked plant species at twelve revegetated study sites. After key testing, each photo interpreter conducted an estimate of ground cover by species and herbage category using a "point line" grid overlay on the color infrared photography. These estimates were then compared statistically to point frame cover estimates that were made on the ground at the time of overflight.

Results indicated that all photo interpreters could readily (>80% correct with varying sample sizes) identify Kentucky bluegrass (*Poa pratensis*), yellow sweet clover (*Melilotus officinale*), alfalfa (*Medicago sativa*), knotweed (*Polygonum* sp.), and Russian thistle (*Salsola kali*), due to their unique crown shapes, textures, shadows and/or spectral signatures. In addition, other common herbaceous species including intermediate wheatgrass (*Agropyron intermedium*), crested wheatgrass (*Agropyron cristatum*) thistle (*Cirsium* spp.) and summer cypress (*Kochia iranica*) could often (>65% correct) be identified. All life forms (grasses, forbs, shrubs) could be separated from each other, although a higher error of commission was evident between forbs and shrub seedlings. A non-parametric statistical comparison (Friedman Index) between the six photo interpreters indicated that familiarity with the vegetation and/or study area, or previous experience with small scale black and white and CIR photography did not significantly ($p>0.05$) improve their abilities to correctly identify the dominant species present.

Statistical comparisons were made between ground and aerial estimates of dominant plant species cover and of various more general ground cover categories. Paired t-tests were used to identify significant trends in over or underestimation of cover from the aerial perspective. Linear regression-correlation was used to assess the accuracy of the aerial estimates.

The paired t-tests by species indicated that the aerial estimates of the three dominant grasses (smooth bromegrass (Bromus inermis), intermediate wheatgrass and crested wheatgrass) were significantly higher ($p<0.05$) than their corresponding ground estimates, while the alfalfa estimates were not significantly different ($p>0.05$). The regression analysis showed that the cover by species for three dominant grasses could not be estimated from the aerial photos using the line point technique. In contrast to this, good results ($r^2 = 0.85$) were obtained for alfalfa cover estimation from the large scale CIR aerial photographs.

Aerial estimation of the various herbage (grasses, forbs plus shrubs, total vegetation) and other ground cover categories (rock, litter, total ground cover) yielded good results. Paired t-tests showed that grass cover and total vegetation cover were generally overestimated ($p<0.05$), while litter and total ground cover were underestimated ($p<0.05$). Forb (plus shrub) and rock cover air photo estimates were not significantly ($p>0.05$) different from the corresponding ground estimates. Regression analyses indicated that forb cover was generally the most reliably estimated cover category, followed by total vegetation and total ground cover. Removal of influential outlier data points improved the relationships for grass and rock cover estimation. Litter cover could not be reliably estimated from the large scale CIR aerial photographs.

While the total costs for ground and aerial surveys were similar, the cover information available from the aerial photography was lower; primarily due to lower species differentiation and litter cover estimation capabilities. However, the use of the aerial survey becomes attractive where a larger number of sample sites or a larger study area are to be studied, when an accurate spatial representation of ground cover is required and where a permanent record of conditions is beneficial. A permanent visual record is invaluable for monitoring revegetation success. Also, the rapid access of data is important to avoid effects of phenological confounding. More efficient year-round utilization of manpower could also be realized by using aerial photo interpretation in the winter to complement information gathered on the ground during the previous growing season. Finally, government regulatory requirements should be closely examined to determine whether the growth form information available from the aerial photography (cover by form) is sufficient, provided that life forms present represent acceptable species.

ACKNOWLEDGEMENTS

This paper summarizes research completed in partial fulfillment of the requirements for Mr. Dams' Master of Science Degree in Range Science at Colorado State University. Funding for the program was provided by the North Atlantic Treaty Organization - Committee on the Challenges of Modern Society Fellowship; Colorado School of Mines - Mining and Mineral Resources Research Institute Fellowship; and a Colorado State University - Forestry and Natural Resources Scholarship. The contributions of Mr. Clem Parkin, Colorado Yampa Coal Co., Steamboat Springs, and the staff of the Rocky Mountain Forest and Range Experiment Station, Fort Collins are greatfully acknowledged.

DIGITAL ANALYSIS OF HIGH ALTITUDE PHOTOGRAPHY

Paul Snook
USDA Forest Service
Rocky Mountain Experiment Station
240 West Prospect CO 80526

Introduction

Resource managers are continually searching for new methods to increase inventory cost-effectiveness. Satellite imagery and aerial photography have contributed to this goal. In this study a comparison is made between manual photointerpretation and digital classification of aerial photography. The object is to estimate land cover for a portion of the Tanana River Basin in Alaska with 20-acre sample plots.

Methods

Duplicate color infrared (CIR) aerial photographic transparencies (1:60,000 scale) were obtained during August, 1978. The film used was Aerochrome CIR SO-103 with a wratten 12 filter. Twenty-acre circular plots (0.2-inch) were located on each transparency. A 1-inch square area encompassing each 20-acre plot was then digitized on a scanning microdensitometer. The aperture setting of 50 micrometers resulted in a ground resolution of 3 meters. Green, red and infrared reflectance and a infrared/red ratio band were used in the classification. In order to minimize spectral noise, spatial resolution was reduced by averaging values of 3x3 pixel blocks. Unsupervised clustering was then performed on each image. A pattern was assigned to each cluster and a final cluster map printed via a color graphics printer. The plot center was ocularly registered on the cluster map. Each pattern was identified by the cover type it represented on the ground. Photo interpretation of low altitude photography (LAP) served as reference.

Results

Acreage estimates of land cover accumulated from 18 plots are given in Table 1. Manual and digital classification of HAP at level 1 provided accurate estimates for forest and non-vegetated classes. Digital classification recognized shrub cover from grassland more consistently than did manual interpretation. At level 2 classification confusion areas within level 1 classes can be identified. Separation of mixed forest classes from deciduous and coniferous forest types could not be done accurately with digital classification. The spatial and spectral resolution of the digital data allowed pure pixels to be recorded as individual components of the mixed stand. When this occurred the classifier separated the mixed stand into deciduous and conifer forest types. It appears that the misclassified pixels were labeled as deciduous forest. This is due to the pixel being dominated by the spectral signature of the larger crowns of deciduous trees. Shrub understory in open mixed stands can also cause digital misclassification as a deciduous forest class because of the similar spectral response of shrub cover and deciduous trees. Manual interpreters can correctly assimilate the 2 forest types within a mixed stand into a single class more consistently than digital classification. Level 2 manual classification results indicate that the confusion between shrub and grassland at level 1 is caused by the misclassification of low shrubs as grassland. This confusion is not apparent in digital classification.

Conclusions

The results obtained in this study indicate that digitized analysis is a viable method for determining most land cover classes. Digital analysis did consistently as well as manual interpretation for determining land cover acreages. Misclassifications that do occur in digital classification are caused by spectral overlap among classes and limitations in spatial awareness inherent in computer analysis of spectral data.

Although digital classification proved to be an accurate method for determining land cover, the use in this study of a small plot size would favor manual interpretation. Digital analysis is time consuming and costly if only a small portion of the image is to be classified. Manual interpretation of small areas is more efficient. However, if wall-to-wall classification of an image covering extensive ground area was needed, digital classification may be a cost-effective alternative to manual interpretation.

TABLE 1. Level 1 and level 2 classification of total acres for each cover type.

LEVEL 1	MANUAL HAP	DIGITAL HAP	MANUAL LAP
FOREST	218.6	208.8	211.0
SHRUB	51.6	77.2	72.0
GRASSLAND	44.6	29.6	31.0
NON-VEGETATED	45.2	44.4	46.0
TOTAL ACRES	360.0	360.0	360.0

LEVEL 2	MANUAL HAP	DIGITAL HAP	MANUAL LAP
CONIFER FOREST	150.4	152.6	149.8
DECIDUOUS FOREST	55.6	54.8	44.0
MIXED FOREST	12.6	1.4	17.2
TALL SHRUB	18.0	22.8	20.0
LOW SHRUB	33.6	54.4	52.0
GRASSLAND	44.6	29.6	31.0
BARREN	38.8	37.2	39.0
WATER	6.4	7.2	7.0
TOTAL ACRES	360.0	360.0	360.0

INDIAN INTEGRATED RESOURCE INVENTORY PROGRAM

J.R. Getter
Bureau of Indian Affairs
1951 Constitution Avenue, N.W.
Washington, D.C. 20245

W.J. Bonner, Jr.
Bureau of Indian Affairs
Lakewood Office Plaza, 730 Simms Street
Golden, Colorado 80401

The Bureau of Indian Affairs (BIA) is responsible for the natural resource management of approximately 52 million acres (21 million hectares) of Trust Lands in the contiguous United States. These lands are distributed in a "patchwork fashion" throughout the country. Management responsibility on these areas include: Range, forest, fish and wildlife, water, minerals, cultural, agricultural, and archeological resources.

In an age of decreasing natural resources and increasing natural resource values, effective multiple resource management is critical. To achieve this, BIA requires information about land status and the activities which impact this status. This information must be accurate and easily accessed. In addition, it must provide both historical as well as current status, and must be interrogable for interactive analysis. Furthermore, this information must be updated on a regular basis; thus, a considerable amount of data handling is implied.

To reduce costs, improve efficiencies, and make management decisions in a more timely fashion, BIA is adopting a "systems approach" to natural resource management which utilizes Geographic Information System (GIS) technology. The GIS will encompass a continuum of spatial and relational data elements, and include functional capabilities such as: data collection, data entry, data base development, data analysis, data base management, display, and report generation. Currently 5 of the 13 BIA Area Offices have initiated their data base construction activities. The remaining 8 offices will follow suit in calendar years 1986 and 1987. In the program, BIA Area Offices provide basic operational funding, while the Office of Trust Responsibilities (OTR) is providing "seed monies" to initiate the technology and conduct appropriate support activities throughout the organization. A total of one million dollars in such support is planned at the OTR level for each of the next five years. Generally these monies will be directed in support of the following activities: data base construction, technical support, applications development, hardware and software, and training.

HYDRAULIC DATA ACQUISITION USING NON-METRIC CLOSE-RANGE PHOTOGRAMMETRY

Kevin G. Coulton
Graduate Student, Washington State University
Albrook Hydraulics Laboratory, Pullman, WA 99164-3001

INTRODUCTION

Field measurements are necessary to define a stream channel environment. Usually data are obtained by stream gaging and channel transect measurements. Both of these methods measure channel hydraulic geometry; i.e., channel width, depth, bed roughness, water slope and velocity. Acquiring data with these conventional methods is labor intensive and expensive. A method that provides hydraulic geometry data with reasonable accuracy, lower cost and a shorter aquisition time would be a valuable alternative. In this paper, procedures for applying photogrammetric field methods for measurement of channel hydraulic geometry are investigated.

HYDRAULIC PRINCIPALS

The hydraulics of a natural channel are described by expressions for free surface open channel flow. These expressions permit the estimation of stream discharge.

Hydraulic Geometry

Hydraulic geometry elements describe the morphology of a channel and the hydraulics of the water flowing through it. A key objective of the research is to define these elements which are amenable to direct remote sensing, or can be indirectly inferred. Channel width can be determined directly. Water slope and flow depth can be inferred from in-stream staff gages. Water velocity can be determined by the displacement of floating objects between sequential photographs, or by creating and analyzing surface waves, similar to those created by dropping a pebble into water.

Discharge Method

The slope-area method (Smith, 1974) determines the discharge from a channel by incorporating hydraulic geometry and water surface slope. The criteria for the application of the method is particular to long, uniform stream reaches. This guides the location and orientation of the ensuing photography.

PHOTOGRAMMETRIC PRINCIPALS

To reduce costs and increase flexibility, non-metric (35-mm) camera equipment is used to photograph channels from close-range vantage points.

Close-range

The desire to provide a practical method to a large user group dictates the use of terrestrial photography from a close range ($\leq$300 meters). Bridges and cliffs are used to attain a good perspective for subsequent photographic measurement of hydraulic geometry.

Non-metric Equipment

The flexibility of 35-mm equipment is established by interchangeable lenses which provide coverage for channels of various configurations. Due to lens distortions and film deformations--inherent with non-metric equipment--an analytical data reduction method, the Direct Linear Transformation (DLT) method (Marzan and Karara, 1975), is being used to compute the orientation parameters of the photography.

PHOTOGRAMMETRIC METHODS

Two basic photogrammetric methods may be employed: direct control or indirect control. Control refers to the location and sophistication of visible reference points, necessary for the analytical reduction of photographic data.

Direct Control Method

This method requires initial field work for the placement and survey of control stakes and staff gages in and around the channel reach of interest. However, subsequent photographs of the site, showing the control network, are taken quickly and permit the determination of unknown float locations and staff gage water heights.

Indirect Control Method

Limited site accessibility dictates the use of indirect control. A reference control grid is set up adjacent to the camera. With the camera mounted on a theodolite, a stereo "backsight" to the control grid establishes the camera position(s). In the subsequent photography of the channel, locations of objects in the channel are determined relative to one of the camera positions.

SUMMARY

Procedures for the measurement of stream channel hydraulic geometry using non-metric close-range photogrammetry are being investigated. Accuracy aspects are being analyzed at the photographic and data reduction stages. Applications of the research to other fluidic phenomena, such as avalanches, lava and mud flows, may be beneficial.

REFERENCES

Marzan, G.T. and Karara, H.M. 1975, A Computer Program for Direct Linear Transformation Solution of the Colinearity Condition, and Some Applications of It: *Technical Papers from the Symposium on Close-Range Photogrammetric Systems*, pp. 420-476.

Smith, A.G. 1974, *Peak Flows by the Slope-Area Method*, Technical Bulletin No. 79, Inland Waters Directorate, Water Resources Branch, Ottawa, Canada.

HARDWOOD CROWN COVER OVER IMPERVIOUS URBAN SURFACES

Howard G. Halverson
USDA Forest Service
University Park, PA 16802

Frederick P. Weber
USDA Forest Service
Salt Lake City, UT 84119

INTRODUCTION

Communities often develop within forested areas and many retain or establish trees to gain environmental benefits. As a result of these activities, many towns have significant crown cover within their boundaries. Marotz and Coiner (1973) found that one-third of the surface area of several communities they studied was composed of tree crowns. As communities develop, some areas are dedicated to roads, parking lots, and other impervious surfaces. Some of these impervious areas are beneath tree crowns.

Precipitation chemistry is altered on contact with some of the surfaces. Precipitation falling directly on a residential street or parking lot is little changed chemically while runoff from major roads and business districts is degraded. Precipitation falling on a hardwood tree crown and appearing as throughfall generally has a decreased hydrogen ion concentration but increased levels of plant nutrients (Halverson et al. 1982). If the surface beneath the crown is impervious, then the throughfall will become part of the runoff. The purpose of this work was to determine what fraction of the impervious surfaces was beneath a tree crown.

METHODS

Four communities located in the Ridge and Valley province were selected as test sites. The towns and surrounding areas were photographed in 1981 with a panoramic camera and color infrared transparency film to test the use of the imagery for mapping gypsy moth defoliation (Ciesla and Acciavatti 1982). After scale and resolution determinations were made, equal area grids were attached to the photographs and the extent of each community determined.

All grid intersection points inside each community were examined and assigned to one of 10 categories of natural or modified surface. Major roads and the central business districts, residential streets, and parking lots were identified separately. Points on a tree crown were further examined to determine if any or all of the surface below the crown was in one of the three impervious categories. Roofs, sidewalks, and other small impervious areas were excluded because they usually drain to surrounding grassed areas and not directly into the drainage network.

After photointerpretation was completed, the total impervious area in each category was computed from the sum of points falling on those surfaces plus the number of points falling on a crown over one of the surfaces. Finally, the proportion of crown coverage over impervious surfaces was computed.

RESULTS

Tree cover in the four communities was close to the value found by Marotz and Coiner (1973), ranging from 31 to 37 percent of the total community area. However, the different impervious areas were not equally shielded by tree crowns (Table 1). Major highways and business districts had little impervious surface beneath a tree crown, possibly due to safety concerns and a lack of suitable sites. The same situation prevailed with existing parking lots.

Residential streets, in contrast, show a substantial tree cover. An average of 22.75 percent of the streets were beneath a tree crown. For residential streets, runoff quality could be altered by the chemical changes that occur during the throughfall process.

Table 1. Total number of sample points and tree crown cover over different impervious surfaces in four communities.

	Community			
	1	2	3	4
Sample size,number	354	547	483	1714
	------number (percent of area)------			
Highway and central business district	0	4(17)	2(14)	6(18)
Parking lot	0	0	0	0
Residential street	13(28)	10(17)	13(20)	54(26)
Total impervious	13(19)	14(14)	15(13)	60(17)

REFERENCES

Ciesla, W.M. and Acciavatti, R.E., 1982, Panoramic aerial photography for mapping gypsy moth defoliation: USDA For. Serv., For. Insect and Dis. Manage., Methods Appl. Group, Rep. 83-1, 17 pp.

Halverson, H.G., DeWalle, D.R., Sharpe, W.E., and Wirries, D.L., 1982, Runoff contaminants from natural and man-made surfaces in a non-industrial urban area: In: 1982 International symposium on urban hydrology, hydraulics and sediment control, OES Publ. Univ. of Kentucky UKY BU128, pp. 223-238.

Marotz, G.A., and Coiner, J.C., 1973, Acquisition and characterization of surface material for urban climatological studies: J. Appl. Meteorol. 12,919-923.

DIGITAL ELEVATION DATA IN LARGE-AREA RESOURCE STUDIES

Randy A. McKinley
Technicolor Government Services
Bureau of Land Management Operations
Bldg. 50, D-442T Denver Federal Center
Denver, CO 80225

James Maslanik
Ph.D. Candidate
Department of Geography
University of Colorado
Boulder, CO

POSTER SUMMARY

The Bureau of Land Management (BLM) manages approximately 350 million acres of range and timberland in the United States. In order to assist field personnel in the management of these land resources, the BLM uses a variety of information types to develop digital data bases within a Geographic Information System (GIS). USGS 30 meter digital elevation data in the form of Digital Elevation Models (DEM's) are routinely used by the BLM. This data provides terrain related parameters such as elevation, slope, and aspect, for inclusion in the GIS. The poster provides a description of processing techniques and specific applications which make DEM's a workable component of operational geographic information systems. Techniques developed by BLM to enter, process, and mosaic multiple (projects with greater than 100 7.5 minute quadrangles) DEM files for million-acre data sets is illustrated.

THE PULSED AIRBORNE LASER PROFILER: A NEW SYSTEM TO QUANTIFY AND EVALUATE FORESTS

Robert D. Baker, Dept. of Forest Science
Texas Agricultural Experiment Station
The Texas A&M University System
College Station, Texas 77843

Leonard A. LeSchack, President
LeSchack Associates, Ltd.
P.O. Box 646
Long Key, Florida 33001

Michael C. Whatley, Dept. of Forest Science
Texas Agricultural Experiment Station
The Texas A&M University System
College Station, Texas 77843

Enumeration of forests is a common requirement in their management and environmental monitoring. Due to constraints of time and expense, complete enumeration of forests is impractical. Traditionally, combinations of field measurements and photogrammetric or remote sensing surveys have been the principal system for procurement of forest data. A new remote sensing tool is the laser. A pulsed airborne laser profiler used in topographic mapping and geophysical research to gain accurate terrain data is now being tested in forestry to obtain measurements of forest stand parameters.

A joint experiment by the Department of Forest Science, Texas Agricultural Experiment Station, LeSchack Associates, Ltd., and Photo Science, Inc., was undertaken in 1984 to evaluate the applicability of the PRAM-3 digital airborne laser system to a variety of problems facing resource managers. The objectives of the overflight of a forested area in western Louisiana were (1) to measure terrain profiles beneath the tree canopy and compare with surface profiles from a clinometer-tape survey, (2) to derive forest stand parameters from canopy profiles, (3) to compare laser derived canopy statistics with a standard inventory of selected timber stands, and (4) to compare spatial variations of frequency of reflected pulses with canopy variables due to species change or level of stress.

Three flight lines covering different forest stands, terrain situations and surface water locations were flown several times each to assure complete coverage. Airborne laser profiler data, vertical accelerometer data, percent of data returns, and voice annotation during the flight were acquired simultaneously along with black and white 70 mm aerial photographs.

The small format photography assisted in accurate positioning of the flight lines on large format aerial photographs and on site. Field verification of the site was also conducted to establish the over-the-ground trace of the laser profiler and to differentiate timber stands. In addition, cultural features, water bodies and streams were annotated.

The system has promise to quantify forests. From an inspection of paper prints of profiles along the flight lines, it is clear that measurements of tree heights and crown diameters can be made by an airborne laser in a fashion generally similar to measurements taken on the ground. Exact replication of the airborne data sets is not observed since the laser "senses" the trees from an entirely different viewpoint that a forester sees them from the ground.

Preliminary measurements made on the paper prints of profiles and field plot measurements of two pine plantations indicate close agreement of means of total tree height (19.5 meters laser vs. 21.5 meters field and 8.8 meters laser vs. 10.0 meters field from the older and younger stands, respectively) and of means of tree crown diameter (4.2 meters laser vs. 4.8 meters field and 3.2 meters laser vs. 2.6 meters field from the older and younger stand, respectively). A clinometer-tape survey produced terrain profiles along the flight lines which essentially agree with those determined from the laser profiles. On an 885 meter length of profile the laser elevation for 23 control points averaged 0.8 meters higher than the field-derived elevations.

Separate analyses of the data tapes for automated measurement of terrain and forest parameters are being made at LeSchack Associates and the Department of Forest Science. Refinement of the LeSchack Associates algorithm for double integration of accelerometer data to correct the laser profiler record for aircraft motion will allow the terrain profile to quite accurately track the topography beneath the timber. Also, the percent of laser pulse returning to the transceiver is being analyzed to allow correlation with standing water, soil moisture and/or stress of forest vegetation. In the Department of Forest Science an algorithm (1) to mathematically determine the position (spacing) of each tree on the laser profile, (2) to calculate each tree's total height to the point of impact of the laser signal with the crown and (3) to calculate the diameter of the tree's crown along the laser trace is underway.

The preliminary agreement of laser data with field data is encouraging. Further development of computer algorithms for analysis and quantification of raw recorded profile, accelerometer and percent of pulse data are needed in order to quantitatively and statistically evaluate the PRAM-3 airborne laser profiling system for evaluating timber and forest terrain.

Illustrations on a poster of how the laser profiling system operates, a sample profile and a depiction of the agreement of profile-derived and field-derived measurements from the study allows a perception of the application of airborne laser measurements as substitutes for field enumerations.

Use of LANDSAT Data in Estimating Evapotranspiration Rates of Sierra Nevada Vegetation in Washoe Valley, Nevada

James A. Brass
NASA Ames Research Center
Moffett Field, CA 94035

and

Vincent G. Ambrosia
Technicolor Government Services, Inc.
Moffett Field, CA 94035

As the impact of anthropogenic activity on biological processes increases, several fundamental science concerns come to the fore. One of these is to develop an understanding of man's influence on the hydrologic cycle, and the impact that it in turn has on atmospheric circulation. Precipitation is the major input and driving force in the hydrologic cycle. Consequently, evapotranspiration (ET), one of the many processes of the hydrologic cycle and a major determinant of precipitation, takes on great significance. Add to this the fact that ET is an important factor in atmospheric circulation, surface temperature and plant physiology, and its role becomes even more crucial.

NASA/Ames Research Center, in cooperation with the U.S. Geological Survey, has been conducting scientific research concerning the effects of vegetation type and distribution on evapotranspiration and, indeed, the entire hydrologic cycle. The objectives of the activity have been (1) to determine the applicability of LANDSAT multispectral scanner data and modeling for the quantification of evapotranspiration rates and (2) to investigate the role vegetation type and distribution play in determining the amount of evapotranspiration from a mountain watershed.

Through the combined use of LANDSAT data for vegetation characterization, and hydrologic relationships between precipitation and water yield, estimates of evapotranspiration were obtained for five elevation zones in the western margin of Washoe Valley, Nevada. The LANDSAT digital analysis followed a modified supervised clustering approach and utilized a maximum likelihood classification, stratified by digital elevation data.

The digital analysis produced 121 spectral classes from the supervised and unsupervised clustering. The spectral classes were then grouped into 26 resource categories, providing the basis for the land cover map and vegetation characterization in the study area. As Washoe Valley was only one component of the total study area, the registered data set was subwindowed to exclude all data outside the watershed boundary.

To determine how well the LANDSAT data would explain the vegetation component an evaluation of the LANDSAT Multispectral classification was completed based primarily on a comparison of land cover information collected by ground survey/photointerpretation and LANDSAT classes. A random sample of 100 verification sites were located throughout the area. The sample units were 40.5 ha in size and represented all spectral types determined through the clustering and classification process. Evaluation of the correlation coefficients between ground data and LANDSAT classes within the 100 primary sample units was used to determine how well the LANDSAT data was predicting the ground condition.

Overall ability of the LANDSAT data to describe the variability of land cover within the study area was good. For the information classes of interest correlation coefficients between ground truth and LANDSAT classes were all above the .85 level.

Since no actual evapotranspiration data from Washoe Valley watershed was available, a determination of empirical evapotranspiration rates for vegetation was used. These rates were based on data found in the research literature and a precipitation-water yield relationship developed by Arteaga and Durbin (1978, p.23). To determine the empirical evapotranspiration loss rates for specific vegetation, calculation of the interception loss for snow and rainfall was estimated using the techniques of Kattelman et al, (Water Resource Bulletin 19(3) 395-402). Snow evaporation loss was estimated from total snow storage and weather station data.

The results of the LANDSAT data analysis were used to assign the estimated evapotranspiration rates to vegetation by elevation zone. This approach was found effective only to the degree that vegetative classes are homogeneous within each elevation zone. The less variation in vegetation within each elevation zone the more reliable the evapotranspiration assignment was.

The research in Washoe Valley, Nevada has shown the potential for integrating LANDSAT data with hydrologic variables to provide accurate estimates of evapotranspiration in a Sierra Nevada watershed. Results of this study have provided additional insight into the role that vegetation type and distribution play in determining the amount of evapotranspiration from a watershed. It is concluded that the techniques developed and used during this research could be applied to other watersheds in the Sierra Nevada.

STUDY OF FLOOD MODERATING CAPABILITY OF FOREST COVER USING REMOTELY SENSED DATA

Om Prakash Dubey
University of Roorkee
Roorkee (India)

ABSTRACT

India suffers extensive damages from floods every year. A salient feature of the flood problem in recent years has been its increasing spread and growing flood indices. The flood damages has increased 21 percent in period of 25 years since 1953 to 1977. The progressive increase in flood damage has brought an increased awareness of having suitable flood control measures. Since contruction of flood control structures requires massive investment and very long construction period, as such some other suitable device has to be planned. One of the important parameter which influences the loss of water due to interception and evapotranspiration is forest cover. Depending upon extent and density of forest cover the amount of water available as surface flows can be evaluated using suitable hydrologic model. A study has been carried out in Himalayan catchment, it was concluded that stream flow gets considerably decreased by increasing forest cover.

INTRODUCTION

Forest cover is a time dependent parameter and is responsible for soil erosion, sedimentation of reservoir and floods etc which are generally caused by ill planned depletion of forest cover. Forestry operation now reach into watershed previously immune from power saw and bulldozer. New mines are coming up, agricultural land area is being extended highways are built. Existing urban areas are being extended and new one being developed. Increasing amount of fertilizers and biocides are being applied. All these factors are influencial upon hydrology alongwith other factors. Canopy interception is a function of biomass and spatial arrangement of vegetal cover in addition to other factors. Cover modification influences the amount of precipitation reaching the earths surface and accordingly the magnitude of interception loss. Litter removal may modify the quantity of water entering mineral soil and may have large effect on rain drop impact, the net precipitation can be expressed as $PN = F(CC+P+E)$ where CC, P and E are canopy characteristic precipitation and evaporation during the storm. Canopy characteristics is dependent upon botanical factors, spatial arrangement and thus can be modified by land used. Management procedures such as logging and crop harvesting and catastophies such as forest fire and wind throw have major effect upon canopy characteristics of forest stand and other types of vegetation. The litter layer on the forest floor intercepts rainfall in same manner as the vegetal canopy. During small rainfall events very little precipitation may actually reach the mineral soil.

The forest floor, however, is not a sponge, it does not have an unlimited capacity to absorb rainfall. Where soils have previously been protected by surface litter cover and when they are laid bare through land use, soil compaction and associated effects may result. The kinetic energy of the storm may be of sufficient magnitude to cause soil splash and plugging of surface soil pores thereby reducing infilteration and increasing surface run off.

The forest vegetation is characterised by high absorptivity low albedo and low transmissivity. Absorptivity of conferous cover > hard wood cover > field crops. Conversely conifer albedo < hardwood albedo < field crops. The absorptivity of conifers, hard wood and grass are 0.9, 0.8, 0.5 and < 0.8 approximately of solar radiation. In closed forest little solar energy reaches the soil surface that is its transmissivity is very low. The magnitude of transmitted shortwave radiation depends upon forest structure composition and density. As such cover changes have large effect on energy budget which limit evaporation, transpiration and snow melt. In addition to the absorptivity and albedo differences the major variable influencing differential evapotranspiration losses, according to vegetational cover are the rooting depth of cover and depth of soil mantle. Shallow soils are generally completely occupied by plant roots, regardless of type of cover. Deeper soils may not be totally occupied by roots of one cover type but be much more fully occupied by roots of another cover type. As large differences is evapotranspiration to occur on deep soils carrying different covers, but only small differences from shallow soils. Manipulation of forest cover has large effect upon soil moisture. Logging by removing transpiring tree cover reduces soil moisture depletion particularly in deep soils, there by increasing soil moisture content. Partial cutting of thinning also increases soil moisture content through reduction in stand density. Remotely sensed data provide basic data bank for resource analysis. Remotely sensed data can be analysed for evaluating landuse pattern and vegetal cover classification, percentage canopy cover, leaf count and other related parameters influencing hydrological parameters. Depending upon the density, class and extent of vegetal cover the amount of water available as surface flow or as ground water can be estimated using suitable hydrologic model. As such a study has been carried out for studying the effect of forest cover on surface flows in a mountaneous catchment in Himalayas.

STUDY AREA AND DATA

The catchment area of about 2000 sq. km areal extent lies between latitudes 29°18'57" to 30°51'58" N and longitudes 77°57'13" to 78°15'56" E. Average rainfall over the ores is 1000 mm to 2000 mm. The maximum temperature recorded is 35°C. Stream flow were the measured at lakhwar dam site.

Different types of photo products of Remotely Sensed Data at 1:50,000 and 1:M and 1:250,000 were analysed for preparing land use map and forest cover map. Results are tabulated in Table I.

Table I

Land Use Parameters

Raingauge Station	Land use in km²				
	Forest	Shrubs	Culti-vation	Blank	Snow
Mussoorie	134	81	93	85	08
Minus	6	12	00	00	00
OSLA	131	00	00	137	45
Uttarkashi	56	00	00	58	35
Barkot	192	276	148	51	00
Damta	170	353	113	127	00

DEVELOPMENT OF MATHEMATICAL MODEL

Stream flow is sum of contribution by different types of forest cover qi = di ai ki, where q,d,a and k are flow contribution, depth of precipitation area of forest cover and runoff coefficient for ith type of vegetal cover in influencing zone of a station. It can be expressed as $Q = AK$ where Q and K are column vector and A is coefficient matrix. The stream flow is observed at various instances and then unknown K can be determined by $K = A^{-1}Q$, this requires that A^{-1} exists. The quation can also be written as $\Delta Q = A\ \Delta K$ and may be solved by minimising ΔQ in least square sense. The generalised inverse may also be adopted by weighing or without weighing the data depending upon the requirement. Once runoff coefficient is known, stream flow arising from different storm at various instances, can be computed. At this stage a theoretical study has been carried for predicting flows at various percentages of forest covers and the results are tabulated in Table II.

Table II

Comparision of Stream Flows at different percentage Forest Cover

Month	Flow in cu mec	
	30 (existing)	60
Jan	29.00	16.00
Feb	12.00	7.00
Mar	10.00	6.00
Apr	20.00	10.00
May	70.00	52.00
Jun	175.00	93.00
Jul	235.00	139.00
Aug	265.00	147.00
Sep	89.00	50.00
Oct	32.00	18.00
Nov	9.00	5.00
Dec	0.00	0.00

RESULTS AND CONCLUSION

Remotely sensed data can be analysed for evaluating land-use parameter. Land use parameters are influencial to hydrologic parameters. Land use is one element of hydrologic landscape that can be deliberately manipulated to create water resources characteristics beneficial to society. Other landscape parameters such as physiography, geomorphology and climate are not easily susceptible to purposive treatment. Forestry operations effect all elements of hydrologic cycle. Forest cover has large capability of retarding the flood. In general it can be said that increase in forest cover in catchment has direct impact on reduction in flood volume. If forest cover is doubled the flood peak reduces to half. It may be very interesting to note that this extra forest may be used as an alternate source of energy too.

LITERATURE CITED (SELECTED)

Chow Vente (1964), Hand Book of Applied Hydrology, McGraw Hill Book Co.

Donald M. Gray, Hand Book on the Principles of Hydrology Water Information Centre, Inc. Huntington, N.Y.

Dubey O.P., Sri Niwas and Awasthi A.K. (1984), Analysis of Remotely Sensed Data for Ground Water Studies of Piedmont Zone, Proc. Asian Conference on Remote Sensing held at Kathmandu.

Dubey O.P. (1985), Remote Sensing Techniques a powerful tool for Rapid Assesment of Water Resources SOARS Symposium on Remote Sensing in Agriculture, Ahmedabad.

Lakhwar Vyasi Pariyojana, Vol. I and II, Irrigation Department, Uttar Pradesh, India.

RADIOMETRIC RESPONSES TO STEP-WISE REMOVAL OF VEGETATION FROM A STATIONARY PLOT

D.L. Isaacson, B.J. Schrumpf and W.J. Ripple
Environmental Remote Sensing Applications Laboratory
Oregon State University
Corvallis, Oregon 97331-6703

BACKGROUND

The relationships between measured spectral reflectance of a desert shrub canopy and selected plant and soil variables were studied by making repeated spectral and photographic observations.

On June 26, 1984 five 2.93m^2 plots, located within a dry lakebed on the Squaw Butte Station of the Eastern Oregon Agricultural Research Center near Burns, were selected for observation. On each of the five plots, dominated by *Artemesia cana*, six sets of observations were made. Each observation set included spectral reflectances from a Barnes Model 12-1000 Radiometer, one 35mm color slide, and a bag of biomass clipped from the plot. The initial spectral measurement and photograph were of the undisturbed plot. The second through fifth observations were at successive stages of incrementally reduced biomass: ca. one-fifth was clipped at each step. The sixth observation was of the completely cleared plot. The biomass was placed in sealed bags for later treatment.

Water content of the above-ground plant biomass was calculated from fresh and dried weights after oven drying at 45°C for 96 hours. Dot counts on the 35mm photographs were used to estimate proportions covered by green leaves, shadow, soil and stems. Partial multiple and canonical correlation were used in analyzing two data sets: 1) an original set of 30 observations, and 2) a set of 25 observations derived by taking the difference between successive observations.

RESULTS AND DISCUSSION

Data Interdependence

Data were organized into two sets because of a concern over interdependence between observations: approximately 80% of the canopy in each observation was also present in prior and subsequent observations. Interdependence was removed by taking the difference between observations, creating a data set of *change* in spectral and plot variables. The effect of interdependence is clearly illustrated in both Tables 1 and 2. The apparent strength of relationships between spectral and plot variables is markedly increased by interdependence, and false positive relationships result.

Spectral and Plot Variable Relationships

The middle infrared Bands 6 and 7 best accounted for variation in plot variables, with Band 7 accounting for more than 60% of plot variation (Table 1). Of the visible bands, only Band 3 significantly accounted for plot variation. Of all the plot variables, total plant-held moisture was best accounted for by the radiometry, and dry biomass was also highly significantly accounted for (Table 2). None of the photo-derived plot variables were significantly associated with radiometric observations.

Table 1. Performance of Barnes Multi-modular Radiometric Bands in Explaining Variation in Plot Data in a Biomass Reduction Experiment.

Original Set (n=30)	MultR	MultR²*(a)*	F*(b)*
1. Band 7 (2.08-2.35μm)	0.961	0.923	45.92**
2. Band 6 (1.55-1.75μm)	0.954	0.910	38.69**
3. Band 4 (0.76-0.90μm)	0.898	0.806	15.92**
4. Band 5 (1.15-1.30μm)	0.894	0.799	15.22**
5. Band 3 (0.63-0.69μm)	0.855	0.731	10.42**
6. Band 1 (0.45-0.52μm)	0.776	0.601	5.79**
7. Band 2 (0.52-0.60μm)	0.775	0.601	5.77**
Change Set (n=25)			
1. Band 7	0.779	0.607	4.63**
2. Band 6	0.707	0.500	3.00 *
3. Band 3	0.702	0.493	2.92 *
4. Band 1	0.671	0.451	2.46ns
5. Band 2	0.663	0.439	2.35ns
6. Band 4	0.505	0.255	1.02ns
7. Band 5	0.415	0.172	0.63ns

(a)- MultR² is the proportion of variation in the radiometric bands that can be predicted from plot data.
(b)- **:significant at 1% level, *:at 5%, ns:not significant

Table 2. Plot Variables From a Plot Reduction Experiment Assessed for Explaining Variation in Barnes Multi-modular Radiometric Data.

Original Set (n=30)	MultR	MultR²*(a)*	F*(b)*
1. Total Moisture	0.989	0.979	146.47**
2. Green Cover	0.985	0.970	101.44**
3. Dry Biomass	0.983	0.965	87.88**
4. Stem Counts	0.931	0.866	20.38**
5. Soil	0.876	0.767	10.33**
6. Shadow	0.868	0.753	9.57**
Change Set (n=25)			
1. Total Moisture	0.890	0.791	9.21**
2. Dry Biomass	0.832	0.692	5.45**
3. Green Cover	0.676	0.457	2.04ns
4. Soil	0.587	0.344	1.27ns
5. Stem Counts	0.524	0.275	0.92ns
6. Shadow	0.415	0.172	0.51ns

(a)- MultR² is the proportion of the variation in plot variables that can be predicted from radiometric data.
(b)- **:significant at 1% level, *:at 5%, ns:not significant

A COMPARISON OF X- AND L-BAND RADAR DATA FOR DISCRIMINATING FOREST COVER TYPES

R.M. Hoffer, S.E. Davidson, P.W. Mueller, and D.F. Lozano-Garcia
Department of Forestry and Natural Resources
Purdue University
West Lafayette, Indiana 47907

Foresters and other resource managers have used aerial photography for many, many years as an aid in resource management decisions. During the past decade, multi-spectral scanner data, especially from Landsat, has become fairly common and has been shown to be of considerable benefit in many situations. Radar data has not been as readily available and quite often, when it has been obtained it has only been generated in a single polarization or single frequency. Consequently, relatively little is known about the capabilities and limitations of radar data of different frequencies, or of multi-polarized radar data for discriminating forest cover types or even differentiating forest from other land use classes.

The objective of this study was to assess the capabilities and limitations of HH- and HV-polarized X-band and HH-, HV-, VH-, and VV-polarized L-band airborne synthetic aperture radar (SAR) data for discriminating various forest and other cover types.

The 3 cm X-band data were obtained on June 30, 1980 over a forested site in Kershaw County, South Carolina. The study site contained large areas of mixed bottomland hardwoods (both old growth and second growth), stands of water tupelo (*Nyssa aquatica*) and slash pine (*Pinus elliottii*), pasture, crop land, exposed soil (both crop land and clearcut areas), as well as water and cultural features. The HH and HV dual polarized SAR data were obtained by NASA's APQ-102 side-looking radar, mounted on the RB-57 aircraft, which was flown at an altitude of 18,000 meters (60,000 feet). The data were optically correlated and the imagery was enlarged to 1:80,000 and 1:50,000 scale prints for analysis. Color infrared photography obtained over the study site in June 1979 and again in August 1980 were used in the interpretation of the SAR data.

The longer wavelength (24.6 cm) L-band SAR data were obtained over a somewhat similar study site on September 14, 1984. However, this study area was located in Baker County in northern Florida where the major cover type is slash pine of several age classes. There are also many cypress (*Taxodium* spp.) and tupelo (*Nyssa* spp.) swamps scattered throughout the area, along with pasture and crop lands, some clearcut and agricultural areas of exposed soil, water and cultural features. The L-band SAR data were obtained by NASA's Convair 990, flying at an altitude of 7,975 meters (26,100 feet). An electronic switch enabled both H (horizontal) and V (vertical) polarized signals to be transmitted, so that all four polarizations (HH, HV, VH and VV) could be obtained simultaneously. The data were digitally processed at the Jet Propulsion Laboratory and subsequently analyzed at Purdue University using an IBM 7350 High Level Image Processing digital display at a variety of scales. Color infrared photography obtained during the SAR flight mission, along with National High Altitude Photography (NHAP) and a variety of other reference data were also used in the analysis of these data.

The results of the analysis of both the X- and L-band SAR data showed several very interesting results. The X-band data had distinct differences due to polarization, most notably the fact that hardwoods had a very high backscatter (and therefore a light tone) but conifers had a very low backscatter (and therefore a dark gray tone) on the HH polarized data. However, on the HV polarized data, the deciduous and coniferous forest could not be differentiated -- it was all a similar medium gray tone! Another interesting result was that some of the pasture areas and conifer stands were very similar in tone on both the HH and HV polarized images, and could only be differentiated on the basis of their texture or speckle characteristics. In general, it appeared that the like-polarized data (i.e., HH) had a wider range of tonal contrast and allowed better discrimination than was the case with the cross-polarized data.

The L-band quad-polarized data displayed some of the same characteristics seen on the X-band data set, but some interesting differences as well. The like-polarized images (HH and VV) provided better discrimination between the pine and swamp forest lands than did the cross-polarized images (HV and VH), but the cross-polarized images appeared to provide a higher contrast between forest and non-forest cover types than did the like-polarized images. The cypress swamplands had a particularly strong backscatter on the HH polarized image, causing these areas to have a distinct white appearance. The HV and VH images were very similar, which would be expected, according to radar theory. All four images exhibited significant variation in backscatter as a function of range, which would also be related to incidence angle.

In summary, analysis of both X- and L-band multi-polarized SAR data indicates distinct differences as a function of polarization and incidence angle, and generally similar results between the two frequencies. Such data may prove to be a valuable adjunct to optical wavelength data.

THE USE OF LANDSAT'S THEMATIC MAPPER FOR WETLANDS INVENTORY

Gregory T. Koeln
Ducks Unlimited, Inc.
One Waterfowl Way
Long Grove, IL 60047

John E. Jacobson
Ducks Unlimited, Inc.
One Waterfowl Way
Long Grove, IL 60047

Charles L. Hill
Earth Resources Laboratory
NASA, NSTL, MS 39529

Crucial to the management of our North American waterfowl resource and particularly in guiding Ducks Unlimited's efforts in wetland protection, enhancement, and management, are current wetland inventories throughout North America. The costs of 1) preparing wetland inventories from large regions in a short period of time, 2) upgrading existing wetland inventories, and 3) monitoring and detecting changes in wetlands can be inhibiting by conventional techniques utilizing aerial photography, ground reconnaissance, and photo interpreters. Ducks Unlimited, trying to find a less expensive and more rapid technique for appraising the wetland resouces, initiated a contract with The National Aeronautics and Space Administration, NASA, in June of 1983 to determine the feasibility of using Landsat's Thematic Mapper (TM) for the inventory of wetlands.

Past efforts using Landsat's Multispectral Scanner (MSS) for wetland inventories have only been marginally successful. However, initial results from utilizing TM data for wetland inventories have encouraged Ducks Unlimited to complete by 1990 an inventory of the prairie pothole region (a 350,000 square mile area covering much of the provinces of Alberta, Saskatchewan, and Manitoba, and the states of North and South Dakota, Minnesota, and Montana). Full coverage of this region requires approximately 58 TM scenes. Long-range plans include updating the inventory every three to five years.

Ducks Unlimited has acquired the hardware and software capabilities to process full TM scenes. From the TM data, map overlays are produced that depict all wetland complexes at scales of 1:24,000, 1:50,000, 1:100,000, and 1:250,000. For each wetland complex, the acreages of open water, deep marsh, shallow marsh, and wet meadow; shoreline length; Universal Transverse Merecator coordinate; and a shape index are calculated and reported. Presently, to process a full scene (eight million acres) and generate required maps and statistics takes 10 to 15 working days. We anticipate that only five working days will be required for these processes in the near future.

Detection of open water has been very successful. Water bodies even smaller than the 30x30 meter resolution of the TM sensor are readily detectable. Shallow marsh, deep marsh, and wet meadows are identifiable with little omission or commission errors, except when these wetland types occur in dry basins. Under these conditions it is difficult to separate wetlands from hay fields, pastures, and row crops. Delays in acquiring TM data and lack of available cloud-free images during the summer months have been problems. Overall, the initial results of analyzing wetlands from TM data have been good. Advanced processing techniques utilizing contextual classifiers or artificial intelligence may be required for improved accuracies.

MILLIMETER-WAVE IMAGING SENSOR FOR FIRE MAPPING

W. J. Wilson, R. J. Howard and G. S. Parks
Jet Propulsion Laboratory
Pasadena, CA 91109

A scanning 3-mm radiometer has been built for use on a small aircraft to produce real-time high resolution images of the ground during atmospheric conditions (smoke, dust, and clouds) in which IR and visual sensors are not usable. This sensor could be used to find fires obscured by smoke or low clouds. A variety of remote sensing applications such as measurements of snow cover and snow water equivalent, precipitation mapping, and vegetation type and extent are also possible. The advantages of millimeter-waves for cloud penetration and the ability to observe different physical phenomena, make it an attractive supplement to visible and infrared remote sensing systems.

The configuration of the mm-wave imaging sensor is shown in Figure 1. A crosstrack scanning antenna system is used and a lightweight flat reflector is scanned ±22-1/2 degrees at a 4 Hz rate. A two-axis scanner, with gyro stabilization, is used to correct for aircraft motions due to turbulence. This provides line of sight stabilization without the use of a conventional stabilized platform. Dual 16-bit microprocessors, with inputs from both rate and angle gyros, transform coordinates and pointing commands to maintain a 0.1° pointing stability to eliminate pixel smear and image distortion. The two-axis scanner also allows use of a butterfly scanning pattern to compensate for the aircraft forward motion and produce a raster scan pattern on the ground. An offset parabolic antenna is used to focus the received signal on the radiometer's feed horn. At an altitude of 750 m (2460 ft) with a forward velocity of 41 m/s (80 kts), a 0.6 x 1.3 km area is scanned in 32 seconds with a spatial resolution of 7 m.

Dual-polarized 98 GHz superheterodyne radiometers have been used to provide polarization information and their outputs are added to improve the sensitivity. The radiometers use balanced mixers with beam-lead diodes and one Gunn diode local oscillator is used for both radiometers. The rms noise per ground resolution element is <0.8 K when both channels are added. A calibrated signal is injected into the feed horn at the end of each scan to provide temperature calibration. A relative accuracy of ±1 K and absolute accuracy of ±5% is expected.

The first test flights were made this summer with the sensor mounted on a commercial helicopter. This provided the flexibility to fly a wide range of flight plans and the freedom to seek out atmospheric conditions of interest, e.g. fog, dust, light rain. In the initial testing, the imaging data was recorded on magnetic tape and then later played back and displayed on the ground for the analysis and evaluation. In the final aircraft configuration the radiometer imaging data will be displayed, along with accurate scene location data, on a TV display. The TV will show each line of data in false colors and the display will scroll down as the aircraft flies along. It will be possible to select different color combinations for the best data interpretation and selected images can be stored on command for later detailed analysis. Preliminary mm-wave images from our tests will be presented and discussed.

This work was supported by the U. S. Army Materiel Command (AMC) for the Army Development and Employment Agency (ADEA).

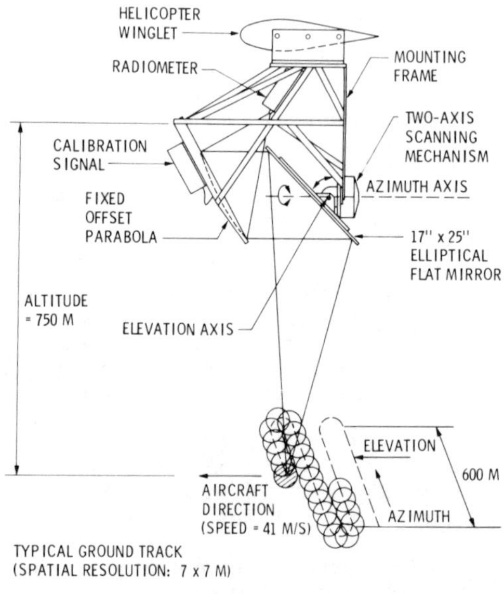

Figure 1. Millimeter-wave Imaging Sensor Configuration

FIRE DETECTION USING NOAA POLAR-ORBITING SATELLITE DATA

M. Matson, G. Stephens
NOAA/National Environmental Satellite, Data, and Information Service
Washington, DC 20233

Fire detection in remote locations is hampered by high cost and inaccessibility. Due to increased concern over the climatic and economic impacts of deforestation and slash and burn agriculture, most of which occurs in underdeveloped parts of the world, it is necessary to find ways to effectively monitor such activity. One promising technique involves the use of the 3.8 micrometer channel onboard the NOAA-series polar-orbiting satellites.

Currently, there are two NOAA polar-orbiting satellites in operation. Each satellite carries the Advanced Very High Resolution Radiometer which has five spectral channels; one visible channel, one near-infrared channel, and three thermal channels, including the 3.8 micrometer channel. Resolution at the satellite subpoint is 1.1 km and a typical satellite image covers a ground swath width of 2600 km. With both satellites, it is possible to cover any part of the world four times daily, twice during the day and twice at night.

The 3.8 micrometer channel is very sensitive to high temperature sources. This is because for high temperature sources the maximum blackbody radiance shifts away from longer thermal wavelengths (e.g., the 11 micrometer area) and towards the 3.8 micrometer wavelength. On a 3.8 micrometer image, high temperature areas will appear as small white or black spots, depending on the enhancement used. It is relatively easy to discriminate these "hot spots" from other features on the image, especially by comparing the image to an image from one of the longer wavelength thermal channels on which the "hot spots" will not appear. It is also important to note that a "hot spot" need not be as large as 1.1 km to cause a response in the 3.8 micrometer channel. Even subresolution "hot spots" will cause a noticeable response.

One type of high temperature target that the 3.8 micrometer channel can detect is fires. Several case studies have been analyzed to demonstrate the utility of this channel for fire detection. These include controlled burning in the forests of the northwest United States, tundra fires in Alaska, forest fires in Siberia, slash and burn agricultural fires in Mexico and Mozambique, and forest clearing fires in Brazil. In some of these cases, several hundred fires are visible on an image.

Figure 1 is a visible band image taken of southern Mexico and northern Guatemala on April 18, 1984. A number of large smoke plumes and smoke covered areas are visible. The more globular white areas are meteorological clouds. Figure 2 is the 3.8 micrometer image of the area. The well defined white spots are fires. Most of the fires seen here are associated with land clearing for agriculture. Scientists familiar with this section of Mexico have identified several distinct areas and the type of burning associated with each (Cloud, personal communication). The Grijalva Basin, the light area near the center of the picture, is a major agricultural center; most of the burning here is for clearing of previously cultivated land. In contrast, the area around Veracruz, in the upper left, and the area around the Guatemala border, in the lower right, are primarily virgin forest, being cleared for agricultural use at great ecological cost. The roughly horizontal cluster of fires near the lower right border is thought to be associated with the camps of Guatemalan refugees.

With the fire detection capability of the 3.8 micrometer channel and the daily coverage of the NOAA-series satellites with their large ground swaths, it is possible to operationally monitor fire areas in any part of the world. Such monitoring would allow scientists to more accurately assess the extent of regional deforestation and the contribution of burning to atmospheric carbon dioxide. In addition, personnel involved with fire control can use the satellite data as an effective supplement to conventional ground and aircraft data for monitoring fire activity over large areas.

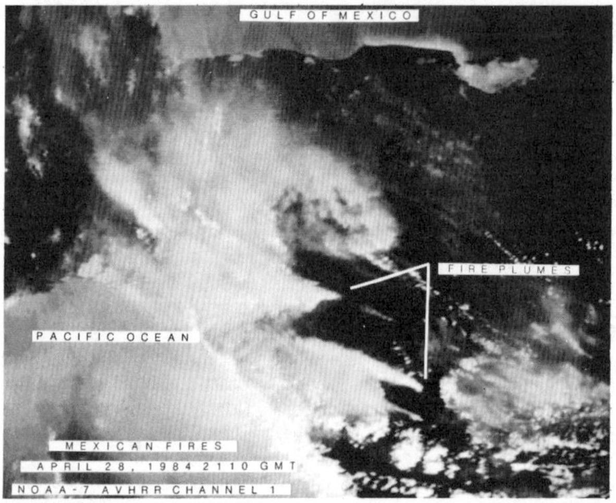

Figure 1. Visible channel image of southern Mexico and northern Guatemala.

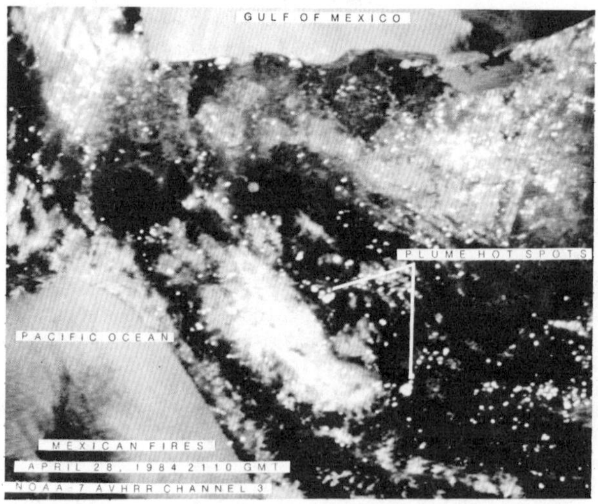

Figure 2. 3.8 micrometer channel image of southern Mexico and northern Guatemala.

FORUM SESSIONS

Coordinator: G. Johnson

INTERFACING REMOTE SENSING WITH SPATIAL DATA SYSTEMS

Moderator
Duane A. Asherin
USDI Fish & Wildlife Service/WELUT
2627 Redwing Road
Fort Collins, Colorado 80526

Panel Members: Jeffery C. Eidenshink - Technicolor-EDC
James W. Merchant - University of Kansas
Edgar A. Work - BLM-DSC

This forum discussion was attended by 75 participants and discussion notes were recorded by Barbara M. White. Participants represented numerous Federal and State natural resource management agencies, universities, and the private sector. The forum topic and panel member comments elicited a lively discussion of topic items; including computer hardware/software considerations, data formats and types, and agency/user problems and needs.

Panel presentations provided the following perspectives, positions, issues, and/or needs. In the 1970's, the utilization and application of remotely sensed data from airborne and satellite devices motivated the development and initial use of spatial data systems or geographic information systems (GIS's). In the 1980's, research and development continues in the interpretation and utilization of remotely sensed data; but increased emphasis is being placed on the development and application of new methods for integrating remotely sensed data, from all sources, with numerous other spatial resource data themes. Federal, State, and private agencies/organizations responsible for natural resource management and assessment activities are placing increasing priority on the use of these technologies and datasets in an integrated manner.

Several data integration problems still remain between remote sensing and GIS technologies. These problems include data registration accuracy, merging and transforming raster and vector data from one type to the other, lack of standardized data exchange format(s), and lack of data supply standards. Another challenge in interfacing datasets and technologies is the development of spatially consistent statistical measures of map accuracy and involves the development of mathematical and statistical operations of advanced analytic systems.

Computer hardware/software systems designed for the analysis of digital, remotely sensed data should include fully integrated capabilities for processing other spatial data types. Spatial data systems (i.e., GIS's), on the other hand, do not necessarily need to be interfaced to systems designed for processing and classifying remotely sensed digital data. Most natural resource managers/

biologists/specialists do not desire to be specialists in the techniques of remote sensing or related analysis methods but they do need the information (e.g., a surface cover map) that can be obtained. Conversely, these end users of spatial resource data are requiring local access to GIS technology and consider "hands-on" capability and experience essential. Resource management specialists need "hands-on" access to the GIS and pertinent databases to examine the physical, biologic, and political interrelationships of data themes and to model potential or real outcomes of alternative management decisions. Alternative results must be evaluated using the specialized knowledge of the resource manager and the various constraints.

Additional concerns and needs discussed in audience-to-panel and audience-to-audience interactions included:

* Need user needs/workload assessments at both the specialist and decisionmaking levels.

* Centralized versus decentralized (distributive) data processing and system access is an issue.

* GIS users demand local access to systems, with interactive, "hands-on" capabilities.

* Management needs to provide adequate time for staff utilization of technology and for training.

* Need to also consider the interfacing of textual/ tabular data with systems, not just spatial.

* Standardized data exchange format(s) is needed.

* Raster to vector transformation(s) is needed.

* Standardization of digital data supply is needed.

* Vendors of remote sensing/spatial data analysis systems for resource management need to consider true user needs regarding the manner in which systems are packaged and marketed.

* Federal, State, and private level needs for remote sensing/spatial data systems are, in many cases, very different; perhaps even within the same agency/organization at different levels.

* Issue five years ago was spatial data systems credibility; today the issue is systems accessibility.

Forum participants expressed their appreciation for this assembly discussion opportunity at the Symposium, indicated that they had benefitted from the topic discussion, and stated their desire that similar forums become standard agenda items in future symposia.

PRACTICAL IMPLICATIONS OF OPERATIONAL IMAGE PROCESSING

Moderator
G. Robinson Barker
Champion Paper Company
435 Clark Road, Suite 411
Jacksonville, Florida 32218

Panel Members: Stanley A. Morain - University of New Mexico
Richard P. Mroczynski - EOSAT
Edgar P. Odenwalder - Coffman Technology

It has been over 10 years since the first earth's resource technological satellite (ERTS 1) was launched into low polar sun-synchronous orbit by the National Aeronautics and Space Administration (NASA). This experimental remote sensing satellite system provided spectacular imagery of the earth's surface. Early investigators and NASA both extolled the virtues of the multi-spectral scanner, and the data analysts became almost euphoric with this satellite, now called Landsat, and its technologically new sensor. By 1975, when Landsat 2 was launched, the original perception of an experimental system had all but vanished, and the byword in Government and academic circles was "applications". This led to a plethora of applications investigations variously called Application Systems Verification and Test (ASVT's) and Applications Pilot Test (APT's) funded, at least in part, by NASA. The new world of "high tech" remote sensing had indeed reached a pinnacle from which it has been falling ever since. Some of the fall can be attributed to natural euphoric cooling, typical of new technology, and still more to Government indecision/inaction, which has seriously impinged on the systems credibility. Much of the dulling of the bloom, however, has been caused by the nature of the technology itself, and its relationship to the general user community. Such an overwhelming inundation of data, sophisticated machinery and processing techniques, access options and capabilities presented a mind boggling problem of assimilation. The users initial feeling of seeing something "nice but not available" was unhappily borne out by reality. Data production was sporatic, unreliable, and slow, if at all, processing and georeferencing was costly and not locally available, interactive access at a stand-alone user facility was essentially not even addressed, and the availability of low cost output product generation capability did not and does not now exist.

The total inadequacy of data collection, handling, and distribution under R&D circumstances is well documented, known throughout the user community and by itself is a major factor in the non-acceptance of digital image processing from Landsat data. Time between accession and archive and archive and user are important considerations in the business community. Information needs are driven by the decision making process, both in a strategic and tactical sense, and bear little on matters of geomorphology, rotation lengths, or other feature-specific characteristics. The question is, does the current system represent a viable market ready product in the first place, and if not, can the commercialization of civilian space, with its market driven philosophy, overcome the problem? To define a product, it is necessary to define the user of the product. The level of user sophistication is measured by the degree of data processing he is prepared to do, from the intial reduction to computer compatible tapes, to hand held output products used for on-line production decisions. To get from one end of the user spectrum to the other takes several iterations. This is the function of a value added digital information business. The presence of such a business at this stage is more conceptual than factual.

Unlike the non-renewable resource user, whose needs can be largely filled by one good image per season over the areas of interest, the renewable resource users needs are seasonal, annual, and continual. It is probably safe to say that the future of civil remote sensing depends to a large degree on the cultivation of a viable market within this user community. Agri-business and forestry alone represent a huge potential market for satellite and related technologies. The needs of this market cannot be met by a designed experimental system such as Landsat. As a result of the current systems operational difficulties, the renewable resource community has sat by with benign, if not hostile indifference for over 10 years while the technology has been developing. To change this attitude is a matter of perception: first, that data needs will be met in terms of content and delivery; second, that the data source will be dependable, continuous, and not part of the problem. These are perceptions required by the end user in developing plans and schedules, and the value added and resale business user who cannot afford to "tool up" for less. Thus, the owner operator must have an attractive product for the value added business to even develop, and this yet to be developed business must generate a cost effective product for end use.

The value added community will be largely a data user community generating information of various descriptions through computer based technology. Unlike traditional computer applications such as accounting, where the data input and user needs have been well defined for years, digital image data are just now beginning to be understood and the user has not even perceived of his needs. Defining an information system based upon digital imagery at a cost the user can afford is the real challenge of the value added sector. Unfortunately, such service companies will require personnel of high skill and substantial capitalization to get started. It will take a broad user base indeed to sustain such an operation. Even if remote sensing data and information could be made available to the various user constituencies at an attractive cost, remotely sensed data products depend upon other products for operational use. Such products as geographical referencing capabilities escalate the user costs in many cases beyond their ability to justify them. It would seem that large industrial users with high stake interests in resource information and with technical staffs capable of further data reduction, would be the logical initial users of this technology. But how about the small or individual user representing such a large potential? He needs the same data and geographical control, but on small segments and he has no interest whatever in any form of data processing and analysis. He simply wants to query the system. How much will this group be willing to pay for information and what will his turn around requirements be?

This brings us to the final issue, that of user interaction and access. This is of vital importance to the developing of this market. With the advent, indeed explosion, of microprocessing capabilities and availability to the general public of low cost microcomputer systems, the matter of user interaction does not present the impossible task envisioned just a few years ago. Remote, inexpensive work stations based around these "micros" have already been configured and a wide variety of program libraries have been established and are available. The major issue separating the user community from digital based information systems, are user awareness and training.

With innovation and imagination, commercial space remote sensing will work. The tools to exploit the power of digital resource information systems are in place, and to a large degree already in the hands of the potential user. Data handling and distribution methodology and the development of a value added service industry are waiting in the wings for commercialization to take place, which seems at last, imminent. Perhaps it is time to repackage and reintroduce remote sensing technology to the resource user community, but this time, in the context of operational production.

REMOTE SENSING FOR DETECTION AND MONITORING OF FOREST DAMAGE CAUSED BY ATMOSPHERIC DEPOSITION OF POLLUTANTS

Moderator
William M. Ciesla
USDA Forest Service, FPM/MAG
3825 East Mulberry
Fort Collins, Colorado 80524

Panel Members: Peter Murtha - University of British Columbia
Lew McCreery - USDA Forest Service
Charles Dull - USDA Forest Service

The potential impact of atmospheric pollutants on forest ecosystems is becoming a major nationwide concern. Recently it has been suggested that the occurrence of several forest declines in North America and Europe are, in part, due to atmospheric deposition of pollutants. The objective of this forum was to explore opportunities to assess forest vegetation damage, which may be associated with atmospheric deposition of pollutants, using remote sensing.

This technology has a long history of successful use in assessment of forest damage caused by insects, fungi, virus, animals, and other biotic causal agents. Experience gained from those examples can serve as a basis for its appliction to assessment of damage caused by air pollutants. In addition, the characteristic appearance of forest vegetation damage associated with specific atmospheric pollutants such as sulfur dioxide, ozone, and fluroides from point sources have been described. These characteristics include damage types, the progression of damage, and distribution of damage with respect to such variables as slope, elevation, and proximity to pollutant source.

Remote sensing provides a means to develop baseline data on the status of vegetation damage at a specific point in time. Through repeated surveys, trend data can be obtained.

Work has been either completed or is underway in many parts of the world to assess forest damage in areas where atmospheric deposition of pollutants is suspected to be part of the causal agent complex. Examples of work in Canada, Mexico, and the United States, where some type of remote sensing was used as an intermediate sampling stage, was presented. In each case, the need to examine records of past episodes of damage, the need to identify the role of insects, diseases, climate, and other factors in the damage complex, and the need for a spatial data base to interface data on location of damage with other resource and topographic data, was emphasized.

RESEARCH AND DEVELOPMENT PRIORITIES IN REMOTE SENSING FOR FOREST AND RANGE RESOURCES MANAGEMENT

Moderator
Dr. Paul T. Tueller
Department of Range, Wildlife and Forestry
University of Nevada Reno
1000 Valley Road
Reno, Nevada 89512

Panel Members: Ray Jackson - USDA
Fred Batson - BLM
Roger Hoffer - Purdue University

Research and development for forest and rangeland remote sensing was characterized as science driven, lacking national goals, bandwagon oriented and underdeveloped. There has been significantly less development than research. Basic research has lagged behind applied efforts; thus, there is a great need for development, but in the United States at least, no institution or group has been willing to take on the task. There is also need for additional high risk research, i.e., innovative research that may or may not lead to immediate applicability, but upon which we must build the science of remote sensing as it relates to forests and rangelands. The need for a greater understanding of scene characteristics, including spectral characteristics, was very strongly emphasized in the forum.

Several important research needs were revealed: (1) Further development and understanding of GIS capabilities. (2) Continued evaluation and development of relatively inexpensive data processing with hardware/software systems based on microcomputer technology. (3) Re-evaluation of the utility of "photographic" tools. (4) The quantitative assessment of the amount and condition of vegetation biomass. (5) Increased capability for merging various kinds of data sets. (6) Laser disk technology to assist in the storing and retrieval of large data sets. (7) Spectral characteristics of scene components and of the entire pixel. (8) A better understanding of the information content of data obtained in the microwave region.

A number of specific problem areas were identified: (1) The cost of TM data. (2) A lack of national goals, particularly as related to the question of continued availability of Landsat data. (3) Vegetation canopy structure influences on spectral responses are little understood. (4) Path radiance questions require additional careful consideration. (5) Be cautious in the use of vegetation indices. (6) The resistance to change factor is a real and significant deterrant to development and application. (7) Training and research in photo interpretation must not be neglected. (8) There is a general lack of research support including support of graduate students in remote sensing. (9) Interdisciplinary research should be fostered. (10) A lack of understanding of vegetation phenelogy influences and the effects of vegetation stress. (11) Approaches to vegetation monitoring are not clear. (12) The lack of open and honest communication between researchers and managers.

The need for considerable rigorous highly quantitative research was made clear. Fresh, new outlooks and approaches are required along with a careful sifting-through of existing technology in order to develop and apply the best of this technology to forest and rangeland management problems.

GLOBAL RESOURCE ASSESSMENT

Moderator: G. Thorley

MONITORING THE EARTH - TOO MANY PLAYERS?

Gene A. Thorley
U.S. Geological Survey
Reston, Virginia 22092

BIOGRAPHICAL SKETCH

Gene Thorley is the Senior Advisor for Remote Sensing, National Mapping Division, U.S. Geological Survey, Reston, Virginia. He received his B.S. and Ph.D. degrees in Forestry from the University of California-Berkeley. Dr. Thorley is a member of the Society of American Foresters and the American Society of Photogrammetry and Remote Sensing, and participated in the development of the first and second editions of the Manual of Remote Sensing.

ABSTRACT

A number of organizations are currently engaged in, or proposing to embark on, worldwide measurement/monitoring programs. Program objectives vary in type and complexity, including a form of technical library (the Global Environment Monitoring System of the United Nations Environment Program), and an ambitious experiment to validate algorithms to derive land surface climatological parameters (International Satellite Land Surface Climatology Project).

Modeling and understanding the Earth as an integrated system is an immense undertaking and will require at least as many measurements as are currently acquired. Improved coordination is required among the diverse national and international measurement/monitoring programs to ensure that sensors, launch dates, orbits, and supporting research are complementary to the maximum extent possible. Development of a long-term program of Earth system monitoring/modeling, such as the International Global Change Program proposed by the International Council of Scientific Unions, could provide the necessary focus and structure for effective international coordination and cooperation.

INTRODUCTION

Remote sensing from satellites provides a unique tool to measure the parameters of the Earth on a worldwide scale. These measurements assist in improving our understanding of such complex Earth systems as the atmosphere, biosphere, lithosphere, and hydrosphere. Recent research suggests that satellite measurements, coupled with

---

Publication authorized by the Director, U.S. Geological Survey, August 26, 1985.

appropriate ancillary data, may enable scientists to interrelate these traditionally separate Earth systems in models which treat the Earth as a "holistic" (integrated) system (Bretherton, 1985). Modeling the Earth as an integrated system through the use of satellite measurements may also lead to new scientific insights and fundamental knowledge heretofore not possible due to the limitations of regional and *in situ* measurement capabilities. In addition, an integrated system approach may provide greater understanding of important global issues, such as the cause and effects of changes in ozone, methane, and carbon dioxide in the atmosphere.

GLOBAL MEASUREMENT/MONITORING PROGRAMS

Man has been measuring/monitoring the Earth from space since the early 1960's, initially with sensors designed to acquire data to improve weather prediction, and currently with sensors to determine a wide variety of parameters, including atmospheric constituents and temperatures, reflectance and emittance from land surfaces, and ocean characteristics. A number of organizations are engaged in, or proposing to embark on, worldwide measurement/monitoring programs. A recent publication by the National Research Council (1983) described over 20 ongoing or proposed international measurement/monitoring programs, and the total number is likely to be in the hundreds if national programs are considered. Five examples of international measurement/monitoring programs (two ongoing and three proposed) are provided below.

Global Environment Monitoring System

The Global Environment Monitoring System (GEMS) is an element of the United Nations Environment Program (UNEP). Established in 1974 with headquarters in Nairobi, Kenya, GEMS aspires to collect all available information relevant to monitoring the environment of the globe (analogous to a technical library). A small staff of administrators and scientists acts as a catalyst to encourage cooperation between U.N. agencies and member nations in acquiring and sharing data in standardized formats (Reader, 1982). Subject areas include monitoring climate, atmospheric pollution, renewable resources, and ocean environments. Recently, GEMS proposed the establishment of a Global Resource Information Database (Croze, 1983).

The Global Resource Information Database (GRID) is conceived as a distributed system, with nodes around the world connected by telecommunication links. Initially, two nodes will be established: GRID-Control at the UNEP/GEMS headquarters in Nairobi, and GRID-Processor at UNEP facilities in Geneva, Switzerland. GRID-Control will manage operations worldwide and serve as the principal center for the demonstration and transfer of technology to interested organizations. GRID-Processor will be responsible for data acquisition, analysis, and distribution. The establishment of GRID-Processor in Geneva will enable synergism with neighboring computer facilities in U.N. organizations and universities, and entry to equipment manufacturers for system improvement and maintenance. In addition, GRID-Processor will have convenient access to baseline data sets such as the World Meteorological Organization's Glacier Monitoring Service and Background Air Pollution Monitoring Network (Fanshawe, 1985).

International Satellite Land Surface Climatology Project

The International Satellite Land Surface Climatology Project (ISLSCP) is an ambitious experiment to validate algorithms to derive land surface climatological parameters. Two elements of the International Council of Scientific Unions (ICSU), the Committee on Space Research and the International Union of Geology and Geophysics, are cooperating with the UNEP to sponsor the project. Three international workshops were held in 1983 to define the character and scope of the project. Over a dozen parameters to be obtained through satellite measurements have been identified for input to mesoscale and global models, including leaf area index, total surface albedo, and surface infrared emittance (ICSU, 1983).

Planning is underway for the first ISLSCP field experiment (FIFE) in the 1986/87 time period. A 10 km by 10 km test area, located in the great plains of the United States, is proposed with four intensive 2-week measurement periods during the annual cycle. Validation of algorithms to obtain land surface climatological properties such as rainfall, snow cover, vegetation index, and surface skin temperature would be attempted (Ohring and Sellers, 1984).

BIOMAP Project

The objective of the BIOMAP project is to geographically locate existing plant associations or bioecosystems and to identify them by the scientific names of their characteristic species. The project is sponsored by the International Union of Biological Sciences (IUBS), an element of ICSU. The project is distinguished from related projects (for example, ISLSCP) by its recognition of geographical distributions of species, which is often not considered when measuring leaf area index or biomass, and its emphasis on complete areal coverage of the biosphere, rather than sampling or extrapolations of ground observations (Baer and Jones, 1984). Activities to identify the character, scope, and funding requirements for BIOMAP are underway for review at the September 1985 IUBS General Assembly.

Global Habitability Program

The Global Habitability Program is an extension of a United States initiative presented at the 1982 United Nations Conference on Exploration and Peaceful Uses of Outer Space (UNISPACE). The United States proposed an international cooperative research program to (1) address changes, either natural or of human origin, affecting the habitability of the Earth, and (2) to acquire the scientific understanding of the Earth as a system (NASA, 1984). Subsequent to the UNISPACE meeting, the National Aeronautics and Space Administration (NASA) has solicited support from other Federal agencies for the program, including naming individuals to serve on a national steering committee. Currently, the term "Global Habitability Program" denotes interdisciplinary research performed under the auspices of the NASA Associate Administrator for Space Science and Applications on the interactions between the major Earth systems, such as the atmosphere, lithosphere, hydrosphere, and biosphere (Tilford, 1984).

International Global Change Program

The International Global Change Program (IGCP) is an emerging program sponsored by the ICSU. An outgrowth of a 1983 U.S. National Research Council Workshop, IGCP is conceived as a long-range, cooperative, interdisciplinary scientific effort to describe the subtle and complex connections that tie the geosphere and biosphere together (NRC, 1983). A resolution to begin program planning was approved at the recent ICSU General Assembly meeting in Ottawa in September 1984. Implementation of the program is targeted for the early 1990's, with research activities perhaps spanning the entire decade.

A U.S. committee to support the program planning process has been formed and has suggested global studies in six areas: (1) studies of lake sediments, ice cores, and tree-rings to recover the record of past environmental change; (2) investigations of the physical environment of the Earth, including the atmosphere and oceans, snow and ice fields, and the solar-terrestrial system; (3) studies of global ecology and the interactions of biota with the physical environment of the Earth; (4) studies of the cycling of chemicals important to life in the air, on land and water, and in living forms, including compounds such as carbon dioxide, methane, and other greenhouse gases; (5) studies of the global cycling of water and the reservoirs such as icecaps and underground aquifers; and (6) studies of the soil reserves of the Earth, their erosion and formation, and the impact of intensive agriculture on the human carrying capacity of the planet (Eddy, 1985).

TOO MANY PLAYERS?

Modeling and understanding the Earth as an integrated system is an immense undertaking and will require at least as many measurements as are currently obtained. Viewing the Earth as a holistic system (paying particular attention to the interface boundaries and feedback loops of the individual subsystems) may allow the identification of gapsin current knowledge, theories, and measurement capabilities. Thus, rather than "too many players," additional measurements and research will be required to augment current programs and strengthen activities where gaps appear to exist.

In addition, improved coordination is required among the diverse national and international measurement/monitoring programs to ensure that sensors, launch dates, orbits, and supporting research are complementary to the maximum extent possible. Presently, such coordination is principally carried out by a number of satellite mission-specific groups, including the Coordination Group on Geostationary Meteorological Satellites, the International Polar Orbiting Meteorological Satellite Group, the SPOT Station Operators Group, and the Landsat Ground Station Operators Working Group. An important shift from the mission-specific approach to interdisciplinary coordination of multimission programs was recently made with the formation of the Committee on Earth Observation Satellites (replacing the groups for Coordination of Land Observation Satellites and Coordination of Ocean Remote Sensing Satellite) (National Environment Satellite Data and Information Service, 1984).

Most importantly, a focus and structure are needed which can build upon the existing national and international measurement/monitoring programs and coordinating groups, and interrelate the measurements, systems, and research to provide the foundation for a scientific program for modeling and understanding the Earth as an integrated system. The ICSU-sponsored International Global Change Program has the potential to provide this necessary focus and structure for effective international coordination and cooperation.

ICSU comprises 20 scientific unions representing all scientific disciplines and 71 national academies, and is recognized as the world's nongovernmental organization of scientists. Representatives from nations with differing political alliances (Soviet, Western, Third World, etc.) meet, communicate, and cooperate in scientific endeavors with a minimum of political overtones and influence.

During the past quarter century this world community of scientists has carried out a number of international global-scale research programs, including the International Geophysical Year, the Global Atmospheric Research Program, the International Lithosphere Project, and the International Biological Program. These programs, however, assessed a single Earth system (for example, lithosphere, atmosphere) and were of limited duration. The proposed global change program, a multidisciplinary program focusing on the interrelationship of the numerous Earth systems over a period of a decade or more, will represent a considerable challenge to plan, coordinate, and implement.

REFERENCES

Baer, D. F., and Jones, L. T., 1984, BIOMAP Project Summary: Rockwell International Space Transportation Systems Division, Downey, CA, 7 p.

Bretherton, F. P., 1985, Earth system science and remote

sensing: National Center for Atmospheric Research, Boulder, CO, 25 p.

Croze, H., 1983, GEMS - How the Global Environment Monitoring System works, in International Conference on Renewable Resource Inventories for Monitoring Changes and Trends, 1983, Proceedings: Corvallis, Oregon State University, p. 520-524.

Eddy, J. A., 1985, International Geosphere-Biosphere Program: Statement before the Subcommittee on Space Science and Applications, U.S. House of Representatives, February 26, 1985, 12 p.

Fanshawe, J., 1985, Global Resource Information Database: Brochure, United Nations Environment Program, Nairobi, Kenya (in press).

International Council of Scientific Unions, 1983, Development of the implementation plan for the International Satellite Land Surface Climatology Project (Phase I): Report to the United Nations Environment Program, Project Number FP/103-83-03 (2426), 113 p.

National Aeronautics and Space Administration, 1984, Global habitability: Brochure, National Aeronautics and Space Administration, Washington, D.C., 7 p.

National Environment Satellite Data and Information Service, 1984, Terms of reference on the Committee on Earth Observation Satellites: National Oceanic and Atmospheric Administration, Washington, D.C., 6 p.

National Research Council, 1983, Toward an International Geosphere-Biosphere Program: Report of Commission on Physical Sciences, Mathematics, and Resources, National Academy Press, 81 p.

Ohring, G., and Sellers, P., 1984, Report of the Design Workshop for the First ISLSCP Field Experiment (FIFE), International Satellite Land Surface Climatology Project: National Center for Atmospheric Research, Boulder, CO, 56 p.

Reader, J., 1982, Global Environment Monitoring System (GEMS): Brochure, United Nations Environment Program, Nairobi, Kenya, 32 p.

Tilford, S. G., 1984, NASA's role in the International Geosphere-Biosphere Program: Statement before the Subcommittee on Space Science and Applications, U.S. House of Representatives, September 13, 1984, 8 p.

THE RELATIONSHIP OF GLOBAL GREEN LEAF BIOMASS TO ATMOSPHERIC CO_2 CONCENTRATIONS

C. J. Tucker
NASA/Goddard Space Flight Center
Greenbelt, Maryland 20771

I. Y. Fung
NASA/Goddard Institute for Space Studies
New York, New York 10025
and
Lamont-Doherty Geological Observatory
of Columbia University, Palisades
New York, New York 10964

C. D. Keeling
Scripps Institution of Oceanography
La Jolla, California 92093

R. H. Gammon
NOAA/Pacific Marine Environmental Laboratory
Seattle, Washington 98115

ABSTRACT

Advanced Very High Resolution Radiometer (AVHRR) data from the National Oceanic and Atmospheric Administration's polar orbiting meteorological satellite have been obtained globally for a 19-month period, processed to produce a green leaf biomass spectral vegetation index for the entire terrestrial surface by month, zonally aggregated by latitude, and compared to atmospheric CO_2 concentrations from observing stations. A strong inverse association was found between the monthly Pt. Barrow CO_2 concentrations and the vegetation index measurements from 50°N - 80°N, between the monthly Mauna Loa CO_2 concentrations and the vegetation index measurements from 10°N - 30°N, 10°N - 80°N, and the global total, and between the globally averaged CO_2 concentrations and the globally averaged vegetation index. No relationships between atmospheric CO_2 concentrations for the South Pole station and the vegetation index measurements from any latitude zone or combinations of zones were found.

QUANTITY AND AREAL DISTRIBUTION OF SOIL ORGANIC CARBON RELATED TO VEGETATION COVER CLASSES FROM LANDSAT DATA

Paul J. Zinke
Edward Katibah
Alan Stangenberger
College of Natural Resources
University of California-Berkeley
Berkeley, California 94720

ABSTRACT

The soil carbon storage of a large region in the central Sierra Nevada was determined using a combination of data obtained from laboratory analyses of soil samples typical of remote sensing derived strata of the landscape based upon soil related vegetation cover classes. A landsat scene was censored to include only granitic rock, keeping soil parent material constant. The pattern of vegetation types related to local soil conditions and their areal extent were used as the parameters to be quantified through remote sensing data. The dominate vegetation cover classes were classified into band 7 to band 5 ratio value increments. Organic carbon analytical data from numerous soil profiles in the area were used to calculate weight of soil carbon representative of the vegetation cover classes. Based upon the areal distribution and the mean carbon storage of each class, the total soil carbon storage for the class was calculated. The regional carbon storage was calculated from the sum of the class amounts. Global soil carbon content estimates would be determined by a similar summation of regional storage estimates. (Project supported by National Science Foundations Ecosytem Studies program, Interagency Agreement No. DEB 81-15316 with the U.S. Department of Energy under contract W-7405-eng-26 with Union Carbide Corporation, and by Project 2937ms California Agricultural Experiment Station, University of California-Berkeley.)

INTRODUCTION

There is a need to determine total soil organic carbon storage on a global basis in models of the global carbon cycle to answer questions regarding ultimate sinks for increasing atmospheric CO_2. The remote sensing data from orbiting satellite instruments offers global coverage and the potential of accurate estimates of soil conditions correlated with vegetation cover classes. Total soil organic matter storage is the result of the balance between the fixation of CO_2 as organic carbon by vegetation, and subsequent respiration of CO_2 by the soil macroflora (vegetation) and microflora. Organic carbon contents of soils have been determined for years in the conduct of local soil surveys. These data can be grouped in strata which are related to soil cover conditions identifiable by remote sensing data. Remote sensing data are based upon reflected radiation from the soil cover of vegetation, vegetal detritus or bare ground. The soil cover as idicated by these data offers a potential for stratifying landscape units related to soil conditions.

This paper presents the results of a study in which remote sensing data coupled with field and laboratory data of soil carbon contents were used to determine the total carbon storage of a large region in the central Sierra Nevada of California.

METHODS

The study site chosen was a full scene of Landsat data for an area extending diagonally from Fresno, California, to the vicinity of Mono Lake, California. (Landsat Scene ID 11495-1715, 26 August 1976.) The objective was to estimate total soil organic carbon storage for one geologic soil parent material stratum in the scene.

Soil Carbon Storage from Field and Laboratory Data

The basic approach for estimating total soil mantle organic carbon storage is essentially to multiply the carbon storage per unit soil area, the intensity factor, by the capacity factor, the areal extent of the soil mantle concerned as follows:

$$Qc = Ic * Ac$$

where:

Qc = Quantity of soil organic carbon $kg/m^2/m$

Ic = Intensity factor; soil carbon as kg per sq. meter

Ac = Capacity factor of areal extent of cover class

The considerable variation in total soil mantle carbon for such a region was subdivided by making the determination for individual cover classes. The total carbon storage for each area is the sum of the carbon contents calculated for each cover class stratum, as follows:

$$\overline{Qc} = \sum_{i=1}^{n} Ic_i \cdot Ac_j$$

$\overline{Qc}$ = Total carbon in entire soil mantle Ic_i = Carbon storage intensity in stratum i Ac_i = Areal extent or soild capacity in stratum i n = number of strata i = stratum index

The intensity factor of carbon storage in each stratum was obtained in utilizing data from field studies and from soil surveys in the Sierra Nevada for soils derived from granite. Data for 165 soild profiles sampled only on granite parent rock in the area, and analyzed for carbon content were used. The carbon contents of the soils were calculated on a square meter basis one meter deep, for each sampling horizon in the profile. The soil profiles were classified according to the associated vegetation cover class and central tendencies for soil carbon storage in each class were calculated means as presented in table 1. Probability distributions (Weibull, 1949) were fitted to the soil carbon storage for each cover class and are shown in figure 1.

Landscape Strata from Remote Sensing Data

The estimation model was expanded by using remote sensing data to provide areal extent estimates of landscape classes within specific strata as follows:

$$Qc = \sum_{i=1}^{n} \sum_{j=1}^{m_i} \cdot \; Ic_{ij} \cdot Ac_{ij}$$

where:

Qc = total soil mantle carbon

Ic_{ij} = soil carbon storage intensity of jth cover class in the ith stratum

Ac_{ij} = areal extent of the jth cover class in the ith stratum

n = number of strata

i = stratum index

m_i = number of cover classes within the ith stratum

j = cover classification index

Classification of Sample Landsat Blocks

The landsat data within the granitic area in the chosen scene was obtained by censoring out all data outside boundaries determined from the State Geology Map of California, thus confining the assessment to one soil parent material. Sample blocks were randomly selected within this area for detailed classification. Nineteen sample blocks were chosen identified using Landsat pixel coordinates; each sample block represented an area of 15,000 Landsat pixels. These represented 8% of the total area of granite in the image, or 3,608,679 pixels.

Sample blocks were identified on Landsat Multispectral Scanner (MSS) digital data for classification into vegetation cover classes for which soil carbon distributions had been made. The entire set of Landsat data representing the sample blocks was processed by a linear classification algorithm, chosen for speed and ease in interpretation. The linear model is as follows:

$$CSV = \frac{2 \times SR_{mss}\ 7}{SR_{mss}\ 5}$$

where:

CSV = classification spectral value

SRmss7 = spectral response of Landsat MSS Band 7

SRmss5 = spectral response of Landsat MSS Band 5

This is the common '7-5' ratio model. It was used as a first order classification method to rapidly classify the data for each random block. The sample blocks were individually analyzed for vegetation classes. Based upon the analyst's experience with the area, and existing California Soil-Vegetation Survey maps, groups of '7-5' ratio values were identified for the six vegetation and land cover classes: 1) bare granite, 2) high density conifer, 3) low density conifer, 4) hardwood and grass, 5) brush, and 6) water and shadow. Meadowland and seep areas although represented in the soil categories were not of enough areal extent to be represented.

Table 2 shows the distribution of pixels for each of the 19 sample blocks in each of the six land cover classes, based upon classification of the Landsat digital data.

Total Regional Soil Carbon Storage

The data for total storage in granitic portion of the central Sierra Nevada are shown in table 3. The total storage was 125 X 10^6 metric tons for a mean areal carbon storage of 7.62 kg/m^2 to a meter depth. The confidence interval of the estimate was calculated from the sample plot data to be plus or minus 3.54% of the mean.

This is a lower carbon storage than was obtained in a separate study conducted by H. Jenny in connection with this study in which no remote sensing data were used, and the average of all field soil data were used to calculate the carbon storage for the same area. These data are also presented in table 3a and estimated a greater storage amount with a high variability.

The carbon storage value of 7.62 kg/m^2 is lower than a worldwide estimate of 10.9 kg/m^2 we (Zinke, et al, 1979) made from 2356 granitic soil profile data assembled from the world literature on soils analyses. However, we feel that the lower value presented for the remote sensing weighted data take into account the large glaciated area along the crest of the Sierra Nevada where carbon storage is at a minimum due to juvenile soils or bare rock and lack of vegetation.

This study had two main applications of remote sensing data using the 7/5 data from Landsat multispectral scanner; one indentifying soil cover classes, and the other in determining the areas of the classes. The addition of an array of new instruments in various polar orbiting satellites giving many more spectral channels as proposed in the Earth Observing System for the 1990's (U.S. NASA, 1984) may give data which will allow a fine tuning of the classifications that can be applied to soil cover classes.

Ultimately, estimates of soil carbon storage from remote sensing data will be of most value when it will be possible to compare soil storage values obtained with a time base of at least a decade so as to estimate the direction of change in soil carbon storage as an element on the global carbon cycle.

BIBLIOGRAPHY

California State Division of Forestry. 1969, Soil Vegetation Surveys in California Quadrangles 89, 90, 91, 99, and 100 Fresno County: Individual Map Sheets 16 per Quadrangle Resources Agency, State of California, Sacramento, California

U.S. National Aeronautics and Space Administration. 1984, Earth Observing System: NASA Tech Memo 86129, Vol. 1, parts 1, 2, 58 p., Washington, D.C.

Weibull, W. 1949, A Statistical Analysis of the Size of Cytoideae in Albatross Cores from the East Pacific Ocean: Nature 164, pp. 1047-1048

Zinke, P.J., Stangenberger, A.G., Post, W.M., Emanual, W.R., Olson, J.S. 1984, Worldwide Organic Soil Carbon and Nitrogen Data: Oak Ridge Nat. Lab. Envir. Sci. Div., Oak Ridge, Tennessee, Publ. no. 2212, 141 pp.

Figure 1. Weibull Cumulative Density Function
Soil Carbon Cumulated to 1 Meter
Fresno Test Site Legend Units

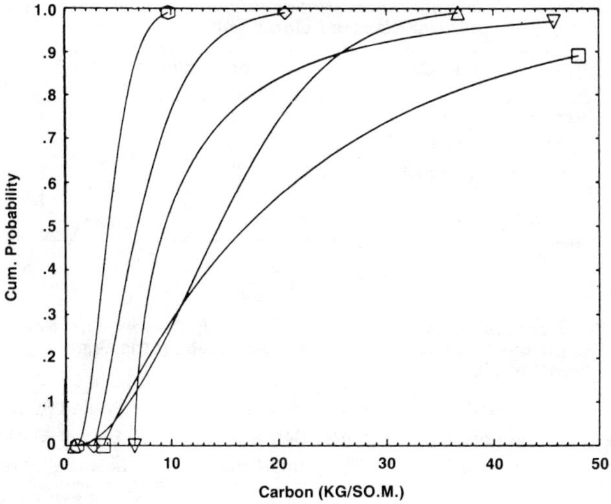

△ High density conifer n = 66
◇ Low density conifer n = 45
⬡ Hardwoods. Oak-grass n = 22
▽ Brush soils n = 7
□ Meadows. Seeps. etc.

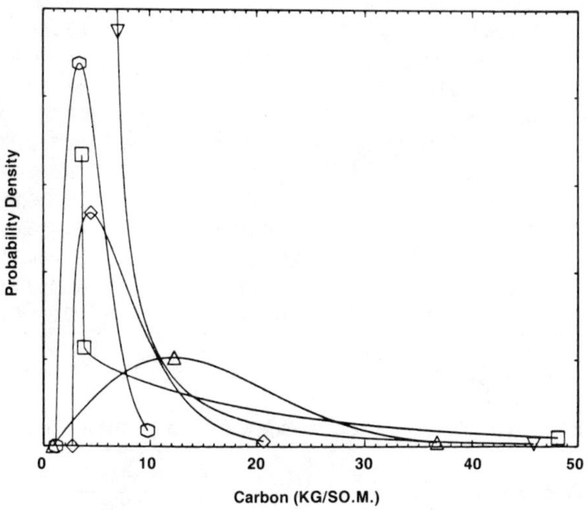

△ High density conifer
◇ Low density conifer
⬡ Hardwoods. Oak-grass
▽ Brush soils
□ Meadows. Seeps. etc.

Table 1. Means of Carbon Storage Amounts in Soils Associated with Cover Classes Determined with Landsat Data

Cover Class a	Carbon Kg/m²/m depth MODE	MEAN
High density conifer	12.28	15.39
Low density conifer	4.39	7.74
Hardwoods-oak-grass	3.38	4.35
Brush	6.56	11.63
Meadows-seeps	3.61	23.23

Table 2. Summary of Estimates of Terrestrial Organic Carbon in the Soil by Sample Block, Including Expanded Estimates for the Granitic Parent Material Stratum

Sample Block	Mean Carbon Storage (Kilograms)	Mean Carbon Storage Per Unit Area (Kilograms/Meter²)	Pooled Standard Error (SE) of the Mean (Kilograms)
4	288,876,167.2	4.33	43,519,730.46
7	375,399,985.8	5.62	35,155,737.68
10	351,144,198.7	5.26	36,879,137.90
16	553,822,077.5	8.29	28,096,670.17
23	667,226,849.5	9.99	30,042,249.02
24	560,044,994.0	8.39	26,281,973.85
27	695,427,064.1	10.41	33,287,569.11
30	254,073,235.9	3.80	48,300,025.18
48	762,625,863.7	11.42	35,855,547.65
51	759,714,879.0	11.38	34,246,274.13
55	164,077,881.2	2.46	57,668,906.71
56	599,898,897.4	8.98	28,304,400.41
58	717,500,374.2	10.74	32,582,574.97
63	486,544,477.4	7.29	35,367,796.04
69	656,440,499.4	9.83	28,818,827.60
84	280,589,570.5	4.20	46,498,053.44
88	616,196,957.0	9.23	27,644,317.39
110	192,208,199.4	2.88	51,150,436.22
113	683,459,642.9	10.23	30,371,616.26

Σ 9,665,271,788 $\qquad \sqrt{\Sigma SE^2}$ 162,816,787.1

Expanded Estimates of Mean Carbon Storage and Pooled Standard Error of the Mean:

$$\text{Expansion Factor: } \frac{3{,}680{,}679}{285{,}000} = \frac{\text{Total Pixels in Stratum}}{\text{Total Pixels in Sample Blocks}}$$

Expanded Estimate of Mean Carbon Storage: .12482373 X 10^{12} Kilograms

Expanded Estimate of Standard Error of the Mean: .00210272 X 10^{12} Kilograms

Table 3. Remote Sensing-Aided Estimate of Terrestrial Organic Carbon in the Soil by Vegetation Class and Cumulatively for the Granitic Parent Material Stratum

Vegetation Class	Percent Areal Extent of Stratum	Estimated Pixel Count	Estimated Areal Extent ($X 10^{10}$ meters^2)	Mean Carbon Storage (Kilograms/Meter^2)	Estimated Total Mean Carbon Storage ($X 10^{12}$ Kilograms)
Bare Granite Soils	29.5	1,086,446	0.4836857761	2.00	.009673715
High Density Conifer	25.1	925,129	0.4118674222	15.49	.063798264
Low Density Conifer	24.4	897,995	0.3997874957	7.76	.031023510
Hardwood	10.5	384,702	0.1712693242	4.39	.007518723
Brush	8.3	306,090	0.1362714601	9.40	.012809517
Water and Shadow	2.2	80,317	0.0357568128	—	—
TOTALS	100.0	3,680,679	1.6386383		.12482373

Table 3a. A Summary Comparison Between Remote Sensing and Non-Remote Sensing-Aided Estimates of Terrestrial Organic Carbon on the Soil for the Granitic Parent Material Stratum

		Non-Remote Sensing-Aided Approach		Remote Sensing-Aided Approach	
		Overall (Kilograms)	Unit Area (Kilograms/Meter^2)	Overall (Kilograms)	Unit Area (Kilograms/Meter^2)
	Estimate of Mean Terrestrial Organic Carbon Storage	$.17861157 \times 10^{12}$	10.90	$.12482373 \times 10^{12}$	7.62
95% Confidence Interval	"Low End" Estimate of Terrestrial Organic Carbon Storage	$.15852613 \times 10^{12}$	9.67	$.12040591 \times 10^{12}$	7.35
95% Confidence Interval	"High End" Estimate of Terrestrial Organic Carbon Storage	$.19869702 \times 10^{12}$	12.13	$.12924155 \times 10^{12}$	7.89
95% Confidence Interval	Confidence Interval Fluctuation Around the Estimate of the Mean (Expressed as a Percentage of the Mean)	± 11.25		± 3.54	

MONITORING TROPICAL FORESTS FROM SATELLITE AND AIRCRAFT PLATFORMS: SOME LIMITATIONS AND NEW APPROACHES

Steven A. Sader, Armond T. Joyce, Robert B. Waide*, and William T. Lawrence*
NASA - Earth Resources Laboratory
National Space Technology Laboratories
NSTL, MS 39529

ABSTRACT

Tropical forests have great importance with respect to the global energy balance, climate, geochemical cycling and various beneficial resources they provide to mankind. The complexity of mixed tropical forest has little parallel with the coniferous forests of North America where much of the previous remote sensing applications have been demonstrated. Frequent cloud cover, inaccessible terrain, data acquisition problems, and closed, diverse forest canopies with high leaf area indices create challenges for remote sensing studies in tropical regions. Studies in Puerto Rico with aircraft acquired TMS have shown spectral differences related to forest successional stages and mountain life zone communities. In consideration of the persistent cloud cover over tropical regions, different polarizations of aircraft L-band, Synthetic Aperture Radar (SAR) data were analyzed to determine the relationship between SAR backscatter and forest structure parameters. A multisensor approach combined with field studies are presented to address ecological studies in tropical environments. Remote sensing techniques are viewed as technology with great potential to contribute new information and scientific understanding of critical global environmental issues in tropical regions.

INTRODUCTION

Remote sensing investigations in tropical regions are not a direct transfer of technology or extension of routine developed from temperate or boreal forest environments. The physical and cultural environment in tropical regions and the distance from U.S. work stations can have a major impact on data acquisition and analysis phases of a remote sensing study. Although there are a few difficulties that may be encountered when conducting remote sensing surveys and research in tropical environments, there is great opportunity for advances in tropical forest and global monitoring techniques through remote sensing.

SOME PITFALLS ENCOUNTERED IN TROPICAL FOREST REMOTE SENSING INVESTIGATIONS

Consider the optimum conditions for conducting a remote sensing investigation: level terrain; homogeneous cover

* Dr. Waide and Dr. Lawrence are with the Terrestrial Ecology Division, Center for Energy and Environmental Research, University of Puerto Rico, San Juan, Puerto Rico.

type and soil surface; clear, dry atmosphere; easy access to the site; good communication systems; good baseline data (historical photo archives, inventory data, good maps). Now consider the other extreme or conditions that make a remote sensing investigation more difficult: broken terrain; multistoried forest canopies; perpetual cloud cover; poor access; difficult logistics; few maps or baseline data; high humidity; an array of poisonous plants, insects, and snakes. Welcome to the tropics!

Cloud Conditions

The most troublesome feature for remote sensing studies in tropical wet and rain forest life zones is the ubiquitous nature of cloud cover. A recent computer search of Landsat 4 and 5 Multispectral Scanner (MSS) and Thematic Mapper (TM) scenes available for the American Tropics and Caribbean countries (about 140 possible scene locations covering land) indicated an average cloud cover of 23% for the best available scenes. Cloud shadows further diminish the percentage of useable data. The data with the least clouds were collected during the "dry" season that extends roughly from December to early April.

Satellite Data Acquisition

Another practical problem concerning Landsat data acquisition occurs when the study area is outside of ground receiving station range. Both Puerto Rico and Costa Rica, the locations of NSTL tropical forest research, are not within range of any receiving station. Consequently, the data must be transmitted through the tracking data relay satellite system (TDRSS-EAST). The TDRSS was in an intermittent data transmission mode for several months during 1984 and 1985 to accommodate an overburdened load of data coming through all METSAT and some military satellites. The probability of obtaining good quality data (low cloud %) for many tropical land areas was further reduced because each overpass was not being transmitted to TDRSS. At the present time, the only way to guarantee Landsat TM data acquisition for areas outside the range of receiving stations is to place a standing purchase order at EROS Data Center, and the first scene with less than 30% cloud coverage will be purchased and mailed. Data acquisition may also be arranged through other governments (e.g., Brazil- INPE).

The advanced very high resolution radiometer (AVHRR) data is not normally archived for tropical regions. Special arrangements must be made with NOAA, to secure the imagery. The AVHRR data is very inexpensive compared to Landsat, however specialized software is required to process it. Several corrections techniques may be necessary to correct for variation in look angle, sun angle effects, cloud contamination, atmospheric attenuation and non-lambertian reflectance (Piwinski, et al., 1983; Tucker, et al., 1985).

Aircraft Data Acquisition

Aircraft missions are very expensive because tropical study sites are usually thousands of miles from flight operation centers. Aircraft missions to unfamiliar countries must be carefully planned and coordinated far in advance. Communications can be difficult for flight crews if there is not at least one crew member who can speak the native language. All flight clearances for within-country flights should be verified prior to the day of the mission. Assistance in obtaining flight clearance through the U.S. Embassy should be solicited. In any event, the State Department will want to be informed of program goals and principal U.S. and foreign agencies involved.

Access to Study Areas

Posner, et al. (1983) estimated that 52 percent of the American Tropics (excluding Mexico, Belize, Brazil, Cuba and Guyana) have slopes of 8% or greater or are located in intermontane valleys and highland plateaus. Considerable travel time is needed due to the narrow, winding, poor quality, secondary roads especially in mountainous areas. In Costa Rica, scientists traveling to or from La Selva field station to San Jose are occasionally delayed for several hours on roads blocked by landslides and mudflows in the lower montane and montane rain forest life zones along the route. In Puerto Rico, in the Maricao Commonwealth Forest, 3 hours of steady hiking and hacking a line through secondary forest were required for Forest Service and NASA crew members to reach a Forest Service inventory plot only 1-1/2 km from the road access point. Average slope gradient in the Maricao Forest was estimated to be greater than 60 percent (Birdsey and Weaver, 1982). Large expanses of lowland tropical forest (eg., amazon basin) are accessible only by river or foot trail. For practical reasons, field plots are often biased to more accessible areas.

REMOTE SENSING APPROACHES TO TROPICAL FOREST MONITORING

Remote sensing techniques coupled with computer technology (geographic data bases and modeling approaches) offer the best opportunity to systematically monitor changes and trends in the tropical (frost-free) environment. It is unlikely that any one type of sensor could provide all of the data needed to address important ecological processes in tropical regions. To accommodate the prodigious land area included in tropical surveys requires coarse resolution sensors (to reduce data volume) that have high temporal data acquisition rates (e.g., AVHRR). Higher resolution sensors (MSS, TM, mapping cameras) are required, to capture the spectral and spatial detail needed to link intensive ecological field studies to the forest community and biome levels. In perpetually cloud-shrouded life zones, where sensors that operate in visible and infrared wavelengths have limited utility, microwave sensors may provide information about the land surface and forest canopy that would otherwise be unobtainable. Sun angle effects and topographic shadows cast in steep topography are apparent in Landsat data because the satellite pass occurs around 9:30 local time when sun angles are low. Misclassification of surface cover will result if topographic corrections are not taken into account. Some recent approaches to the investigation of the tropical forest environment utilizing remote sensing technologies are presented.

Deforestation Studies

There are many uncertainties concerning rates and trends of tropical "deforestation". Deforestation can be defined as the temporary or permanent removal of forest cover for agricultural or other purposes (Grainger, 1980). Tropical forest clearing rates differ by more than 100% (Brown and Lugo 1980). There is no quantifiable basis to determine the reliability of the "best available data" concerning forest clearing rates. There is no standardized system as yet, however, the 1981 FAO survey incorporated remote sensing interpretations derived from Landsat, radar systems and aerial photography in an inventory conducted for 52% of the tropical forest biome (Grainger, 1982). Several authors have demonstrated the utility of remotely sensed data for detecting tropical forest change (Williams et al. 1979; Sader, 1980; Green, 1982; and Tucker et al., 1984).

Studies in the Amazon basin (Tucker, et al., 1984) have shown that recent forest clearing can be detected with AVHRR data. Forest classification using AVHRR performed as well as Landsat MSS data in large homogeneous areas, but detection of smaller parcels were unsuccessful as a result of the coarse resolution of local area coverage (LAC - 1 km) AVHRR data (Graven, et al., 1985). However, the AVHRR data may become an important tool to locate tropical forest areas as the first stage in a nested multistage sample design. More detailed measurements of forest change and characteristics could be made at second and tertiary stages with higher resolution sensors.

In Costa Rica, clearing of primary forest was related to various biophysical landscape characteristics and transportation network. The total area of primary forest decreased from approximately 68% in 1940 to only 18% in 1983. Quantitative information on these relationships will form the basis for predictive modeling of changes in the forest environment in tropical regions (Sader and Joyce, 1985).

Secondary Forest Succession Studies with Near Infrared/Red Ratios

One aspect of tropical forest research for which remote sensing could make a contribution is to improve the information needed to model the exchanges of terrestrial carbon with the atmosphere. Existing global carbon models that treat biomass accumulation linearly in the early successional stages may not be an accurate depiction in secondary tropical forests where growth is faster than in temperate zones. There is general agreement in the literature that definite stages of early tropical forest succession exist and their composition and structure are relatively easy to define (Budowski, 1965; Ewel 1980). The importance of small successional gaps in the dynamics of tropical forests has generated a considerable amount of interest among tropical ecologists (Schulz 1960; Budowski 1965; Webb et al;, 1972, Knight 1975; Whitmore 1975, 1978; Hartshorn 1978, 1980;

Denslow 1980; Janos 1980; Foster and Brokaw 1982; Brokaw 1982). However, the frequency and distribution of gaps is poorly known for most tropical forests (Denslow 1980). The development of remote sensing tools could be of potentially great benefit in understanding the importance of gaps in the structure of tropical forests.

Preliminary studies of secondary forest succession at Luquillo experimental forest in Puerto Rico indicate that near IR/red ratios (10 meter TMS data) offered better separation of forest successional age classes than did individual TMS channels and multichannel combinations. Older age classes had higher ratios than early successional stands (Fig. 1). Research is underway to determine if the higher ratios in older successional stages can be attributed to higher leaf area index (LAI) or other characteristics of the canopy. Maximum LAI in the Tabonuco forest in Puerto Rico is around 7.3 with a mean of 6.3 (Brown and Lugo, 1983). These investigations conducted by the authors of this paper, through the NASA Landsat TM research program will test the capabilities of TM and TMS data to detect spectral differences between early successional stages and the manner in which those changes can be related to above-ground biomass accumulation in tropical secondary forests.

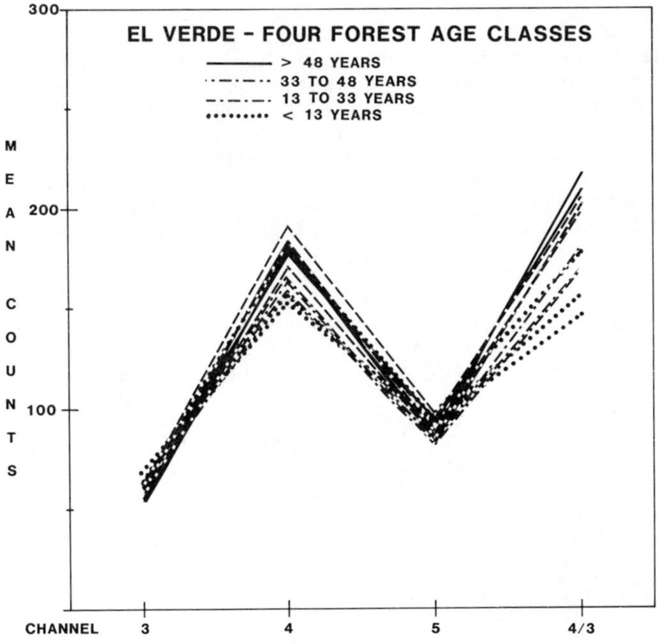

Figure 1. Mean Spectral Response of Thematic Mapper Channels for Subtropical Forest Successional Age Classes - Puerto Rico.

Ecological Stratification Using Near IR/Red Ratios and Terrain Data

The near IR/red ratio technique also improved the delineation of four subtropical forest communities in another Luquillo Forest research site. Sierra Palm (Prestoea montana) had the highest ratios followed by the Tabonuco type, Palo Colorado type and the drawf community on the ridge tops. These forest communities represent ecological gradients that are strongly influenced by elevation, climate and soils. Tree height, number of species, average tree diameter, basal area and above-ground phytomass, all decrease with increasing elevation and only number of stems increases. Topographic stratification approaches are being tested along with channel ratio techniques to identify zones of similar biological productivity and response.

Microwave Studies of Forest Structural Parameters

Investigations using microwave sensors are of particular interest for tropical forest monitoring because active microwave systems can penetrate clouds and record the backscatter returned from surface features. There is little evidence that Synthetic Aperture Radar (SAR) systems are capable of separating species differences except for where there are morphological differences in vegetation types (e.g., conifer vs. hardwood) or where site characteristics (i.e., moisture regime) are major determinate of forest type.

The challenge is to deconvolve the backscatter coefficient (σ) into amounts that can be attributed to vegetation structure from the amount influenced by physiography and moisture content at the surface. Plant height, density, geometry and presence or absence of standing water affect the SAR response. Also incidence angle and polarization can have a significant effect on the radar return signal (Ormsby et al., 1985; Wu, 1984; Hoekman, 1985 ; Pampaloni and Paloscia, 1985).

L-band SAR data was acquired over dense, southern Mississippi, bottomland hardwoods. Green leaf area and above ground biomass in these forests approach the levels that are present in some tropical forest life zones. Digital values using different polarizations and ratioed polarizations were compared to stand structure measurements (average height, average basal area, average diameter breast height, number of stems/acre) and biomass estimates collected in the field. Multipolarized L-Band SAR data has also been acquired over study sites in Costa Rica, and analysis of these data are underway.

Multisensor Approach to Ecological Studies in Tropical Forest Environments

There are limitations in the use of greenness indices (near IR/red ratios) when applied to some tropical wet and rain forests. First, the indices reach an asymptotic or saturation response at higher green leaf densities (Tucker et al., 1985), and the greenness indices and other reflectance measures do not account for respiration losses: maintenance and night respiration. Nor does greenness account for changing non-leaf biomass, and allocation of photosynthate above and below ground which are important variables in the estimation of net primary productivity (NPP) rates.

The respiration and allocation measurement may be addressed with the combination of three remote sensor systems and ground studies. First, studies might be conducted to determine if there is a linear relationship between greenness (using Barnes radiometer* measurements and the TM or TMS near IR/red ratio) and photosynthesis or photosynthetically active radiation (using the Li Cor gas exchange system on several dominant canopy species). Because temperature (day, night, and monthly) is important in determining respiration rate, TIMS, a thermal scanner operating in six narrow bands partitioned in the 8 to 12 micron region, could be used to study diurnal changes in radiant energy. There is evidence in predominately tropical evergreen forests that water vapor content of canopy-air is slightly higher than above the canopy at night, including unusually open stomata in darkness (Aoki, et al., 1974). This phenomenon indicates that the vegetation may be transpiring water vapor and is important in understanding nighttime respiration. Nighttime measurements of leaf respiration with a temperature controlled gas exchange system (from canopy towers or platforms) would provide the field verification.

Another factor in understanding the maintenance-respiration partitioning of energy relates to carbon allocation above and below ground. This is especially important in the tropics because maintenance costs for support of photosynthesis systems is relatively constant throughout the year, but in temperate regions the respiration losses are reduced to low levels in winter. In the tropics, the efficient production of structural biomass may be less advantageous than a large supply of carbohydrates available for maintenance costs (Jordan, 1984). A determination of tree height over time is needed to understand the allocation of carbon above and below ground. The height measurements could be conducted by operating a laser in the profiling mode.

Preliminary results of NASA-NSTL research in Costa Rica and results from other investigations (Arp, et al., 1974; Neilson and Aldred, no date) indicate that laser profilers (and radar altimeters) can penetrate small openings in dense tropical canopies and measure tree height with good accuracy compared to ground measurements. It is reasonable to think that the degree to which the laser pulses can penetrate the canopy may also afford a method for measuring light penetration, and its relation to photosynthesis rates at different levels of the canopy. Laser measurements of the canopy coupled with day and night temperature, gas exchange and nutrient content (phosphorus and nitrogen) and cycling will provide a basis for understanding respiration and allocation in the context of NPP estimates. This approach to esti-

---

* Use of trade names do not constitute an endorsement of the products by NASA or the U.S. Government.

mating NPP in tropical environments was incorporated into a joint NASA - University of Puerto Rico research proposal that is currently under review.

CONCLUSIONS

Remote sensing investigations in tropical regions are not a direct transfer of technology and routine developed from temperate or boreal forest environments. Recent improvements in satellite and aircraft sensor design and data analysis techniques offer the potential to monitor important tropical forest ecological processes that could provide information to improve global carbon and climate models. Within the next decade, remote sensing from various platforms operating in several wavelength regions combined with computer technology, sampling design, and modeling approaches will likely become the major components of a truly systematic tropical forest and global monitoring system.

REFERENCES

Aoki, M.K. Yabuki, and H. Koyama, 1974, Micrometerology of Pasoh Forest: In IBP synthesis meeting, Kuala Lumpur, Aug. 12-18, 1974, 14 pp.

Arp, H., and J. P. Burns, 1982, Mapping in tropical forests: A new approach using the laser APR: Photogrammetric Engineering and Remote Sensing, Vol. 48, No. 1, pp. 91-100.

Birdsey, R. A. and P. L. Weaver, 1982, The forest resources in Puerto Rico: Resource Bulletin SO-85, Southern Forest Exp. Sta., New Orleans, LA.

Brokaw, N.V.L., 1982, Tree falls: frequency, timing, and consequences: In the Ecology of a Tropical Forest, Seasonal Rhythms and Long-term Changes. E.G. Leight, Jr., A.S. Rand, and D.M. Windsor, eds. Smithsonian Institution Press, Washington, D.C.

Brown, S., A. E. Lugo, S. Silander and L. Liegal, 1983, Research history and opportunities in the Luquillo Experimental Forest: USDA-Forest Service, Gen. Tech. Rep. SO-44, New Orleans, LA, 128 pp.

Brown, S., and A. E. Lugo, 1980, Preliminary estimate of the storage of organic carbon in tropical forest ecosystems: In S. Brown, et al. (ed.) The Role of Tropical Forests on the World Carbon Cycle: CONF-800350, UC-11, NTIS, 5285 Port Royal Rd., Springfield, VA 22161.

Budowski, G., 1965, Distribution of tropical American rain forest species in the light of successional processes: Turrialba 15(1) 40-42.

Denslow, J.S., 1980, Gap partitioning among tropical rainforest trees: In J. Ewel (ed.). Tropical/Succession. Biotropica 12(2), pp. 47-55.

Ewel, J., 1980, Tropical succession: Manifold routes to maturity: In J. Ewel (ed.), Tropical Succession. Biotropica Vol. 12(2), 95 p.

Foster, R.B. and N.V.L. Brokaw, 1982, Structure and history of the vegetation of Barro Colorado Island: In the Ecology of a Tropical Forest, Seasonal Rhythms and Long-term Changes. E.G. Leigh, Jr., A.S. Rand, and D.M. Windsor, eds. Smithsonian Institution Press, Washington, D.C.

Gervin, J.C., A.G. Kerber, R.G, Witt, Y.C. Lu and R. Sekhon, 1985, Comparison of level 1 land cover classification accuracy for MSS and AVHRR data: Int. Jour. of Remote Sensing, Vol. 6, No. 1, pp 47-57.

Grainger, A., 1980, The state of the world's tropical forest: The Ecologist, 10, 6-54.

Grainger, A., 1982, Improving the monitoring of deforestation in the humid tropics: In Tropical Rain Forests: Ecology and Management, British Ecological Society, No. 2, Blackwell Scientific Publications, pp. 387-395.

Green, K.M., 1982, Using Landsat to monitor tropical forest ecosystems: realistic expectations of digital processing technology: In Tropical Rain Forests: Ecology and Management, British Ecological Society, No. 2, Blackwell Scientific Publications, pp. 397-409.

Hartshorn, G.S., 1980, Neotropical forest dynamics: In J. Ewel (ed.). Tropical Succession. Supplement to Biotropica 12(2), pp. 23-30.

Hartshorn, G.S., 1978, Tree falls and tropical forest dynamics: In P.B. Tomlinson and M.H. Zimmerman (eds.). Tropical Trees as Living Systems. Cambridge Univ. Press, London.

Hoekman, D.H., 1985, Radar backscattering of forest stands: Int. J. Remote Sensing, Vol. 6, No. 2, 325-343.

Janos, D.P. 1980. Mycorrhizae influence tropical succession: In J. Ewel (ed.). Tropical Succession. Supplement to Biotropica 12(2), pp. 56-64.

Jordan, C.F., 1984, Total Tropical Ecosystem Productivity and Productivity useful to man: In Tropical Eclolgy, ed. by C.F. Jordan, Editors comments, p 226 Benchmark papers in Ecology V. 10, Hutchinson Ross Pubblishing Co., Shroudsburg, PA.

Knight, D.H., 1975, A phytosociological analysis of species-rich tropical forest on Barro Colorado Island, Panama: Ecol. Monog. 45(3): 259-284.

Nielsen U. and A.H. Aldred, (no date), New developments for tropical surveys prove successful: Forest Management Institute, Canadian Forestry Service, Ottawa, Canada.

Ormsby, J.P., B.J. Blanchard, and A.J. Blanchard, 1985, Detection of lowland flooding using active microwave systems: Photogrammetric Engineering and Remote Sensing, Vol. 51, No. 3, pp. 317-328.

Pampaloni, P. and S. Paloscia, 1985, Experimental relationships between microwave emission and vegetation features: Int. J. Remote Sensing, Vol. 6, No. 2, 315-323.

Piwinski, D.J., L.B. Schoch, M.J. Duggin, V. Whitehead and E. Ryland, 1983, Dependence of NOAA-AVHRR recorded radiance of scan angle, atmospheric turbidity and unresolved cloud: In proc. of Seventeenth Int. Sym. on Remote Sensing of Env., Ann Arbor, Michigan.

Posner, J.L., G.A. Antonini, G. Montanez, R. Cecil and M. Grigsby, 1983, Land systems of hill and highland tropical America: Revista Geografica, No. 98, pp. 5-22.

Sader, S.A., 1980, Renewable resources analysis and monitoring in Costa Rica: In 14th Int. Sym. on Remote Sensing of Environment, April 23-30, San Jose, Costa Rica, ERIM, Ann Arbor, Michigan, Vol. 2, pp. 1251-1259.

Sader, S.A. and A.T. Joyce, 1985, Relationship between forest clearing and biophysical factors in a tropical environment: In IX World Forestry Congress, Special Document of Technical Commission I, Subtheme 6.1.B., July 1-12, 1985, Mexico City, In press.

Schulz, J.P., 1960, Ecological studies on rain forest in northern Suriname: North-Holland, Amsterdam. 267 p.

Tucker, L.J. J.R.G. Townsend, and T.E. Goff, 1985, African Land - Cover classification using satellite data: Science, vol. 227, no. 4685 pp. 369-375.

Tucker, C.J., B.N. Holben, and T.E. Goff, 1984, Intensive forest clearing in Rondonia, Brazil, as detected by satellite remote sensing: Remote Sensing of Environment 15: 255-261.

Webb, L.J., J.G. Tracey, and W.T. Williams, 1972, Regeneration and pattern in the sub-tropical rain forest: J. Ecol. 60(3): 675-695.

Whitmore, T.C., 1978, Gaps in the forest canopy: In P.B. Tomlinson and M.H. Zimmerman (eds.). Tropical Trees as Living Systems. Cambridge Univ. Press, London.

Whitmore, T.C., 1975, Tropical Rain Forests of the Far East. Clarendon Press, Oxford. 282 p.

Williams, D.L. and L.D. Miller, 1979, Monitoring forest canopy alteration around the world with digital analysis of Landsat imagery: NASA-GSFC, Greenbelt, Maryland, 45 pp.

Wu, S.T., 1984, Analysis of data acquired by shuttle imaging radar and Landsat-4 thematic mapper over Baldwin County, Alabama: NASA-Earth Resources Laboratory, Report No. 228, NSTL, MS, 27 pp.

AN ECOREGION-CONTINUUM APPROACH TO GLOBAL VEGETATIVE BIOMASS ESTIMATION

Thomas L. Logan
Jet Propulsion Laboratory
California Institute of Technology
Pasadena, California 91109

BIOGRAPHICAL SKETCH

Thomas L. Logan is a geographer with ten years of experience in the Cartographic Applications Group of the Image Processing Applications and Development Section at the Jet Propulsion Laboratory. His emphasis has centered on the remote sensing and image processing of forest and range applications, as well as general Geographical Information Systems technology, classification techniques, database development, and Cartographic applications. Dr. Logan received his B.A. from the California State University, Northridge, in 1976, and his M.A. and Ph.D degrees from the University of California at Santa Barbara in 1978 and 1983, respectively.

ABSTRACT

The hypothesis has been proposed that the total vegetative biomass of the Earth's surface can be accurately estimated from environmental satellite imagery using an "Ecoregion-Continuum" approach. The basic tenets of the approach are 1) that the globe can be divided into regions that are ecologically consistent; 2) that a unique spectral/biomass relationship exists for each ecoregion; and 3) that the unique ecoregion spectral/biomass relationships are actually part of a spectral/biomass "relationship continuum" from agriculture (linear) to tropical forests (non-linear). Once the "Ecoregion-Continuum" is fully defined, it reduces to a simple look-up table of satellite spectral values---making ecoregion biomass estimation a relatively easy task.

INTRODUCTION

The global estimation of vegetative biomass involves a complex spatial problem that may be broken into smaller components for intensive analysis and comprehension of fundamental relationships. Considering the large differences in canopy expression and biomass quantities between forest and monocultural environments, the use of a regional approach to biomass estimation would appear reasonable. This paper draws upon the results of monocultural (and near-monocultural) biomass estimation research in the literature, a statistical comparison evaluating several different vegetation environments, and a research project in which the biomass of a coniferous forest was accurately estimated from satellite imagery. The results suggest the hypothesis that different biomass/spectral relationships exist for each vegetative environment (or ecoregion), and when the biomass/spectral relationships of each ecoregion are reviewed in succession, they represent a continuum of linear (agriculture)

* Part of the research described in this paper was carried out at the Jet Propulsion Laboratory, California Institute of Technology, under NASA contract NAS7-918. Presented at the Tenth W. T. Pecora Memorial Remote Sensing Symposium, Fort Collins, Colorado, August 20-22, 1985.

to non-linear (tropical) relationships across the various vegetative environments (agriculture-grassland-deciduous-coniferous-tropical).

In 1983, research was performed (Logan and Strahler, 1983; Logan, 1983a, 1983b) that developed a table look-up procedure for accurately estimating the total standing biomass (carbon) of a coniferous forest from NOAA-6 AVHRR-LAC spectral imagery. The look-up table was derived from a graph which bears resemblance to a two-dimensional "Rooster's Tail," with well defined crescentic layers of biomass plotted against AVHRR channels 1 (RED) and 2 (NIR). The approach differs from conventional regression techniques by providing a simple mathematical mechanism for compensating for the non-linearities in the biomass/spectral relationships.

Numerous researchers have established linear and near-linear regression relationships for agriculture and grassland environments. However, as the vegetative environment becomes heterogeneous (mixed species, increasing trees, progressing towards forestation), the biomass/spectral relationships become progressively more non-linear, reducing the reliability of regression techniques. This has led to the observation that different biomass/spectral relationships exist for each vegetative environment (or ecoregion). It is hypothesized that the biomass/spectral relationships are actually a continuum across the various vegetative environments with the "Rooster's Tail" being condensed for agricultural environments, and 'fanning' out as heterogenity increases. This occurs much like the natural environment which is a continuum from monocultural to heterogeneous forested conditions. If reliable biomass/spectral trends can be established across the progression of ecoregions, then approximate biomass (and carbon) quantities (and change) can be easily predicted on a global scale using the table look-up technique.

The "Ecoregion-Continuum" approach to global biomass estimation provides a mechanism for recognizing the vast differences in vegetative environments. A linear biomass/spectral relationship, for example, exists for agricultural crops (±r-square=.95) because virtually the entire plant is composed of chlorophylic leaves that are fully exposed to the satellite's view. The grassland environment, however, contains more variance and therefore has a less linear biomass/spectral relationship (±r-square=.8) than the agricultural environment. The higher variance results from the greater proportion of non-chlorophylic trunk and branch parts, heterogenity of mixed species, varying vegetation heights, shadow, subsurface roots, and variable vegetation-to-soil ratios. As the number of trees increases, the biomass/spectral linearity decreases. Very poor regression correlations have been found between the total standing biomass (carbon) of the coniferous forest environment and satellite spectral data (±r-square=.61). This is because most of the tree's biomass is located in its branches, stem, and roots, which are not sensed by satellite sensors. Considerable leaf chlorophyl is also obscured by the deep shadows cast by neighboring trees. Other research (Gholz, 1982) has established an allometric relationship between healthy coniferous biomass canopy, leaf area index (LAI), and total tree biomass, that exists under ideal circumstances. However, this relationship does not usually persist in the natural environment where the majority of trees are stressed by crowding, sun-deprivation, water saturation/deprivation, and soil/mineral deficiencies (Waring, 1983). The "Ecoregion-Continuum" approach can compensate for biomass which can not be directly sensed by using a statistical averaging procedure based on extensive ground-truth.

The "Rooster's Tail" graph which relates AVHRR spectral imagery with coniferous forest biomass is the product of a statistical averaging approach. It assumes that direct biomass sensing of a heterogeneous forest is not possible, and that total biomass (including trunk, stem, and roots) can be statistically estimated by calibrating the known total biomass (derived from extensive ground-truth) with the spectral data. Thus, the deduced relationship is only suitable for the specific ecoregion involved, and site-specific biomass estimation will be variably accurate, but considerably more accurate than from regression analysis (about 16x more accurate for the coniferous forest studied).

If the "Rooster's Tail" continuum can be fully validated, a considerable potential exists for developing table look-up techniques which will quickly and accurately measure total vegetative biomass (carbon) from satellite spectral imagery for each of the world's ecoregions. This will provide a new, more accurate procedure for spatially estimating vegetative biomass, provide a new perspective of the carbon parameter for the study of global bio-chemical cycles, including assistance in the understanding of carbon pathways and rates of carbon exchange.

ECOREGION CONCEPT

The concept of an ecoregion is generally defined to be a continuous geographical area in which the presence of one or more important ecological associations is distinctly different from that of adjacent regions. In other words, for some selected criteria, the within region variance is considerably less than the between region variance. For biomass considerations, the ecoregion should be based on more than simple vegetative associations, since a given plant community's biomass will vary according to a variety of topographic and ecological conditions. Thus, an ecoregion is characterized not only by distinctive floral associations, but also the region's fauna, climate, landforms, topographic conditions, and soil properties. Within an ecoregion, the ecological relationships between all selected criteria should be similar enough that identical management practices anywhere in the region would produce comparable results.

The use of satellite imagery for biomass estimation is consistent with the ecoregion concept. Taken within an ecoregion, the digital number (DN) of a satellite image can have an internally consistent meaning, whereas the same DN across two or more ecoregions may correspond to different ground features. This is one form of the spectral signature extension problem which occurs because of atmospheric and spacecraft sensor variations. Sensor and atmospheric perturbations are known to occur within single images, but are particularly significant between different satellite images and dates. The chief causes of the variability include: changes in platform altitude, pitch, yaw, roll, crab; and unequal distribution of atmospheric contaminants, particularly haze. The difficult task of compensating for the variations caused by these effects has typically been ignored in the majority of remote sensing projects because the studies usually involved such small portions (sub-samples) of the total available satellite imagery that all degradation effects could reasonably be considered negligible if not uniform. In the few satellite processing studies that have involved large geographic areas, ecoregions have typically been utilized.

In 1980, a study (Mcleod and Johnson, 1980) was completed that mapped the vegetation, biomass, and soils of the California Desert Conservation Area (Mojave Desert of Southern California). In that research, unsupervised clusters were obtained from two-pixel-wide transects that crossed over eleven separate Landsat frames which had been digitally mosaicked together. In the processing of those clusters, "...the existence of a systematic drift in signature characteristics from north to south and from east to west was shown" (p. 267). The solution to this problem was to organize their test areas into four separate regions which were then stratified and labeled separately.

In a second large area remote sensing project (Strahler, 1981), the problems of local forest ecological consideratins and satellite/atmosphere spectral variations were considered to be related problems best solved simultaneously. In that project, the study area was divided into "...natural regions, in which ecological relations and (possibly) signatures [would] be stable" (p. 21). Subsequent classification was then performed on an ecoregion basis. Specific atmospheric corrections were not attempted in either of the two studies because of the unavailability of requisite atmospheric calibration data.

The results of the above research indicate that when a regional approach has not been utilized in studies involving large geographic areas, regional classification errors can be expected. An important distinction must be made however, between the estimation of total vegetative biomass and estimation of the photosynthesizing biomass of the vegetative canopy. Several researchers (ERIM, 1984; Tucker, et al., 1983) for example, have generated "green wave" maps of North America and Africa from AVHRR data using the Normalized Difference (ND) transform [(NIR-RED)/(NIR+RED)]. Theses maps typically cross many ecoregions and are useful for providing an indication of the photosynthetic activities of the canopy. However, these maps are not providing estimates of total vegetative biomass. It is likely that these maps contain some regional "green wave" errors because the spectral signal between different ecoregions is the product of different optical interactions and vegetative structures. A forest canopy, for example, may require comparatively more photosynthetic activity than a grasslands area to show the same spectral reflectance. This is likely because the forest canopy interactions are much more complex (particularly with regard to shadow) than are grassland canopies, which tend to be very open. Thus, it appears that large area remote sensing projects should consider a regional approach defined by similarity of ecological characteristics.

The regional concept is complementary with satellite-based research in that the effects of sensor and atmospheric perturbations can be reduced. Some atmospheric variations are related to local regional situations such as haze over urban sites and dust over semi-arid areas. This suggests that atmospheric variations may be more uniform (and therefore less problematic) within ecoregions than between ecoregions.

ECOREGION DELINEATION

The regional concept has been a traditional theme in the discipline of Geography (Whittlesey, 1954), and is still considered to be central to that discipline by some (Hart, 1982). However, despite the emphasis placed upon regional study by Geographers and the practitioners of other disciplines, there has been little work in the development of large scale integrated ecological delineations of world regions. Traditionally, most global delineations have been based on a single criteria such as climate (e.g., Koppen-Geiger, 1954), or vegetation (e.g., Brockmann-Jerosch, 1951). Where highly detailed systematic ecosystem classifications have been developed (Vinogradov, et al., 1962; Brink, et al., 1966; Lacate, 1969) the definitions tend to be so detailed as to preclude the actual delineation of geographic boundaries (e.g., Brown, et al., 1980). Recently, however, Bailey (1981) has developed an ecoregion delineation of North America, where the approach is to: "reconcile the biotic and abiotic classifications into a single geographic classification that is relatively objective. The scheme is based on a major division of all lands into lowlands and highlands. Within the classification, bioclimatic criteria are used to determine the upper rungs of the heirarchy, and geologic and geomorphic criteria are taken into account mainly at the lower levels" (Bailey, 1978).

The classification scheme used by Bailey is based on the work of Thornwaite (1931), Koppen (Trewartha, 1943), Kuchler (1964), and Crowley (1967), and includes four levels of generalization: A subcontinental area of broad climatic similarity is termed a "Domain." Within a Domain is a "Division," which is defined by isolating areas of differing vegetation and regional climates. The next level down is the "Province," which is delineated on the basis of uniform regional climate and the type of zonal rocks. The finest level of subdivision is the "Section," which is characterized by a single climax association and is usually defined by Kuchler's (1964) potential vegetation divisions. Bailey's ecoregion "Province" appears to be consistent with the scale to which conventional ground-based biomass estimates are considered valid. Within such delineated areas, satellite signature extension problems do not provide a significant source of error.

A logical implementation of the ecoregion approach to global scale biomass estimation might begin with an internally consistent delineation system, from which biomass could be accurately calculated on a region-by-region basis, and then aggregated to the global scale. At present, Bailey's system appears to provide the necessary ecoregion requirements for global scale biomass estimation. Bailey's system was originally developed for the United States (1978, 1980), but has recently been expanded to the entire North American continent (1981). The system's underlying principles could be applied towards defining world-wide region delineations.

BIOMASS RELATIONSHIPS IN A CONIFEROUS FOREST ECOREGION

The "Ecoregion-Continuum" approach is hypothesized on the basis of three spectral/biomass relationships. The first two, agriculture and grassland, are well established in the literature as virtually linear and near-linear, respectively. The suggested continuum towards non-linearity becomes visible when the third biomass/spectral relationship, coniferous forest, is graphed. The spectral/biomass relationship for the coniferous forest was developed in a research study (Logan and Strahler, 1983; Logan, 1983a, 1983b) conducted in Bailey's Sierran Forest Province (Figure 1) using NOAA-6 AVHRR-LAC spectral reflectance. This relationship was described above as a "Rooster's Tail" (Figures 2 and 3) and provides an accurate and convenient tool for predicting total vegetative biomass from AVHRR spectral channels 1 (RED) and 2 (NIR) in the Sierran Forest Province.

Development of the "Rooster's Tail" relationship was a large and complex task. The study area consisted of nearly 1.2 million acres completely encompassing the Eldorado National Forest (Figure 1) in California. This forest was chosen because it provided an example of the type of coniferous needleleaf forest that dominates most of western North America, and it represents a vegetative regime very different from agriculture and grassland. The vegetation types found in this area are representative of a large portion of the world's biomass.

The technical approach involved with developing the "Rooster's Tail" relationship began with preparation of a ground-based biomass map of the 1.2 million acre Eldorado study area, and relating it to the spectral properties of the AVHRR using correlation, regression, and multispectral classification techniques. A biomass ground-truth map was prepared by adaptation of a Forest Type Map of the Eldorado National Forest which had been created in a previous research project (Strahler, et al., 1983) from the modelling of Landsat imagery, digital elevation data, texture, and ground-truth information. Timber stratums in the Type Map were converted to oven-dry biomass (in kilograms) by relating the timber biomass of standard size trees (generated from biomass regression coefficients available in the literature) to their basal area. That relationship was then used as a calibration multiplier applied to basal area statistics generated by USFS software from 1982 inventory ground-truth. The non-timbered stratums in the Type Map were converted to biomass using a variety of techniques based on an earlier inventory of the Eldorado National Forest, field experience, and consultation with USFS personnel.

The Eldorado study area biomass data plane (used as the 'Control,' or ground-truth base for all subsequent analysis) was thus determined to contain an approximate 135.8 billion kilograms of oven-dry biomass. A precise estimate of the error associated with the biomass calculation could not be determined because of the variety of data sources used in its preparation. The accuracy, however, was confidently placed at plus or minus eight percent. The consistency in the pattern of subsequent results provided a positive indication of the integrity and reliability of the ground-based biomass data plane.

The Eldorado biomass sample base (originally created at a 63.62 meter/pixel/square resolution) and the corresponding AVHRR imagery were placed in a common map projection and resampled to a joint spatial resolution of 827.06 meters/pixel/square. This size was an exact multiple (13x) of the biomass sample base, permitting (13x13=) 169 biomass pixels to be aggregated to form one new biomass pixel at the AVHRR scale. This technique assumed that most errors in the biomass data were random rather than systematic, which is a reasonable assumption considering the variety of data sources employed in its making. Thus, existing errors in the calculation of the Eldorado biomass data plane were averaged out when the aggregation process converted the data to AVHRR spatial scale. The averaging of errors over large areas has also been reported in a related remote sensing application by Hanuschak, et al. (1979).

Several popular channel transformations of the AVHRR sensor system were created and compared with the biomass sample data in a correlation analysis. Results revealed a generally non-linear relationship between biomass and the Transformed Vegetation Index (TVI), Normalized Difference (ND), Gray-McCrary Index (GMI), Eigenvector Images 1 and 2, and all four AVHRR channels. The highest correlation of -.7813 between AVHRR1 (RED) and AVHRR2 (NIR) indicated that only 61% of total variance was explained by the linear function. A multiple regression of AVHRR1 and AVHRR2 on biomass produced a nearly identical result. Multispectral classification of AVHRR channels 1 and 2 were then performed, and polygon overlay technqiues utilized to determine the mean biomass per pixel in each classification. A two-dimensional plot of spectral class means with average class biomass values as the Z-value (Figure 2), revealed the biomass-to-spectral channel relationship as a rather well-behaved non-linear relationship somewhat analogous to a "Rooster's Tail." Multiple regression and multispectral classification techniques were also used in a 'Train and Test' situation to determine their comparative predictive capabilities. Both techniques produced correct total biomass estimates, but the classification technique did so with a residual standard error considerably less than that for the regression technique.

ECOREGION-CONTINUUM HYPOTHESIS

The homogeneous to heterogeneous trend of the agriculture-to-grassland-to-coniferous forest environment, suggest a linear to non-linear continuum across ecoregions. Development of spectral/biomass relationships for intervening ecoregions is a time consuming task that is yet to be performed. However, the hypothesis suggests that those intervening relationships need only be developed once from representative ecoregions.

To establish the relative placement of the three ecoregions in the continuum as well as the location of deciduous and tropical forests, a surrogate procedure was performed. From three to eight sites (averaging 100 pixels each) were selected in each of the five ecoregions from AVHRR-LAC satellite imagery. The digital values of the RED and NIR spectral channels from each site where calculated and ratioed. The ratios from each ecoregion were then averaged and graphed (Figure 4). The results show the ecoregion ordering as: agriculture, grassland, deciduous forest, coniferous forest, then tropical forest. A hypothetical graphing of the five ecoregions representing the "Ecoregion-Continuum" is also shown (Figure 5).

CONCLUSION

The hypothesis has been proposed that the total vegetative biomass of the Earth's surface can be accurately estimated from environmental satellite imagery using an "Ecoregion-Continuum" approach. The basic tenets of the approach are 1) that the globe can be divided into regions that are ecologically consistent in terms of vegetative associations, climate, landforms, topographic conditions, and spectral signature extension; 2) that a unique spectral/biomass relationship exists for each ecoregion; 3) that the spectral/biomass relationship represents a "statistical average" in which accurate regional biomass estimates are produced at the expense of relatively higher local (site) biomass variance; 4) that the unique ecoregion spectral/biomass relationships are actually part of a spectral/biomass "relationship continuum" which resembles a "Rooster's Tail;" 5) that the continuum is from agriculture (linear spectral/biomass relationship; condensed "Rooster's Tail") to tropical forest (non-linear spectral/biomass relationship; fanned-out "Rooster's Tail"); and 6) that the nominal ordering of the ecoregions is: agriculture to grassland to deciduous forest to coniferous forest to tropical forest.

The "Ecoregion-Continuum" hypothesis is currently based upon the known spectral/biomass relationships of the agriculture, grassland, and coniferous forest ecoregions. Development of relationships in the other ecoregions is necessary to further substantiate and refine the hypothesis. Adjustments for seasonal and spacecraft radiometric variations will also need to be modelled (although these will probably manifest as relatively minor compressions/expansions of the "Rooster's Tail"). However, once the "Ecoregion-Continuum" is fully defined, it reduces to a simple look-up table of satellite spectral values---making ecoregion (and global) biomass estimation a relatively trival operation.

Validation of the "Ecoregion-Continuum" will provide several important insights into fundamental science and technology issues confronting contemporary civilization. In particular, the approach could provide significant perspectives of near and far term global vegetative/ecological change, resource management, carbon cycling, and carbon dioxide release and absorption. This will enhance the prospects of providing civilization with the technical capability of monitoring (and perhaps managing) the quantity of carbon in the atmosphere (and dependent global temperatures).

REFERENCES

Bailey, R., 1978. Ecoregions of the United States, RARE 11 Map B, USFS/ Forest Service (map).

Bailey, R., 1980. Description of the Ecoregions of the United States, USDA/Forest Service, Ogden, Utah, 77p.

Bailey, R., 1981. Ecoregions of North America, USDA/Forest Service (map).

Brink, A., J. Mabbutt, R. Webster, and P. Beckett, 1966. Report of the Working Group on Land Classification and Data Storage, MEXE Report No. 940, Christchurch, NZ., 97p.

Brockmann-Jerosch, 1957. Vegetation of the Earth, Justus Perthes:Gotha, West Germany (map).

Brown, D., C. Lorre, and C. Pase, 1980. A Digitized Systematic Classification for Ecosystems with an Illustrated Summary of the Natural Vegetation of North America, USDA/Forest Service General Technical Report RM-73, Rocky Mountain Forest and Range Experiment Station, Fort Collins, CO., 93p.

Crowley, J., 1967. "Biogeography in Canada," Canadian Geographer, 11: 312-326.

ERIM, 1984. "Cover Photo," Photogrammetric Engineering and Remote Sensing, L(12):1647-1653.

Gholz, H., 1982. "Environmental Limits on Aboveground Net Primary Production, Leaf Area, and Biomass in Vegetation Zones of the Pacific Northwest," Ecology, 63(2):469-481.

Hanuschak, G., R. Sigman, M. Craig, M. Ozga, R. Luebbe, P. Cook, D. Kleweno, and C. Miller, 1979. "Crop-Area Estimates from Landsat; Transition from Research and Development to Timely Results," Proceedings of the Symposium on Machine Processing of Remotely Sensed Data, LARS/Purdue University, West Lafayette, IN., 86-94.

Hart, J., 1982. "The Highest Form of the Geographer's Art," Annals of the Association of American Geographers, 72(1):1-29.

Koppen-Geiger, 1954. Klima der Erde, Justus Perthes:Darmstadt, West Germany; American Distributor: A.J. Nystrom and Co., Chicago (map).

Kuchler, A., 1964. Potential Natural Vegetation of the Conterminous United States, American Geographical Society Special Publication No. 36; 116p.

Lacate, D., 1969. Guidelines for Biophysical Land Classification, Dept. of Fisheries and Forestry, Canada Forestry Service, Publication No. 1264.

Logan, T., and A. Strahler, 1983. "Surface Vegetative Biomass Modelling From Combined AVHRR and Landsat Satellite Data," Proceedings of the 17th International Symposium on Remote Sensing of Environment, ERIM, Ann Arbor, MI.

Logan, T., 1983a. "Regional Forest Biomass Modelling From Combined AVHRR and Landsat Data," CERMA Energy Resource Management Conference, San Francisco, CA., August 22-26.

Logan, T., 1983b. Regional Biomass Estimation of a Coniferous Forest Environment From NOAA-AVHRR Satellite Imagery. Dissertation, University of California at Santa Barbara. University Microfilms International, Ann Arbor, MI., 283p.

McLeod, R., and H. Johnson, 1980. "Resources Inventory Techniques Used in the California Desert Conservation Area," Arid Lands Resources Inventory Workshop, USDA/Forest Service, p260-271.

Strahler, A., 1981. "Stratification of Natural Vegetation for Forest and Rangeland Inventory Using Landsat Digital Imagery and Collateral Data," International Journal of Remote Sensing, 2(1):15-41.

Strahler, A., C. Woodcock, and T. Logan, 1983. "Spatial Inventory Integrating Raster Databases and Point Sample Data," ACSM-ASP Annual Convention, American Congress on Surveying and Mapping, Washington, D.C.

Thornthwaite, C., 1931. "The Climate of North America According to a New Classification," Geographic Review, 21:633-655.

Trewartha, G., 1943. An Introduction to Weather and Climate, 2nd Edition, McGraw Hill:New York.

Tucker, C., 1983. "Satellite Remote Sensing of Total Dry Matter Production in the Senegalese Sahel," Remote Sensing of Environment, 13(6): 461-474.

Vinogradov, B., K. Gerenchuk, A. Isachenko, K. Rahm, and Y. Tseselchuk, 1962. "Basic Principles of Landscape Mapping," Soviet Geography---Review and Translation, 3(6):15-29.

Waring, R., 1983. "Estimating Forest Growth and Efficiency in Relation to Canopy Leaf Area," Advanced Ecological Research, (in press).

Whittlesey, D., 1954. "The Regional Concept and the Regional Method," in: James and Jones (eds.), American Geography, Inventory and Prospect, Association of American Geographers, Syracuse University Press, p19-69.

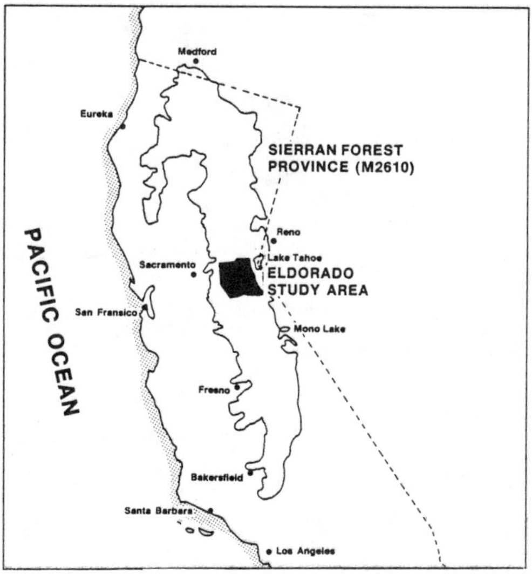

Figure 1: Location of Bailey's Sierran Forest Province and the 1.2 million acre Eldorado study area used for generating the coniferous forest spectral/ biomass relationship.

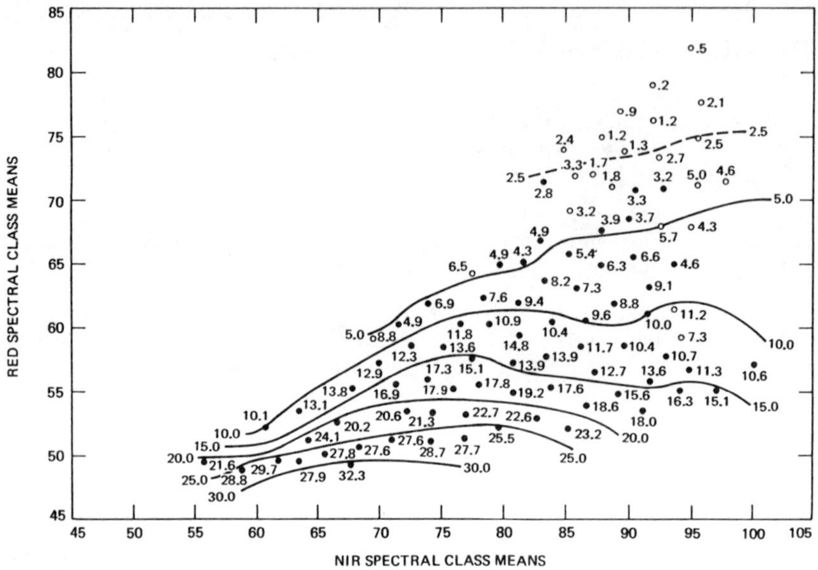

Figure 2: The relationship between biomass and forest spectral properties is shown as a function of AVHRR channels 1 (RED) and 2 (NIR) class spectral means. The average biomass per AVHRR pixel of each class is provided next to its plotted spectral location. Classes with 14 or less pixels (low-confidence) are identified with open circles. Contours are drawn at 5 million kilogram intervals with liberties taken near low confidence classes.

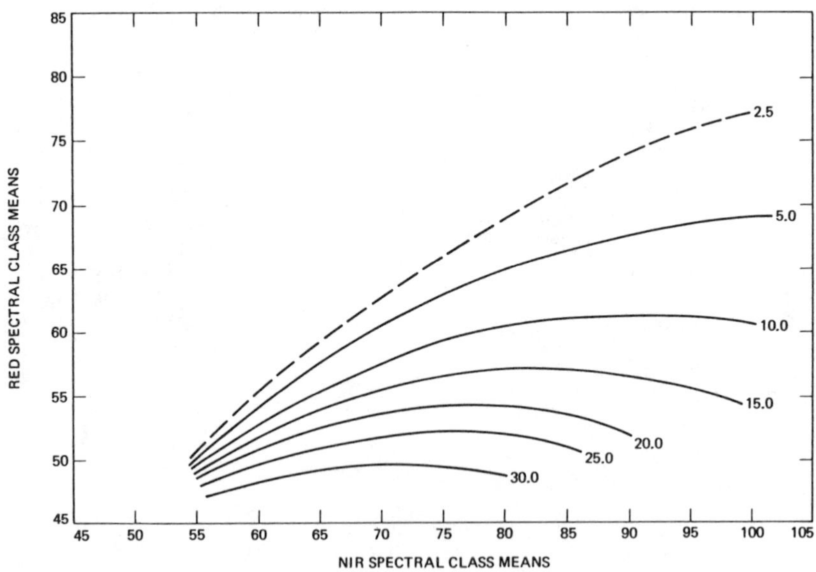

Figure 3: Theoretical stylization of Figure 2.

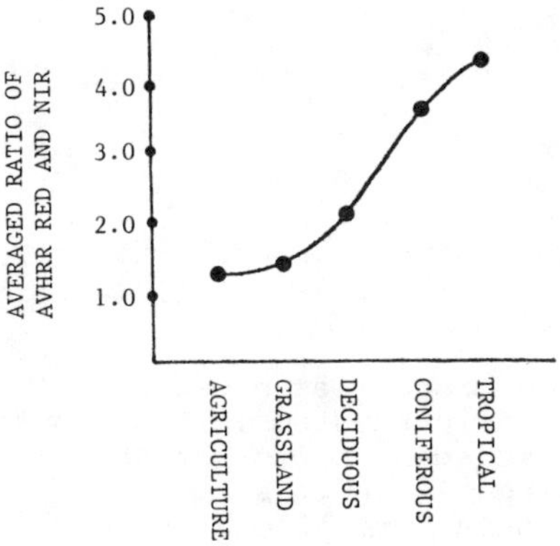

Figure 4: Ordering of ecoregions in the spectral/biomass continuum as inferred from AVHRR imagery. Three to eight sites (averaging 100 pixels each) were selected in each of the five ecoregions from AVHRR-LAC satellite imagery, their digital values ratioed, and averaged within ecoregions.

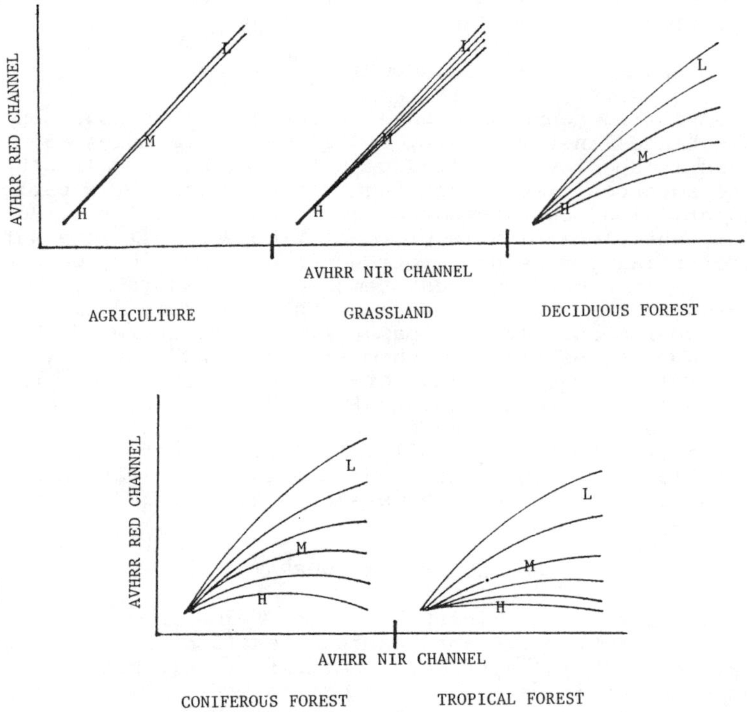

Figure 5: Hypothetical interpolation of "Rooster's Tail" spectral/biomass relationship demonstrating the theorized form of the "Ecoregion-Continuum" in each of the five discussed ecoregions. H='High' biomass; M='Medium' biomass; L='Low' biomass.

PRECURSORS OF CHANGE IN TERRESTRIAL ECOSYSTEMS

R.H. Waring
School of Forestry
Oregon State University
Corvallis, Oregon 97331

ABSTRACT

Plant canopies reflect underlying properties of ecosystems. With a variety of remote sensing systems, it may be possible to estimate changes in seasonal photosynthesis, the cost of maintaining conducting tissue in stems and branches, the proportion of photosynthate allocated to roots, the susceptibility of plants to insects and disease and the rate of litter decomposition. Large scale changes in ecosystem properties are harbingers of change in the structure and composition of ecosystems. An experimental approach to manipulating mixed forests of hardwoods and conifers is proposed to verify principles and the possible application of remote sensing technology.

INTRODUCTION

Remote sensing techniques have proven useful in assessing major changes in the density of vegetative canopies and changes in species composition (Botkin et al. 1984). With recent advances, remote sensing technology may be capable of identifying ecosystems subject to change at local, regional, and global levels of resolution. To fully evaluate the potential for using remote sensing to predict, we need to know more about how ecosystems operate and then to test the new remote sensing technology in rigorously designed field experiments. In this paper I shall first review how we believe terrestrial ecosystems operate and identify key structural and chemical characteristics of the canopy that may be useful. With knowledge of these important variables, we can then focus on which kinds of sensors are appropriate for testing. Finally, I shall suggest an example of a field experiment that might provide verification of both our understanding of ecosystem operation and remote sensing applications.

HOW ECOSYSTEMS OPERATE

Plants, animals, and microbes each performs unique functions in ecosystems. Plants capture carbon dioxide through the process of photosynthesis and make organic matter. Organic matter in various forms offers habitat and supplies food for animals. Animals prepare the seedbed, pollinate flowers, and disseminate seeds over long distances. Microbes decay animal and plant remains into soil organic matter and release nutrients in the process. The organic matter incorporated in soils helps maintain a store of water and nutrients for plant roots. Thus a cycle of life and interdependency exists in every ecosystem.

Few ecosystems exist in a state where conditions are optimal for all groups. Normally, one group is favored over the others. But conditions are unstable, changing with the seasons, and as vegetation develops. We see this most notably in temperate regions where plants grow in one season and shed leaves, fruits, and some roots during the dormant season. Occasionally, plants reduce leaf production because of late frosts. At other times, insects defoliate the vegetation in mid-summer. Usually, these kinds of variations do not persist. If they do, they are harbingers of change in the kind of vegetation that will inhabit the area.

Ecologists are discovering that even before leaves are shed or defoliation occurs, a variety of structural and chemical alterations usually become appparent in the vegetation canopy. These alterations indicate whether an imbalance is developing that can make the vegetation susceptible to change. Coupled with information about the weather and soils, we can often interpret what are the causes of the observed changes in canopy structure and chemistry.

Why should the canopy be so sensitive to these structural and chemical alterations? Certain properties of the canopy indicate the physiological status of the vegetation. The physiological status of the vegetation, in turn, also defines the nutritional quality of plants for animals. In addition, the chemical composition of foliage and other vegetative parts is indicative of how rapidly organic materials will be decayed by microbes.

Climatic change, polluted environments, and biogenic emission of trace gases all impact terrestrial ecosystems. If alterations in the atmospheric environment become widespread, then recognizing symptoms of change in vegetative canopies has practical significance. Coupling remote sensing to specific physiological and structural features is a prerequisite for widespread application.

PREDICTING DRY MATTER PRODUCTION

For many annual agricultural crops, dry matter production is closely correlated with the amount of short-wave radiation (400-700 nm) absorbed by a changing canopy throughout a growing season (Monteith 1977). Unfortunately, perennial vegetation is often restricted in carbon uptake by freezing conditions, soil drought, and low atmospheric humidity so that growth is less than that predicted from the amount of short-wave radiation absorbed (Berry and Downton 1982, Jarvis and Leverenz 1983). Moreover, when both annual and perennial vegetation are grown with different amounts of water and nutrients, the proportion of growth that goes into roots and shoots will change (Fig. 1). In such cases, the measured above-ground production is not a constant fraction of the carbon captured by the canopy. Finally, as perennial vegetation grows in stature, the number of living cells in non-photosynthetic tissues increases in the conducting tissues between roots and leaves. These non-photosynthetic cells require carbon resources for maintenance processes that otherwise might go toward growth.

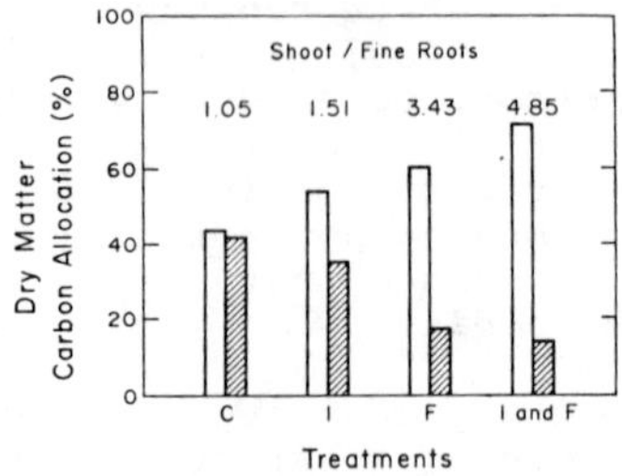

Fig. 1. When pine trees in Sweden were irrigated (I), fertilized (F), or received both treatments (I and F), they allocated increasing proportion of carbon into shoots (clear bar) vs. fine roots (cross-hatched bar) compared to untreated trees (C). Data from Axelsson (1981).

In a pine forest, the amount of non-photosynthetic living cells in the stems may increase over a 50-year period sufficiently to account for an 80% reduction in observed above-ground growth (Fig. 2). This helps to explain why older forests never match the production rates of young forests, even when they have similar canopies (Fig. 3).

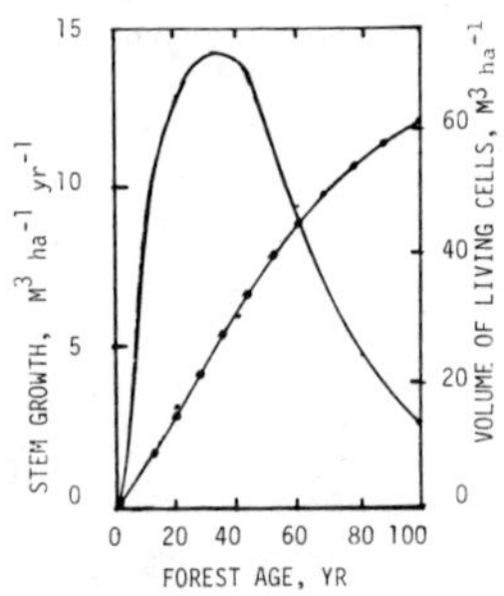

Fig. 2. Stem growth of pine forests in Norway peak on good sites at about 30 yr, corresponding with attaintment of maximum canopy and photosynthetic uptake. Thereafter, stem growth decreases in proportion to the increase in volume of living cells in the stem. Trees are 8 m tall at 20 yr and 26 m at 100 yr. Data from forestry yield tables of Brautseg (1969), analysis by Waring and Schlesinger (1985).

What canopy characteristics might allow us to predict carbon uptake, maintenance costs, and changing patterns in allocation of above-ground and below-ground production? For vegetation which sheds leaves during periods of stress, it may be possible to correlate carbon uptake directly with the amount of leaf area and the duration

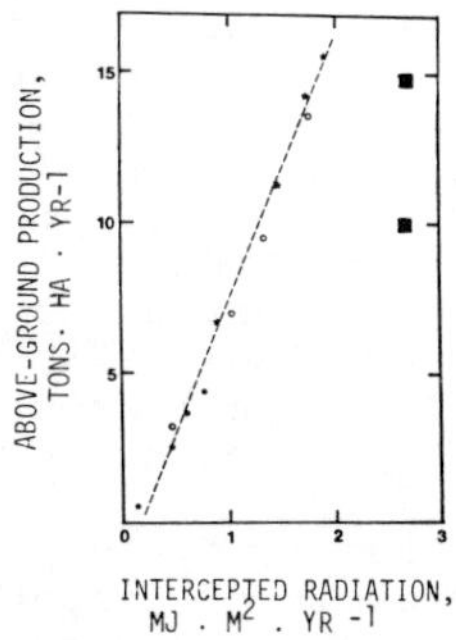

Fig. 3. Measured above-ground production by eucalyptus plantations in Australia closely corresponds to the estimated short-wave radiation intercepted by their crowns during the year. The circles and stars represent three different plantations fertilized to varying degrees and compared at 2, 4, and 10 yr in age. The two squares represent much older and taller forests of eucalytus with rates of growth below that of younger plantations with similar canopy leaf area indices. After Linder et al. (1985).

of leaf display (Fig. 4). The normalized ratio of infrared to red radiation reflected by vegetation appears to correlate fairly well with seasonal changes in leaf area (Jordan 1969, Tucker 1980, Botkin et al. 1984, Williams et al. 1985, Wardley and Curran 1984, Spanner et al. 1984). The assumption that chlorotic foliage will reflect more red light than chlorophyll-rich foliage and thus decrease the normalized ratio of infrared/red light is not fully confirmed. In fact, in one experiment where foliage was allowed to become chlorotic through drought, the ratio proved insensitive to alternations in chlorophyll concentration per unit area (Curran and Milton 1983). Chlorophyll concentrations do appear directly correlated with the wavelengths between 680 and 750 nm where red light reflectance changes most abruptly (Horler et al. 1983, Schwaller et al. 1983). Thus it should be possible to discriminate between green and yellow, less photosynthetically active foliage.

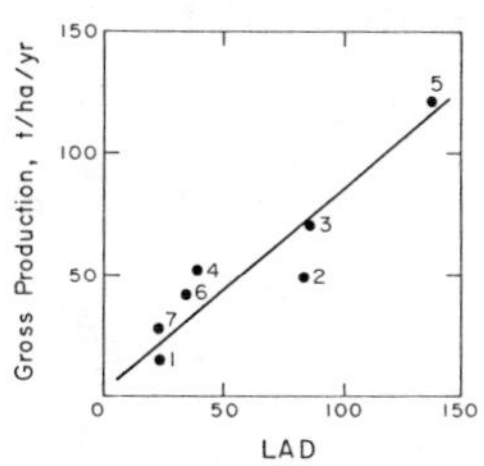

Fig. 4. Annual carbon uptake (gross production) by a wide variety of deciduous and evergeen tropical forests is correlated with the amount of leaf area and the months for which it is displayed (LAD index). After Kira and Shidei (1967).

For evergreen species it may be necessary during part of the year to reduce estimates of carbon uptake derived from light reflectance data when canopies are frozen or drought-stressed. Surface temperature may be measured to within 0.5° K with infrared sensors (Idso 1982) and atmospheric water vapor concentration at the canopy surface, an index of water stress, has been estimated within 10% above corn and wheat using microwave (36 GHz) sensors (Paloscia and Pampaloni 1984, Pampaloni and Paloscia 1985). Idso (1982) indicates that humidity may also be measured with infrared thermometry.

With a variety of sensors it may be possible to estimate carbon uptake by an actively photosynthesizing canopy throughout the year. The clear-day estimates of normalized infrared/red reflectance ratios integrated for the growing season or longer periods are a first step toward this goal (Tucker et al. 1985, Goward et al. 1985).

Maintenance cost has not yet been assessed by remote sensing but should be important in tall forests. The amount of living cells in stems and branches is a function of wood type, total biomass, canopy leaf area, and total height (Waring 1983, Waring and Schlesinger 1985). The major wood types (softwoods, ring-porous hardwoods, and diffuse-porous hardwoods) can sometimes be distinguished by observing seasonal patterns in foliage development and duration of foliage display (Lechowicz 1984). Tree heights can be measured within 1 meter using an airborne infrared laser system (Schreier et al. 1985, Nelson et al. 1984). Similar laser systems have shown promise in estimating merchantable timber volume (closely related to total above-ground biomass) in forests made up of trees with varying shapes (Maclean and Krabill 1985). Assuming the proportion of living cells in above-ground biomass could be estimated, maintenance cost of these cells is a function of temperature, doubling about every 10°C rise above freezing, up to a lethal temperature around 40 to 50°C (Amthor 1984).

Root growth is difficult to measure directly but indirectly, it might be estimated by substracting from estimates of annual photosynthesis that portion of carbon used in maintenance respiration and in production of new foliage and other above-ground biomass. Laboratory experiments suggest it may be possible also to detect major changes in shoot to root growth by measuring lignin/nitrogen ratios in the foliage (Waring et al. 1985). More will be said about these chemical ratios latter.

SUSCEPTIBILITY OF VEGETATION TO INSECTS AND PATHOGENS

A wide variety of plants growing under cultivation as well as in more natural conditions appear more sensitive to insect and disease attacks when conditions for growth are less than optimal. Nutritional imbalances, unfavorable water relations, and generally poor adaptation to a specific environment can result in predisposing plants to insect and disease attack (Bryant et al. 1983, Matson and Waring 1984, Tuomi et al. 1984, Cox and McEvoy 1983, Lambert and Turner 1977, Gibbs 1967, Schoeneweiss 1975).

Any combination of environmental conditions that drastically reduced photosynthesis below normal could cause vegetation to become susceptible to insects and pathogens. Thus periodic reductions in canopy leaf area or reductions in photosynthetic activity are often followed by some sort of disturbance. The scale is important. If large areas of vegetation exhibit stress, the likelihood of disturbance is much greater than if only isolated areas are involved. Using some of the techniques previously described, remote sensing offers a particularly effective means of assessing stress over large areas.

Plants under stress tend to allocate less to above-ground dry matter production per unit of leaf area (Waring 1983, Waring and Schlesinger 1985). Thus by periodically evaluating growth in above-ground biomass and the canopy leaf area, it may be possible, from the ratio of these variables, to risk-rate large units of vegetation.

Plants have evolved many chemical defenses. Often the specific chemical reflects the relative availability of resources. Plants growing in nitrogen-rich environments or nitrogen-impoverished environments develop different kinds of defenses. With low nitrogen availability plants often grow slowly and have thick cell walls rich in lignin (Lainson and Thornley 1982). The opposite is true when nitrogen is readily available. Furthermore, with high nitrogen availability, the canopy is often dense and light is limiting (Gartlan et al. 1980). The lignin to nitrogen ratio in the foliage may be worth close examination. Under conditions where the ratio is noted to change dramatically over time, the vegetation may be predisposed for attack by insects or disease and eventual replacement by younger, smaller plants or by species better adapted to the changed environment.

DECOMPOSITION BY MICROBES

Decomposition by microbes is a very complex process. Mechanical and digestive activities performed by animals also play an important role in hastening the rate of decomposition (Reichle 1977, McBrayer et al. 1977). The amount of material available for decomposition depends on foliage and root production, as well as defoliation, water stress, and tree mortality (Jarvis and Leverenz 1983). Remote sensing may be able to account for seasonal changes in above-ground production of dead organic material.

Under a particular climate, it may also be possible to predict decomposition rates as a function of lignin/nitrogen ratios (Fig. 5). These ratios can easily be changed by varying the relative availability of nitrogen (Table 1). Lignin, protein nitrogen and possibly other mineral and organic compounds have not yet been assayed using airborne remote sensors. However, in the laboratory there has been some success distinguishing these compounds by their unique absorption characteristics in the infrared portion of the spectrum (Barton and Burdick 1979). D.L. Peterson at NASA Ames Research Center now heads a project to evaluate whether remote sensing with an instrument such as the Airborne Visible and Infrared Imaging Spectrometer (Goetz et al. 1983, Vane et al.(1983) might provide a means of assaying foliage chemistry from space.

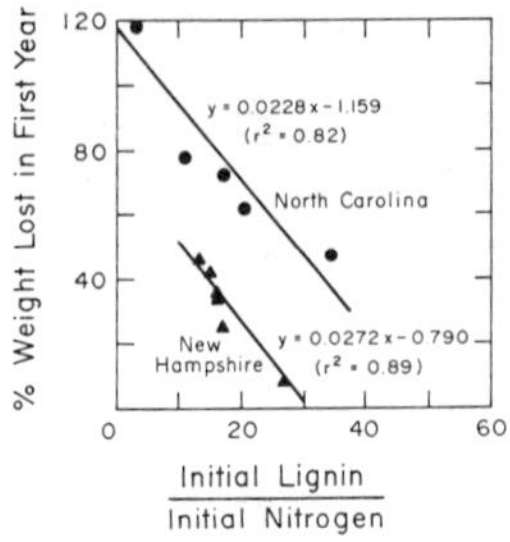

Fig. 5. Decomposition of leaves, as measured by weight loss, decreases progressively as the fraction of lignin/nitrogen increases. Decomposition is about 50% faster in North Carolina than in New Hampshire because of differences in temperature. After Melillo et al. (1982).

Table 1. Increasing fertilization with nitrogen in Swedish pine forests produced litter with progressively lower lignin/nitrogen ratios which allowed decomposition rates to increase. After Berg and Staaf (1980).

Nitrogen Added kg N/ha/yr	Lignin/Nitrogen in leaf litter	% weight loss in 1 yr
0	75	21
30	60	22
60	50	23
90	41	25

Under a specified climate, and with knowledge of litter chemistry it should be possible to estimate how rapidly litter will decompose and nutrients will be released (Meentemeyer 1978, Fogel and Cromack 1977, Berg et al. 1984, Melillo et al. 1982). If litter production exceeds decomposition, organic matter will accumulate in the system. When the reverse is true, a net loss of carbon from litter will result. Of course harvesting vegetation also results in major reductions in the amount of carbon stored in the ecosystem.

In summary, it may be possible to indirectly assess most of the major components of an ecosystem carbon budget from a variety of above-ground measurements made on vegetation and environment throughout the year. Remote sensing may offer a means of monitoring many of the key variables.

CRITICAL EXPERIMENTS

Before remote sensing technology can be applied to assess potential or actual shifts in ecosystem operation, some large integrated experiments will be required. These experiments must eventually encompass a wide range in vegetation types, climates, and soil conditions to confirm the generality of basic principles and the reliability of the sensors to distinguish key variables.

Most experimental ecosystem research to date has focused on vegetation dominated by a single species. To predict changes in vegetative composition requires that ecosystems be studied with a variety of species present. There have been broad surveys across regions but these studies lack adequate ground truth and have involved only a few types of remote sensing systems.

To test the ability of remote sensing to adequately estimate seasonal photosynthetic activity, changes in canopy chemistry, and the importance of maintenance costs in reducing growth, experiments need to be done in areas where temperature or drought limits photosynthesis, where vegetation is tall, and where the composition of the vegetation can be clearly changed. We find these prerequisites in a variety of temperate forests. For relative simplicity, we might look first at the northern hardwood - mixed coniferous forest dominated by maple, basswood, spruce, and fir. This type of forest is well developed in the Lake States and New England States on relatively uniform glacially-derived soils with modest variations in topography. A large number of experimental forests used by universities in the United States contain this vegetation type. An opportunity exists to develop an integrated experiment on one or more of these experimental forests.

What kind of experiment would provide the most insight? An experiment where the relative availability of a critical resource favored either conifers or hardwoods should be most insightful. The availability of nitrogen has been suggested as such a resource (Linder and Rook 1984). Excess nitrogen is difficult for conifers to assimilate. Conifers only increase photosynthetic rates slightly with increased nitrogen available and cannot easily store the element in an inorganic form. Many hardwoods, on the other hand, may increase photosynthetic rates dramatically when supplied with adequate nitrogen and can store the mineral in inorganic form (Linder and Rook 1984, Waring and Schlesinger 1985).

The availability of nitrogen can be increased by fertiliation and decreased by application of sawdust and sugar which cause nitrogen to be tied up by decomposer microbes and made less available to plants (Turner 1977, Johnson and Edwards 1979, Waring and Pitman 1985). By adding nitrogen, we would predict a lowering in the lignin to nitrogen ratio of foliage and litter, an increase in decomposition rates, more growth of hardwoods relative to conifers per unit of leaf area, and increasing susceptibility of conifers to pests and pathogens. In adding sawdust and sugar, the ratio of lignin/nitrogen in foliage is expected to rise, the decomposition rate to fall, and conifers to be favored over hardwoods.

Such experiments need to include an assessment of nitrogen availability in the soil, plant, and litter, and any change in the relative availability of other nutrients. Seasonal changes in canopy leaf area, by species, and photosynthetic activity must all be coupled with temperature and light measurements, above and below the canopy. Short-term carbon dioxide flux above canopies might be assessed from aircraft (Desjardins et al. 1982). Maintenance respiration needs to be accounted for by estimating living cells in trees and its importance then evaluated by comparing growth rates of large and small trees with similar photosynthetic rates per unit of foliage. Some estimate of root growth also is required to evaluate changes in the allocation of photosynthate.

A variety of experiments would be desirable in addition to the one proposed. Balanced and imbalanced nutrient supply, drought and flooding are other obvious treatments that should alter ecosystem function and lead to changes in the composition of vegetation. A wide range of treatments is necessary to evaluate how well the general approach of using canopy structure and chemistry applies to a variety of ecosystems. A wide range in remote sensors is also desirable. In this way, we may be able to develop a capability for distinguishing characteristics not directly measureable by any one type of sensor. With a proven experimental base, developed at the local level we can safely and effectively expand to the regional and global scale.

ACKOWLEDGMENTS

Many of the ideas expressed in this paper were developed through discussion at two NASA meetings of the Land Processes Terrrestrial Biology Working Group chaired by Berrien Moore, III. In particular, I thank John Aber and Jerry Melillo for providing special insights in regard to linking canopy characteristics to decomposition. I also thank Hank Margolis for technical editing.

REFERENCES

Amthor, J.S. 1984. The role of maintenance respiration in plant growth. Plant, Cell and Env. 7:561-569.

Axelsson, B. 1981. Site differences in yield differences in biological production or in redistribution of carbon within trees. Res. Report No.9., Dept. Ecol. Environ. Res., Swedish Univ. Agric. Sci., Uppsala.

Barton, F.E., and D. Burdick. 1979. Preliminary study of the analysis of forages with a filter-type near-infrared reflectance spectrometer. J. Agric. and Food Chem. 27: 1248-1252.

Berg, B., P.-E. Jansson, and V. Meentemeyer. 1984. p.389-404. In: G.I. Agren, ed. State and change of forest ecosystems. Swedish Univ. Agr. Sci., Dept. of Ecology and Env. Res., Report No. 13.

Berg,B.,and H. Staaf. 1980. Decomposition rates and chemical changes of Scots pine needle litter. II. Influence of chemical composition. Ecol. Bull. (Stockholm) 32:373-390.

Berry, J.A. and W.J.S. Downton. 1982. Environmental regulation of photosynthesis. p. 262-343. In: Govindjee, ed. Photosynthesis, development, carbon metabolism, and plant productivity. Vol. II. Academic Press, New York.

Botkin, D.B., J.E. Estes, R.M. MacDonald, and M.V. Wilson.1984. Studying the Earth´s vegetation from space. BioScience 34:508-514.

Brantseg, A. 1969. Furu sonnafjells. Produksjonstabeller. Medd. Nor. Skogforsoeksves 26 (84).

Bryant, J.P., F.S. Chapin, III, and D.R. Klein. 1983. Carbon/nutrient balance of boreal plants in relation to vertebrate herbivory. Oikos 40:357-368.

Cox, C.S. and P.B. McEvoy. 1983. Effects of summer moisture on the capacity of tansy ragwort (Senecio jacobaea) to compensate for defoliation by cinnabar moth (Tyria jacobaeae). J. Appl. Ecology 20: 225-234.

Curran, P.J., and E.J. Milton. 1983. The relationship between the chlorophyll concentration, LAI and reflectance of a single vegetation canopy. Int. J. Remote Sensing 4:247-255.

Desjardins, R.L., E.J. Brach, P. Alvo, and P.H. Schuepp. 1982. Aircraft monitoring of surface carbon dioxide exchange. Science 216:733-735.

Fogel, R., and K. Cromack, Jr. 1977. Effect of habitat and substrate quality on Douglas-fir litter decompotion in western Oregon. Can. J. Bot. 55:1632-1640.

Gartlan, J.S., D.B. McKey, P.G. Waterman, C.N. Mbi, and T.T. Struhsaker. 1980. A comparative study of the phytochemistry of two African rain forests. Biochem. Syst. Ecol. 8:401-422.

Gibbs, J.N. 1967. The role of host vigour in the susceptibility of pines to Fomes annosus. Annuals of Botany

31:803-815.

Goetz, A.F.H., B.N. Rock, and L.C. Rowan. 1983. Remote sensing for exploration: an overview. Economic Geology 78:573-590.

Goward, S.N., C.J. Tucker, and D.G. Dye. 1985. North American vegetation patterns observed with the NOAA-7 advanced VHRR. Vegetatio (in press).

Horler, D.N.H., M. Dockray, and J. Barber. 1983. The red edge of plant leaf reflectance. Int. J. Remote Sensing 4:273-288.

Idso, S.B. 1982. Humidity measurements by I.R. thermometry. Remote Sens. Environ. 12:87-91.

Jarvis, P.G., and J.W. Leverenz. 1983. Productivity of temperate, deciduous and evergreen forests. p. 233-280. In: O.L. Lange, P.S. Noble, C.B. Osmond, and H. Ziegler, eds. Encyclopedia of Plant Physiology Vol. 12D. Springer-Verlag, Berlin and New York.

Johnson, D.W., and N.T. Edwards. 1979. The effects of stem girdling on biogeochemical cycles within a mixed deciduous forest in eastern Tennessee. II. Soil nitrogen mineralization and denitrification rates. Oecologia 40: 259-271.

Jordan, C.F. 1969. Derivation of leaf area index from quality of light on the forest floor. Ecology 50:663-666.

Kira, T., and T. Shidei. 1967. Primary production and turnover of organic matter in different forest ecosystems of the western Pacific. Jpn. J. Ecol. 17:70-87.

Lainson, R.A. and J.H.M. Thornley. 1982. A model for leaf expansion in cucumber. Annual of Bot. 50:407-425.

Lambert, M.J., and J. Turner. 1977. Dieback in high site quality Pinus radiata stands -- the role of sulphur and boron deficiencies. New Zealand J. For. Sci. 7:333-348.

Lechowicz, M.J. 1984. Why do temperate deciduous trees leaf out at different times? Adaption and ecology of forest communities. Amer. Naturalist 124:821-842.

Linder, S., R. McMurtrie, and J.J. Landsberg. 1985. Crop physiology of forest trees. I.U.F.R.O. S2.01-00. Symp. Univ. of Helsinki, July 1984. (in press).

Linder, S. and D.A. Rook. 1984. Effects of mineral nutrition on carbon dioxide exchange and partitioning of carbon in trees. p. 211-236. In: G.D. Bowen and E.K.S. Nambiar, eds. Nutrition of forest trees. Academic Press, New York.

Maclean, G.A., and W.B. Krabill. 1985. Estimating gross

merchantable timber volume using an airborne lidar system. Can. J. Remote Sensing (in press).

Matson, P.A., and R.H. Waring. 1984. Effects of nutrients and light limitations on mountain hemlock: susceptibility to laminated rootrot. Ecology 65:1517-1524.

McBrayer, J.F., J.M. Ferris, L.J. Metz, C.S. Gist, B.W. Cornaby, Y. Kitazawa, T. Kitazawa, J.G. Wernz, G.W. Krantz, and H. Jensen. 1977. Decomposition invertebrate populations in U.S. forest biomes. Pedobiologia 17: 89-96.

Meentemeyer, V. 1978. Macroclimate and lignin control of decomposition rates. Ecology 59:465-472.

Melillo, J.M., J.D. Aber, and J.F. Muratore. 1982. Nitrogen and lignin control of hardwood leaf litter decomposition dynamics. Ecology 63:621-626.

Monteith, J.L. 1977. Climate and efficiency of crop production in Britian. Philos. Trans. Royal Soc. (London). Ser. B. 281:277-294.

Nelson, R., W. Krabill, and G. MacLean. 1984. Determining forest canopy characteristics using airborne laser data. Remote Sensing of Environ. 15:201-212.

Nihlgard, B. 1985. The ammonium hypothesis -- an additional explanation to the forest dieback in Europe. Ambio 14:2-8.

Paloscia, S., and P. Pampaloni. 1984. Microwave remote sensing of plant water stress. Remote Sensing of Environ. 16:249-255.

Pampaloni, P., and S. Paloscia. 1985. Experimental relationships between microwave emission and vegetative features. Int. J. Remote Sensing 6:315-323.

Reichle, D.E. 1977. The role of soil invertebrates in nutrient cycling. p. 145-156. In: Soil organisms as components of ecosystems. Ecol. Bull. (Stockholm) No. 25.

Schoeneweiss, D.F. 1975. Predisposition, stress, and plant disease. Annual Review of Phytopathology 13:193-211.

Schreier, H., J. Lougheed, C. Tucker, and D. Leckie. 1985. Automated measurements of terrain reflection and height variations using an airborne infrared laser system. Int. J. Remote Sensing 6:101-113.

Schwaller, M.R., C.C. Schnetzler, and P.E. Marshall. 1983. The change in leaf reflectance of sugar maple (*Acer saccharum* Marsh) seedlings in response to heavy metal stress. Int. J. Remote Sensing 4:93-100.

Spanner, M.A., K.B. Teuber, W. Acevedo, D.L. Peterson, S.W. Running, and D.A. Mouat. 1984. Remote sensing of leaf area index of temperate coniferous forest: Early results. Tenth Int. Symp. on Machine Processing of Remotely Sensed Data, Purdue Univ., June, 1984. (in press).

Tucker, C.J. 1980. A critical review of remote sensing and other methods for non-destructive estimation of standing crop biomass. Grass and Forage Sci. 35:177-182.

Tucker, C.J., J.R.G. Townshend, and T.E. Goff. 1985. African land-cover classification using satellite data. Science 227:369-375.

Tuomi, J., P. Niemela, E. Haukioja, S. Siren, and S. Neuvonen. 1984. Nutrient stress: an explanation for plant anti-herbivore responses to defoliation. Oecologia 61:208-210.

Turner, J. 1977. Effect of nitrogen availability on nitrogen cycling in a Douglas-fir stand. Forest Sci. 23:307-316.

Wardley, N.W., and P.J. Curran. 1984. The estimation of green-leaf-area index from remotely sensed airborne multispectral scanner data. Int. J. Remote Sensing 5:671-679.

Waring, R.H. 1983. Estimating forest growth and efficiency in relation to canopy leaf area index. Adv. Ecol. Res. 13:327-354.

Waring, R.H., A.J.S. McDonald, S. Larsson, T. Ericsson, A. Wiren, E. Arwidsson, A. Ericsson, and T. Lohammar. 1985. Differences in chemical composition of plants grown at constant relative growth rates with stable mineral nutrition. Oecologia 66:157-160.

Waring, R.H., and G.B. Pitman. 1985. Modifying lodgepole pine stands to change susceptibility to mountain pine beetle attack. Ecology 66:889-897.

Waring, R.H., and W.H. Schlesinger. 1985. Forest ecosystems: concepts and management. Academic Press, New York. (in press).

Williams, D.L. R.F. Nelson, and C.L. Dottavio. 1985. A georeference LANDSAT digital database for forest insect-damage assessment. Int. J. Remote Sensing 6:643-656.

Vane, G., A.F.H. Goetz, J.B. Wellamn. 1983. Airborne imaging spectrometer: A new tool for remote sensing. IGARSS´83, Vol. 2, FA-4. p. 6.1-6.5. San Francisco.

NEW AND EMERGING TECHNOLOGY - DATA ACQUISITION

Moderator: W. Barnes

AN AIRBORNE IMAGING SYSTEM FOR MEASURING ANISOTROPIC REFLECTANCE PATTERNS

by W. L. Barnes, F. G. Huegel, and J. R. Irons
Goddard Space Flight Center
Greenbelt, Maryland 20771

ABSTRACT

An airborne system to image anisotropic reflectance patterns from plant canopies in 32 spectral bands between 0.4 and 0.9 micrometers has been developed and tested at Goddard Space Flight Center. The system consists of the Advanced Solid-state Array Spectroradiometer (ASAS), a 32-channel imaging spectrometer, and a semi-automated pointing system that allows the ASAS to be pointed 45° fore or aft of nadir. The system has been flown on an SC-7 Skyvan at an altitude of 3.0 kilometers. At this altitude the ASAS had a nadir spatial resolution of 2.6 meters and a swath of 1.3 kilometers. The pointing system allowed the acquisition of 150 lines of data at 15° fore-aft pointing increments. Thus, seven angular views of 32 spectral channels each were acquired in a single pass.

Initial experiments will be to quantify atmospheric effects on multiple angle measurements of reflectance and to examine the anisotropic reflectance patterns of forest canopies.

INTRODUCTION

Advances in remote sensing of the Earth's surface from space over the last twenty years have resulted from improvements in spatial and spectral resolution, signal-to-noise, number and location of spectral channels, and temporal resolution. A parameter that has not been examined in detail is the angular dependence of reflected solar radiance from plant canopies. Current attempts to model this dependency require data sets containing carefully measured surface and remote parameters in order to verify the models. An airborne system has been developed at the Goddard Space Flight Center that measures the reflected spectral radiance from a scene on the order of 500 x 200 pixels at seven or more view angles and 30 spectral channels in a single pass of the aircraft. The system has been successfully flown and will be utilized operationally starting in August 1985. The following sections will discuss the scientific objectives of the data from this system, outline the system including a description of the Advanced Solid-State Array Spectroradiometer (ASAS) and the sensor pointing system, and briefly discuss the results from the initial flights of the system.

OBJECTIVES

The scientific rationale for operating the pointable ASAS system is to understand, characterize and utilize anisotropic reflectance patterns from terrestrial surfaces. The non-Lambertian character of most terrestrial reflectance patterns is well known from emperic measurements (e.g., Deering 1985;

Kriebel 1978) and analytic modeling (e.g., Kimes, 1984; Norman, 1979). Research is currently establishing the relationships between the biophysical characteristics of terrestrial surfaces, particularly plant canopies, and the anisotropy of light reflected from the surfaces.

Several compelling reasons for observing and characterizing anisotropic reflectance are apparent. Knowledge of land albedos is limited by current remote sensing capabilities from space (Barrett, 1974). A single observation from nadir or other view angle is inadequate for describing the anisotropic reflectnce from the entire hemisphere about a surface. The assumption of Lambertian reflectance from a terrestrial surface can lead to either the over estimation or under estimation of the actual albedo (Eaton and Dirmhirn, 1979). Multiple angle observations and knowledge of reflectance anisotropy are required for improved albedo determinations.

Another justification for multiple angle observations is the estimation of biophysical parameters. The work of Goel and Thompson (1984) has shown that plant canopy characteistics (e.g., leaf area index and leaf angle distribution) can be accurately estimated by model inversion if data from several viewing and illuminating geometries are available. Additionally, a study by Smith and Oliver (1974) indicated that off-nadir view zenith angles may be preferable for discriminating levels of plant cover density.

The off-nadir pointing of future spaceborne sensors presents another reason for studying reflectance anisotropy. The Systeme Probatoire d'Observation (SPOT), for example, is scheduled to carry pointable sensors into orbit beginning in 1986. The use of spaceborne sensors with wide FOV's is also increasing in global and continental scale terrestrial studies. For instance, anisotropic surface reflectance is a consideration in analyses of Advanced Very High Resolution Radiometer (AVHRR) data (Holben et al., 1984). All of these remote sensing endeavors establish a need for characterizing the anisotropic reflectance from terrestrial surface materials.

APPROACH

The basic requirements for the science tasks under consideration are multispectral measurements of a well-instrumented test site at a number of sensor view angles between $\pm$ 45° in a time period short enough to preclude any significant change in incident radiation. The system being used to make these measurements consists of a 32 channel visible and near-infrared imaging spectrometer mounted on a platform that points the radiometer at predetermined angles along the path of the aircraft under microcomputer control. The paragraphs below will discuss the performance parameters of the imaging spectrometer and of the pointing system and the following section will discuss recent test flights.

ASAS

The Advanced Solid-State Array Spectroradiometer (ASAS) is modified version of the Scanning Imaging Spectroradiometer (SIS) that was developed in the early 1970's for NASA's Johnson Space Center (JSC) by TRW and later modified by substitution of a silicon detector array for the original vidicon to become the Solid-state Array Spectroradiometer (SAS). The SAS was modified and renamed ASAS in 1983 after installation of a new CID detector array developed for the Naval Ocean Systems Center (NOSC) by General Electric (Stewart et al, 1985). Subsequently the ASAS was transferred to GSFC.

A conceptual drawing of the ASAS optical train is shown in Figure 1. The reflected solar radiance from a line of 512 pixels at right angles to the aircraft direction of flight is imaged by the collecting optic onto the entrance slit of a spectrometer with a 75 line/mm transmission grating blazed at 530 nm. Located at the focal plane of the spectrometer is the 512 x 32 element silicon CID detector array. Thus, each of the 512 surface pixels is dispersed into 32 spectral channels. It should be emphasized that Figure 1 is idealized in that the ASAS optics actually consist of 38 optical elements plus the grating and detector array.

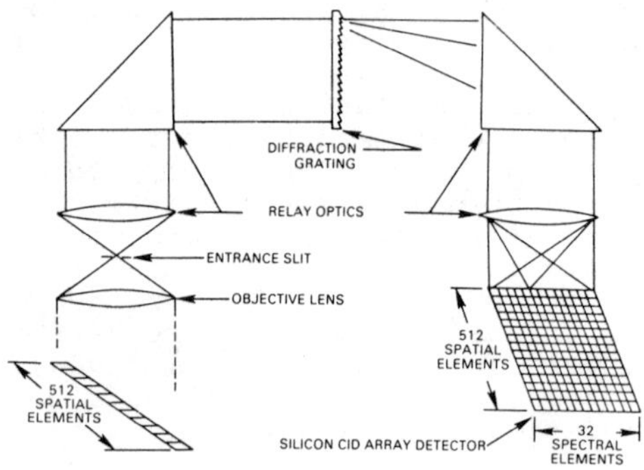

FIGURE 1. ASAS Optical System

The ASAS image is formed from the motion of the aircraft and sampling of the detector array at the proper times to ensure contiguity of the 512 pixel lines. This type of image formation is commonly designated as pushbroom and has the advantage that there are essentially no moving parts and the dwell time for each pixel is (in this case) at least 512 times greater than can be achieved with a scanning system having a single detector element.

The output from the CID array is read out and digitized to twelve bit accuracy at a rate variable between 3 and 64 frames per second for a data rate of 0.7 to 14.4 megabits per second. The ASAS system includes two interchangable objective lens with 10° and 25° field-of-view (FOV).

Therefore, it is possible to use the ASAS from a variety of platforms with widely differing altitudes and ground speeds.

To illustrate the system's flexibility, Table 1 lists the footprints, swaths and data rates for both objectives as viewed from a SC-7 Skyvan at 3 kilometer altitude, the NASA U-2 at 20 kilometers, and the NASA Shuttle at 283 kilometers. The system has flown on both the Skyvan and the U-2, but there are presently no plans to include it as part of an STS mission.

Table 1 - ASAS Performance Parameter

Platform	Alt (km)	FOV (deg)	IFOV (mr)	Footprint (m)	Swath (km)	Frame Rate (sec^{-1})	Data Rate (Mb/sec)
Skyvan	3	25	0.85	2.6	1.3	26	5.85
	3	10	0.34	1.0	0.5	64	14.4
U-2	20	25	0.85	17.0	8.7	12	2.7
	20	10	0.34	6.8	3.5	30	6.75
STS	283	25	0.85	241.0	123.0	30	6.75
	283	10	0.34	96.2	49.3	64	14.4

At present, only 30 of the 32 spectral channels of ASAS are operable. The spectral range covered by the 30 channels is 455 to 873 nanometers with each channel having a full width at half maximum of 14 nanometers. The spectral range is adjustable between missions by moving the detector array in the focal plane. Such an adjustment requires complete recharacterization of the system. The spectral response of the ASAS has been measured by observing the analog output of a single element as the wavelength of a 10 micrometer diameter spot is varried between 400 and 900 nanometer. This results in the spot traveling across the array in the 32-element spectral direction. The results from three of these scans while observing spectral channels 8, 9, and 10 are shown in Figure 2. The overlap between channels is

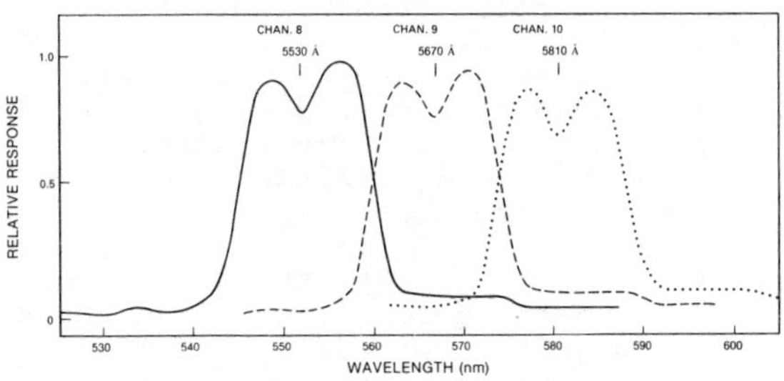

FIGURE 2. Spectral response of ASAS channels 8-10

considerably less than shown previously (Stewart et al, 1985) due to correction of an error in array timing. The dip near the middle of each channel is due to a 12-micrometer wide lead of aluminum and thick polysilicon that runs through the middle of each 81-micrometer wide element. The integrated response at wavelengths greater than two half-widths from the band center is typically less than five percent of the total response. Total crosstalk between channels is of the order of 20 percent.

Video gain is adjustable by factors of one, two, four or eight. In combination with an offset capability, near optimum signal-to-noise can be obtained for all combinations of scene contrast and noise. Laboratory tests for a variey of gains, frame rates and different detector elements have shown saturation to be about 3400 digital counts (4096 count range) and typical rms noise to full-scale signal of 1000:1.

Pointing System

During the last 18 months, a pointable mount has been developed at GSFC that allows the ASAS to view an area on the surface at several angles in a single pass of the aircraft. The ASAS camera and a color TV camera are boresighted on a gimballed platform that is driven by a DC motor via a worm gear and worm wheel as shown in Figure 3.

The total airborne system is shown in block form in Figure 4. The heart of the system is a Compaq Model 286 portable microcomputer that controls the system during a data pass. Prior to a data pass the operator enters the start and stop angles (typically +45° and -45°), the angular increment between acquisitions, the number of scan lines to be acquired at each angular step and the speed and altitude of the aircraft. The system determines the ASAS frame rate required for contiguity of the scan lines at nadir and the time allowed for the operator to visually reacquire the test site between acquisitions (typically 2-7 seconds). To start a data acquisition, the pilot and operator align the test site along the aircraft path using separate video monitors (see Figure 4), the operator starts the acquisition as the ASAS reaches the edge of the site, the Compaq automatically counts the desired number of scan lines and moves the platform to the next angle; the operator waits for the ASAS to again reach the edge of the site, pushes the button, and the Compaq starts counting again. As shown in Figure 4, the output from the video camera and time code are recorded continuously on a video tape recorder while the ASAS data together with time code, roll and pitch information, and a data start pulse at each acquisition angle are recorded by a 14 track Ampex recorder.

At the conclusion of each flight, the quality of the data can be examined using the quick-look system shown in Figure 5 and by reviewing the imagery from the video recorder. The data is then returned to GSFC where the aircraft tapes are converted to 6250 BPI computer compatible tapes (CCT's) using an HP-1000 minicomputer. Each hour of aircraft flight time results in 26-28 CCT's.

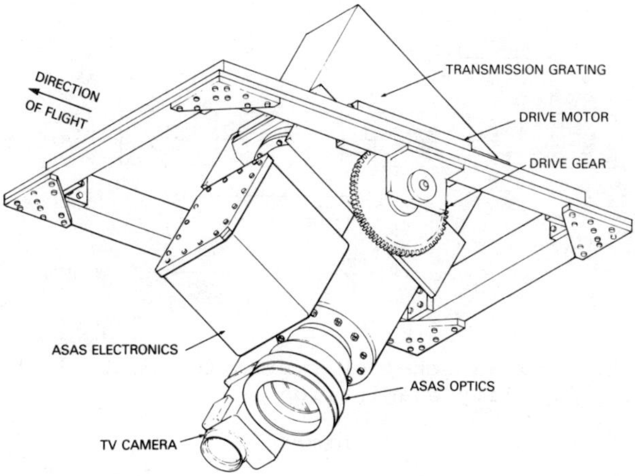

Figure 3. ASAS pointing system

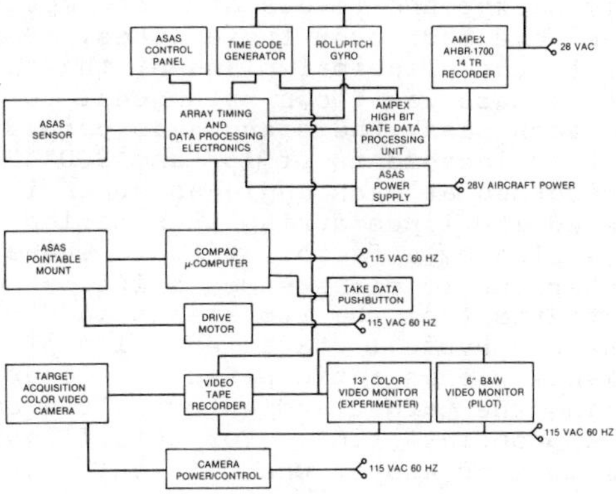

Figure 4. ASAS pointing system-airborne configuration

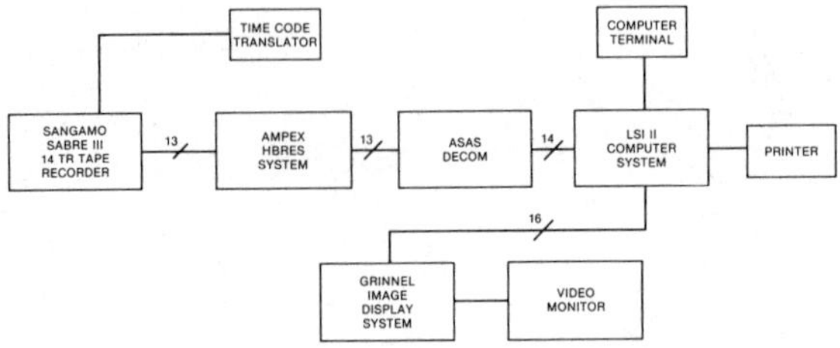

Figure 5. ASAS quick-look system for post flight data quality examination

DATA

The initial flight of the ASAS pointing system was on June 17, 1985. The sensor was flown on an SC-7 Skyvan over a geobotany research area at an altitude of 10,000 ft. Two tree stands were of interest within the area. One stand is growing on soil containing high concentrations of lead, zinc, and copper. The other stand is growing on soil containing only background levels of these metals. Two aircraft passes were made over these sites. One pass was flown parallel to the principal plane of the sun and the sites. The other pass was flown orthogonal to the principal plane. During each pass, the sensor was pointed from 45° fore to 45° aft at increments of 15° and 300 ASAS frames (lines) were recorded at each angle as shown in Figure 6. The SC-7 advanced 150 lines during this period, so the data was oversampled by a factor of two. The data will be analyzed to determine if differences exist between the reflectance patterns from the two stands and to relate the patterns to the biophysical characteristics of the stands. System performance was as anticipated; the operator had no problems locating the target and entering the data start signal at the appropriate time. For this flight the pilot did not have a monitor and it was difficult for him to maintain the required flight line in a cross-wind once the target was no longer in front of the aircraft. Therefore, a monitor for the pilot will be added for subsequent flights as shown in Figure 4.

SUMMARY

A pointable airborne imaging spectrometer system has been developed and successfully tested. The system records 30 contiguous, 14 nanometer wide spectral channels between 455 and 873 nanometers while being pointed at predetermined angles fore or aft of the aircraft. The system has demonstrated its ease of use while recording the solar spectral reflectance from an area 0.8 x 1.4 kilometer with 2.6 meter spatial resolution at 7 angles from +45° to -45° in a single pass of the aircraft. The system will be flown operationally in August and September 1985 to examine reflectance patterns from forests exposed to acid depositions and to study atmospheric effects on remote, multiple angle observations.

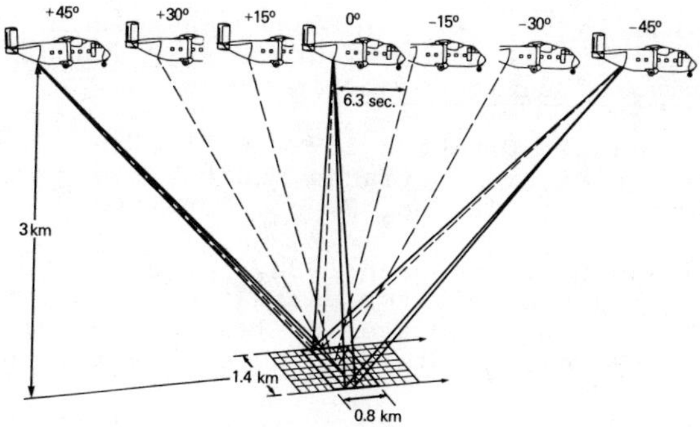

Figure 6. ASAS pointing configuration for June]7,]985 test flight.

REFERENCES

Barrett, E.C. 1974, Climatology from Satellites, Methuen, London, p. 418

Deering, D.W., and P. Leone 1985, A sphere-scanning radiometer for rapid directional measurements of sky and ground radiance, Remote Sensing of Environment (in press)

Eaton, F.D., and I. Dirmhirn 1979, Reflectance irradiance indicatrices of natural surfaces and their effect on albedo, Applied Optics, Vol. 18, No. 7, pp. 994-1008

Goel, N.S., and R.L. Thompson 1984, Inversion of vegetation canopy reflectance models for estimating agronomic variables IV: Estimation of LAI and average leaf angle using measured canopy reflectances, Remote Sensing of Environment (in press)

Holben, B.N., D.S. Kimes, and R.S. Fraser 1984, Directional reflectance response in AVHRR red and near-IR bands for three cover types and varying atmospheric conditions, Remote Sensing of Environment (in press)

Kimes, D.S. 1984, Modeling the directional reflectance from complete homogeneous vegetation canopies with various leaf-orientation distributions, Journal of the Optical Society of America A, Vol. 1, pp. 725-737

Kriebel, K.T. 1978, Measured spectral bidirectional reflection properties of four vegetated surfaces, Applied Optics, Vol. 17, No. 2, pp. 253-259

Norman, J.M. 1979, Modeling of complete crop canopy in Modification of the Aerial Environment of Plants, B.G. Barfield and J.F. Gerber, eds. ASE monograph No 2, St. Joseph, Michigan, pp. 249-277

Smith, J.A., and R.E. Oliver 1974, Effects of changing canopy directional reflectance on feature selection, Applied Optics, Vol. 13, No. 7, pp. 1599-1604

Stewart, S.E., R.R. Buntzen, and J. Carbone 1985, Advanced CID array multispectral pushbroom scanner, Laser Focus/Electro-Optics, Vol. 21, No. 6, pp. 88-100

HIGH-SPECTRAL RESOLUTION REMOTE SENSING WITH THE AIRBORNE VISIBLE/INFRARED IMAGING SPECTROMETER (AVIRIS)

Gregg Vane
Jet Propulsion Laboratory/
California Institute of Technology
4800 Oak Grove Drive
Pasadena, California 91109

SUMMARY

The value of high-spectral resolution remote sensing has been thoroughly demonstrated in the recent past through the use of several airborne and spaceborne spectroradiometers and a prototype infrared imaging spectrometer which has been developed at the Jet Propulsion Laboratory (Goetz, et al, 1985; Vane and Goetz, 1985). Imaging spectrometry is defined as the simultaneous acquisition of images in many narrow contiguous spectral bands. This approach to imaging allows a complete reflectance spectrum to be reconstructed for each picture element (pixel) in the scene. If the surface materials in the scene have unique spectral signatures, the information in the reconstructed spectrum will allow these materials to be directly and uniquely identified on a pixel-by-pixel basis. Many naturally occurring surface materials have diagnostic absorption features that are 20 to 40 nm in width at the half power points; it is for these types of materials that imaging spectrometry has proven especially effective. Examples of the successful application of imaging spectrometry to geological and botanical remote sensing will be presented from data acquired with the prototype instrument mentioned above.

A second-generation sensor is now under construction at JPL which will image the 0.4 to 2.4μm region in 224 contiguous 10 nm spectral bands. This instrument, which is called the Airborne Visible/Infrared Imaging Spectrometer (AVIRIS), will be an operational NASA facility instrument to be flown on the NASA U-2 and ER-2 when it is completed in 1987 (Vane, et al., 1984). From an operating altitude of 20 km, AVIRIS will have a ground pixel size of 20 m and swath width of 11 km, giving images that are 550 pixels across. Data will be encoded to 10 bits to take advantage of the high radiometric performance of the instrument. Signal-to-noise ratios of about 200 to 1 and 100 to 1 are expected in the visible and short wavelength infrared, respectively. The foreoptics of AVIRIS consist of a modified Kennedy scanner with a three-sided mirror designed to scan a field of view of 33 degrees. Light collected by the foreoptics is split between four spectrometers which are optically connected to the foreoptics with light fibers. At the focus of each spectrometer is a line array of detectors. The spectrometer for the visible portion of the spectrum uses a 32 element silicon line array while each of the three spectrometers for the near and short wavelength infrared uses a 64 element line array of indium antimonide.

A key element of the AVIRIS system is the data handling facility also under construction at JPL which is based on a VAX 11-780 minicomputer. This facility will provide the archiving and retrieval of all AVIRIS data. The facility has been sized to permit the complete pre-processing of a given data set within two weeks of acquisition. Pre-processing consists first of converting the raw data recorded aboard the U-2 on a high density tape to a computer compatible 6250 bpi tape (CCT) which is the archival medium. The archival CCT will contain the raw image data and the engineering, calibration and aircraft motion data acquired with the image data. Retrieval processing, which will be done only on those portions of the raw data requested by experimenters, will consist of radiometric correction to the raw data using the appropriate calibration data, and geometric corrections based on knowledge of aircraft motion.

AVIRIS experimeters will be chosen by NASA through "the dear colleague" letter process. The first letter will be released in late 1985 or early 1986. For details contact Dr. Robert E. Murphy, Earth Science and Applications Division, NASA Headquarters, Code EE, Washington, D.C. 20546.

REFERENCES

Goetz, A.F.H., Vane, G., Solomon, J.E., and Rock, B.N., 1985, Imaging Spectrometry for Earth Remote Sensing: Science, Vol. 228, No. 4705, pp. 1147-1153.

Vane, G., Chrisp, M., Enmark, H., Macenka, S., and Solomon, J., 1984, Airborne Visible/Infrared Imaging Spectrometer: An Advanced Tool for Earth Remote Sensing: Proc. 1984 IEEE Int'l Geoscience and Remote Sensing Symposium, Pub. No. SP215, European Space Agency, Paris, France, pp. 751-757.

Vane, G. and Goetz, A.F.H. (editors), 1985, Proceedings of the Airborne Imaging Spectrometer Data Analysis Workshop, April 8-10, 1985: JPL Pub. No. 85-41, Pasadena, California, 173 pages.

RESEARCH OPTICAL SENSOR

Carl F. Schueler
Hughes Santa Barbara Research Center
75 Coromar Drive
Goleta, CA 93117

ABSTRACT

Advances in Landsat sensor technology at Santa Barbara Research Center (SBRC) have been sponsored by the National Aeronautics and Space Administration's Goddard Space Flight Center (NASA GSFC). The most recent development in this area is a concept known as the Research Optical Sensor (ROS). ROS is intended for land resource research, and must provide excellent spatial, spectral, and radiometric resolution, plus ground programmable data acquisition flexibility so that the sensor can be tailored on-orbit to specific research requirements. Fundamental imaging requirements that ROS must satisfy include a ground resolution of 10 meters in the visible and near infrared (VNIR) and 20 meters in the short-wave infrared (SWIR), many spectral bands covering the VNIR and SWIR at a spectral resolution of 20 nm, and enhanced pointing capability for bidirectional reflectance distribution function (BRDF) measurement as well as for stereo and cross-track imaging. A variety of sensor concepts were initially examined to arrive at a flexible design for land-resource scientists.

BACKGROUND

The state of the art of pushbroom Earth observation space sensors was last reviewed for the Pecora symposium in 1983 (Mika 1983). Significant technical progress has been made since to prove the feasibility of the pushbroom sensor concepts that underlie the advanced Landsat initiative. In particular, both the required VNIR and SWIR line array detector fabrication have been demonstrated at RCA and at SBRC under NASA GSFC sponsorship (Tower et al. 1984; Cannata et al. 1985). As a result, the goal of building a pushbroom space sensor is now considerably closer that it was in 1983, and a sensor concept that offers a combination of advanced features and reasonable cost and risk using demonstrated technology has been designed by SBRC. This Research Optical Sensor (ROS) concept was specified by NASA in late 1983 to offer the following key features:

- **Progress over TM** - a significant step forward in radiometric, spatial, and spectral performance relative to the Thematic Mapper.
- **Data Versatility** - flexible data acquisition capability under ground control for extensive research experimentation.
- **Growth Potential to Imaging Spectrometry**
- **Reasonable Risk** - use demonstrated focal-plane technology.

The progress beyond Thematic Mapper performance is probably the major item of interest, because ROS is intended as the research and development bridge between current whiskbroom Thematic Mapper technology and future operational instrumentation. It is currently practical to consider a hardware development effort in the near future leading to procurement of an ROS instrument with data acquisition capabilities listed in Table 1.

Table 1. Research Optical Sensor Performance

Spatial Resolution	10m + multiples (VNIR) 20m + multiples (SWIR) 5m + multiples [Panchromatic (PAN)] 40m + multiples [Middle IR (MIR)] 120m + multiples [Thermal IR (TIR)]
Coverage	10-60 km, selectable in 10-km increments from 500-km orbit
Spectral Resolution and Coverage	PAN 200 nm, 1 band at 0.55-0.75 μm VNIR 20 nm, 12 bands over 0.4-1.0 μm SWIR 20 nm, 6 bands over 1.0-2.5 μm MIR 1 band at 3.6-4.0 μm TIR 2 bands at 10.4-11.3 μm and 11.5-12.4 μm
Radiometric Accuracy	2-5% absolute; <2% relative; 0.25% intraband
Sensitivity	0.5% Reflectance (NEΔρ) 0.2K Temperature (NEΔT)
Pointing	Cross-Track ±41° (±4000 steps) Fore/Aft ±0-41° (±1500 steps)

The versatility for research experimentation is the next most important priority for ROS. During the past fifteen years the emphasis in NASA's remote sensing efforts has been on development of sensors to allow compilation of thematic maps and to make crop production predictions. To meet these needs, the Multispectral Scanner (MSS) and the Thematic Mapper (TM) systems were designed to provide wide-area coverage almost twice a month from a sun-synchronous orbit. NASA's priorities have since switched from operational to research applications. Therefore, the ROS mission is focused on the needs of the remote sensing research community. As such, the requirement for global, repetitive coverage is not as important as it was for TM, and flexibility in radiometric, spatial, and spectral capability is more critical.

The other features listed above are also important. Growth potential to imaging spectrometry will ensure that the ROS concept will not suffer technological obsolescence as area detector array technology matures, and risk is to be minimized because the cost of a new space sensor escalates rapidly when unproven technology is part of the concept. SBRC's ROS design meets the NASA objectives, and a comprehensive, 800-page final report fully documents the results (Hughes SBRC 1985).

The Research Optical Sensor concept envisoned by NASA prior to the SBRC design study was not restricted to any one technology or approach, but instead was only defined in terms of the functional requirements presented in Table 1 above. Accordingly, the SBRC study embraced a number of alternate methods to accomplish the NASA performance objectives. Moreover, NASA requested not only a baseline concept definition to meet the stated goals, but also more detailed evaluation of the sensor concept's capability to offer some specific research data. In particular, excellent radiometric accuracy combined with atmospheric correction capability are viewed as crucial to the goal of developing a space sensor that can accurately estimate ground reflectance and temperature. Thus, the study encompassed both sensor design and applications.

DESIGN TRADEOFFS AND BASELINE CONCEPT

Sensor Design Tradeoffs

There are basically two design approaches to sensor implementation, illustrated in Figure 1: the "whiskbroom" approach (a), exemplified by the TM sensor on Landsat 5 (Hughes SBRC 1984), and the "pushbroom" linear array approach (b), with its extension to the use of area arrays (c), which are undergoing technology development at various laboratories.

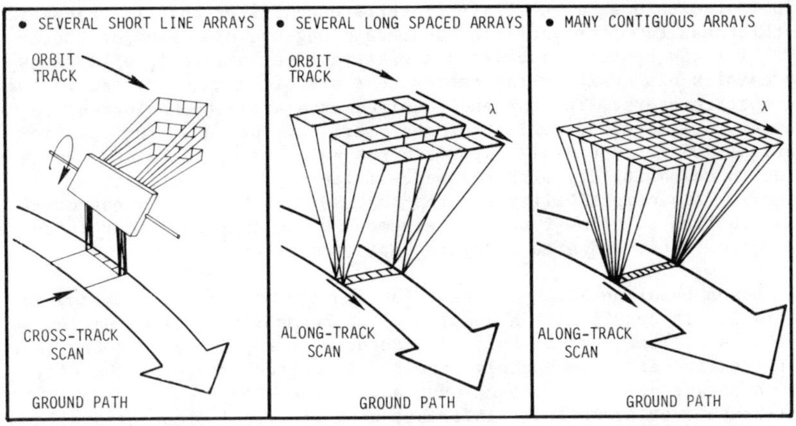

(a) Whiskbroom (b) Pushbroom (c) Area array spectrometer

Figure 1. Fundamental scanning approaches

The chief advantage to the whiskbroom approach is that only a few detector elements are required to produce imagery by virtue of a scanning mirror system. Therefore, technology risk and focal plane cost are at a minimum for this approach. It is ideal for current operational applications, and represents the approach of choice for the 1980's U.S. commercial Landsat initiative. One disadvantage of the whiskbroom approach, however, lies in the very short signal integration times available because a few detectors have to cover an entire ground swath in a very short time (185 km in 60 ms for TM), leading to limitations in signal-to-noise ratio (SNR). A second disadvantage is that the scan-mirror mechanical system is optimized for a specified orbit.

The relatively slow scan pushbroom approach allows better than a factor of 400 increase in signal integration time over the whiskbroom approach. Using pushbroom technology, the ROS concept can provide significantly finer spectral and spatial resolution than TM, and yet with equivalent or better SNR, as required by NASA. Moreover, the pushbroom approach offers the added advantage of multiorbit potential, since the ground scan is controlled purely by spacecraft motion. Only electronic detector readout timing changes need be implemented to accommodate alternate orbits. Therefore, alternate pushbroom concepts were considered that primarily differ in the means by which spectral bands are separated and defined on the focal plane.

Included among the various options was the exotic Fourier transform spectrometer, which offers electronic spectral tunability and perfect band-to-band registration. Although this concept is feasible in principle, mechanical and signal-processing requirements placed it beyond current to near-term state of the art. Another interesting approach, suggested by Mr. Marvin Maxwell of NASA GSFC, was to use an area detector array with a rapidly moving filter wheel to obtain time-multiplexed spectral selection, but again the mechanical requirement for a moving mechanism as well as the need for area array technology makes this concept less attractive than others available.

A third alternate prospect is the linear spectral wedge (LSW) filter spectrometer. This involves the use of area detector arrays that are illuminated through a multilayer interference filter that is graduated in thickness to offer variable bandpass along one dimension of the array. The LSW approach appears quite attractive because it eliminates the complex dispersive-relay optics that are often incorporated in imaging spectrometers. The LSW and required area array technologies are, however, still beyond current or near-term development. The opto-mechanical ROS concept design approach was nevertheless developed so as to maintain compatibility with the wedge filter spectral definition method. This compatibility ensures that ROS has the growth potential NASA requires to allow future development of an imaging spectrometer as the wedge filter and area array technologies mature.

The chosen baseline ROS spectral selection approach is illustrated in Figure 2. It consists of a complex glass beamsplitter illuminating several sets of long, narrow multiline detector arrays. Besides the fact that the line-array technology required is current state of the art, a key advantage of this approach over any of the others considered is object-space overlay of image information on the focal-plane arrays located below the respective beamsplitter exit ports, which define individual spectral zones. Therefore, along-track displacement is limited to the separation of detector lines within each spectral zone. This is important because along-track object-space separation can result in band-to-band image distortion due to terrain relief as well as orbit velocity variation. While such errors can be corrected by ground data post-processing, object-space registration is desirable.

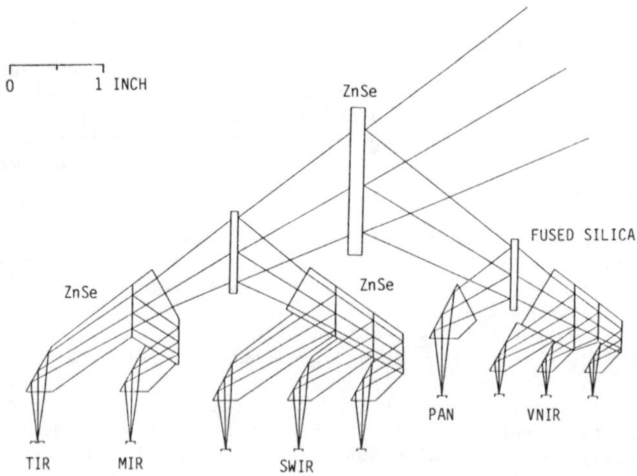

Figure 2. ROS beamsplitter spectral separation concept

The results of these and other tradeoff considerations led to the ROS baseline sensor concept illustrated in Figure 3, which offers the following key features:

- Compact, all-reflective, three-mirror telescope with 6.87° FOV for a 60-km swath from a 500-km polar, sun-synchronous orbit;
- Omnidirectional off-nadir pointing via movable mirrors attached to a fixed instrument, with no image rotation;
- Modular beamsplitter/focal-plane unit permitting in-orbit replacement;
- Use of current or near-term detector technology, featuring RCA silicon VNIR and SBRC HgCdTe SWIR/MIR/TIR arrays with active cooling, permitting space station implementation;
- Electronic spectral selection, selectable swath, instantaneous field-of-view (IFOV) and dynamic range with on-board radiometric calibration, band-to-band registration measurement, and spectral characterization; and
- Growth potential to include an imaging spectrometer.

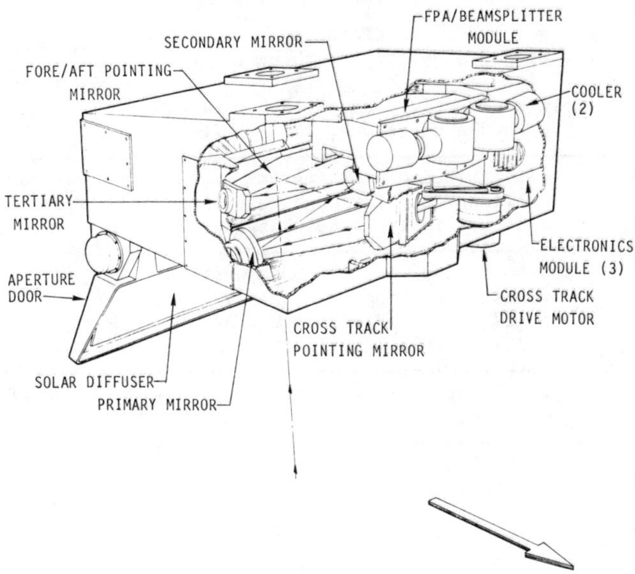

Figure 3. Research Optical Sensor concept cutaway view

ROS Baseline Design Concept

Table 2 summarizes major specifications of the ROS concept shown in Figure 3. A key feature of this design is that simultaneous fore/aft and cross-track pointing may be performed without image rotation, for a view to any point on the Earth within 450 km of the nadir point in any direction from a nominal 500-km orbit, as illustrated in Figure 4. Stereo images at angles as small as 10° from nadir can be acquired due to the rapid slew rate afforded by the mirror motors. A bonus of the two-mirror pointing system is cancellation of polarization effects at high incidence angles.

Table 2. ROS Design Summary

Optics	• 3-mirror all-reflective telescope. 6.87° FOV • 9-way beamsplitter	
EFL/EPD = f/number	750 mm/200 mm = 3.75	[EFL: Effective focal length] [EPD: Effective pupil diameter]
Focal Plane Array (FPA)* (20 km swath segment)	• 40,800 detector elements, 30 mm long, T_{op} = 295K VNIR/PAN = 140K SWIR = 40K MIR/TIR	[T_{op}: Operating temperature]
VNIR	• 12 bands, 2048 elements/band • Si detectors, 15 μm detector center-to-center spacing (pitch)	
PAN	• 1 effective band, 8192 elements • Si VNIR detectors, 7.5 μm pitch (masked and staggered VNIR detectors)	
SWIR	• 6 bands, 1024 elements/band • HgCdTe detectors, 30 μm pitch	
MIR	• 1 band, 512 elements • HgCdTe detectors, 60 μm pitch	
TIR	• 2 effective bands, 1366 elements • HgCdTe detectors, 180 μm pitch	
Electronics	• Distributed architecture with up to three parallel slices, one for each 20 km segment of FPA • 10-bit internal quantization; 8 bits transmitted	
Mechanical	• Monocoque graphite-epoxy optical bench • Active cooling (235 watt VM cooler) • 127 × 94 × 69.9 cm (50 × 37 × 27.5 inches) • 307 kg (675 lb) total weight	
Power (Maximum)†	• 688 watts (20 km FPA) to 1196 watts (60 km FPA) • 300 watts standby	

*Multiply by 2 for 40 km FPA, and by 3 for 60 km FPA.
†Nominal operating loads from 50 to 70% of maximum.

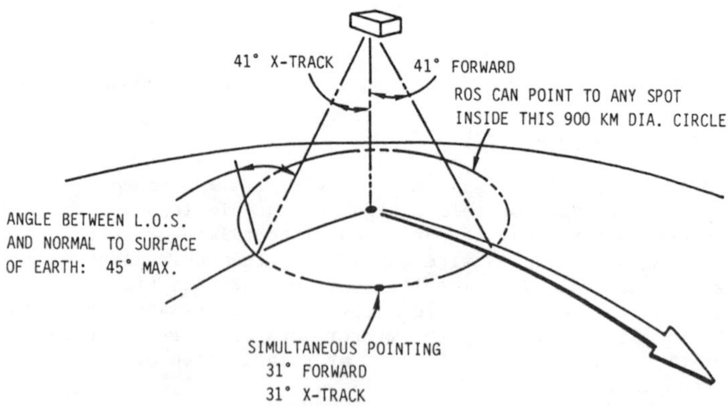

Figure 4. Simultaneous pointing capability (500-km orbit)

Additionally, a solar diffuser mounted on the inside of the aperture door can be positioned to reflect sunlight into the ROS optical chain to provide an end-to-end solar calibration. This diffuser is shielded from contamination except when the aperture door is in the intermediate calibration position. Because of the pivotal importance of data accuracy for the ROS mission, several calibration mechanisms in addition to the solar diffuser are included in the baseline design and are shown in Figure 5. These include radiometric calibration of the entire instrument by an internally-mounted tungsten ribbon filament calibrator, spectral "vacuum-shift" calibration, a thermal blackbody calibration source, and a spatial registration calibration.

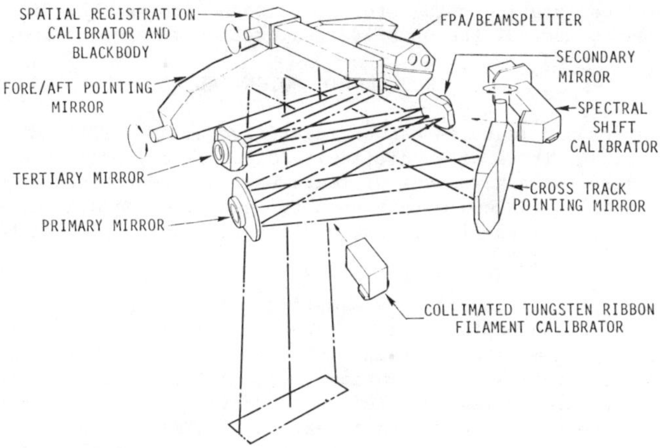

Figure 5. ROS optical schematic

The light leaving the cross-track pointing mirror is reflected off the primary mirror of the three-mirror (triplet) telescope shown in the optical schematic diagram of Figure 5, and then to the focal plane assembly (FPA) module via the secondary and tertiary mirrors. Light entering the FPA module first strikes a VNIR/LWIR dichroic beamsplitter plate (shown in Figure 2) and is then directed to various beamsplitter prism assemblies through two additional beamsplitter plates. One of these latter two is a neutral density 50/10 plate to provide both panchromatic (10%) and VNIR (50%) imaging simultaneously, and the other is a dichroic plate to separate SWIR energy from MIR/TIR energy.

The beamsplitter shown in Figure 2 has a total of nine exit ports: three VNIR, one PAN, three SWIR, one MIR, and one TIR. By using multiple-array-per-chip detector technology, however, ROS obtains 12 VNIR bands and six SWIR bands, along with one PAN, one MIR, and two TIR bands, for a total of 22 electronically selectable spectral bands. Four Si detector arrays are mounted on each VNIR chip, and two photovoltaic (PV) HgCdTe arrays on each SWIR chip, so that multiple bands may be obtained at each VNIR/SWIR beamsplitter exit port. Integral stripe filters are mounted on each detector array to provide final spectral separation at the focal plane.

To minimize risk and cost, the ROS FPA FOV may, if desired, initially be restricted to 2.29°, or 20 km from a 500-km orbit, but the overall optomechanical structure, FPA module clearance, cooling, and electronics systems are all designed to handle the full 60-km FPA-beamsplitter package. With 22 available bands, the signal processing electronics provide the capability to transmit the entire complement of data the ROS can produce at 100 megabits/second (MBS) per 10-km spatial ground swath.

Because the NASA-specified data rate limit is 150 MBS, ground commandable swath or band selection is also available, and a Hughes SBRC developed error-correcting successive-difference pulse code modulation (ESPCM) data compression algorithm can offer data compression without information loss (Schueler 1985). Moreover, ground commands can reprogram on-board processors for a wide range of additional data manipulation, including pixel aggregation, gain ranging, bit selection, quantization encoding, and automatic on-board registration by programmable sample time-delay. Telemetry processing includes commands to perform electronic calibration using stored or ground programmable test signals as well as to control the on-board calibration mechanisms provided.

ROS GROWTH POTENTIAL

While many ground features can be identified using a limited number of spectral bands, it is often difficult to accurately predict which bands are necessary for a given set of features. A research sensor designed to include all bands would be a boon to selecting the minimal sets required for specific applications, as illustrated in a recent NASA report (NASA JPL 1984).

Continuous spectral coverage by sampling adjacent spectral bands can be provided using the approach illustrated in Figure 1(c). An area detector array is illuminated in such a way that the closely spaced rows of detectors receive image information at successive spectral intervals. One method to accomplish this relies on variable bandpass interference coating technology, illustrated in Figure 6. The ROS optomechanical design can incorporate either the current technology focal plane described in Table 2, or an advanced spectrometer focal plane employing the wedge filter concept. The spectrometer would offer nearly continuous spectral coverage with 20 nm or finer bandpass. The required wedge filter technology is approaching fruition in current research and development programs.

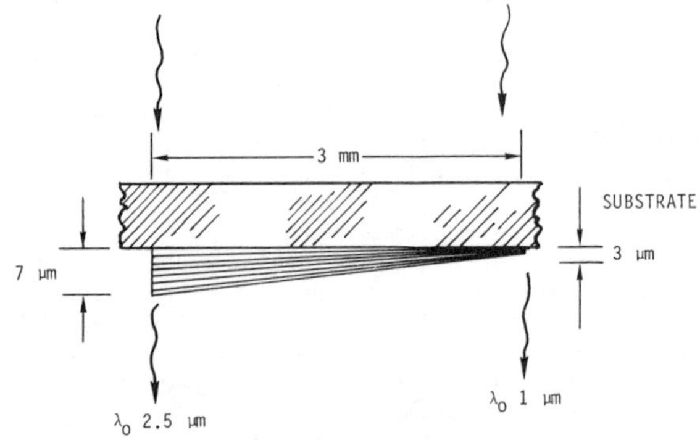

Figure 6. Variable wedge filter technology concept for imaging spectrometry on ROS

APPLICATIONS STUDIES

Key applications research issues that were addressed during the SBRC ROS study included polarimetry, radiometry, atmospheric correction, and spatial sharpening using a fine resolution panchromatic spectral band.

Polarimetry

Ground features sometimes exhibit spectral polarization signature as well as spectral reflectance characteristics, so that it may be of interest to offer polarimetry as an added feature of a multispectral sensor. This issue was addressed during the study to identify ways that polarization may be measured by the ROS instrument, and several candidate methods were suggested that would be feasible within the context of the selected concept.

Radiometry

Highly accurate radiometry is a priority for a research imaging radiometer such as ROS, so that special effort was devoted to identifying various independent means to provide excellent radiometric calibration both on the ground prior to launch, as well as on orbit. The results of the radiometry study, performed in collaboration with Prof. Philip Slater of the University of Arizona, indicate that simultaneous radiometric calibration using several independent calibration techniques will offer the best performance.

Atmospheric Correction

With accurate radiometry, one still only has a measure of the radiance striking the sensor aperture in space, and it is necessary to correct for atmospheric effects to estimate ground radiance. Accordingly, the goal of a cooperative study with Prof. Jeff Dozier of the University of California at Santa Barbara was to create a set of data specifications that would enable ROS data itself to be used to accurately estimate ground radiance, without resorting to ancillary measurements, such as radiosonde information that is commonly used to correct atmospheric effects on TM data. The analysis indicates that correction is feasible if ROS can provide 10 to 12 bit radiometric sensitivity. Based upon further sensor configuration analysis, a goal of 10-bit overall sensitivity appears to be feasible, including both signal-to-noise ratio (SNR) and image blur radiometric degradation effects.

Spatial Sharpening

Finally, the NASA requirements for ROS listed in Table 1 include a 5-meter resolution panchromatic (PAN) spectral band. A major purpose of this band is to provide additional spatial feature identification beyond that available in the coarser-resolution VNIR and SWIR bands. A study was therefore performed to identify on-board signal-processing algorithms that can use the panchromatic sharpening band to identify mixed boundary pixels at the edges of fields, or other features smaller than the VNIR or SWIR bands can discriminate. The results of this study, based on simulations using TM data, show that simple but effective mixed pixel discrimination approaches may be implemented in hardware on the sensor.

SUMMARY

The NASA sponsored Hughes/SBRC ROS study has produced a detailed design analysis that provides a description of the envisioned instrument layout and performance. Moreover, instrument applications have been addressed to the extent necessary to select a preliminary set of spectral bands that will meet research needs. The key results of the Hughes/SBRC design study are:

- Layout and preliminary spectral band selection has been generated.
- Proven focal plane technology can be employed and the preliminary signal processing design has been completed.
- Active cooling will allow space station application.
- Performance prediction includes over 90% reliability for three-year unattended mission.

ROS is intended to allow verification of scientific models of the effects of atmospheric and other environmental conditions on the data retrieved from space sensors. Such model verification leads directly to enhanced understanding of the relationship of sensor data to ground reflectance or emittance, and therefore to enhanced data accuracy and operational utility.

ACKNOWLEDGMENTS

The author acknowledges the fine contributions of the entire Research Optical Sensor team listed in the full ROS report, including the support of John Tower and colleagues of the RCA Advanced Technology Laboratories. Moreover, the ROS study upon which this paper is based was supported by NASA GSFC contract number NAS5-28168 under the technical direction of Mr. Harry Shaw of the Sensor Concepts and Development Branch in the Laboratory for Terrestrial Physics. The author also wishes to thank Dr. William Barnes of NASA GSFC for the invitation to prepare and present this paper at Pecora 10.

REFERENCES

Cannata, R., G. Orias, and F. Brown 1985, High Performance PV HgCdTe Arrays, Proceedings of the IRIS Detector Specialty Group, 6-8 August, Boulder CO (in press).

Hughes Santa Barbara Research Center 1984, Thematic Mapper: Design Through Flight Evaluation Final Report, NASA Goddard Space Flight Center Contract No. NAS5-24200, SBRC Report No. 41741.

Hughes Santa Barbara Research Center 1985, Concept Definition Study for A Multispectral Linear Array Research Optical Sensor and Imaging Radiometer Final Report, NASA Goddard Space Flight Center Contract No. NAS5-28168, SBRC Report No. 50913.

Mika, A. 1983, Multilinear Array Instrument Concepts for Future Land Remote Sensing, Proceedings Pecora 8, Augustana Research Institute, pp. 259-268.

NASA Jet Propulsion Laboratory 1984, Imaging Spectrometry: The Next Step in Remote Sensing.

Schueler, C.F., C. deBoer, B. Marks, and M. Stegall 1985, On-Board Data Compression for Advanced Landsat, Proceedings of the Society of Photo-Optical Instrumentation Engineers, Vol. 534 (in press).

Tower, J.R., S.B. Warren, L. Pellon, R.T. Strong, H. Elabd, A.D. Cope, D.N. Hoffman, W.M. Kramer, and R.W. Longsderff 1984, Multispectral Linear Array Visible and Infrared Sensors, Proceedings of IGARSS, Strasbourg, 27-30 Aug., pp. 759-764.

MODIS: A MODERATE-RESOLUTION IMAGING SYSTEM FOR THE SPACE STATION POLAR-ORBITING PLATFORM

W. L. Barnes
Goddard Space Flight Center
Greenbelt, Maryland 20771

ABSTRACT

The Space Station being developed by NASA includes in its baseline configuration both polar and equatorial unmanned platforms. The Earth Observing System (EOS), a proposed sensor compliment for the polar platform, includes the Moderate-resolution Imaging System (MODIS) as part of its initial orbital configuration. MODIS is designed to generate a 10-year data base for a wide variety of remote sensing research problems in the areas of ocean, atmosphere and land science. The system consists of 52-spectral channels viewing a 1500 kilometer swath in the wavelength region from 0.4 to 12.0 micrometers. The ten thermal channels will image the Earth daily with a spatial resolution of 1.0 kilometer, while the channels measuring reflected solar radiance will survey the Earth once every 2 days with a spatial resolution of 500 meters (12 channels) or 1.0 kilometer (30 channels). MODIS will consist of two sensor packages designated MODIS-T (tilt) and MODIS-N (nadir). In order to avoid specular reflectance from the ocean surface, MODIS-T, which will output 17 of its 64 spectral channels in the visible and near infrared (<1.0 micrometer) region, can be tilted to view fore or aft up to 60° while scanning ± 45° crosstrack. MODIS-N will view nadir and also scan ± 45° crosstrack with a total of 35 spectral channels.

Preliminary sensitivity studies have indicated that a whiskbroom scanner with an imaging spectrometer aft optics system is feasible for MODIS-T while a whisk-broom scanner with 71 individual filter/detector modules in the focal plane will meet the MODIS-N requirements.

INTRODUCTION

In the Spring of 1981, NASA started an intensive in-house study to examine the problem of what the next major advance in remote sensing science of the Earth's surface should be in the context of technology available for flight on a low altitude polar orbiting satellite in the early 1990's. This effort was originally designated as System Z and later changed to the Earth Observing System (EOS), which has become the anticipated payload(s) for the polar platform portion of the Space Station. A Science and Mission Requirements Working Group was formed with representatives from most of the traditional Earth science disciplines to define major Earth science questions for the 1990's and to delineate low Earth orbit observables that would materially address these questions. The results of this group's deliberations (Butler et al., 1984) included a requirement for a multispectral radiometer capable of frequent global surveys at a 1 kilometer spatial resolution. This system was designated the Moderate resolution Imaging Spectrometer (MODIS). The MODIS Instrument Panel (Table 1) was formed in mid-1984 to further define the scientific requirements and generate a set of sensor parameters that would insure achievement of the scientific goals. This paper is a combination of the recommendations of the MODIS Instrument Panel (Esaias, et al., 1985) and the results from two NASA Phase A contracts (NASA a, 1985, and NASA b, 1985)

Table 1 Instrument Panel For
The Moderate-Resolution Imaging Spectrometer

Members Names	Organization
Wayne E. Esaias (Chairman)	*NASA HQS
W. L. Barnes (Executive Secretary)	GSFC
Mark R. Abbott	Scripps/JPL
Steve Cox	Colorado State University
Robert H. Evans	University of Miami
Robert S. Fraser	GSFC
Alexander F. H. Goetz	JPL
Christopher O. Justice	University of MD
E. Paul McClain	National Oceanic and Atmospheric Administration
Marvin S. Maxwell	GSFC
Roert E. Murphy	NASA HQS
Joseph M. Prospero	University of Miami
Barrett Rock	JPL
Steven W. Running	University of MT
Raymond C. Smith	University of CA/ Santa Barbara
Jerry E. Solomon	JPL
Joel Susskind	GSFC
Ex-Officio Members	
Howard R. Gordon	University of Miami
Michael Spanner	Ames Research Center

*Presently at GSFC

OBJECTIVES

Over the last few years, there has developed within the scientific community the belief that the Earth must be treated as a single interacting ecosystem if major advances in our knowledge of the Earth and man's impact are to be achieved. This has resulted in numerous integrated research programs such as the United Nations Environmental Program, International Satellite Land Climatology Program, and the International Geosphere/ Biosphere Program. A common theme of all these programs is the need for long-term global data bases to establish the Earth's present state and to delineate trends. These data bases are vital for the initialization and testing of a variety of models that are presently under development. The natural cycles are such that the Earth's surface must be viewed every few days in order that most dynamic events can be observed and a minimum 10-year data base is required if we are to separate long-term trends from natural cycles. Ten years of multispectral global surveys repeated every few days is a massive data base. Therefore, it is imperative that the spatial resolution of the sensor be kept at the minimum required for the scientific objectives. Fortunately, there has been considerable progress in the last couple of years in utilizing data with spatial resolution near one kilomater. Specifically, the NOAA/AVHRR has been utilized for vegetation studies (Hayes, 1985) and to develop continental and global scale data sets suitable for improved inputs to global models (Tucker, 1985). In addition, the data from the Nimbus-7 Coastal Zone Color Scanner has demonstrated open ocean chlorophyll surveys with accuracies not possible by any other means (Gordon, 1983).

Terrestrial studies amenable to solution via low resolution imaging radiometry (MODIS) include rate of tropical deforestation and type and rate of regrowth, areal distribution and effect of acid rain on the boreal forests of Europe and North America, rate and extent of desertification at the edge of the world's deserts, update of global vegetation maps, extent of freeze or drought damage in croplands and natural communities, land cover change and its effect on terrestrial biophysical systems, continental changes in snow cover and associated changes in albedo, and derived products including standing green biomass, intercepted photosynthetically active radiation and net primary productivity. Computer simulation models that will depend on data from a sensor of this type include global surface climate models, carbon models, hydrologic/energy models, and comprehensive biogeochemical cycling models.

Biological oceanographers use utilize MODIS data to characterize the global distribution of phytoplankton biomass and its temporal and spatial variability. Using in-situ data together with regional scale models, it will be possible to convert biomass to primary productivity. Thus oceanic and global measurements of primary productivity and their temporal variation will be possible. In addition, mesoscale features such as warm and cold core rings and jets can be observed and their development followed for months or years using ocean color data. The effects of riverine fluxes and other forms of sediment transport can be observed and quantified to demonstrate their importance to ocean productivity and global biogeochemical cycles.

Atmospheric constitutents that can be monitored directly by MODIS include clouds and aerosols, both of which have a direct impact on climate as well as on the geochemical and hydrological cycles. In addition, water vapor distribution in the upper troposphere could be mapped using a 6.7 micrometer absorption band.

REQUIREMENTS

Using the science objectives outlined above, one must determine a set of requirements which will define a sensor system capable of supplying the needed data. Constraints on these requirements include sensor and spacecraft technology, resources, and availability of algorithms. Each of the three major discipline groups of the MODIS Instrument Panel (terrestrial, oceans, and atmospheres) outlined the sensor attributes needed for achieving their scientific objectives and these were then combined into a comprehensive set of sensor requirements. The same general approach will be used below.

Terrestrial Studies

The terrestrial studies subgroup proposed a spatial resolution of 500 meters based on a trade-off between adequate descrimination of agricultural and forestry features and the volume of data that can be practically handled. The initial EOS payload will also include the High Resolution Imaging Spectrometer (HIRIS) with 192 spectral channels of 30 meter spatial resolution which could be used to examine sub-pixel structure in selected areas.

Required spectral channels are essentially the same as those for Thematic Mapper with the possible addition of a couple of channels in both the near infrared and shortwave infrared. It has been demonstrated by Gurney et al (1983), Heimburg et al (1982), and Soer (1980) that thermal data is potentially quite useful in studies of soil moisture and evapotranspiration, therefore, a total of 7 bands in the 3.5 to 4.0 and 8.5 to 12.0 micrometer window regions were recommended.

For terrestrial studies, it was determined that an equatorial crossing time between 1 and 2 p.m. was acceptable and that a revisit time of two or three days was sufficient to measure vegetation dynamics.

Ocean Studies

Requirements generated by the ocean color research community are based on six years experience with the Nimubus-7 CZCS data. This has resulted in a desire for at least 17 spectral bands in the wavelength region from 0.4 to 1.0 micrometer with those bands in the visible portion having bandwidths smaller than the 20 nanometers of the CZCS and signal-to-noise in excess of 600:1. The near-infrared bandwidth and signal-to-noise requirements are less restrictive since these channels are principally used to perform aerosol corrections.

The ocean color requirement with the greatest impact on the overall system is the necessity to point fore or aft of nadir over a 20° range. Without this ability, a large portion of the ocean color data would be useless due to specular reflection (sunglint) from the ocean surface. Another unique ocean color requirement stems from the work of Gordon (1981) demonstrating the improvements that can be achieved in chlorophyll concentration accuracy if the system is periodically calibrated by viewing the sun. Therefore, an articulated, well characterized, diffuser plate should be included.

Additional ocean color requirements include one or more polarized channels to enhance atmospheric corrections; and equatorial crossing time as near noon as possible, spatial resolution of one kilometer in coastal regions with degradation to four kilometers in open oceans; minimal sensitivity to incoming polarized radiance at all viewing angles (with the exception of the polarized channel); a continuous, ten year global data base, and two-day revisit.

Atmospheric studies

Atmospheric science requirements for MODIS are embedded to a large extent in the terrestrial and ocean requirements. Therefore, only unique requirements need be discussed. Chief among the latter is a set of channels originally proposed by Kyle et al (1978) for detailed examination of cloud parameters. The most demanding of these are a set of narrow ($\Delta\lambda$ = 1.2 nm) channels located in the oxygen A-band near 0.76 micrometers that are to be used to measure cloud altitude. Other requested channels are more conventional and readily accommodated within the constraints established by the terrestrial and oceanic requirements.

APPROACH

MODIS must be capable of conducting global surveys to support terrestrial, oceanographic, and atmospheric science. The two attributes of the MODIS that are crucial to its mission are the numerous spectral channels in the region between 0.4 and 12.0 micrometers and its revisit time of 2 days for channels viewing reflected solar radiance and 1 day for thermal channels. The primary goal is to develop a sensor system that will address the widest possible variety of research tasks.

In developing the scientific requirements for the MODIS, the following requirements: (a) the ocean color channels must be capable of a view 20° fore and aft of nadir (b) uninterrupted long-term global surveys of ocean chlorophyll; and (c) minimum atmospheric path radiance for terrestrial sensing (i.e., nadir view) were incompatible with a single sensor package. Therefore, it was proposed that MODIS be divided into two packages to be designated MODIS-T (tilt) and MODIS-N (nadir); the former to contain the required ocean color channels and to be pointable fore or aft of nadir, and the latter containing those channels with no requirement for off-nadir pointing. The two packages will be discussed below.

MODIS-T

MODIS-T will address those science requirements that call for viewing the surface at predetermined angles forward or aft of the subsatellite point (nadir). These include:

(a) minimizing the amount of specular reflectance from the surface, (b) examining the bidirectional reflectance distribution function (BRDF) from large homogeneous targets and (3) performing atmospheric studies by examining the spectral signal as a function of optical depth. The ocean color requirements include: fore and aft pointing 17 visible and near infrared region spectral channels, a 10 nanometers spectral width, a visible channel signal-to-noise minimum of 600:1 and a long-term global data set. Spatial resolution of 1 kilometer in the coastal regions is sufficient, but the resolution in open oceans could be reduced to 4 - 10 kilometers. It was decided, however, that any such reduction in resolution could best be done during ground processing. The BRDF requirements include viewing to angles as large as 60 degrees both fore and aft. Other requirements for both BRDF and those for the proposed atmospheric studies are less restrictive and therefore fit within those for ocean color.

The MODIS-T requirements can be satisfied by any of several types of imaging radiometers. A practical system in terms of size, complexity, available technology, and overall utility is that of the imaging spectrometer shown in Figures 1 and 2. The version shown consists of a crosstrack scan mirror, collecting optics, spectrometer, and a 64 x 64 element silicon detector array with the entire system capable of ± 60 degrees rotation about the optics axis to give the required fore-aft tilt. The 90 degrees swath is viewed in 9.5 seconds, the time required for the subsatellite point to advance 64 kilometers.

The spectrometer disperses the beam from the entrance slit along one dimension of the detector array. Therefore, the image of the spectrometer slit on the surface consists of 64 pixels along-track with each pixel being dispersed within the sensor into 64 perfectly registered contiguous spectral bands of approximately 10 nanometers width covering the range from 400 to 1000 nanometers. A long-pass filter will probably be deposited on the last few rows of the detector array to suppress second-order energy.

The parameters used to calculate the MODIS-T signal-to-noise are listed in Table 2. It is worth noting that the required entrance aperture, which sizes the entire system, is only 2.8 centimeters. The optics in Figure 2 are designed around a 5 cm entrance aperture which would result in even better signal-to-noise. Also, note that the signal-to-noise calculations were made using a sensor look angle of 20° and an atmospheric model with total nadir optical thickness of 0.72.

The calculated signal-to-nose (S/N) for sixteen (16) of the sixty-four (64) MODIS-T channels using the parameters listed in Table 2 are tabulated in Table 3. The resulting S/N are in excess of 600:1 from 400 through 640 nanometers and greater than 300:1 for wavelengths less than 870 nanometers.

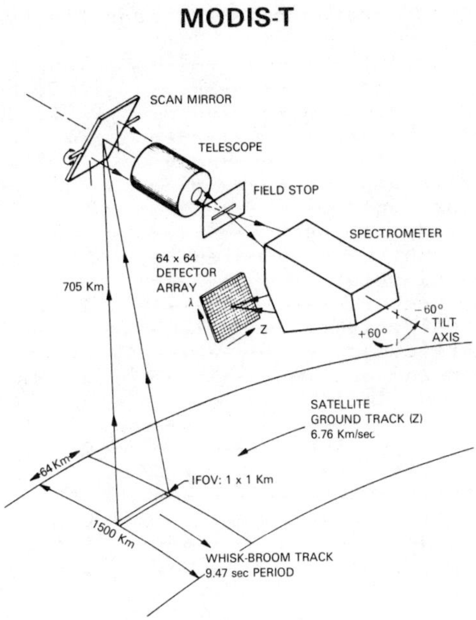

FIGURE 1. Imaging spectrometer concept for MODIS-T

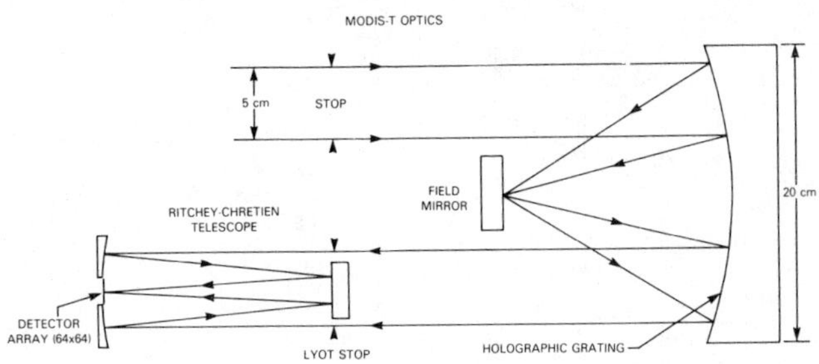

FIGURE 2. MODIS-T optics (scan mirror not shown)
(Shafer 1981)

TABLE 2 - MODIS-T parameters for sensitivity calculations

1.	Satellite height --------------	705 kilometers
2.	Ground resolution -------------	1000 meters
3.	Swath width -------------------	1513 kilometers
4.	Wavelength --------------------	See Table 3
5.	Spectral bandwidth-------------	10 nanometers
6.	Solar zenith angle ------------	22 degrees
7.	Sensor look angle -------------	20 degrees
8.	Optical transmission ----------	0.35
9.	Detector size -----------------	50 micrometers
10.	Telescope diameter ------------	2.82 centimeters
11.	Optical f-number --------------	1.25
12.	Surface reflectivity ----------	See Table 3
13.	Quantum efficiency ------------	See Table 3
14.	Saturation radiance -----------	2.1 mw/cm^2-sr- m
15.	(Integration time)/ (Dwell time) ------------------	1.0
16.	Time to map the earth ---------	2 days
17.	Number of detectors per spectral band--------------	64
18.	Scanning efficiency -----------	0.25
19.	Expected S/N-------------------	See Table 3

MODIS-N

MODIS-N (nadir) will address those scientific tasks that do not require off-nadir pointing. The requirements cover the spectral range from 0.4 to 12.0 μm and include a strong 500-meter resolution justification for several channels in the region from 0.4 to 12.0 μm. Spectral widths vary from 1.2 to 500 nanometers.

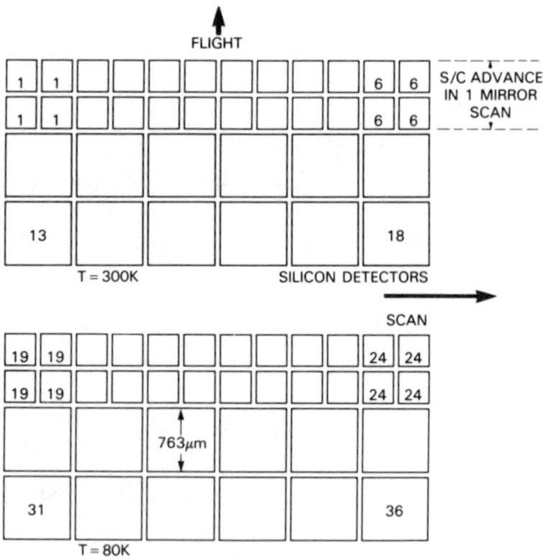

FIGURE 3. MODIS-N focal plane (conceptual)

Table 3 - S/N for sixteen of the sixty-four MODIS-T spectral bands

Band No.	Wavelength (nm)	Detector Q.E.	Water Reflectance (%)	S/N	Comments
1	410	0.43	4.7	812	dissolved org mat
3	430	0.47	5.0	840	dissolved org mat
4	440	0.5	5.1	860	chloro abs max
6	460	0.52	5.1	888	
9	490	0.57	3.5	868	high chloro concen
12	520	0.6	2.8	786	low susp sedi conc
14	540	0.63	2.1	755	
17	570	0.66	1.7	719	chlorophyll min.
22	620	0.67	1.0	632	high susp sedi conc
27	670	0.67	1.0	575	chloro abs max.
28	680	0.67	1.0	563	chloro fluorescence
34	750	0.64	1.0	478	atmos correction
37	780	0.59	1.0	430	atmos correction
47	870	0.39	1.0	301	atmos correction
55	950	0.16	1.0	152	H_2O
60	1000	0.07	1.0	72	

NOTE: Reflectance obtained from Figure 3-124 of Wolfe and Zissis, 1978.

The requirements outlined above result in a system with at least 35 spectral bands. Due to the range of spectral widths and a requirement to measure polarization, it is impractical to use the imaging spectrometer concept of MODIS-T. Instead, a more conventional imaging radiometer concept will be used. This system consists of a cross-track scan mirror and collecting optics similar to those shown in Figure 1 for MODIS-T and a set of individual detector elements with spectral interference filters located in the focal plane of the collector. The layout of the focal plane is illustrated schematically in Figure 3 which shows the detector/filter layout for 36 channels, 18 maintained at a temperature of 300K and 18 at 80K. Each of the twenty-four 1000-meter resolution channels has a single detector/filter module, whereas each of the twelve 500-meter resolution channels has four detector elements with a single spectral filter over all of them. Since the scan speed is 64 times faster than that of MODIS-T, the optics for MODIS-N must be larger in order for the system to collect the required number of photons in the alloted time.

The parameters used to calculate expected signal-to-noise are similar to those of Table 2 with a sensor look angle of 0°. The results for the MODIS-N channels are listed in Tables 4 and 5. Typical but by no means complete applications are also listed for each band.

Signal-to-noise calculations for MODIS-N were made using an optics diameter of 40 centimeters. The optics size could be reduced by increasing the number of detectors for each band so as to scan more than one line for each pass of the

Table 4 - MODIS-N - visible/near-IR channels

Chan No.	λ (nm)	Δλ (nm)	IFOV (meters)	Surface Reflectance (%)		S/N (calc.)	Comments
1	470	20	500	3	(B)	740	soil-veg diff
2	550	20	500	10	(B)	920	green peak
3	670	20	500	6.5	(B)	770	chloro abs
4	710	20	500	9	(B)	830	red-NIR trans
5	880	20	500	25	(B)	850	veg max ref
6	960	20	500	24	(B)	520	H_2O peak
7	435	10	1000	5.1	(C)	1480	low chloro
8	490	10	1000	3.5	(C)	1520	nonlin. chlo
9	520	10	1000	2.8	(C)	1390	high chloro
10	565	10	1000	1.8	(C)	1290	chlo baseline
11	590	10	1000	0.6	(C)	1160	sediment
12	665	10	1000	0.2	(C)	950	atmos/sediment
13	765	10	1000	0.1	(C)	720	atmos correc
14	865	10	1000	0.1	(C)	470	atmos correc
15	754	1.2	1000	30	(D)	920	cloud altitude
16	761	1.2	1000	90	(D)	1550	cloud altitude
17	763	1.2	1000	50	(D)	1160	cloud altitude
18	500	100	1000	2.5	(B)	2880	polarization
19	500	100	1000	2.5	(B)	2880	polarization
20	1080	20	500	25	(B)	830	leaf morphology
21	1131	20	500	10	(A)	520	cloud H_2O abs
22	1240	20	500	24	(B)	750	leaf H_2O abs
23	1550	20	500	14	(B)	480	leaf H_2O abs
24	1640	20	500	10	(A)	375	snow/cloud diff
25	2130	50	500	10	(A)	250	cloud penetra

References for surface reflectances:

(A) Colwell, 1983 (Figure 5-30); (B) Huck et al, 1984; (C) Wolfe and Zissis, 1978 (Figure 3-124); (D) Kyle et al, 1978

TABLE 5 - MODIS-N - thermal channels

Channel NO.	λ (nm)	Δλ (nm)	IFOV (meters)	NE Δ T (°C @ 270K)	Comments
26	3750	90	1000	0.14	clouds & surf temp.
27	3959	50	1000	0.14	clouds & surf temp.
28	4050	50	1000	0.13	clouds & surf temp.
29	8300	500	1000	0.01	strat aero det
30	8550	500	1000	0.01	clouds & surf temp.
31	8810	500	1000	0.01	clouds & surf temp.
32	9900	500	1000	0.01	strat aero det
33	10150	500	1000	0.01	clouds & surf temp.
34	11030	500	1000	0.01	clouds & surf temp.
35	12020	500	1000	0.01	clouds & surf temp.

mirror. A system consisting of 6-element linear detector arrays for each of the 1 km resolution channels and 23 centimeter optics is under consideration. This does, however, increase the mechanical and electrical complexity of the focal plane, the cooling requirements for the detector array and the amount of system testing and calibration.

Calibration

Conversion of digital counts to radiance requires (1) complete characterization of the system prior to launch including response to a standard source, (2) inflight response to known sources, and (3) constant vigilance over the life of the mission to detect changes in the system.

The principal standards for prelaunch radiometric calibration will be a visible and near infrared integrating sphere similar to those used for MSS, TM, CZCS and AVHRR, and a calibrated blackbody. Improved versions of the integrating sphere are in the planning stage. Accuracies commensurate with the noise figures shown in Table 5 will require improved thermal infrared sources. Inflight visible and near-infrared calibration has typically taken the form of monitoring system response to incandescent lamps and referring these changes to the prelaunch values. Thermal inflight calibration usually is based on viewing a blackbody built into the backscan portion of the sensor and the near-zero temperature of space. Inclusion of an aperture filling visible and near infrared source that is external to the sensor is highly desirable for both MODIS-N and MODIS-T, but may be difficult for the former due to aperture size. Inclusion of more than one blackbody in MODIS-N is also desirable.

For many of the planned applications of MODIS, the reflected spectral radiance is compared against the incoming solar spectral radiance. Direct measurement of the solar radiance using a well characterized difuse reflector bypasses many of the difficulties inherent in radiometric calibration (Gordon, 1981). Therefore, inclusion of a deployable calibrated diffuser is required for MODIS-T and highly desirable for MODIS-N.

The third type of inflight radiometric calibration is accomplished using surface targets that are well characterized and located in areas with a generally clear atmosphere. Careful measurements of the surface and atmosphere at the time of spacecraft overpass coupled with radiative transfer modeling, can result in accurate radiometric calibration.

In addition to inflight radiometric calibration, it is imperative that any post launch degradation in the spectral and spatial response of the system be known. Spatial response may best be measured using known surface targets. Inflight spectral characterization may be accomplished using absorption lines in rare earth glasses or input from light emitting diodes.

Use of as many of the calibration techniques described above as is practical for both MODIS-N and MODIS-T will insure the greatest possible scientific return for the resources expended.

Data rates

EOS guidelines call for a 100% duty cycle for MODIS. This implies that the thermal channels will be on at all times and the reflected solar channels will be on for one-third (1/3) of each orbit. Assuming ten bit digitization, 40% over-sampling in the cross-track direction and contiguity at nadir, MODIS-N, with 24 channels having 1000 meter resolution and 12 channels with 500 meter resolution, will output 7.6 megabits/second for one-third of each orbit and 1.2 megabits per second during the remainder of the orbit. MODIS-T using the same assumptions and outputing 17 of its 64 spectral channels has a data rate of 1.2 megabits/second (day only). These rates result in a total of 3.4×10^{11} bits/day. This is roughly equivalent to 200 computer compatible tapes per day (6250 BPI).

SUMMARY

MODIS, as presently conceived, is a system of two imaging spectroradiometers designed for the widest possible applicability to research tasks that require long-term (5-10 years), low resolution (0.5 - 1.0 kilometer), global, multispectral (52 channels between 0.4 and 12.0 micrometers) data sets. The system described in this paper is preliminary and subject to scientific and technological review and modification and it is anticipated that both shall occur prior to selection of a final system configuration. However, the basic concept outlined above is likely to remain unchanged. Present plans call for the MODIS development to be managed at Goddard Space Flight Center as a NASA facility. A NASA Announcement of Opportunity (AO) is planned for release in April 1986 calling for investigations based around MODIS and the other EOS flight instruments. The MODIS Phase B study will be initiated in October, 1987 with a launch planned for late 1992.

REFERENCES

Butler, D.M., et al 1984, Earth Observing System: Science and Mission Requirements Working Group Report, NASA Technical Memorandum 86129, GSFC

Colwell, R.N. 1983, ed. Manual of Remote Sensing, American Society of Photogrammetry, Virginia

Esaias, W.E., et al 1985, Moderate Resolution Imaging Spectrometer (MODIS): MODIS Instrument Panel Report, NASA Technical Memorandum (in review), GSFC

Gordon, H.R. 1981, Reduction of Error Introduced in the Processing of Coastal Zone Color Scanner-Type Imagery Resulting from Sensor Calibration and Solar Irradiance Uncertainty: Applied Optics, Vol. 20, pp. 207-210

Gordon, H.R. and A.V. Morel (1983), Remote Assessment of Ocean Color for Visible Satellite Imagery: A Review, Springer-Verlag, New York, p. 114

Gurney, R.J., J.P. Ormsby, and D.K. Hall 1983, A Comparison of Remotely Sensed Surface Temperature and Biomass Estimates for Aiding Evapotranspiration Determination in Central Alaska: Permafrost, 4th International Conf. Proc., pp. 401-404

Hayes, L. 1985, The Current Use of TIROS-N Series of Meteorological Satellites for Land Cover Studies: International Journal, Vol. 6, No. 1

Heimburg, K.F., L.H. Allen, and W.C. Huber 1982, Evapotranspiration estimates based on surface temperature and net radiation: Development of remote sensing methods. Florida Water Resources Research Center, Publication No. 66, University of FL Gainesville, FL

Huck, F.O., R.E. Davis, C.L. Fales, R.M. Aherron, R.F. Arduini, and R.W. Samms 1984, Study of remote sensor spectral responses and data processing algorithms for feature classification: Optical Engineering, Vol. 23 No. 5, pp. 650-666

Kyle, H.L., J.R. Curran and W.L. Barnes 1978, A Cloud Physics Radiometer, Proceedings of Third Conference on Atmospheric Radiation, Davis, CA, June 28-30, 1978

NASA 1985, Phase-A Study for a Moderate Resolution Imaging Spectrometer-Nadir Viewing (MODIS-N), Final Report NAS5-27145 (in preparation), GSFC

NASA 1985, Phase-A Study for a Moderate Resolution Imaging Spectrometer-Tiltable (MODIS-T), Final Report NAS5-27157 (in preparation), GSFC

Shafer, D. 1981, Simple infrared telescope with stray-light rejection: Modern Utilization of Infrared Technology, SPIE Vol. 304, pp. 171-175

Soer, G.J.R. 1980, Estimation of regional evapotranspiration and soil moisture conditions using remotely sensed crop surface temperatures, Remote Sensing Environ., Vol 9, pp. 27-45

Tucker, C.J., G. Townshend, and T.E. Goff 1985, African Land-Cover Classification using Satellite Data: Science, Vol. 227, pp 369-375

Wolfe, W.L. and G.J. Zissis 1978, ed., The Infrared Handbook, Office of Naval Research, Department of the Navy, Washington, DC

IMAGING SPECTROMETRY: WHAT THE FUTURE HOLDS

Alexander F. H. Goetz
Jet Propulsion Laboratory/
California Institute of Technology
4800 Oak Grove Drive
Pasadena, California 91109

SUMMARY

Imaging spectrometry for Earth remote sensing has developed over the last few years to satisfy the need for more precise remote sensing information than can be gathered by broad-band multispectral scanners. The goal is to create images in which each picture element represents a composite spectrum of all the materials within an instanteous-field-of-view. That is equivalent to a laboratory spectrum both in spectral as well as radiometric resolution. If these criteria are met then all the information theoretically available about the surface composition will be acquired by the sensor.

Research during the last five years, both in the field and in the laboratory, has revealed the potential of high-spectral resolution remote sensing for studies of rocks, soils, vegetation and water, all the components of the Earth's surface. Technological developments in the area of infrared area array detectors, optics and data systems have made it possible to build imaging spectrometer sensors that cover the wavelength region of reflected sunlight from 0.4 to 2.5 μm.

An imaging spectrometry program being implemented at the Jet Propulsion Laboratory and contains four sensor elements. At present the Airborne Imaging Spectrometer (AIS) is the only sensor collecting data. It acquires images in 128 bands simultaneously in the wavelength region 1.2 to 2.4 μm. In late 1986 the Airborne Visible and Infrared Imaging Spectrometer (AVIRIS), now under construction, will be operational. This instrument, to be flown on the NASA ERS-2, will cover a swath 11 km wide and collect image data in 224 spectral bands between 0.4 and 2.4 μm simultaneously. The next step, presently in the planning stage, is the Shuttle Imaging Spectrometer Experiment (SISEX) which will acquire data on a global basis in 128 spectral bands over a 12 km swath. A follow-on to SISEX, called the High-Resolution Imaging Spectrometer (HIRIS), is envisioned for the space platform expected to be launched in the mid 1990's. HIRIS would acquire data similar to SISEX, but over a swath 50 or more kilometers in width.

Technology is now available to develop imaging spectrometer sensors as well as to handle and analyze the vast quantities of data that will be produced. The stage is set for a new era in remote sensing in which a much larger number of questions can be addressed with remote sensing since the full information content of the reflected signal can be acquired and utilized.

The research described in this summary was carried out by the Jet Propulsion Laboratory, California Institute of Technology, under a contract with the National Aeronautics and Space Administration.

NEW AND EMERGING TECHNOLOGY - DATA PROCESSING

Moderator: M. Devirian
J. Urena

THE MASSIVELY PARALLEL PROCESSOR - PROGRAMMING AND APPLICATIONS

H. K. Ramapriyan,
J. P. Strong and J. C. Tilton
Space Data and Computing Division
NASA Goddard Space Flight Center
Greenbelt, Maryland 20771

ABSTRACT

The Massively Parallel Processor (MPP) is a unique computer capable of orders of magnitude higher computational speeds than conventional computers.

The MPP was conceptualized at the Goddard Space Flight Center (GSFC) in 1977 and developed, under contract, by the Goodyear Aerospace Corporation. It was delivered to GSFC in May 1983. Since then, GSFC has been active in integrating it with a VAX 11/780 host computer and developing systems and applications software. This paper describes the facilities available for developing software on the MPP and the use of its high-speed capabilities for NASA's applications. The applications discussed concentrate on processing of remotely sensed images and include data preparation (preprocessing), feature extraction, unsupervised and supervised classification, and post-classification processing.

INTRODUCTION

The large volume of data produced by spaceborne high resolution sensors and the processing and analysis techniques to derive useful information from such data require considerably greater computational capabilities than offered by conventional (sequential) computers. Many array processors (AP's) have been developed in recent years to address this computational problem. A review of AP's, their architecture and use in remote sensing can be found in Ramapriyan and Strong (1983). Among these the Massively Parallel Processor (MPP) at GSFC stands out as a unique, general purpose, easily programmable and significantly faster computer.

The MPP is a Single Instruction Multiple Data (SIMD) computer consisting of 16,384 relatively simple processing elements (PE's) configured as a 128 x 128 array unit (ARU). Each PE has a 1024-bit local memory and is capable of arithmetic and logical operations in a bit-serial fashion. At a given time, all the PE's execute the same instruction. However, it is possible to inhibit a specific subset of PE's from performing the given operation by using a "mask" bit-plane. Also, each PE is connected to its four nearest neighbors (N, E, S, W) thus permitting shifting an entire bit-plane of data by one step in the array in one cycle (100 nsec). This arrangement makes the MPP ideally suited for image processing where point or local neighborhood operations can take advantage of the parallelism. As an example, the MPP can perform eight bit additions at the rate of over 6 billion per second.

In addition the MPP has 2 MB buffer memory (Staging Buffer) which is expandable to 64 MB. This permits data exchange rates of up to 160 MB/sec with the ARU and 40 MB/sec with the host computer. In the present configuration, however, these rates are limited to 20 MB/sec and 6 MB/sec respectively. The slowest part of the system for high data volume applications is the I/O between mass storage devices and the VAX (at less than 1 MB/sec). The Staging Buffer permits highly flexible reformatting of data including byte to bit-plane conversion and in-place transposition of large matrices by simple reordering of dimensions between input and output. More details about the MPP architecture can be found in Batcher (1980) and Schaefer et al (1982).

PROGRAMMING

Various tools are available to facilitate programming the MPP. The best speeds possible are achieved using the MPP's assembly language (MCL). However, the speed gains provided by the MPP are considerable even when it is programmed in the high order language currently available on the MPP: Parallel Pascal. (There are plans to develop a Parallel Fortran and/or Forth for the MPP.) Parallel Pascal is essentially standard Pascal with a few parallel extensions. Developed by Reeves and Bruner (1980, 1982), it was designed to be easy to use and efficient, and to provide good error detection. The extensions are carefully chosen to make Parallel Pascal portable even among the wide variety of parallel architectures. These extensions allow the declaration of parallel arrays, a few parallel manipulation and indexing functions and extended control structures.

Parallel arrays are explicitly declared in Parallel Pascal by affixing the word PARALLEL to the standard Pascal array declaration, viz.:

```
name: PARALLEL ARRAY[index range] OF type ;
```

This explicit declaration of parallel arrays provides the programmer direct control over which arrays are operated on in parallel. The last two dimensions of parallel arrays are restricted in the MPP implementation of Parallel Pascal to 128-by-128 (or the last dimension to 16,384) to conform with the architecture of the MPP.

Operationally, parallel arrays are different from standard Pascal arrays in that parallel arrays can be added, divided, compared, etc. as aggregate units rather than element-by-element. Special indexing mechanisms are provided to operate selectively on portions of a parallel array. Standard elemental functions, such as 'trunc', 'round', 'sqr', 'exp', and 'cos', perform the same operation on each element of a parallel array, independently and in parallel. Special transformational functions, such as 'rotate', 'sum' and 'max', perform transformations on the entire parallel array. A complete list of elemental and transformational functions can be found in Reeves and Bruner (1982).

Parallel Pascal programs can be split between the host VAX and the MPP computers. The programmer specifies which subroutines will run on the VAX, and which subroutines will run on the MPP, through a compiler flag. Typically, the main program would be run on the VAX to allow easier non-parallel array data (parameter) input. This compiler flag also allows the programmer to optimize a program by directing to the VAX subroutines involving only scalar operations. (The VAX generally performs scalar operations more efficiently than the MPP.)

Another compiler flag is the translator flag. With the translator option turned on, all parallel constructs are translated into their serial equivalents, and VAX (rather than MPP) assembly language is generated. This translator option makes it possible to check out programs on the VAX without using the MPP. In this case, one would normally declare parallel arrays to be smaller than the normal 128 X 128 size (e.g. 16 X 16) for convenient debugging. This option permits the development of software for the MPP at locations other than GSFC for eventual installation on the MPP.

It is of interest to compare programming for the MPP in Parallel Pascal to programming a conventional computer or a vector array processor. As typical examples we will use, respectively, a VAX-11/780 minicomputer and an FPS-AP180V array processor. For instance, consider adding two arrays, multiplying by a third, and placing the result in a fourth array. Programming for the MPP in Parallel Pascal tends to be more compact in Parallel Pascal. For the MPP we would write:

```
d := a*b + c ;
```

where a, b, c and d are declared earlier as PARALLEL ARRAY's. Here, in one simple statement, one utilizes the full parallel power of the MPP to perform these operations.

Programming the VAX-11/780 minicomputer in Pascal to perform the same operations we would write:

```
FOR i:=0 TO 127 DO
  FOR j:=0 TO 127 DO
    d[i,j] := a[i,j]*b[i,j] + c[i,j] ;
```

where a, b, c and d are declared earlier as ARRAY's. Two FOR loops are required to do the same operation as was performed on the MPP without looping. Also, 16384 distinct multiplication, addition and assignment operations must be done on the VAX where single parallel multiplication, addition and assignment operations are done on the MPP.

One could use AP Math Library routines to perform the same operations on the FPS-AP180V vector array processor. For convenience and efficiency we would represent our 128-by-128 data arrays as 16384 element vectors when using the AP Math Library. To call the appropriate library routines from AP-FORTRAN we would write:

```
CALL FVMUL(A,1,B,1,D,1,16384)
CALL FVADD(D,1,C,1,D,1,16384)
```

where FVADD is the vector add library routine and FVMUL is the vector multiply library routine. As in Parallel Pascal, no loops are required to perform the specified operations. However, the required programming statements are not at all intuitive as they are in Parallel Pascal.

The MPP programming environment is being gradually enriched by a growing subroutine library. University of Illinois has contributed linear algebra subroutines for the MPP (Gallopoulos, 1985). Also, a versatile subroutine library has been contributed by Cornell University (Reeves, 1985). Besides adding several useful subroutines for use with standard 128-by-128 arrays on the MPP, this subroutine library contains subroutines that conveniently perform operations on "large" arrays with dimensions some small multiple of the standard 128-by-128 array. For example, this library could be used for performing correlations, convolutions, image rotations, or general geometric corrections on 512 x 512 arrays.

APPLICATIONS

For an algorithm to be considered suitable for implementation on the MPP, the following criteria should hold:

1. The algorithm should be significantly compute-bound on conventional machines or of general utility so that it could be part of a subroutine library for use in bigger programs

2. It should be possible to parallelize the algorithm to take advantage of the MPP's architecture.

Several algorithms pertinent to processing and analysis of remotely sensed images and satisfying the above criteria have been implemented on the MPP. These algorithms will be discussed briefly in this section. The mathematical details of the algorithms are beyond the scope of this paper, but can be found in the references cited. The applications covered include data preparation (preprocessing), feature extraction, unsupervised and supervised classification, and post-classification processing.

Data Preparation

Three separate algorithms are covered under this heading: Thematic Mapper Geometric Correction, Synthetic Aperture Radar Signal Processing, and Stereo Image Matching.

Thematic Mapper Geometric Correction: The Landsat Thematic Mapper (TM) produces images in 7 spectral bands amounting to approximately 240 MB of image data per scene. Many corrections are required to assure internal geometric consistency of the images and their rectification to standard map projections. These corrections are derived using a model of the imaging system. The details of the model and its use for geometric correction are provided in Prakash and Beyer (1981) and Sehn (1981).

The major steps in the geometric correction are coordinate computations and interpolations for resampling. These involve over 100 operations per output pixel. The procedures to perform these computations vary depending on the hardware used. There are at least four implementations of this algorithm:

1. The GSFC's Land Analysis System - LAS (Fischel, 1983) with an AP180V;

2. The GSFC's Applications Developmental Data System - ADDS (Mocarsky and Ramapriyan, 1983) with special purpose hardware;

3. The Thematic Mapper Image Processing System (TIPS) in the Landsat Ground Segment (Beyer, 1983) with special purpose hardware; and

4. The Goodyear implementation on the MPP at GSFC (Faiss, 1984).

The Mpp implementation differs significantly from the others in that it handles 128 x 128 image segments at a time and does not use certain linear interpolation approximations for coordinate computations.

Synthetic Aperture Radar (SAR) Signal Processing: Processing of signal (phase history) data from spaceborne SAR's into images representing the target backscatter coefficients is highly compute-bound. In general, it involves correlation with a two-dimensional reference function (point target response). The size of the reference function is approximately 1500 x 4200 for Seasat SAR and 900 x 2400 for the Shuttle Imaging Radar - B (SIR-B). However, it has been shown that with negligible effect on image quality, the processing can be separated into two one dimensional correlation operations (Wu, 1982). Thus, the signal data are first correlated in the range (cross-track) direction. Then, the results are transposed and correlated in the azimuth (along-track) direction. The one dimensional correlations are carried out using the Fast Fourier Transform (FFT). The correlation of each line of data requires a forward FFT, multiplication by the complex (range or azimuth) reference function, and an inverse FFT. In addition to the correlations, it is necessary to correct for range walk (caused by earth's rotation) and range curvature (caused by variation of the distance from a point target to the spacecraft).

The most time consuming operations of the above are the FFT's. The MPP implementation performs sixteen 4096-long complex FFT's simultaneously. The complex multiplications between the forward and the inverse FFT's take negligible time.

Stereo Image Matching: Elevation data for large regions are conveniently obtained by matching pairs of images of a given scene taken from known platform locations. Manual matching of such images is quite laborious. Several automatic matching techniques have been developed (for example: Marr and Poggio, 1979; Gennery, 1980; Norvelle, 1981). These require finding, for every point in one of the images (reference image), the corresponding point in the other image (test image). Hence, it is a very computationally intensive procedure. One method of reducing the computations is to perform the matching over a subsampled grid. However, this reduces the resolution of the result and does not permit taking advantage of continuity constraints on elevations to eliminate significant errors.

An algorithm has been developed on the MPP to perform the above matching without subsampling. This is an iterative algorithm which alternates between matching and local geometric correction. The iterations are carried out until the two images match to within a given tolerance. This process identifies local disparities between the two images. These disparities are used to compute the elevations. This algorithm is being tested on image pairs from the Shuttle Imaging Radar-B (SIR-B) obtained with different viewing angles.

Feature Extraction

There has been a considerable amount of interest in the characterization of texture in images and the use of textural features for image classification. The most commonly used definitions of textural features were given by Haralick and Shanmugan (1973) in terms of the "Spatial Grey-Tone Dependence Matrix (SGTDM)". These features are: Variance, Skewness, Kurtosis, Correlation, Inertia, Homogeneity, Energy and Entropy.

For the textural features to be meaningful, they have to be defined over neighborhoods with large numbers of samples. Typical sizes of neighborhoods are 9 x 9 to 31 x 31. Since the natural boundaries of objects in an image are not necessarily straight or parallel and perpendicular to the scan lines it is often necessary to compute textural features at neighborhoods centered at all possible pixels in the image. However, to reduce computations on conventional machines, these features are calculated over a sparser set of samples than the original image.

The approach to the computation of these features depends on the type of computer used. For the first six features, a recursive summation technique can be used to save computation time on a sequential computer. However, energy and entropy features cannot be computed in this manner because they require calculating the SGTDM and consequently a large amount of computation time. On the MPP, all the textural features can be calculated conveniently since they are obtained by local operations over small neighborhoods.

Classification

Unsupervised Classification (Clustering): The Basic ISODATA Clustering algorithm (Duda and Hart, 1973) was one of the first programs implemented on the MPP. This algorithm classifies the given multispectral image into a fixed number K of clusters. The value of K is an input to the program.

The algorithm starts with an initial guess of the K cluster centroids. Then, each pixel in the image is assigned to the cluster whose centroid is closest to its feature vector. Next, each of the cluster centroids is updated by averaging the feature vectors of all the pixels assigned to that cluster. The pixel assignment and the centroid update steps are alternated iteratively until no further change occurs or a prespecified number of iterations is completed.

Since the K distance calculations are to be performed for every pixel, the assignment step is very time-consuming. Typically, on conventional computers, this algorithm (and similar iterative clustering) are not run to convergence, but limited to about 10 iterations. However, it is practical to run clustering algorithms to convergence on the MPP in a near-interactive manner and explore the effects of changing parameters such as the number of clusters and the initial selection of centroids.

Supervised Classification: One of the first to be developed and most widely accepted and used supervised classification techniques is the per pixel maximum likelihood classifier with the assumption of normally distributed spectral classes. The classification of each pixel requires, for each spectral class, computation of the value of the likelihood function using the pixel's feature vector. The pixel is then assigned to the class with the largest likelihood function. The computation of the quadratic forms in the likelihood functions is the most time consuming step, but on the MPP, this is accomplished by computing them simultaneously for all pixels in a 128 x 128 neighborhood.

Contextual Classification: In the contextual approach to classification of multispectral image data, the probable classifications of neighboring pixels influence the classification of each pixel. Classification accuracies can be improved through this

approach since certain ground-cover classes naturally tend to occur more frequently in some contexts than in others. The contextual classifier presently implemented on the MPP is the algorithm formulated by Swain et al (1981) and further developed by Tilton et al (1982). Here compound decision theory is invoked to develop a classification method which exploits spectral/spatial information. See Swain et al (1981) for a complete derivation.
This contextual classifier was developed earlier at Purdue University, but it was tested only on 50-by-50 pixel data sets. The results produced in these tests were encouraging, but they were of limited value because of the small size of the test data sets. The classifier was not tested on larger data sets because it took too long to run on a VAX-11/780 minicomputer.

Testing the contextual classifier on large data sets has now become feasible with its implementation on the MPP. For example, a test of a version of classifier on the MPP takes a total of 5 minutes to produce a 120-by-120 pixel classification. It would take roughly 12 hours to perform the same classification on a VAX-11/780 minicomputer. This test is performed using a portion of a 7 band Thematic Mapper image. The MPP CPU processing time could be reduced to under 30 seconds by programming in MPP machine code.

Connected Components Labelling

The problem of labelling connected components is a classic one in image processing and has several applications. The problem consists of assigning a unique label to all pixels, within each connected region in an image. Typical applications of connected component labelling are: Identification of individual lakes, fields or forests in a classification map to facilitate area calculations, labelling of homogenous objects as a first step in the Extraction and Classification of Homogenous Objects (ECHO) classification algorithm (Kettig and Landgrebe, 1976), labelling and measurement of contour lines or other linear features, and inclusion of classification maps into a geocoded data structure (Goldberg, 1984).

The approach to implementation of the connected component labelling algorithm depends significantly on the computer architecture. Lumia et al (1983) have analyzed implementations on sequential machines with virtual memory. On parallel computers like the MPP, the algorithm is implemented by shrinking each of the connected components into isolated points, assigning a unique label to each such isolated point and growing the labels bounded by the input boundary points. The shrinking and growing are handled in parallel. However, to assure consecutive labels, it is necessary to use a sequential method to label the isolated points after shrinking.

TIMING SUMMARY

In this section we present a summary of the timings on the MPP for the applications algorithms discussed above. These timings are shown in table below.

With the exceptions noted below, the image sizes used were 512 x 512. The TM geometric correction time is for a full scene (5984 x 6320 input and 5965 x 6967 output). The SAR signal processing time is for single look processing of a full SIR-B scene (25000 x 4096). The contextual classification test was performed on a 128 x 128 image. The contextual classifications CPU timings include a VAX timing as

ALGORITHM	CPU TIME (SEC.)	I/O TIME 500 KB (SEC.)	I/O TIME 10 MB (SEC.)
TM Geometric Correction (7 bands; full scene)	20*	1080	56.00
SAR Signal Processing (SIR-B; Single look)	240	600	30.00
Stereo Image Matching	12*	1.5	0.01
Textural Feature Extraction (4 bands; 32 x 32 window)	0.7	4	0.01
Basic ISODATA Clustering (4 bands; 100 iterations; 16 Classes)	19	2.6	0.13
Maximum Likelihood Classification (4 bands; 16 Classes)	3.5	2.6	0.13
Contextual Classification (7 bands; 6 Classes)	150 (MPP) 90 (VAX)	0.25	0.02
Connected Components Labelling	0.9	1.6	0.08

* Estimates

well as MPP timing because substantial number of serial calculations are performed on the VAX. TM geometric correction and the Basic ISODATA algorithms have been coded in assembly language. All the other algorithms are in Parallel Pascal. The typical speed gains a 3 to 5 can be achieved by coding in assembly language (or by optimizing the assembled code manually). The I/O times indicated here are estimates based on the data volume being transferred and the slowest link in the data transfer pipeline. However, the effective data rates are dependent on other overheads associated with subroutine calls and data blocking and have not yet been optimized for the programs discussed here.

CONCLUSIONS

The use of the MPP's unique architecture for high-speed computation have been demonstrated for various applications of interest to the remote sensing community. Even with the easy-to-use high level language, Parallel Pascal, it has been shown that the MPP can solve applications problems orders of magnitude faster than conventional sequential machines. The actual (wallclock) times achieved for high data-volume problems is currently limited by the I/O hardware.

REFERENCES

Batcher, K. E., 1980, "Design of a Massively Parallel Processor", IEEE Transactions on Computers, Vol. 29, No. 9.

Beyer, E. P., 1983, "Thematic Mapper Geometric Correction Processing", Proceedings of the Seventeenth International Symposium on Remote Sensing of Environment, Ann Arbor, MI., pp 319-334.

Duda R. O., and Hart P. E., 1973, Pattern Classification and Scene Analysis, John Wiley and Sons, New York, p. 201.

Faiss, R. O., 1984, "Landsat Thematic Mapper Data Processing with the Massively Parallel Processor", Proceedings of the ASP-ACSM Convention, March 11-16, Washington, D.C., p. 263.

Fischel, D., 1983, "Some Current Uses of Array Processors for Preprocessing of Remote Sensing Data", Proceedings of the Seventeenth International Symposium on Remote Sensing of Environment, Ann Arbor, MI., pp. 185-196.

Gallopoulos, E., 1985, Processor Arrays for Problems in Computational Physics, Ph. D. Thesis, Department of Computer Science, University of Illinois, Urbana.

Gennery, D. B., 1980, "Modeling the Environment of an Exploring Vehicle by Means of Stereo Vision", Stanford Artificial Intelligence Laboratory, Memo AIM-339.

Goldberg, M., et al, 1984, "Topological Grid Structure: A Data Structure for Earth Science Modelling", Proceedings of the Pecora IX Symposium, Sioux Falls, SD, pp. 83-86.

Haralick, R. M., Shanmugan, K. S., and Kinstein, I., 1973, "Textural Features for Image Classification", IEEE Transactions on Systems, Man and Cybernetics, Vol. SMC-3, pp 610-621.

Kettig, R. L. and Landgrebe, D. A., 1976, "Classification of Multispectral Image Data by Extraction and Classification of Homogeneous Objects", IEEE Transactions on Geoscience Electronics, Vol. GE-14, pp. 19-26.

Lumia, R., Shapiro, L., and Zuniga, O., 1983, "A New Connected Components Algorithm for Virtual Memory Computers", Computer Vision, Graphics, and Image Processing, Vol. 22, pp 287-300

Marr, D. and Poggio, T., 1979, "A Computational Theory of Human Stereo Vision", Proceedings of the Royal Society of London, pp. 301-328.

Mocarsky, W. L. and Ramapriyan, H. K., 1983, "The Applications Developmental Data System", Proceedings of the Seventeenth International Symposium on the Remote Sensing of Environment, pp. 1353-1362.

Norvelle, F. R., 1981, "Interactive Digital Correlation Techniques for Automatic Compilation of Elevation Data", Engineering Topographic Laboratories, Report No. ETL-0272.

Prakash, A. and Beyer, E. P., 1981, "Landsat-D Thematic Mapper Image Resampling for Scan Geometry Correction", Seventh International Symposium on Machine Processing of Remotely Sensed Data, pp. 189-200, LARS, Purdue University.

Ramapriyan, H. K. and Strong, J. P., 1983, "Applications of Array Processors in the Analysis of Remote Sensing Images", Proceedings of the Seventeenth International Symposium on the Remote Sensing of Environment, Ann Arbor, MI., pp. 197-216.

Reeves A. P. and Bruner J. D., 1980, "High Level Language Specification and Efficient Function Implementation for the Massively Parallel Processor", Interim Report TR-EE 80-32, School of Electrical engineering, Purdue University, W. Lafayette, IN.

Reeves A. P. and Bruner J. D., 1982, "The Language Parallel Pascal and other Aspects of the Massively Parallel Processor", Final Report for NASA Grant 5-3, School of Electrical Engineering, Cornell University, Ithaca, NY.

Reeves, A. P, 1985, Parallel Pascal Development System, School of Electrical Engineering, Cornell University, Ithaca, NY.

Schaefer, D. H., Fischer, J. R., Wallgren, K. R., 1982, "The Massively Parallel Processor", AIAA Journal of Guidance, Control and Dynamics, Vol. 5, No. 3,pp. 313-315.

Sehn, G. J., 1981, "Thematic Mapper Imagery Geometric Correction - Payload Correction Subsystem Description", General Electric Program Information Release, 1K50-LSD-1014, pp. 1-20.

Swain, P. H., Vardeman S. B., and Tilton, J. C., 1981, "Contextual Classification of Multispectral Image Data", Pattern Recognition, Vol. 13, No. 6, pp. 429-441.

Tilton, J. C., Swain, P. H., and Vardeman S. B., 1982, "Estimation of Context for Statistical Classification of Multispectral Image Data", IEEE Transactions on Geosciences and Remote Sensing, Vol. GE-20, No. 4, pp. 445-452.

Wu, C. et al, 1982, "Modeling and a Correlation Algorithm for Spaceborne SAR Signals", IEEE Transactions on Aerospace and Electronics Systems, Vol. AES-18, pp. 563-575.

CONCURRENT PROCESSING TECHNOLOGY FOR LAND REMOTE SENSING

Jerry E. Solomon and Meemong Lee
Jet Propulsion Laboratory
Mail Stop 168/514
4800 Oak Grove Drive
Pasadena, California 91109

ABSTRACT

As the available data for land remote sensing becomes increasingly larger and more complex, e.g., thematic mapper, imaging spectrometer, and synthetic aperture radar imagery, the need for cost effective powerful computational resources becomes greater. Many scientists find their ability to utilize these new types of data severely limited by locally available computer facilities since most do not have easy access to super-computer class machines. Help appears to be on the way in the form of high performance/cost ratio machines built around the concepts of concurrent processing utilizing low cost microprocessors as the basic building blocks. The field is a rapidly expanding one with a great diversity in the architectures being proposed and built. The purpose of this paper is to present a short overview of the current state of the field and an in depth look at a specific concurrent processor implementation. A detailed discussion is presented of results obtained using the California Institute of Technology (Caltech) nearest neighbor concurrent processor (NNCP), sometimes referred to as the "hypercube", for image processing and analysis problems. Efficient utilization of concurrent processor machines requires a very different approach to algorithm design and implementation than in the case of serial machines. The machine architecture is described and a discussion of the implications of this architecture for image processing algorithm developed is given. Finally, the potential for constructing a super-computer "workstation" is described in light of the experience gained from the NNCP machine.

EMPLOYING GEOGRAPHIC REASONING IN LANDSCAPE MAPPING

James W. Merchant
Kansas Applied Remote Sensing Program
University of Kansas
Space Technology Center
Lawrence, Kansas 66045

ABSTRACT

"Geographic reasoning" is defined as the systematic application of geographic knowledge and understandings in the analysis of data acquired via remote sensing. Visual image interpreters routinely use geographic reasoning in preparing maps of land use and ecoregions. A strategy for employing such reasoning in classification of Landsat Thematic Mapper digital data is proposed. An experiment is designed to test the proposed strategy on a Kansas site. Thematic Mapper data are initially stratified into water, vegetated and non-vegetated pixels. A region-growing algorithm is then used to define "fields" of similar land cover composition. Fields are characterized by size and neighborhood attributes. Hierarchical clustering is employed to test the hypothesis that the diversity, field sizes and spatial interspersion of land cover types are useful discriminants of landscape regions. The hypothesis is shown valid. A major focus of current research is upon implementation of these results.

INTRODUCTION

Classification of landscapes and mapping of landscape regions (e.g., ecoregions, land use regions) are important foci of study in geography and other earth sciences (Spence and Taylor, 1970; Gardiner and Gregory, 1977). The concept of the geographic region is based upon the premise that there are identifiable physical and/or cultural landscape elements which serve to characterize an area and differentiate it from surrounding areas. Remotely sensed data are commonly used to aid in classification and mapping of landscape regions.

Conventional computer-assisted classification of digital multispectral scanner data may be considered monothetic in that it is usually based upon essentially one type of variable - spectral reflectance (though this variable may be comprised of individual measurements taken in several spectral bands). Spectrally-based classification algorithms have been found to be effective in distinguishing many types of "land cover" (e.g., grass, coniferous forest, water). They have been less successfully employed to identify and map landscape regions (Merchant, 1984a, b). Classification of ecoregions and land use requires a polythetic procedure (Spence and Taylor, 1970). Land cover composition, which can often be estimated via multispectral classification, is an important variable, but it alone does not permit differentiation among, or demarcation of, such regions. Other variables must be employed. Bailey, et al., (1978) observe that "the relationship between components of landscape and physical and biological process is almost always through spatial pattern or structure rather than through composition alone."

Landscape regions are comprised of mosaics of individual parcels or "patches" of land cover. Forman and Godron (1981) propose that "the structure of a landscape is primarily a series of patches surrounded by a matrix . . . the numbers of patches of each patch origin [natural or cultural], biotic patch type, size and shape determine in part the landscape structure. However, the spatial configuration among the patches present may be just as important as the numbers." Landscape structure is infrequently random (Grigg, 1965; Forman and Godron, 1981). Randomness implies that each event has an equal probability of occurrence. Landscapes, particularly those resulting from human decisions, are likely to reflect some spatial order.

The spatial character of the landscape may arise from a variety of phenomena and processes. In the "natural" environment where man is not a significant factor, spatial structure may reflect the pattern of vegetation communities and/or soils which tend to develop in a particular mosaic pattern under a given set of geologic, topographic (e.g., elevation, slope, aspect, drainage) and climatic circumstances. Where man is a significant organism in the environment, the spatial character of the surface will be related to segmentation of the landscape resulting from land ownership, land use (including settlement, cropping, grazing, and mineral extraction practices), transportation and energy resource development, and urbanization. In most instances, these phenomena will still reflect, to a substantial extent, the natural environment (e.g., soils, slope, drainage, relief, climate), since the environment imposes some constraints (at least economic) on the practical use of the land.

Visual image interpreters routinely employ spatial cues evident in a scene to aid in data stratification, classification and mapping. Such cues may include parcel size, parcel shape, pattern, texture, context and associations of cover types. Optimal visual image interpretation requires a systematic, logical approach to data analysis, and usually employs deductive reasoning and inference. Visual image interpretation requires the use of geographic reasoning and spatial logic. "Geographic reasoning" is defined as the systematic application of geographic knowledge, understandings and expectations in the analysis of data acquired via remote sensing for the purpose of accomplishing a specific objective or set of objectives. "Spatial logic" is defined as the formal expression and systematic application of decision rules which are formulated to assist in the attainment of a particular analytic objective (e.g., preparation of a 1:250,000 scale land use map portraying classes X, Y and Z) and are based upon characteristics and relationships evident in a particular landscape as depicted in a particular set of remotely sensed data. Spatial logic is developed as the image analyst transforms observations, understandings and geographic knowledge of the landscape into decision rules which can be applied in data analysis to segment that landscape into regions of interest. Such rules typically require consideration of both spectral <u>and</u> spatial properties of the landscape, and may require inference and deduction.

This research is directed towards examining means for applying such reasoning in the classification of digital multispectral data. The research has two specific objectives:

1. To formulate and test a data analysis strategy, founded upon spatial logic and emulating some aspects of visual image interpretation, which can be employed to prepare maps of landscape regions from digital multispectral data; and
2. To evaluate the utility of Landsat Thematic Mapper (TM) data for this application.

DEVELOPMENT OF A DATA ANALYSIS STRATEGY

A data analysis strategy, founded upon geographic reasoning and spatial logic, was formulated. The strategy, described at length previously (Merchant 1984 a, b), is here only summarized.

It was observed that visual image interpreters creating land use maps appear to rely on analysis of the spatial distributions of rather generalized land cover types in order to demarcate regions of land use. For example, an image interpreter can identify "cropland" without first having to identify corn, wheat, oats and soybeans. Cropland can be defined on the basis of the size, shape and spatial arrangement of "fields" (parcels, "patches") of "crop-like" stuff, bare soil and other less frequently occurring cover types (e.g., water, trees).

It was proposed that a greenness/brightness data transformation be used initially to stratify the data into, what might be termed, "least common denominator" cover classes - water, organic (vegetated) and inorganic (non-water, non-vegetation) pixels. A "region-growing" technique would then be employed to define "fields" of similar cover composition within each stratum. Fields would be subsequently characterized in a more refined manner (e.g., crop-like, grass-like). Such cover classes should comprise the fundamental constituents of all land use classes, should have "physical significance," and should be readily and consistently identifiable by any data analyst.

Having "mapped" the distribution of fundamental cover types, the task would then be to "classify" land use on the basis of differences in the spatial structure of the landscape so depicted. Three properties of the landscape would be employed, initially, in an attempt to differentiate landscape regions:

1. Fundamental cover composition;
2. Cover diversity (i.e., the number and types of cover per unit area); and
3. Cover interspersion (i.e., parcel size, arrangement and context).

These three variables, selected from a broad range of possible measures of landscape structure, were thought to characterize important elements of spatial structure employed by visual image interpreters in the definition of land use regions. A complex supervised iterative contextual classification algorithm was developed to assign final land use/land cover labels to each field.

INITIAL EXPERIMENTS

The strategy outlined above was tested on a 640 x 480-pixel scene selected from a Landsat TM image acquired September 3, 1982 (ID 40049-16273). The study area includes the City of Topeka, Kansas, and vicinity. The success of classification was evaluated with respect to (1) contemporary aerial photography and (2) a conventional supervised multispectral classification of the same data set carried out by other researchers.

Initial classification results were unsatisfactory. Several adjustments and retrials resulted in little improvement. Consequently, it was decided to examine class statistics to determine if problems lay within the logic of the proposed strategy or were within the complex final classification algorithm employed.

HIERARCHICAL CLUSTERING OF SAMPLE DATA

It was hypothesized that the problems experienced in classification were related to the complex manner in which the classification was implemented, rather than to the logic of the overall proposed strategy. An experiment was conducted to test this hypothesis. The experiment had two major objectives:

1. To determine if the proposed measures of cover composition, field size and interspersion, computed for the landscape defined by the data processing carried out heretofore, could be used to discriminate between land use classes; and
2. To determine if all measures were of equal value and to identify useful measures.

If the measures were shown to be valid and useful, then one might assume that the problems experienced with the classification program were likely due to the complexity of implementation rather than to basic flaws in spatial logic.

Forty samples of twelve land use/land cover classes were selected (Table 1). Each of the forty samples was characterized by a set of eleven statistical variables which were designed to characterize land cover composition, field size attributes and land cover interspersion (Table 2).

Various combinations of the eleven variable sets were entered into a hierarchical clustering program. Hierarchical clustering (HC) is a tool used to cluster together (or link) objects of interest according to selected measures of their similarity (or, if desired, their dissimilarity) (Davis, 1973; Mather, 1976). The intent of clustering was to determine whether any of the statistics computed served as useful discriminants for separating samples of different classes, and, if so, which served as the best discriminants.

RESULTS

A 40-sample x 40-sample error matrix was constructed for each HC run. The error matrix portrays linkages between all samples at the twelve-cluster level. A perfectly accurate twelve-cluster linkage would have resulted in all samples of a single land use class being linked with one another and no linkages between different land use classes (Figure 1).*

A relative accuracy (RA) score was computed for each of the twelve classes and for each run. The within-class relative accuracy score was computed as:

$$\frac{\text{actual within-class links achieved}}{\text{maximum within-class links possible}}$$

The overall score for the entire run was computed as:

$$\frac{\text{actual "correct" within-class links achieved over all classes}}{\text{maximum within-class links possible}}$$

---

*Note that, since the error matrices are symmetrical, only one-half of the matrix is shown.

TABLE 1. LAND USE/LAND COVER CLASSES

Class	Number of Sample Sites (Sample Numbers)
1. Residential - Urban Core	4 (1-4)
2. Residential - Older	4 (5-8)
3. Residential - Newer	4 (9-12)
4. Central Business District	3 (13-15)
5. Secondary/Neighborhood Business Districts	4 (16-19)
6. Industrial	2 (20-21)
7. Institutional	3 (22-24)
8. Urban Open Land	3 (25-27)
9. Cropland and Pasture - Upland	4 (28-31)
10. Cropland and Pasture - Floodplain	4 (32-35)
11. Rangeland	3 (36-38)
12. Woodland	2 (39-40)

TABLE 2. VARIABLES COMPUTED FOR EACH SAMPLE

1. The relative frequency of occurrence (RF) of each of the nine cover types of which any sample is comprised (expressed as the percent of the total pixels in the sample):

   (1) Water
   (2) Dark Inorganic Material
   (3) Medium-Dark Inorganic Material
   (4) Medium Inorganic Material
   (5) Medium-Bright Inorganic Material
   (6) Bright Inorganic Material
   (7) Crop-like Material
   (8) Tree-like Material
   (9) Grass-like Material;

2. The relative frequency of occurrence of Water, Inorganic Material (the sum of cover classes 2-6) and Organic Material (the sum of cover classes 7-9);
3. The mean field size of each cover type present in a sample;
4. The mean field sizes of Water, Inorganic and Organic Material;
5. The standard deviation of field sizes of each cover type present;
6. The standard deviation of field sizes of Water, Inorganic and Organic Material;
7. The modal field size of each cover type present;
8. The modal field size of Water, Inorganic and Organic Material;
9. The percent of the sample in each of nine field size categories:
10. The mean, standard deviation and modal field size for <u>all fields</u> in the sample; and
11. An index of spatial complexity (interspersion) for the entire sample (SCI), where

$$SCI = \frac{\text{fields}}{\text{pixels}} * 100$$

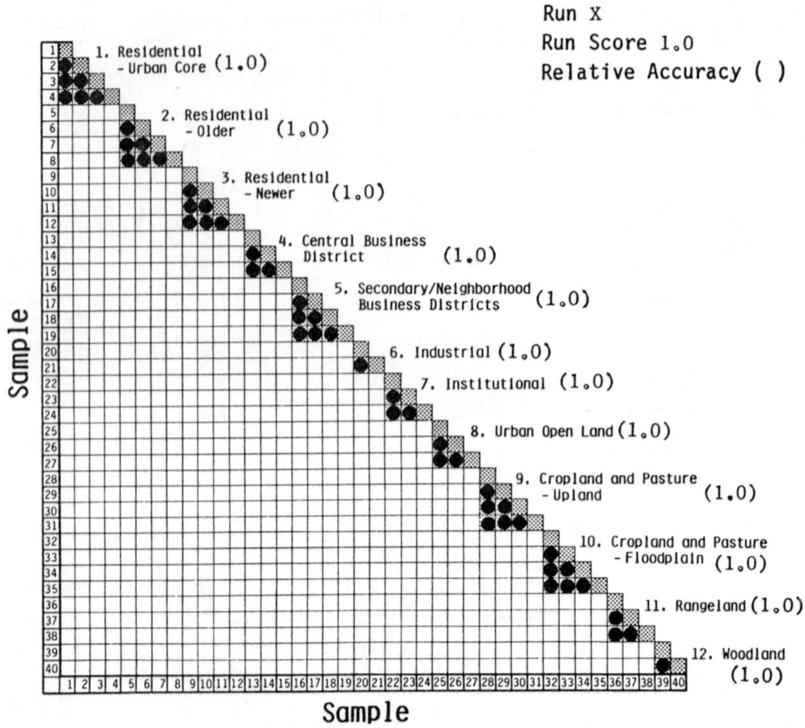

Figure 1. Example of an error matrix portraying perfect linkage between all samples within each land use class and no confusion between classes. Linkage indicated by •.

In either case, the score ranges from 0 (indicating no "correct" links) to 1.0 (indicating that all samples of a given class were linked perfectly). Note, however, that this score cannot be interpreted as percent correct. A class that has a score of 1.0 may still have links with other different classes. That is, all samples of a given class may have been linked together (giving an RA score of 1.0) but all may have also been linked, in a particular run, to samples of a completely different class. Such "confusion" is graphically portrayed on the error matrices. The RA scores and error matrices provided a guide to interpretation and comparison of the results of each run.

Because of limited space, only the best outcome of hierarchical clustering is discussed in detail below. Complete results are presented in Merchant (1984b).

The most satisfactory result (Run 7) was obtained by clustering on four variables:

(1) The relative frequency of occurrence (RF) of each of the nine cover types of which any sample could be comprised;
(2) The mean field size of each cover type present in a sample;
(3) The standard deviation of field sizes of each cover type present; and
(4) The modal field size of each cover type present.

Analysis of the error matrix for Run 7 produced an overall RA score of 0.56, a substantial improvement over other runs (Figure 2). This was attributed to the addition of one or more of the field size descriptors. Subsequent addition of the spatial complexity index (SCI) and percent of the sample in each of nine field size categories (PCF) to the variable set produced no discernible increase in success of clustering.

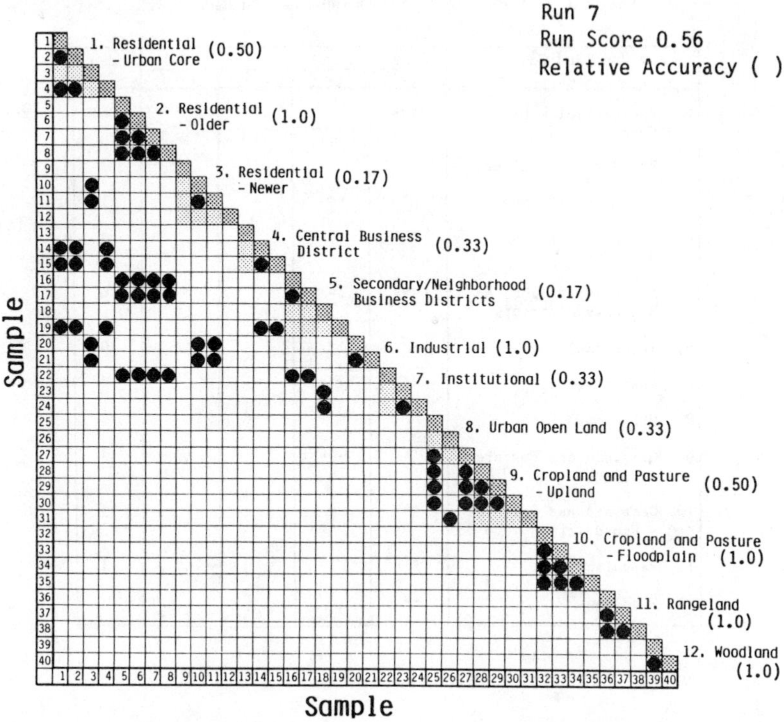

FIGURE 2. Error matrix for best HC run.

The result of Run 7 was compared with the result of the next best run, Run 2 (which was based on the RF alone). The success of within-class linkage was noted to have improved in Classes 2, 5 and 10, but decreased in Classes 3 and 4.

Rural classes were generally quite well-defined. All samples of Classes 10, 11 and 12 (Cropland and Pasture-Floodplain, Rangeland and Woodland, respectively) were linked with one another, and were linked with no other classes. The confusion between Class 10 and Sample 18, Class 5 (White Lakes Mall), which was present in Run 2, was absent in Run 7. Field size apparently aided in sorting these classes. It was noted that confusion between Classes 9 and 10, observed in other HC runs, was also resolved favorably.

Class 6 (Industrial) was perfectly defined in Run 2, but was confused in Run 7 with Sample 3 (Class 1) and Samples 10 and 11 (Class 3). This confusion of industrial and residential land uses was believed to be an artifact of the 3-to-1 weighting applied to field size attributes. It was noted that the SCI, properly weighted, might have been used to help segregate these samples at least to the extent of separating industrial from residential land uses (Table 3).

TABLE 3. RELATIONSHIPS BETWEEN REGIONAL SPATIAL COMPLEXITY INDEX (SCI) (NUMBER OF SAMPLES) AND LAND USE/LAND COVER

Land Use/Land Cover Classes	Spatial Complexity Index (SCI)				Mean SCI
	1-10	11-20	21-30	31-40	
1. Residential - Urban Core	***	*			6
2. Residential - Older			**	**	31
3. Residential - Newer		***	*		21
4. Central Business District	***				3
5. Secondary/Neighborhood Business Districts	**	*		*	16
6. Industrial	**				10
7. Institutional	**		*		12
8. Urban Open Land			***		24
9. Cropland and Pasture - Upland	*	***			13
10. Cropland and Pasture - Floodplain	***	*			10
11. Rangeland	***				3
12. Woodland	**				6

Note: $SCI = \frac{\text{fields}}{\text{pixels}} * 100$

SCI = 100 = maximum interspersion; as SCI approaches 0 it indicates decreasing interspersion

Overall, urban classes (Classes 1-8) and rural classes (Classes 9-12) were well distinguished from one another. Only Classes 8 and 9 (Urban Open Land and Cropland and Pasture-Upland, respectively) were confused. In terms of RF and field size, these might be considered "transitional" classes. It was observed, again, that even this confusion might have been reduced by use of a properly weighted SCI.

The results of the hierarchical clustering experiments may be summarized as follows:

1. The results showed that the relative frequency of occurrence of each of the nine cover types of which a sample could be comprised and field size statistics could be profitably employed to separate land use/land cover samples into a logical classification.
2. Two other variables tested were shown to have little value. These were:
   (a) the percent of the sample in each of nine field size categories; and
   (b) the spatial complexity index (SCI) of the sample site.

3. Classification wherein cover was labeled only as Water, Inorganic and Organic material was not as successful as classification wherein cover was broken into nine types.
4. Rural and urban land use/land cover classes were well-separated from one another in the best run. Most confusion between classes was logical and explainable.
5. There was evidence that results might be improved if field size, RF and SCI variables were more equitably weighted.
6. Comparison of the results of these experiments with conventional spectral classifications of the study area produced from TM and MSS data shows that the addition of spatial information aids in definition of land use and in separation of certain land use classes.

It should be noted that all of these findings reflect the manner in which fields were defined (i.e., through greenness-brightness transformations and region-growing) and labeled, the decisions that were made during these steps of analysis (e.g., thresholds selected), and the way in which spatial structure and cover composition were measured (e.g., RF, $\bar{x}$ field size, SCI). They also reflect the manner in which hierarchical clustering was conducted. Nevertheless, it seems clear that the type and diversity of cover and field size are important variables that can be employed to help classify and identify landscape regions in digital TM data.

CONCLUSIONS

A case has been made for the more rigorous application of geographic reasoning and spatial logic in analysis of digital remotely sensed data. A strategy for mapping landscape regions, founded upon spatial logic, has been proposed and tested. The strategy was judged to be valid, but full implementation will require additional effort.

The major conclusions of this research to date are as follows:

1. Spatial logic and geographic reasoning provide useful frameworks for formulation of digital data analysis strategies;
2. Greenness and brightness data transformations and region-growing are useful for defining landscape structure in a cogent manner;
3. Several measures of the landscape structure so-defined are effective for classification of land use. These include:
   (a) the relative frequency of occurrence of land cover types;
   (b) the mean field size for each cover type that occurs; and
   (c) the standard deviation of the field sizes of each cover type that occurs;
4. Other measures of landscape structure, such as the spatial complexity index, may be useful discriminants if weighted appropriately; and
5. The results of the hierarchical clustering experiments suggest that the overall approach to data analysis proposed is valid. The failure of the contextual classification algorithm to perform successfully is believed to stem from the internal complexity and logic of the program and not from the spatial logic which guided the overall strategy. Further research is warranted.

ACKNOWLEDGMENTS

The constructive critiques and helpful suggestions of T. H. Lee Williams and George F. Jenks have been important contributions to this research. Tshow Chu and Chien-Chung Chan provided expert and innovative software development. The financial support of the National Aeronautics and Space Administration (Grant NGL 17-004-024) is gratefully acknowledged.

REFERENCES

Bailey, R. G., R. D. Pfister and J. A. Henderson, 1978, Nature of Land and Resource Classification - A Review: Journal of Forestry, Vol. 76, pp. 650-655.

Davis, J. C., 1973, Statistics and Data Analysis in Geology, J. Wiley, New York.

Forman, R. T. T. and M. Godron, 1981, Patches and Structural Components for a Landscape Ecology: Bioscience, Vol. 31, pp. 733-739.

Gardiner, V. and K. J. Gregory, 1977, Progress in Portraying the Physical Landscape: Progress in Physical Geography, Vol. 1, , Edward Arnold, London, pp. 1-22.

Grigg, D., 1965, The Logic of Regional Systems: Annals of the Association of American Geographers, Vol. 44, pp. 465-491.

Mather, P. M., 1976, Computational Methods of Multivariate Analysis in Physical Geography, J. Wiley, London.

Merchant, J. W., 1984a, Using Spatial Logic in Classification of Landsat TM Data: Proceedings of the Pecora IX Symposium, Institute of Electrical and Electronic Engineers, Piscataway, NJ, pp. 378-385.

Merchant, J. W., 1984b, Employing Spatial Logic in Classification of Landsat Thematic Mapper Data, Ph.D. Dissertation, Department of Geography, University of Kansas, Lawrence, KS.

Spence, N. A. and P. J. Taylor, 1970, Quantitative Methods in Regional Taxonomy: Progress in Geography, Vol. 2, Edward Arnold, London, pp. 1-64.

ANALYSIS OF IMAGE PROCESSING ALGORITHMS FOR CLASSIFYING THE FORESTS OF NORTHERN MINNESOTA

S.R. Yool, J.L. Star, J.E. Estes, and D.B. Botkin
University of California
Santa Barbara, CA 93106

ABSTRACT

It is important to test the limits of space-based sensor data and associated processing algorithms to understand their potential for the production of world forest maps of verified accuracy. Such maps are needed to evaluate concerns that massive forest clearing will result in net carbon dioxide emissions to the atmosphere, promoting global warming and related environmental effects. Landsat data from a forest test site in northeastern Minnesota are used to evaluate waveband ratios, statistical filters and principal components algorithms for large-area forest classification. These algorithms were used with Landsat data to produce spectral variables. Performances of the spectral variables were ranked based on their accuracy in classifying vegetation at the Minnesota test site. Analysis of performance indicates that no single spectral variable is best always for forest classification, but performance appears to depend on the vegetation class and the spectral variable used. Differences in performance appear to be related to variations in background reflectance, surface illumination and spatial pattern by class. These results underscore the complexity of the relations between the land cover regime, remotely-sensed data, and the algorithms used to process these data.

INTRODUCTION AND OVERVIEW

Available forest maps are of insufficient accuracy, currency and scale to evaluate the effect massive forest cutting is having on global carbon balance (Houghton and Woodwell, 1981). Scientists are concerned that an imbalance in the global carbon cycle will promote a "greenhouse effect" by the middle of the next century (Sagan et al., 1979). To help evaluate this concern, they have advocated the use of space-based remote sensing to produce accurate maps of the world's forests (Botkin et al., 1984; Woodwell et al., 1983). The accuracies of such maps, particularly in areas of high spatial complexity, may depend upon both the environmental characteristics of the area, and the data processing algorithms used to generate these maps. This paper analyzes the relative performances of data processing algorithms for distinguishing forest type classes within a complex Landsat scene. Landsat data were selected for these tests because of the sensitivity of the Landsat bands to vegetation, and because the large quantity of experimental data involving Landsat serves as a good basis for comparison.

Surface spectral variations are needed to classify forest types. Undesirable spectral variations related to the angles of the sun and sensor, the state of the atmosphere, and the type of background soil, however, typically have an adverse effect on classification accuracy. Algorithms have been developed to reduce these undesirable spectral variations in agricultural and rangeland vegetation (Kriegler et al., 1969; Carneggie et al., 1974; Richardson and Wiegand, 1977; Tucker, 1979; Holben and Justice, 1980). We test whether such algorithms offer advantages over unprocessed data when large and complex forested areas are classified. First, we describe our test scene, then review the algorithms developed for information extraction from remotely-sensed data. Next, we describe the methods we used to test the performances of select algorithms. Finally, we analyze performance results, and speculate about the physical bases for these results.

TEST SCENE

The test scene selected for this work is located in the north-central United States near Ely, Minnesota and contains portions of the Superior National Forest and Boundary Waters Canoe Area (Figure 1).

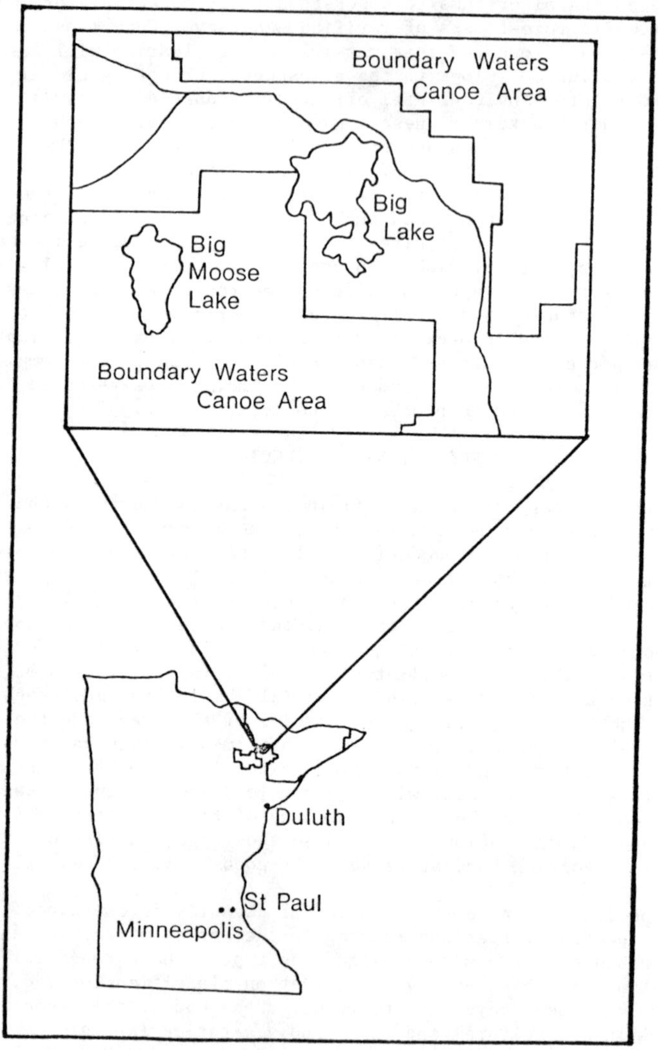

Figure 1. Northern Minnesota Test Scene.

The test scene lies within the Superior Upland physiographic province. The forest is part of the Laurentian Peneplain, and is underlain by Pre-Cambrian rock. Numerous lowland ridges and locally steep scarps (30-150m) are drained by a grid of interconnected lakes and streams, occupying the granite or slate troughs and basins carved by repeated glaciation during the Pleistocene. These features have produced a flora similar to the boreal forest. Nine vegetation classes are found in the scene and were used for testing. These classes were identified using the United Nations Educational, Scientific, and Cultural Organization (UNESCO) scheme for global vegetation mapping (UNESCO, 1973). This scheme defines vegetation in terms of its physiognomy and structure, which makes it compatible with space-based remote sensing technology. UNESCO designates canopies having >80% crown cover as "forest," and canopies with <80% crown cover, as "woodland" (UNESCO, 1973). Table 1 lists the UNESCO classes identified within the test scene.

Class A = Evergreen needle-leaved woodland with round crowns, ENLWRC (Pinus banksiana/Jack pine; Pinus resinosa/Red pine; Pinus strobus/White pine)

Class B = Evergreen needle-leaved forest with round crowns, ENLFRC (Pinus banksiana/Jack pine; Pinus resinosa/Red pine; Pinus strobus/White pine)

Class C = Grass

Class D = Dwarf Shrub/Bog (Picea mariana/Black spruce <5m and >10% cover; Sphagnum, Typhus)

Class E = Evergreen needle-leaved woodland with conical crowns, ENLWCC (Picea mariana/Black spruce; Picea glauca/White spruce; Abies balsamea/Balsam fir)

Class F = Evergreen needle-leaved forest with conical crowns, ENFLCC (Picea mariana/Black spruce; Picea glauca/White spruce; Abies balsamea/Balsam fir)

Class G = Deciduous broad-leaved forest, DBLF (Populus spp./Aspen; Betula papyrifera/Paper birch; Acer rubrum/Red maple)

Class H = Mixed evergreen needle-leaved and deciduous broad-leaved forest, MF (Classes B, F, and G)

Class I = Water

Table 1. UNESCO classes identified within the Northern Minnesota Test Scene.

ALGORITHMS FOR INFORMATION EXTRACTION

Despite considerable evidence that reflectance in the visible and near-infrared wavelengths is correlated with physical parameters of vegetation (Tucker, 1979; Curran, 1980), the strength of these correlations is affected by changes in sun-surface-sensor angles (Duggin and Ellis, 1980), and by varying atmospheric conditions. The capacity to gain information about vegetation from these wavelengths depends therefore on the amount of spectral variability angular and atmospheric factors contribute to the reflectance data.

Selection of cloud-free scenes acquired at times of high solar elevation serves as a reasonable "first-order" reduction of spectral variations imposed by changing sun-surface-sensor angles and the atmospheric screen (Holben and Justice, 1980). Data processing has the potential to increase classification accuracies by reducing noise beyond the "first order," and to decrease data volume and associated processing costs (Maxwell, 1976). A variety of data processing algorithms have been developed to reduce noise, and thereby increase the amount of signal coming from surface features such as vegetation. Three major groups of algorithms in common use are waveband ratios, statistical filters, and principal components (Kriegler et al., 1969; Haralick, 1979; Moik, 1980).

The ratio of the reflectance values from certain spectral wavebands may increase signal-to-noise ratios and compress the data volume for some applications (Maxwell, 1976). Prior work on crop classification demonstrates that ratios reduce undesirable radiometric variations and tend to enhance vegetation-related signals for scenes free of haze. We decided to test several ratios developed for agriculture and rangeland applications (Rouse et al., 1973; Carneggie, 1974). Statistical filters are typically moving windows within which measurements of spatial variation in pixel brightness are made. Spatial or statistical filtering can be used to produce image regions having uniform spatial texture (Haralick, 1979) that represent the spatial variations in pixel brightness (Pickett, 1970). Texture data can be useful in vegetation classification if (1) the spatial sampling rate of the sensor is compatible (from a sampling theory perspective) with variations chacterizing the vegetation classes (Jenkins and Watts, 1968) and (2) if different classes have unique textures. Tonal variations in a forest are produced by the shapes and patterns of different tree species and stand densities. Thus, texture may be a useful discriminant of different vegetation classes. Principal component analysis has been reported useful for data compression (Jenson and Waltz, 1979; Kourtz and Scott, 1978). Signal-to-noise ratio is enhanced because in-phase radiometric variations that correlate among the spectral bands are discarded from the set of orthogonal linear combinations produced in the analysis (Moik, 1980). Further background on these three algorithms is given in Yool et al. (1985).

METHODS

We tested the performances of seventeen spectral variables of Landsat data. These variables were derived using three standard image processing algorithms reported in the literature (Table 2). The original Landsat spectral variables, MSS4,MSS5,MSS6, and MSS7, were included in the tests as standards for comparison.

Table 2. Spectral Variables, Northern Minnesota Test Scene.

Number	Variable	Type
1	MSS4	ORIGINAL (O)
2	MSS5	ORIGINAL (O)
3	MSS6	ORIGINAL (O)
4	MSS7	ORIGINAL (O)
5	MSS7/MSS5	RATIO (R)
6	MSS6/MSS5	RATIO (R)
7	(MSS6 - MSS5)/(MSS6 + MSS5) Vegetation Index Using MSS6 "VI6"	RATIO (R)
8	(MSS7 - MSS5)/(MSS7 + MSS5) VI7	RATIO (R)
9	SQRT(VI6 + 0.5) Transformed Vegetation Index Using VI6 "TVI6"	RATIO (R)
10	SQRT(VI7 + 0.5) TVI7	RATIO (R)
11	Local-Mean (M) Texture Of MSS5	TEXTURE (T)
12	Local Standard-Deviation (SD) Texture Of MSS5	TEXTURE (T)
13	Local Standard-Deviation (SD) Texture Of MSS7	TEXTURE (T)
14	MSS5.MSS7	ORIGINAL (O)
15	MSS5.MSS6	ORIGINAL (O)
16	Local Standard-Deviation (SD) Texture Of MSS7/MSS5	RATEX (RT)
17	Local-Mean (M) Texture Of MSS7/MSS5	RATEX (RT)
18	(MSS7/MSS5)/PC2	COMPLEX (C)
19	(Local SD Texture Of MSS7/MSS5)/PC2	COMPLEX (C)
20	(Local SD Texture Of MSS7)/(Local SD Texture Of MSS5)	RATEX (RT)
21	(MSS7/MSS5)/(Local SD Texture Of MSS7/MSS5)	RATEX (RT)
22	First Principal Component PC1	COMPLEX (C)
23	Second Principal Component PC2	COMPLEX (C)

The spectral variables split into five types.

* The ORIGINAL (O) spectral variables (1-4, 14, 15) consist of the unprocessed Landsat bands.
* The RATIO (R) variables (5-10) consist of linear combinations of MSS6 and MSS5 (MSS6/MSS6, "V16," and "TV16"), and MSS7 and MSS5 (MSS7/MSS5, "V17," and "TV17"). The amount of change in ratio values related to vegetation should exceed the changes in bands before "ratioing" them.
* The TEXTURE (T) variables (11, 12, and 13), are produced by passing a 3x3 moving window over an image. By this process the local mean and standard deviations were computed for each pixel in Landsat MSS5, and the local standard deviation was computed for MSS7. The texture produced by computing the local mean tends to enhance edges of forest type classes and improve their discrimination (Maxwell, 1976). The local standard deviation filter may enhance differences in gray-tone patterns if unique patterns are produced by different forest types.
* The RATEX (RT) variables (16, 17, 20, 21), are combinations of RATIO and TEXTURE variables. They are a measure of the texture or spatial variability in the ratio variables. RATEX variables may increase the feature recognition qualities possible using RATIO or TEXTURE variables separately.
* The COMPLEX (C) variables (18, 19, 22, 23) use the first or second principal components (PC1 or PC2) of the four original Landsat MSS variables. The first principal component is interpreted as differential illumination in the scene that may be associated with topographic preferences of certain forest cover types. The second

principal component is uncorrelated with the brightness variations associated with the first principal component, and may provide more vegetation-specific information. Principal components could enhance vegetation discrimination when used singly. Noise may be reduced however, and vegetation class separations enhanced if a principal component is combined with another type of spectral variable, such as a waveband ratio. The processing used to reduce noise present in the original data involves a different algorithm to generate waveband ratios than to derive the principal components, and their combination may have an additive effect (Logan and Strahler, 1983). To test these ideas, we assessed the performances of the first and second principal components singly, and in combination with select RATIO variables.

Performance Assessment Approaches

In previous work we examined divergence and classification accuracy approaches for ranking and analyzing the performances of spectral variables of Landsat data (Yool et al., 1985). Divergence is a measure of the statistical separability between spectral classes (Latty and Hoffer, 1980). Classification accuracy quantifies the ability of training statistics to assign pixels to the correct class. Our original hypothesis was that the results from both divergence and classification accuracy approaches would agree and could be used as converging lines of support to validate the rankings of the spectral variables. We found agreements among the original, unprocessed MSS variables. The unprocessed MSS variables were normally distributed. Disagreements among the processed data apparently resulted because they are not normally distributed and therefore violate the Gaussian assumption required by the divergence algorithm. (Though the classification algorithm we used also assumes a normal distribution, it invokes a maximum likelihood rule to resolve ambiguities caused by overlapping statistical distributions between classes.) Because no divergence algorithm was available to handle non-Gaussian distributions, we evaluated approaches to classification accuracy assessment, which included a "train-test" approach.

The "train-test" approach to classification accuracy assessment is based on the classification of testing sites using vegetation class statistics from training sites. A Monte Carlo technique was used to simulate the effect of random training and testing site selection. Results demonstrated that the inherent intraclass variability in natural vegetation influences strongly the statistical separability of a given class when training and testing sites are selected at random. Due to the uncertainty associated with the train-test approach, we selected a backclassification approach, and have used this approach in the performance tests discussed below.

Backclassification Approach

The backclassification approach uses the same sites for testing as used for training, and is not considered a legitimate measure of accuracy (Logan and Strahler, 1983). Backclassification results do however, provide insight into the *relative* effectiveness of the spectral variables to discriminate between classes. This approach is particularly efficient when many variables are being tested. Accordingly, we adopted this approach for the performance evaluations, using a parallelepiped-maximum likelihood classifier for the backclassifications. This classifier stores class decision boundaries in a look-up table. Each pixel is assigned to the class within whose decision boundaries its spectral signature falls. A Bayesian maximum likelihood algorithm resolves ambiguities should a pixel's spectral signature fall within the decision boundaries of two or more classes. We produced backclassification accuracies and 95% confidence limits for all spectral variables on each of nine classes. Confidence limits on the estimated proportion correct were calculated from an approximation to the binomial distribution (Moik, 1980; Winkler and Hays, 1975), then used to determine whether the performances of the spectral variables were significantly different. If confidence limits overlapped between variables, their performance differences were deemed insignificant.

RESULTS

We produced backclassification accuracies, with 95% confidence limits, for all spectral variables on classes A through I. Table 3 summarizes the results, and gives the relative rankings of the variables. We restrict our discussion to extremes in performance.

Table 3. Performances by Vegetation Class, Northern Minnesota Forest Region: <25% = poor (.); 26-50% = fair (*); 51-75% = good (**); >75% = excellent (***); T = "tie."

Rank	Score (%)	Type	A	B	C	D	E	F	G	H	I
1	48	14 O	*	*	*	**	**	**	**	.	***
2	47	17 RT	*	**	***	**	*	.	**	.	***
3	46	15 O	*	*	*	**	**	**	**	.	***
4T	45	7 R	*	*	**	**	**	.	**	.	***
4T	45	10 R	*	*	**	**	*	.	**	.	***
5T	44	5 R	*	*	**	**	*	.	*	.	***
5T	44	8 R	*	**	**	**	*	.	**	.	***
5T	44	9 R	*	**	*	*	*	*	**	.	***
6	43	6 R	*	*	**	**	*	.	**	.	***
7T	42	3 O	*	**	.	.	**	**	**	.	***
7T	42	4 O	.	**	.	.	**	**	**	.	***
7T	42	18 C	*	.	***	**	*	*	.	.	***
8	38	11 T	.	.	*	**	*	**	.	*	***
9	37	1 O	.	*	*	*	*	**	*	.	***
10	36	2 O	.	.	**	**	*	**	.	*	***
11T	28	16 RT	*	.	.	**	*	.	.	.	**
11T	28	19 C	*	.	.	***	*	.	**	.	.
12	26	21 RT	**	.	.	.	.	.	.	.	***
13T	22	20 RT	.	.	***	*	.	.	.	**	.
13T	22	13 T	.	.	.	.	*	*	.	*	***
14	20	12 T	.	.	*	.	.	.	.	.	**
15	10	22 C	.	.	**	.	.	.	.	.	.
16	9	23 C	.	.	**	.	.	.	.	.	.

DISCUSSION

Mean performance scores for the Landsat spectral variables ranged from a low of 9% to a high of 48% (Table 3). Natural breaks in these scores occur in two places namely, between ranks 10 and 11, and ranks 14 and

15. To insure performance differences were indeed coupled to the spectral variables, we tested several variables believed to contain little information. These included variable 20, a ratio of MSS7, SD Texture to MSS5, SD Texture and variable 21, a ratio of ratios. As expected, variables 20 and 21 had low scores. Several observations can be made on the results in Table 3.

(1) The two-band ORIGINAL variables and RATIO variables occupy the top one-half of the rankings. The range of their overall mean scores indicates that performance differences between them were insignificant.

(2) The RATIO variables are all close-ranked, separated by only two percentage points. This result strengthens the idea that ratios are very similar in their capacity for discrimination between vegetation classes (Perry and Lautenschlager, 1984; Jensen and Hodgson, 1985; Yool et al;, 1985).

(3) RATEX variable 17 far outperformed other RATEX variables. The performances of RATEX variables using a mean filter appear to outperform the majority of variables using the standard deviation filter because the different dynamic ranges of the different classes are preserved by the mean-filtering procedure. Modifying the scaling to enlarge the dynamic range of variables using the standard deviation filter enlarges the variance, and does not improve separability.

(4) The one-dimensional ORIGINAL spectral variables appear in the middle one-third of the rankings. Because these variables are the standards for comparison, it would appear that some processing has merit for discrimination between vegetation classes within this test scene.

(5) The RATEX and COMPLEX variables performed poorly overall. The high intraclass variances typical of natural vegetation make class separation difficult, particularly for the RATEX variables, because the texture algorithm used for the RATEX variables was based on a standard deviation filter, which performed poorly overall.

(6) ORIGINAL variables 14 and 15, and all RATIO variables, performed comparably on the non-conifer classes (i.e., classes C, D, G, and I). Though the performances of these spectral variables are not comparable to the others because they are two-dimensional, it is noteworthy that the RATIO variables performed as well as the two-dimensional variables.

(7) With the exception of RATEX variable 17, which used a moving-average filter, the performances of the TEXTURE and RATEX variables using a moving standard deviation filter were significantly lower than the other variables. Their performances were poorest overall because they produced similar statistical distributions for most of the vegetation classes. The similarity in the statistical distributions of the vegetation classes for these variables results because the moving standard deviation filter measures the spatial variations in pixel brightness levels, and does not preserve the differences in dynamic ranges between the classes. Accordingly, vegetation classes with different dynamic ranges could have similar SD textures.

(8) The moving-average algorithm used to produce TEXTURE variables 11 and 17 performs significantly better than any of the SD TEXTURE variables. The moving-average algorithm is essentially a low-pass filter. This filter acts to sharpen class boundaries and reduce intraclass variances compared with the SD TEXTURE variables, while preserving differences in the dynamic ranges between the vegetation classes. RATEX variable 17 outperformed all other ratios produced from MSS5 and MSS7. Furthermore, it outperformed

slightly all the ORIGINAL Landsat bands, including the unfiltered MSS7/MSS5 (variable 5) and the individual ORIGINAL bands MSS5, and MSS7.

(9) The differences in performance between ORIGINAL MSS6 and MSS7, and the RATIO variables, are insignificant overall. MSS4 and MSS5, which due to their intercorrelation are ranked rogether, perform poorest of the datasets in the top one-third of the rankings. These results are in contrast with prior work, which suggested that MSS4 and MSS5 were the best overall single-band data for natural vegetation classification (Logan and Strahler, 1983; Yool et al., 1985).

(10) Though most RATEX variables performed poorly, they outperformed most others on select classes. RATEX variable 21 for example outperformed all others on Class A, while RATEX variable 20 excelled on the mixed forest class (H). The overall low performances of TEXTURE and RATEX and COMPLEX variables may stem from the differential responses of the Landsat bands used in these variables to wavelength-dependent variations in background reflectances, and from high intraclass variances encountered in natural vegetation. A further source of poor performance due to high within-class variance may result from differential illumination, which has not been "ratioed out" in SD TEXTURE variables.

SUMMARY AND CONCLUSIONS

This research tested the performances of different data processing algorithms for natural vegetation classification. We analyzed performances of waveband ratios, statistical filters, and principal components algorithms using Landsat MSS data from a complex forested area in northern Minnesota. Differences in performances of the spectral variables apparently result in part from the relative capacities of data processing algorithms to reduce undesirable variations in background reflectance and surface illumination, and to enhance the spectral reflectance of different classes.

Performance differences between two-dimensional ORIGINAL variables and the RATIO variables were insignificant overall. RATIO or "vegetation index" variables appear to be functionally equivalent, and do not perform significantly better than the individual ORIGINAL Landsat variables. Overall classification accuracy for any given RATIO variable is <50%. ORIGINAL variable MSS6 was the best unprocessed one-dimensional spectral variable overall, while a local-mean filter of MSS7/MSS5 performed best of both processed and unprocessed one-dimensional spectral variables. RATEX and COMPLEX variables performed poorest overall, but RATEX variables 20 and 21 performed best on Classes A and H, respectively.

The results of these tests suggest the complexity of relations between the physical expression of natural vegetation, environmental variability and data processing techniques, that is likely to exist when large geographic areas are surveyed by satellite-borne sensors. Based on these results, the processing of the original Landsat MSS bands does not result in significant increases in performance for the majority of classes. We believe that the excellent performances of certain variables on certain classes however, is a sound scientific basis for the use of processed data for classification, particularly if general vegetation class information is available a priori. Accordingly we have been performing additional tests on other areas having high spatial complexity in order to test this idea out, and to understand further the limits of remote sensing and image processing algorithms for global forest surveys.

ACKNOWLEDGMENTS

We wish to acknowledge the assistance of Gloria Fletcher in the production of graphics. This research was supported in part by NASA contracts NAGW-455 and NAGW-323.

REFERENCES

Botkin, D.B., J.E. Estes, R.M. MacDonald, and Mark V. Wilson (1984); Studying The Earth's Vegetation From Space; Biosci; vol. 34, no. 8; pp. 508-514

Carneggie, D.M., S.D. deGloria, and R.N. Colwell (1974); Usefulness of ERTS-1 and Supporting Aircraft Data for Monitoring Plant Development and Range Conditions in California's Annual Grassland; BLM Final Report 53500-CT3-266

Curran, P. (1980); Multispectral Remote Sensing of Vegetation Amount; In Progress in Physical Geography vol. 4, no. 3, pp. 315-341

Duggin, M.J. and P.J. Ellis (1980); Relating Ground Level Reflected Radiance Measurements In The Landsat MSS Bandpasses To Satellite Altitudes; Proc., ASP Fall Technical Meeting; Niagara Falls, N.Y.; pp. RS-1-G to 10

Haralick, R.M. (1979); Statistical and Structural Approaches To Texture; Proc., IEEE; vol. 67, no. 5

Holben, B. and C. Justice (1980); An Examination of Spectral Band Ratioing To Reduce The Topographic Effect On Remotely Sensed Data; Tech. Memorandum 80604; NASA Goddard Space Flight Center

Houghton, R.A. and G.M. Woodwell (1981); Biotic Contributions To The Global Carbon Cycle: The Role Of Remote Sensing; Proc., Seventh Int'l Symp. Mach. Proc. Rem. Sens. Data; Purdue Univ.; West Lafayette, IN; pp. 593-600

Jackson, R.D., R.J. Reginato, P.J. Pinter, Jr. and S.B. Idso (1979); Plant canopy information extraction from composite scene reflectance of row crops; Appl. Opt. vol. 18, no. 22, pp. 3775-3782

Jenkins, G.M. and D.G. Watts (1968); Spectral Analysis and Its Applications; (San Francisco: Holden-Day, Inc.); 525p.

Jensen, J.R. and M.E. Hodgson (1985); Remote Sensing Forest Biomass: An Evaluation Using High Resolution Remote Sensor Data And Loblolly Pine Plots; The Prof. Geog.; vol. 37, no. 1; pp. 46-56

Jenson, S.K. and F.A. Waltz (1979); Principal Components Analysis And Canonical Analysis In Remote Sensing; Proc., Amer. Soc. Photogramm.; pp. 337-348

Kourtz, P.H. and A. J. Scott (1978); An Improved Image Enhancement Technique And Its Application To Forest Fire Management; Proc., Fifth Can. Symp. Rem. Sens.; Victoria, Aug. 1978; pp. 72-78

Kriegler, F.J., W.A. Malila, R.F. Nalepka and W. Richardson (1969); Preprocessing Transforms and Their Effects on Multispectral Recognition; Proc., Sixth Int'l Symp. Rem. Sens. Environ.; Environmental Research Institute of Michigan (ERIM); Ann Arbor, MI; pp. 97-121

Latty, R.S. and R.M. Hoffer (1980); Waveband Evaluation of Proposed Thematic Mapper In Forest Cover Classification; Proc., Amer. Soc. Photogram.; Niagara Falls, N.Y.; pp. RS-2-D-1 to RS-2-D-12

Logan, T.L. and A.H. Strahler (1983); Optimal Landsat Transforms For Forest Applications; Proc., Symp. on Mach. Proc. of Rem. Sens. Data; LARS/Purdue University; West Lafayette, Inc.

Maxwell, E.L. (1976); Multivariate System Analysis of Multispectral Imagery; Photo. Eng. & Rem. Sens. vol. 43, pp. 1173-86

Moik, J.G. (1980); Digital Processing of Remotely Sensed Images; NASA SP-431; 330p.

Perry, C.R. and L.F. Lautenschlager (1984); Functional Equivalence of Spectral Vegetation Indices; Rem. Sens. Environ.; vol.14,pp.169-182

Richardson, A.J. and C.L. Wiegand (1977); Distinguishing Vegetation From Soil Background; Photog. Eng. & Rem. Sens. vol. 43, pp.1541-52

Rouse, J.W., R.H. Haas, J.A. Schell and D. W. Deering (1973); Monitoring Vegetation Systems In The Great Plains With ERTS; Third ERTS Symp., NASA Special Publication 351 pp. 309-317

Sagan, C., O.B. Toon, and J.B. Pollock (1979); Anthropogenic Changes and the Earth's Climate; Science vol. 206, pp. 1363-68

Tucker, C.J. (1979); Red and Photographic Infrared Linear Combinations for Monitoring Vegetation; Rem. Sens. Environ. vol. 8, pp. 127-150

United Nations Educational, Scientific and Cultural Organization (UNESCO; 1973); International Classification And Mapping Of Vegetation; Series 6, Ecology And Conservation, United Nations Educational, Scientific And Cultural Organization; Paris

Winkler, R.L. and W.L. Hays (1975); Statistics: Probability, Inference, and Decision (2nd ed.); (New York; Holt, Rinehart and Winston); pp. 374-375

Woodwell, G.M., J.E. Hobbie, R.A. Houghton, J.M. Melillo, B.J.Peterson, G.R. Shaver, T.A. Stone, B. Moore, and A.B. Park (1983); Deforestation Measured by Landsat: Steps Toward a Method; U.S.D.O.E.; DOE/EV/10468-1; 62p.

Yool, S.R., D.W. Eckhardt, J.L. Star, T.L. Becking, and J.E. Estes (1985); Image Processing For Surveying Natural Vegetation: Possible Effects On Classification Accuracy; 51st ASP Annual Meeting; Amer. Soc. Photogram., Wash., D.C.; pp. 595-603

INTERPRETATION OF FOREST COVER ON MICROWAVE AND OPTICAL SATELLITE IMAGERY

P.W. Mueller, D.F. Lozano-Garcia, and R.M. Hoffer
Department of Forestry and Natural Resources
Purdue University
West Lafayette, Indiana 47907

ABSTRACT

The two major portions of the electromagnetic spectrum utilized in remote sensing are the optical and microwave regions. The first Thematic Mapper (TM) -- a new generation optical scanner -- was launched in 1982, which obtained multispectral data in the visible, near infrared, middle infrared, and thermal infrared wavelengths. In October 1984, the Shuttle Imaging Radar-B sensor (SIR-B) was operated from the cargo bay of the Space Shuttle orbiter Challenger and collected L-band synthetic aperture radar imagery at various incidence angles. The SIR-B sensor operated at a 23.5 cm wavelength (microwave region) with an HH polarization. Both SIR-B and TM data were acquired for a predominantly forested area of northeastern Florida. On October 9, 10, and 11, 1984, three different sets of radar data were acquired, each at a different incidence angle -- 28°, 45°, and 58°. The TM data utilized in the study were acquired October 12, 1984.

This study investigated the information content of the various TM Bands and the three SIR-B incidence-angle data sets acquired for this site. Digital techniques were used to produce enhanced imagery which was studied to determine the utility and importance of the various bands and look-angles as they pertain to forest resources. The results indicate that both the optical and microwave portions of the electromagnetic spectrum provide a variety of information about the forest and other cover types present.

INTRODUCTION

With the launch of Landsat-1 in July 1972, man entered a new era for obtaining information about earth resources. Landsat-1 was the first unmanned satellite designed specifically for collecting data about earth resources on a global, repetitive, multispectral basis. Since 1972, numerous studies -- including those by Hoffer et al. (1973), Kalensky and Scherk (1975), and Strahler et al. (1978) -- have demonstrated the effectiveness of computer-aided analysis of optical Multispectral Scanner (MSS) data acquired by the Landsat satellites. In 1982, a new generation optical scanner -- the Thematic Mapper (TM) -- was launched, thereby providing multispectral data with greater spectral dimensionality, finer spectral resolution, and finer spatial resolution than the MSS sensor.

Two major regions of the electromagnetic (EM) spectrum --

optical and microwave -- can be utilized for remote sensing of the Earth's surface. The optical region includes ultraviolet (UV), visible, reflective infrared, and thermal infrared energy, and ranges from approximately 0.3 µm to 14 µm. The microwave region is characterized by much longer wavelengths -- 1 mm to 1 m (Lillesand and Kiefer 1979). Usually, only the region from approximately 0.8 cm to 25 cm is utilized with radar sensor systems. Remote sensing systems operating in the microwave region of the electromagnetic spectrum, such as synthetic aperture radar (SAR), offer some unique advantages over the traditionally utilized optical systems. Because SAR systems are active sensors -- both transmitting and receiving the microwave signal -- they are not dependent on solar illumination. Additionally, the radar signal can penetrate cloud cover which often hinders the acquisition of optical data for forest cover mapping in tropical regions. The Shuttle Imaging Radar-B (SIR-B), which operated in October 1984, was the third spaceborne SAR, being preceeded by the Seasat SAR in 1978, and SIR-A in 1981.

Data acquired by satellite remote sensors of the Earth's surface can provide valuable information to resource managers. The synoptic view provided by these sensors enables the user to detect macro-details for large geographic areas which would be missed when utilizing lower-altitude sensor systems. With the introduction of satellite sensors, multi-parameter data sets were possible. These parameters include date, spectral bands, polarization, incidence angle, etc. By obtaining measurements of the energy reflected or emitted in various wavelength regions (i.e., spectral bands) for earth surface materials, broad classes can be discriminated. Analyzing the data in a digital format allows one to enhance the imagery and also to statistically characterize and classify the data.

OBJECTIVE

The purpose of this study was to investigate the utility of various optical wavelength regions and L-band synthetic aperture radar data acquired at different incidence angles for identifying broad cover types. Interpretation of the individual optical waveband and radar incidence-angle images were conducted and various cover types were digitally characterized.

STUDY SITE

The study site is located in Baker County, Florida (Figure 1). The dominant cultural feature in the study site is the highway intersection of Interstate 10 and U.S. 90 which is located approximately 30 km south of the Okefenokee Swamp and 55 km west of Jacksonville. This highway intersection is evident in Figures 2 and 3. Soils of the area are predominantly sandy and the topography is relatively flat with the elevation ranging from approximately 27 to 53 meters above mean sea level.

The study site is largely forest, with ownership principally divided between Champion International Corporation,

Figure 1. Location of the Forestry Test Site in the State of Florida

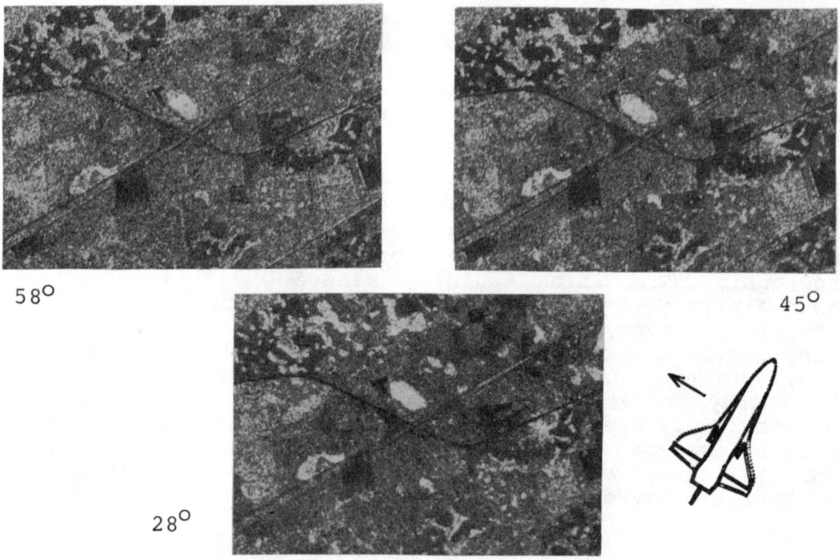

Figure 2. Example of Shuttle Imaging Radar-B Imagery acquired at three different incidence angles (58°, 45°, and 28°) for the Florida Forestry Test Site. See Figure 3.

Owens-Illinois Corporation, Southern Resin and Chemical Company, and the U.S. Department of Agriculture Forest Service (Osceola National Forest). The major forest cover types are pine flatwoods, cypress swamps, and cypress-tupelo swamps. The pine flatwoods are composed of slash pine (*Pinus elliottii*) and/or longleaf pine (*P. palustris*), the former often in plantations. Pine is managed for sawtimber with a rotation age of 50 years on the Osceola National Forest, while it is managed for pulpwood with a rotation age of 25 to 30 years on land owned by the three forest product companies. The cypress and cypress-tupelo swamps largely consist of such deciduous species as pondcypress (*Taxodium distichum* var. *nutans*), baldcypress (*T. distichum*), swamp tupelo (*Nyssa sylvatica* var. *biflora*), water tupelo (*N. aquatica*), black tupelo (*N. sylvatica*), sweetbay (*Magnolia virginiana*), red maple (*Acer*

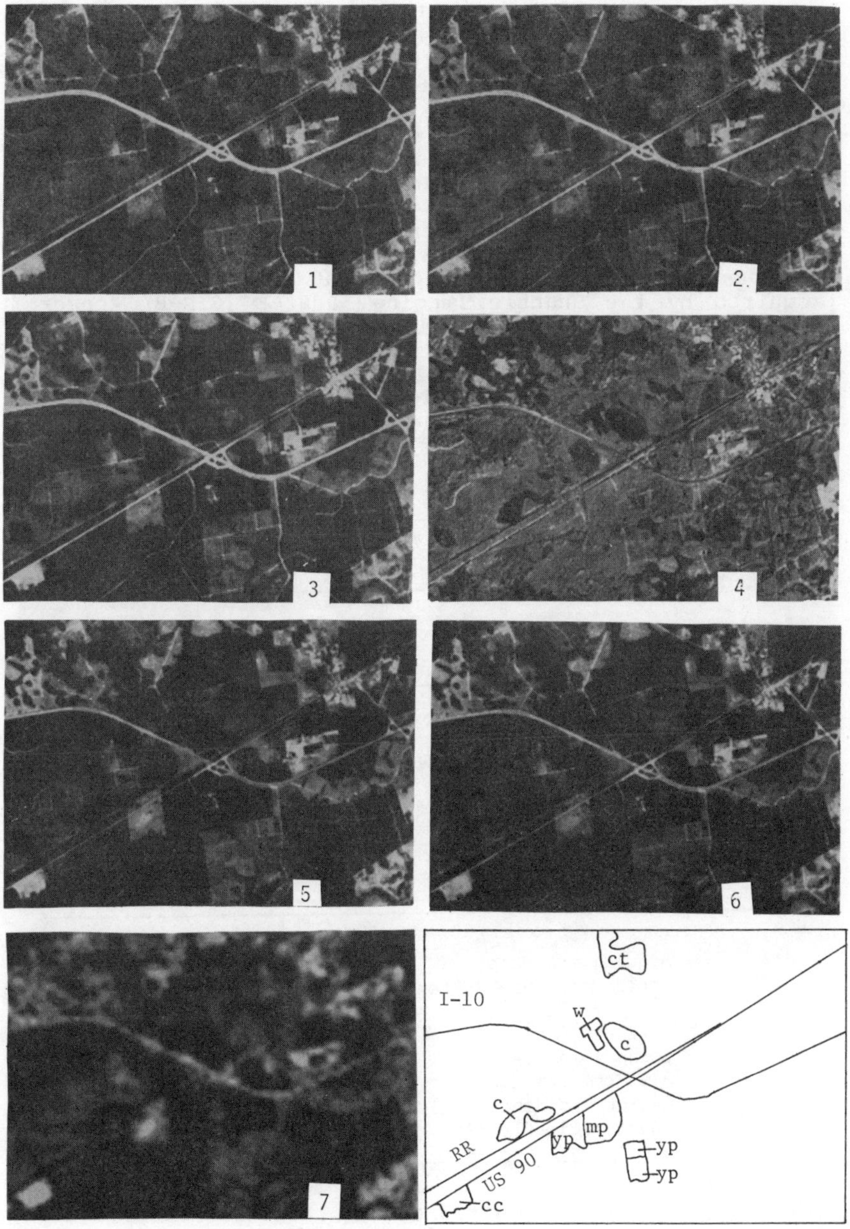

Figure 3. Example of Landsat-5 Thematic Mapper Imagery for the Florida Forestry Test Site. Key: c = cypress swamp; cc = recent clearcut; ct = cypress-tupelo swamp; mp = medium-age pine; rr = railroad tracks; yp = young pine; w = water. TM Bands as indicated.

rubrum), and sweetgum (Liquidambar styraciflua), often with some scattered slash pine and/or pond pine (P. serotina). The cypress swamps are distinguished from the cypress-tupelo swamps in that the former are predominantly pondcypress and baldcypress.

MATERIALS AND METHODS

The optical data utilized in this study was obtained by the Landsat-5 Thematic Mapper sensor which collects data in seven spectral bands as outlined in Table 1. The data obtained in the microwave portion of the EM spectrum was acquired by the Shuttle Imaging Radar-B (SIR-B) sensor which was aboard Space Shuttle Flight 41-G. The TM sensor collected data at different optical wavelengths, whereas the SIR-B sensor data was all collected at the same wavelength, but with different viewing geometries.

Table 1. Characteristics of the Landsat Thematic Mapper sensor and the digital data for the Florida forestry test site. (Adapted from Freden and Gordon, 1983.)

Spectral Band Designation	Wavelength Range (Micrometers)	Spectral Region	Ground IFOV	Pixel Size
TM 1	0.45 - 0.52	Visible blue	30 m	28.5 m
TM 2	0.52 - 0.60	Visible green	30 m	28.5 m
TM 3	0.63 - 0.69	Visible red	30 m	28.5 m
TM 4	0.76 - 0.90	Near infrared	30 m	28.5 m
TM 5	1.55 - 1.75	Middle infrared	30 m	28.5 m
TM 6	2.08 - 2.35	Middle infrared	30 m	28.5 m
TM 7	10.40 - 12.50	Thermal infrared	120 m	28.5 m

Acquisition date	12 October 1984
World Reference System	
Path	17
Row	39
Scene ID number	5022 515311 FL
Platform altitude	705 km
Center incidence angle	0 degrees
Quantization levels	256

The SIR-B data were acquired on the three consecutive days, October 9-11, 1984. The characteristics of the SIR-B data utilized in this study are summarized in Table 2. The repeat cycle for imaging an area was approximately 24 hours with a slight westward drift of the orbital track.

The SIR-B sensor had a tiltable antenna which enabled the same area to be imaged on different passes. Consequently, the viewing geometry of the SIR-B sensor was altered with each orbital pass. Figure 4 illustrates the generalized

Table 2. Characteristics of the digitally correlated SIR-B data for the Florida forestry test site.

Parameter	Value		
Center incidence angle	58.4^{o}	45.3^{o}	28.4^{o}
Acquisition date	9 OCT 84	10 OCT 84	11 OCT 84
Center acquisition time: GMT	0934	0917	0900
EDT	0534	0517	0500
Data take scene number	AK-064.2-003	AK-080.2-003	AK-96.2-003
Orbital track (from true north)	45.0^{o}	45.0^{o}	45.6^{o}
Platform altitude	229.45 km	230.12 km	225.67 km
Slant range to near edge	409.81 km	311.37 km	249.46 km
Center resolution (ground range x azimuth)	16.5 m x 31.5 m	19.8 m x 34.1 m	29.6 m x 25.5 m
Pixel size	12.5 m x 12.5 m	12.5 m x 12.5 m	12.5 m x 12.5 m
Quantization levels	256	256	256
Wavelength	23.5 cm	23.5 cm	23.5 cm
Polarization	HH	HH	HH

viewing geometry of the SIR-B sensor for the three data acquisitions. The three SIR-B data sets were obtained at center incidence angles of 58°, 45°, and 28°. Incidence angle is measured at the target and is defined as the angle the radar beam makes with the surface vertical, where the surface vertical is an imaginary line that passes through the center of the earth. The SIR-B data were digitally correlated and the three data sets were registered at the NASA Jet Propulsion Laboratory.

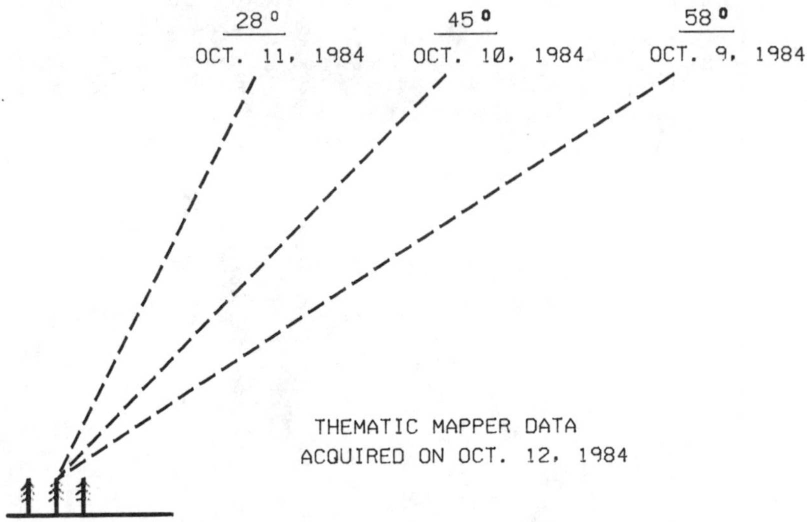

Figure 4. SIR-B multiple incidence angle data set for the Florida Forestry Test Site.

The characteristics of the TM data are listed in Table 1. The acquisition date was October 12, 1984, just one day after the last SIR-B overpass. Data were obtained in four major portions of the EM optical region -- visible, near infrared, middle infrared, and thermal infrared. No clouds were present over the study site.

Qualitative interpretations of the ten images -- three SIR-B and seven TM -- were conducted to characterize the data. The interpretation of digital imagery is similar to aerial photo interpretation using tone, texture, shape, pattern, size, shadow, and association as interpretive clues. Tone and texture are the major factors to consider when developing interpretive keys for forest typing of radar images (Knowlton and Hoffer 1981; Morain 1980). Because these two characteristics are also important for interpretation of optical images, both tone and texture were noted for several broad cover type classes when interpreting the ten images. The broad cover types (non-exhaustive) of interest included water, two classes of deciduous forest, three classes of pine, clearcut areas, pasture, bare soil, and roads (Table 3). Tone was classified as white, light gray, gray, dark gray, or black; and texture was classified as smooth, medium, rough, or mottled.

Table 3. Description of cover type classes utilized in both the qualitative and quantitative investigations.

Cover Type Class	Description[1]
WATER	Exposed water body
DECID1	Cypress swamp
DECID2	Cypress-tupelo swamp
PINE1	Young pine plantation, 3 - 5 years old
PINE2	Medium-age pine, 25 - 30 years old
PINE3	Older pine, 51 - 52 years old
CLEARCUT	Recently harvested forest area
PASTURE	Non-forested grazing area
SOIL	Bare soil

[1]Forest types described in text under Study Site.

The images that were interpreted were displayed on the IBM 7350 High Level Image Processing System digital display. The images were digitally enhanced using different histogram techniques. Imagery from two sensors are shown in Figures 2 and 3. These images are a sampling of the entire set of data analyzed in this study.

Secondly, to develop a quantitative understanding of the digital data, first order statistics were generated for the nine broad cover type classes. At least three data blocks were selected for each cover type class in a supervised manner for each data type. For each of the ten image data sets, the mean and standard deviation were calculated for each class. These results were then used to quantitatively verify the qualitative analysis results.

Reference data for the interpretation of the SIR-B and TM images included aerial and ground photography, field notes, and forest stand inventory information. The aerial photography included both oblique and vertical photographs. Complete vertical color infrared photographic (1:100,000) coverage of the study site was obtained in September 1984. Additionally, 1:58,000 color infrared vertical photography was obtained in 1983 and 1984 as part of the National High Altitude Photography program. In October 1984, when the SIR-B sensor was operating aboard the orbiter Challenger during Space Shuttle Flight 41-G, 35 mm format color and color infrared photographs of the study site were taken from light aircraft and on the ground as additional reference data. Field trips were made to the study site in August and October 1984, and notes on stand conditions were taken. Detailed stand maps and inventory data were provided by the timberland owners.

RESULTS AND DISCUSSION

Qualitative Interpretation

The results of the image interpretation are summarized in Table 4, where the tone and texture of the nine cover type classes for the ten images are listed. Even though the actual resolution of the SIR-B imagery is higher than that of the Thematic Mapper data, the TM images, except for Band 7 (10.40 - 12.50 μm), were much sharper and features could be recognized more easily than on the SIR-B images. This is primarily due to the speckle characteristics of the SIR-B data and also the side-look view of the radar.

SIR-B. The 28° incidence-angle image provided the maximum amount of differentiation between the general forest classes -- all five classes could be differentiated. This image also had the greatest overall contrast. The 45° incidence-angle image allowed the PINE2 class to be differentiated from PINE3 and DECID2, however, these latter two classes still could not be differentiated effectively. The 58° incidence-angle image had the poorest overall contrast and the three forest classes PINE2, PINE3, and DECID2 were easily confused.

Deciduous swamp forests were highlighted, particularly on the 28° incidence-angle imagery, by a tone of white with light gray or white with gray. Wherever water was standing beneath a forest canopy -- composed of deciduous forest species in all cases observed -- a very bright radar backscatter resulted. This result is similar to the findings of Krohn et al. (1983), MacDonald et al. (1980), and Ormsby et al. (1985) with L-band imagery from other SAR sensors. However, it is not clear from this study whether L-band SAR can be used to discriminate between pine and deciduous forest species when the underlying water is not present, since that situation did not occur to any great extent on this study site; this was not discussed by the authors of the other studies.

In general, relatively smooth surfaces such as WATER, SOIL, and PASTURE were very similar in tone and texture. This is not unexpected since such surfaces act as mirrors, reflecting most of the radar signal away from the sensor resulting in a dark tone.

In some cases, the edges of forest stands were highlighted by a bright strip. This occurred when the trees along the edge of one stand were taller than the trees or other feature in front of them, and the resulting edge was exposed to the radar signal. Thus, the edge of trees formed a "wall" and consequently a corner reflector. This is most evident when forest stands are adjacent to clearings of water bodies, and was most evident on the 58° incidence-angle image, and least apparent on the 28° image. As the incidence angle is increased, the signal is striking the boles of the trees more directly and the return consequently becomes greater.

Table 4. Visual interpretation results for the ten digital images.

	Data Type																			
	SIR-B 28°		SIR-B 45°		SIR-B 58°		TM1		TM2		TM3		TM4		TM5		TM6		TM7	
Cover Type[1]	TON[2]	TEX	TON	TEX	TON	TEX	TON	TEX	TON	TEX	TON	TEX	TON	TEX	TON	TEX	TON	TEX	TON	TEX
WATER	BL	SM	BL	SM	BL-GR	SM-MD	DG	RG-MT	DG-BL	MD	BL-DG	RG	BL	SM	BL	MD	DG-BL	MD	BL-DG	M
DECID1	WH+LG	SM-MD	WH+GR	MD	WH-GR	MD	DG	RG-MT	GR-DG	RG	GR	SM	DG-BL	SM-MD	GR-DG	MD	GR-DG	MD	DG-GR	SM-MD
DECID2	WH+GR	MD	LG	MT	GR	MT	GR-DG	RG	DG	MD	GR-DG	SM-RG	GR-BL	MD-RG	LG-DG	MD	LG-GR	MD	DG-GR	SM-MD
PINE1	BL+GR	SM-MD	BL+LG	SM-MD	BL+GR	MD-RG	GR	RG	LG	RG	LG	MD-RG	GR	MD-RG	LG	MD	LG	RG-MD	LG-GR	SM
PINE2	GR-LG	MT	GR	MT	GR	MT	DG	RG-MT	GR-DG	MD-RG	DG	MD-RG	GR	MD	GR-DG	MD	DG-GR	MD	GR	SM
PINE3	LG	MT	LG	MT	GR	MT	DG	RG-MT	GR-DG	MD-RG	GR	SM-MD	GR	MD	GR-DG	MD	DG-GR	MD	GR	SM
CLEARCUT	BL-GR	RG-MT	BL+GR	MD	BL-GR	MD-RG	LG	RG	WH-LG	RG	WH-LG	SM-MD	DG-BL	RG	WH-LG	MD	WH-LG	SM-MD	LG	SM-MD
PASTURE	BL	SM	BL	SM	BL+GR	MD	GR	MD-RG	LG	MD-RG	LG	MD-RG	WH-LG	MD	WH-LG	MD	LG	MD	LG	MD
SOIL	BL	SM	BL	SM	BL+GR	MD	WH	SM	WH	SM	WH	SM	WH-LG	SM-MD	WH	SM	WH	SM	WH-LG	RG

[1]See Table 3 for explanation.

[2]Key: TON = tone, WH = white, LG = light gray, GR = gray, DG = dark gray, BL = black
TEX = texture, SM = smooth, MD = medium, RG = rough, MT = mottled
\- to (a range of tone or texture)
\+ with

Thematic Mapper. The relatively smooth cover types so easily confused in the SIR-B imagery were easily differentiated using the TM data. In all the TM Bands, WATER could be differentiated from SOIL and PASTURE. Furthermore, SOIL and PASTURE could also be separated in all the TM Bands, except for Band 4 (near infrared).

Differences were noted between pine and deciduous forest types. However, the swamps were not as strikingly highlighted as they were on the SIR-B imagery. TM Band 4 (0.76 - 0.90 um) was useful for detecting exposed water, but also, water standing beneath forest cover. Many of the forested swamps had a very dark tone on the Band 4 image. Since water has minimal reflectance and vegetation high reflectance in the near infrared, the dark tone may indicate that the trees in these areas were widely spaced or seasonal leaf fall had begun.

The blue visible (Band 1) image had an overall rough texture, which was probably due to atmospheric haze. For this reason, this image was the least desirable of the visible wavelength bands. Cultural features such as roads were most apparent in imagery from Bands 3 (0.63 - 0.69 µm) and 6 (2.08 - 2.35 µm). Even narrow dirt forest roads were readily apparent. On the SIR-B imagery, many of these could not be delineated, even when they cut a swath through forested areas. This inability was probably due to the side-looking geometry of the SIR-B sensor. There is slightly more detail for cultural features provided by Band 3. In addition, Band 3 provides information regarding turbidity of water that Band 6 does not provide. For these reasons, Band 3 was the most valuable visible wavelength Band.

From a visual interpretation standpoint, the Band 7 (thermal infrared) imagery is difficult to analyze due to its poor spatial resolution of 120 m. However, this does not mean that the information is useless. Mueller et al. (1985) report that Band 7 appeared to provide useful information when digitally classifying TM data of forest lands.

Quantitative Analysis

The mean and standard deviation corresponding to each of the nine cover types are shown graphically in Figures 5 and 6, corresponding to the SIR-B and TM data, respectively. One of the advantages of working with digital data is that such statistics can be easily generated. Since the different channels of data from a particular sensor are not absolutely calibrated, interpretation of these plots must be done on a channel by channel basis.

As the digital number value increases for a given channel, the tone on the corresponding image will become brighter and vice versa. When the qualitative interpretation results are compared with these plots, it can be seen that these two methods of characterizing the data gave similar results.

Figures 5 and 6 vividly illustrate the greater variance in

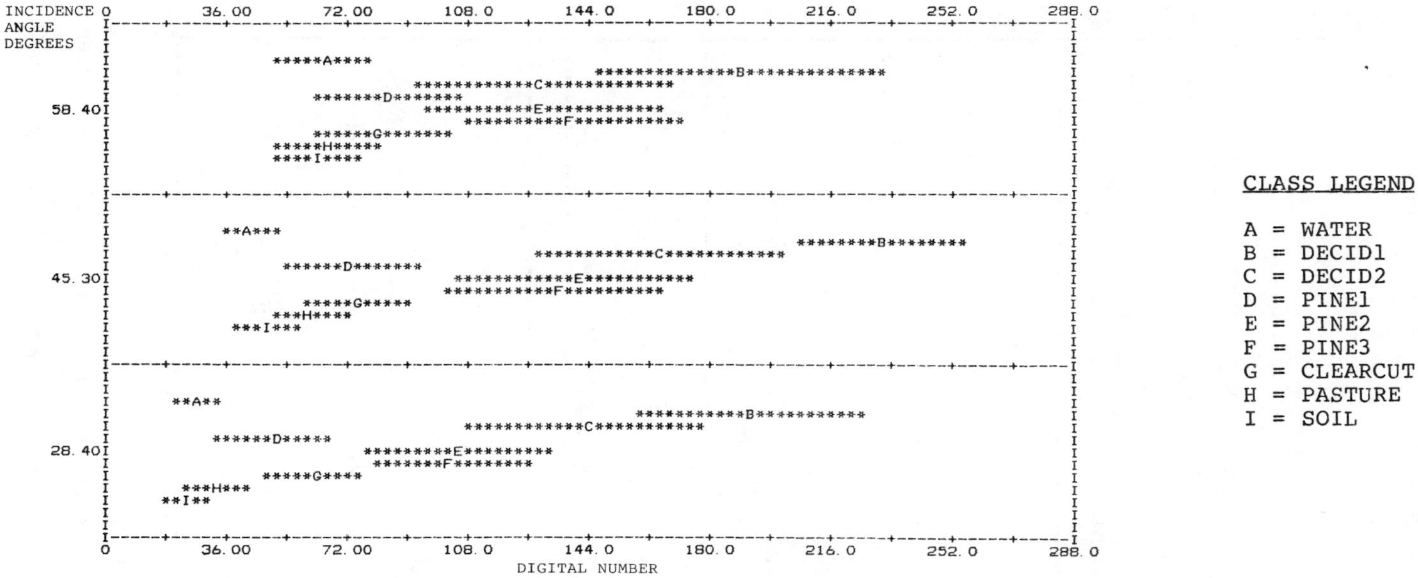

Figure 5. Multiple incidence angle backscatter plot for the SIR-B data. Mean (letter) plus and minus one standard deviation (asterisks) are indicated for each class. Class descriptions in Table 3.

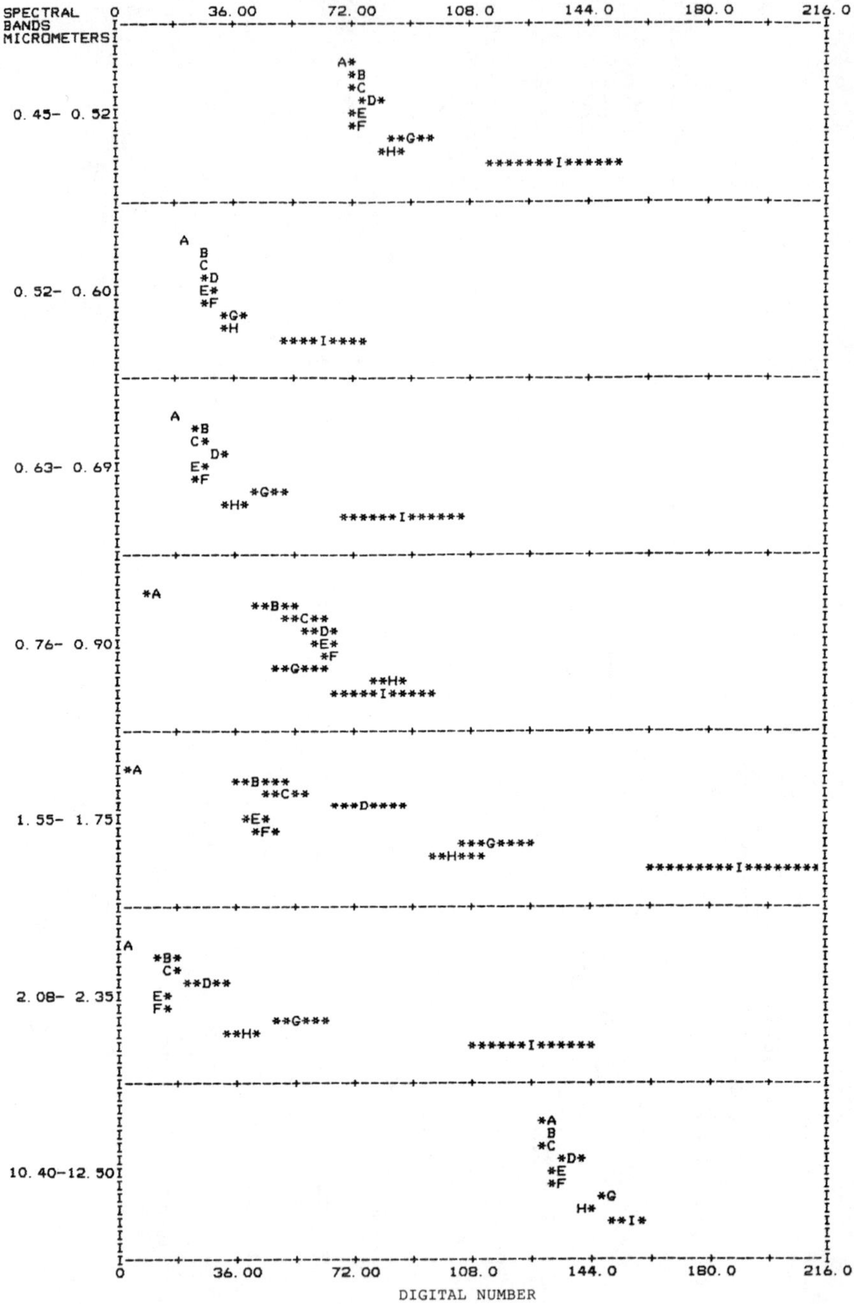

Figure 6. Coincident spectral plot for the TM data. Mean (letter) plus and minus one standard deviation (asterisks) are indicated for each class. See Figure 5 for legend.

the SIR-B data for a single cover type class. This is largely due to the inherent speckle noise, which causes the general lack of clarity in the SIR-B images.

Working with the digital data in a quantitative manner is advantageous because more precise results can be obtained. In further studies, we hope to investigate these two data sets in more detail using computer-aided digital analysis techniques.

CONCLUSIONS

- Even though the SIR-B system resolution was better than that of the TM sensor (except for TM Band 7), the TM imagery had greater detail and clarity.
- On the SIR-B imagery, relatively smooth surfaces such as bare soil, pasture, and water, could not be effectively differentiated, however, with most of the TM images, these cover types could be easily separated.
- Of the ten images analyzed, cultural features were most apparent on the TM Band 3 (0.63 - 0.69 μm) image.
- Of the three SIR-B images, the 28° incidence-angle image provided the maximum amount of differentiation between the general forest classes, present in the study site: young pine, medium-age pine, older pine, cypress swamps, and cypress-tupelo swamps.
- L-band SAR imagery, particularly if obtained at an incidence angle near 28°, can be used to detect standing water beneath forest vegetation.
- Forest stand edges were highlighted on the SIR-B imagery when the exposed edge was facing the sensor and the trees were of sufficient height to form a "wall".

REFERENCES

Freden, S.C., and F. Gordon, Jr. 1983, Landsat Satellites. R.N. Colwell, ed. Manual of Remote Sensing. Falls Church, VA: American Society of Photogrammetry. pp. 517-570.

Hoffer, R.M., and LARS Staff. 1973, Techniques for Computer-Aided Analysis of ERTS-1 Data, Useful in Geologic, Forest and Water Resource Surveys. Proc. Third Earth Resources Technology Satellite-1 Symposium; December 10-14, 1973, NASA Goddard Space Flight Center, Washington, D.C. Vol. 1, Sect. A. pp. 1687-1708.

Kalensky, Z., and L.R. Scherk. 1975, Accuracy of Forest Mapping from Landsat Computer Compatible Tapes. Proc. 10th Int. Symp. Remote Sensing of Environment; October 6-10, 1975, Ann Arbor, MI. pp. 1159-1167.

Knowlton, D.J., and R.M. Hoffer. 1981, Radar Imagery for Forest Cover Mapping. Proc. 7th Int. Symp. Machine Processing Remotely Sensed Data; June 23-26, 1981, Purdue University, W. Lafayette, IN. pp. 626-632.

Krohn, M.D., N.M. Milton, and D.B. Segal. 1983, Seasat Synthetic Aperture Radar (SAR) Response to Lowland Vegetation Types in Eastern Maryland and Virginia. J. Geophys. Res. 88(C36):1937-1952.

Lillesand, T.M., and R.W. Kiefer. 1979, Remote Sensing and Image Interpretation. New York: John Wiley & Sons. 612 p.

MacDonald, H.C., W.P. Waite, and J.S. Demarcke. 1980, Use of Seasat Satellite Radar Imagery for the Detection of Standing Water Beneath Forest Vegetation. Tech. Papers American Society of Photogrammetry Fall Tech. Meeting; October 7-10, 1980, Niagara Falls, NY. pp. RS-3-B-1 to RS-3-B-13.

Morain, S.A. 1980, Use of Radar for Vegetation Analysis. Proc. Seminar on Remote Sensing; October 1980, South Dakota State University, Brookings, SD. pp. 40-58.

Mueller, P.W., R.M. Hoffer, and J.E. Jacobson. 1985, Evaluation of Landsat Thematic Mapper Data for Classifying Forest Lands. Proc. Indiana Academy of Science, Vol. 95 (In Press).

Ormsby, J.P., B.J. Blanchard, and A.J. Blanchard. 1985, Detection of Lowland Flooding Using Active Microwave Systems. Photogramm. Eng. Remote Sensing 51(3):317-328.

Strahler, A.H., T.L. Logan, and N.A. Bryant. 1978, Improving Forest Cover Classification Accuracy from Landsat by Incorporating Topographic Information. Proc. 12th Int. Symp. Remote Sensing of Environment; April 20-26, 1978, Ann Arbor, MI. pp. 927-942.

PHOTOGRAPHS

Program Cochairperson, Dick Driscoll (L) with welcoming speakers, Barbara Rutstein, Mayor of Fort Collins, and Dr. Philip Austin, President of Colorado State University.

Mrs. Pecora receives an enlargement of a Shuttle Large Format Camera scene of the Bearpaw Mountains from Symposium Cochairpersons, Bill Ciesla (L) and Dick Driscoll.

Larry Pettinger moderating the Plenary Session.

Chuck Poulton discussing the evolution of remote sensing for range management during the Plenary Session.

Bob Heller discussing the state-of-the-art of remote sensing for forestry at the Plenary Session.

Don Lauer contemplating new horizons of remote sensing for forest and range resource management at the Plenary Session.

An attentive audience at the opening session.

Coffee break at the Colorado State University Student Center.

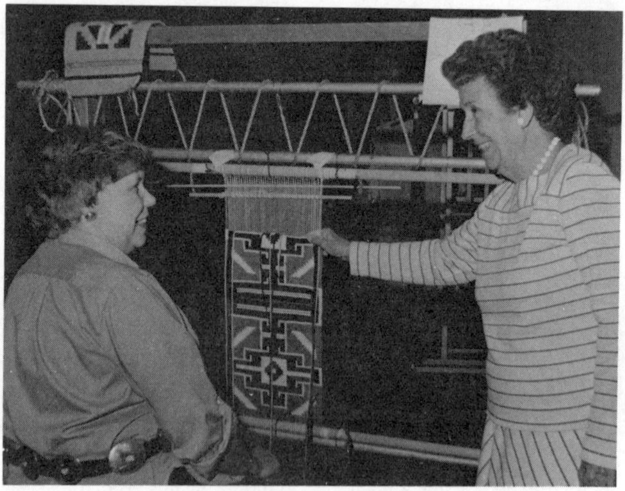

Mrs. Pecora gets a demonstration of home weaving techniques from Barbara Burridge of the Northern Colorado Weavers Guild.

Mrs. Pecora addresses the Awards Banquet.

C.B. (Chuck) Williams, President of EOSAT addresses the Pecora Awards Banquet.

Guy Baier, Bureau of Land Management Deputy for Renewable Resources addresses the luncheon.

Discussion at the Poster Session.

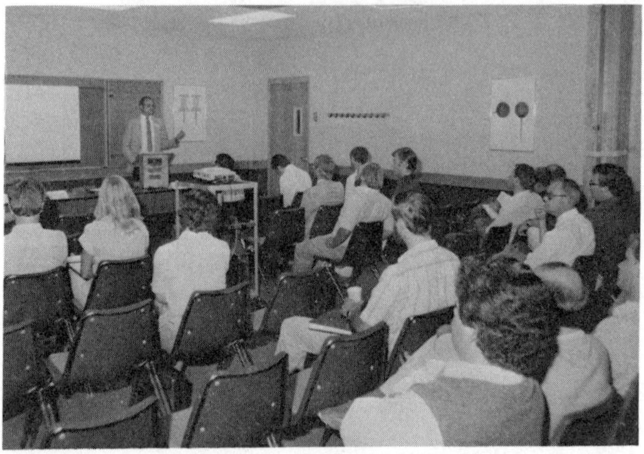

Bill Ciesla leading the forum session on remote sensing for detecting and monitoring forest damage caused by atmospheric pollutant deposition.

Rex McHail demonstrates a new model of the zoom transfer scope.

Dr. Dianne Evans (R), Jet Propulsion Laboratory, receives the Pecora Award for Dr. Charles Elachi from Ann Bradley, NASA Associate Deputy Administrator.

Part of the head table at the Awards Banquet. Left to Right: Dick Driscoll, Joyce Driscoll, Wynn Pecora, C.R. Williams, Ann Bradley, and Dianne Evans.

Pecora 10 Planning Committee: L-R—Upper row: Gregg Johnson, Bob Haas, Bill Ciesla, Dick Driscoll, Ray Byrnes, Dick Mroczynski, Larry Pettinger, Dick Myhre; Lower row (L-R): K.C. Wehde, Mrs. Pecora, Pat Ciesla

ATTENDEES

ATTENDEES

Barry W. Adams
Bag 3014, 530-8 St. So.
2nd Floor Sun Centre
Lethbridge, Alberta
Canada T1J 2J8
(403) 329-5486

Sean Ahearn
University of Wisconsin
1711 Madison St.
Madison, WI 53703
(608) 263-5964

Francis Ahern
Canada Ctr. for Remote Sens.
2464 Sheffield Road
Ottowa, Ontario
Canada K1A 0Y7
(613) 993-9884

Ray Albrandt
Bureau of Indian Affairs
Northern Cheyenne Agency
Lame Deer, MT 59043
(406) 477-6211

David Alerich
USDA Forest Service
P.O. Box 1775
Media, PA 19063
(215) 461-3037

Vincent G. Ambrosia
Tech. Govt. Services, Inc.
MS 242-4 NASA-AMES Res. Cen.
Moffett Field NAS, CA 94035
(415) 694-6184

Hal Anderson
Idaho Dept. Water Res.
Statehouse
Boise, ID 83720
(208) 334-4492

Kenneth Andresen
Bureau of Land Management
Bldg. 50-DFC, P.O. Box 25047
Denver, CO 80225
(303) 236-0102

Peter Archibald
Tamat Graphics, Ltd.
717 Belfast Rd.
Ottowa, Ontario
Canada K1A 0Y7
(613) 993-9879

Raymond E. Arndt
U.S. Geological Survey
6903 Essex Ave.
Springfield, VA 22150
(703) 860-6276

Duane Asherin
US Fish & Wildlife Service
2627 Redwing Rd.
Creekside One
Fort Collins, CO 80526
(303) 226-9402

Ghassem Asrar
Kansas State University
Waters Annex Bldg.
Manhatten, KS 66506
(913) 532-5731

Dale Baer
USDA FS, PNW, FIA
P.O. Box 3890
Portland, OR 97208
(503) 231-2249

Guy Baier
Bureau of Land Management
18th and C Street NW
Washington, D.C. 20240
(202) 343-4896

Stan Bain
USDA Forest Service
P.O. Box 147
Kalispell, MT 59901
(406) 755-5401

Robert Baker
Texas A&M University
Dept. of Forest Science
College Station, TX 77843
(409) 845-5033

G. Robinson Barker
Champion International Corp.
435 Clark Rd., Ste. 411
Jacksonville, FL 32218
(904) 764-0545

W.L. Barnes
NASA/Goddard Space Flt. Ctr.
GSFC Mail Code 625
Greenbelt, MD 20771
(301) 344-8101

Mehdi Bassiri
Isfahan Univ. of Tech.
28 Negarstan Ln., Azar St.
Isfahan, Iran
031-62690

Fred Batson
US Bureau of Land Management
Bldg. 50, P.O. Box 25047
Denver, CO 80225-0047
(303) 236-6376

Marvin Bauer
University of Minnesota
110 Green Hall
St. Paul, MN 55108
(612) 376-8179

Jean Beaubien
Canadian Forest Service
1055 Rue Du, PEPS Box 3800
Ste-Foy, Quebec
Canada G1V 4C7
(418) 648-5822

Don Becker
EROS Data Center
USGS/Mundt Federal Bldg.
Sioux Falls, SD 57198
(605) 594-6171

Stephen Berwick
Int. Inst. Environ. & Dvlpmt.
1717 Massachusetts Ave. NW
Suite 302
Washington, DC 20036
(202) 462-8918

Robert Bewley
Bureau of Land Management
P.O. Box 6770
Albuquerque, NM 87197-6770
(505) 766-2455

Merv Bilbrey
Charlotte Cnty. Prop. Office
118 W. Olympia Ave.
Punta Gorda, FL 33950
(813) 637-2250

Jim Blohm
Iowa State University
6A Schilletter Village
Ames, IA 50010
(515) 294-1221

William J. Bonner, Jr.
Bureau of Indian Affairs
Lakewood Office Plaza
730 Simms St.
Golden, CO 80401
(303) 236-2667

J. David Born
USDA Forest Service
507-25th St.
Ogden, UT 84401
(801) 625-5385

Bruno Boucek
Sir Sanford Fleming College
P.O. Box 8000, Albert St.
Lindsay, Ont., Canada K9V 5E6
(705) 324-9144

Harry Bowlin
USDA Forest Service
630 Sansome Street
San Francisco, CA 94111
(415) 556-2654

John Breckenridge
Bureau of Land Management
708 Prairie Ave., Apt. C
Cheyenne, WY 82009
(307) 772-2264

John A. Brockhaus
N. Carolina State University
3707 Burt Dr.
Raleigh, NC 27606
(919) 737-2891

Victor Buczkowski
Lockheed Eng. & Mgt. Co.
7500 Greenway Center Dr.
Greenbelt, MD 20770
(301) 982-5500

Gregory S. Burns
NASA-NSTL-Earth Res. Lab.
National Space Tech. Lab.
NSTL Station, MS 39560
(601) 688-1911

Raymond Byrnes
USGS/EROS Data Center
Mundt Federal Bldg.
Sioux Falls, SD 57198
(605) 594-6546

James Carr
Soil Conservation Service
2125 The Knolls
Lincoln, NE 68512
(402) 741-5315

Hugh Carr, Jr.
Lockheed Eng. & Mgt. Serv. Co
Nat Spc tech Lab, Bldg 1100
NSTL, MS 39529
(601) 688-3665

Clifford W. Carter
Soil Conservation Service
7104 Wind Chime
Fort Worth, TX 76133
(817) 334-5282

George Casely
Department of Commerce NOAA
11420 Rockville Pike
NBOC1/300
Rockville, MD 20852
(301) 443-3925

"Sky" Chamard
E. Coyote Enterprises, Inc.
Box 10761
Eugene, OR 97440
(503) 683-2504

Henry Cheatham
USDA Forest Service
324-25th Street
Ogden, UT 84401
(801) 625-5532

Edward Chine
Technicolor Gov. Ser. Inc.
Denver Fed. Ctr. Bld. 50
Denver, CO 80225
(303) 236-6492

William Ciesla
USDA Forest Service
3825 E. Mulberry
Fort Collins, CO 80524
(303) 224-1788

William A. Clerke
USDA Forest Service
1720 Peachtree Rd.
Atlanta, GA 30367
(404) 881-2592

Warren Cohen
Colorado State University
719 Colorado
Fort Collins, CO 80524

Robert O. Coleman
Dept. of Forest & Wood Sci.
Colorado State University
914 W. Lake St., #220
Fort Collins, CO 80521
(303) 491-5255

Mike Cornelison
SD Dept. of Sch. & Pub. Land
500 East Capitol Ave.
Pierre, SC 57501
(605) 773-3303

Thomas Costello
Bureau of Land Management
DFC, Bld. 50, Box 25047
Denver, CO 80225-0047
(303) 236-0144

Kevin Coulton
Washington State University
Albrook Hydraulics Lab.
Pullman, WA 99164-3001
(509) 335-4546

Margaret Courain
Dept. of Commerce, NOAA
Washington, DC 20235
(202) 634-7318

Eugene Cox
NM Inst. of Mining & Tech.
Box 2328
Socorro, NM 87801

Wallace Crisco
Bureau of Land Management
12282 W. Mississippi Ave.
Lakewood, CO 80228
(303) 236-6382

Robert Crippen
Univ. of CA, Santa Barbara
Remote Sensing Research Unit
Santa Barbara, CA 93106
(805) 961-3845

Mathew Cross
Colorado State University
3898 S. Bannock
Englewood, CO 80110

Dale Cummins
Bureau of Land Management
1037-20th Street
Denver, CO 80202
(303) 294-7016

Paul Cuplin
Bureau of Land Management
12198 W. Ohio Dr.
Lakewood, CO 80228
(303) 236-0162

Ray Czaplewski
USDA Forest Service
240 W. Prospect St.
Fort Collins, CO 80526
(303) 224-1877

Robert Dana
USDA Forest Service
240 W. Prospect Ave.
Fort Collins, CO 80256
(303) 224-1284

Steve Davidson
Evangel College
Springfield, MO 65802
(417) 865-2811

Harlan DeGarmo
USDA Soil Cons. Service
100 Centennial Mall
N. Room 345
Lincoln, NE 68508-3866
(402) 471-5355

Clarence DeMars
USDA Forest Service
1960 Addison St.
Berkeley, CA 94704
(415) 486-3576

Andy Dimas
Bureau of Land Management
P.O. Box 1449
Sante Fe, NM 87501
(505) 988-6279

Alice Doremus
USDA Forest Service
1415 W. Rost St.
Walla Walla, WA 99362
(509) 522-6277

Charles W. Dull
USDA Forest Service
3620 IPS NE Room 2103
Doraville, GA 30340
(404) 221-4796

Robert Dale
KD^2
330A Iolani Ave.
Honolulu, Hawaii 96813
(808) 521-5159

Bob Dams
Intera Technologies Ltd.
1200-510 5th St. S.W.
Calgary, Alb. Canada T2P 3S2
(403) 266-0900

Dennis Darling
USADA Soil Cons. Service
P.O. Box 6567
Fort Worth, TX 76115
(817) 334-5292

Charles Davis
Forest Res. Consultants-SRP
P.O. Box 68
Montmorenci, SC 29839
(803) 725-6517

Leslie DeGroff
Colorado State University
491 S. Wheeling St.
Aurora, CO 80012

Stephen Dewhurst
University of Minnesota
1234 St. Helena Ave.
Santa Rosa, CA 95404

Roy Dixon
Dept. of Natural Resources
1007 Century St.
Winnipeg, Manitoba
Canada R3H 0W4
945-6594

Richard Driscoll
2217 Sheffield
Fort Collins, CO 80526
(303) 484-4470

Michael Dwyer
USDI Bureau of Land Mgmt.
2020 Arapahoe
Denver, CO 80202
(303) 294-7100

David Eckhardt
770 A 29th St. #723
Boulder, CO 80303
(303) 447-4648

Jeffery Eidenshink
Tech. Govt. Ser., EROS Data
Mundt Federal Bldg.
Sioux Falls, SD 57198
(605) 594-6114

Thomas Elder
Arizona State Land Dept.
Views From Above, Box 7731
Phoenix, AZ 85011
(602) 255-4061

Ull M. Erich
Dornier System
POB 1360
D-799 Friedrichshafen
West Germany

Bob Evans
SW Gla. Water Mgmt. Dis.
2379 Broad St.
Brooksville, FL 33512-9712
(904) 796-7211

David Evans
Louisiana State University
School of Frstry. Wildlife
Baton Rouge, LA 70803
(504) 388-6826

Collin Fallat
WY Dept. of Agriculture
2219 Carey Ave.
Cheyenne, WY 82002
(307) 777-6576

Donna Field
University of Wyoming
Entomology, Plant Sci. Dept.
P.O. Box 3354
Laramie, WY 82071
(307) 766-2298

Gloria Fletcher
Univ. of CA, Santa Barbara
Remote Sensing Unit
7341 Hillsboro
Goleta, CA 93117
(805) 635-1752

Gary Folger
5 South Pearl St.
Denver, CO 80209
(303) 722-0412

George Fosdick
Cathedral Blf. Shale`Oil Co.
P.O. Box 2687
Grand Junction, CO 81502
(303) 244-3231

Larry Fox
Humboldt State University
1783 Hyland St.
Bayside, CA 95524
(707) 826-4280

Janet Franklin
University of California
Dept. of Geography
Santa Barbara, CA 93106
(805) 961-3663

Bob Gerber
CA State Off. Emergency Ser.
2800 Meadowview Rd.
Sacramento, CA 95832
(916) 427-4216

Don Gipe
4633 E. Flower
Mesa, AZ 85206
(602) 832-5157

Mark Godfrey
Timberline
302-958 W. 8th Ave.
Vancouver, Br. Canada V5Z 1E5
(604) 733-0731

Marty Goldblatt
MN Dept. of Natural Res.
1201 E. Highway 2
Grand Rapids, MN 55744
(218) 327-1749

Bill Grauel
Bureau of Indian Affairs
General Delivery
Dulce, NM 87528
(505) 759-3661

Glen Green
WA University in St. Louis
Box 1169
St. Louis, MO 63130
(314) 889-5679

Steven Gregonis
Bureau of Land Management
1037-20th St.
Denver, CO 80202
(303) 294-7017

R.H. Griffin, II
SCS/USDA
P.O. Box 6567
Ft. Worth, TX 76115
(817) 334-5231

John Gulliksen
DMA
494 Bright Spur Lane
Ballwin, MO 63011
(314) 263-4991

Bob Haas
Tech. Govt. Serv. EROS Data
Mundt Federal Building
Sioux Falls, SD 57198
(605) 594-6114

Donald M. Hall
Colorado State University
419 Whedbee St.
Fort Collins, CO 80524
(303) 226-9389

Jay Hart
Nat. Research Ecology Lab.
Colorado State University
Fort Collins, CO 80523
(303) 491-1976

Patricia Hall
USDA Forest Service, PNW
3625-93rd Ave. SW
Olympia, WA 98502
(206) 753-9494

Wendel Hann
USDA Forest Service
Box 7669, Federal Bldg.
Missoula, MT 59807
(406) 329-3214

Jon Hanson
USDA-ARS
Crops Res. Lab., CSU
Fort Collins, CO 80521
(303) 484-8777

Claire Hay
Private Consultant
3614 Mountain View Ave.
Longmont, CO 80501

Frederick Henderson, III
The Geosat Committee
153 Kearny, Suite 209
San Francisco, CA 94108
(415) 981-6265

Edward Gryczan
USDA Forest Service
3825 E. Mulberry
Fort Collins, CO 80524
(303) 224-1807

Preston L. Guthrie
Bureau of Indian Affairs
1500 NE Irving, Rm. 530
Portland, OR 97208
(503) 231-2219

Ron Hacker
Utah State University
340 East 300 North
Logan, UT 84321
(801) 750-2471

Dorothy K. Hall
NASA/Goddard Spact Fl. Cen.
Code 624
Greenbelt, MD 20771

Lawrence Hipps
Utah State University
UMC 48
Logan, UT 84322
(801) 750-2009

Howard Halverson
USDA Forest Service
104 Academic Projects
University Park, PA 16802
(814) 863-1933

Don Hansen
Colorado State University
3041 LaPorte Ave.
Fort Collins, CO 80521
(303) 226-9294

Sharon Hatch
Lockheed Eng & Mgmt Ser Co
Nat Space Tech Lab, Blg 1100
NSTL, MS 39529
(601) 688-3665

Robert Heller
21 Eastwood Dr.
Orinda, CA 94563

Dennis Herlocker
Louis Berger Internation, Inc
P.O. Box 2954
Mogadishu, Somalia
(206) 425-1957

Karl Hermann
NC Div of Land Resources
1204 Watauga St.
Raleigh, NC 27604
(919) 733-2090

Kathie Hirsch
ND Pub. Ser. Commission
13th Floor State Capitol
Bismarck, ND 58505
(701) 224-2298

Bruce Hiserote
USDA FS PNW FIA
P.O. Box 3890
Portland, OR 97208
(503) 231-2249

Holly Hixson
WESTVACO
Box 1950, Timberlands Div.
Summerville, SC 29484
(803) 571-5000

Roger Hoffer
Purdue University
Dept of For & Nat Res
West Lafayette, IN 47907
(317) 494-3605

Thomas Holm
USGS/EROS Data Center
Mundt Federal Bldg.
Sioux Falls, SD 57198
(605) 594-6540

Emil Horvath
Tech Gov Ser, EROS Data
Mundt Federal Bldg
Sioux Falls, SD 57198
(605) 594-6114

W. Howard
Dept. of Commerce-NOAA
EROS Data Center
Sioux Falls, SD 57198
(605) 594-2261

Don Hunter
US Fish & Wildlife Service
2627 Redwing Rd, Creekside I
Fort Collins, CO 80526
(303) 226-9382

David Hyatt
Defense Mapping Agency
Attn: Toas 6500 Brookes Ln
Washington, DC 20315-0030
(202) 227-2639

Joji Iisaka
Canada Ctr Remote Sensing
2464 Sheffield
Ottawa, Ont. Canada K1A 0Y7
(613) 998-9183

Joy Ikelman
NOAA-Dept. of Commerce
325 Broadway
Boulder, CO 80303
497-6419

Ben Jackson
Louisiana State University
School of Forestry
Baton Rouge, LA 70803
(504) 388-4131

Ray Jackson
US Water Conservation Lab
4331 E. Broadway Road
Phoenix, AZ 85040
(602) 261-4356

Suresh Jain
Moniteq Ltd.
630 Rivermede Rd.
Concord, Ont. Canada L4K 2H7
(416) 669-5334

Dave Jamieson
Timberline
302-958 West 8th Ave.
Vancouver, BC Canada V5Z 1E5
(402) 471-2081

James Jaspers
Texas Tech University
P.O. Box 4035
Lubbock, TS 79409
(806) 742-1982

Kathie Jewell
Bureau of Land Management
P.O. Box 38600
Billings, MT 59107

Janet Johnson
USDA Forest Service
P.O. Box 7669
Missoula, MT 59807
(406) 329-3141

Gregg Johnson
Technicolor Gov. Ser. Inc.
Denver Fed. Cen. Blg. 50
Denver, CO 80225
(303) 236-6492

David Johnston
Tech. Gov. Ser, EROS
Mundt Federal Bldg.
Sioux Falls, SD 57198
(605) 594-6114

Leonard Jurgens
Soil Conservation Service
2490 W 26th Ave, Bld A 3rd
Denver, CO 80211
(303) 964-0466

Marilyn Keller
Pacific Gas & Electric Co
3400 Crow Canyon Road
San Ramon, CA 94583
(415) 820-2000

Michael Kerr
Mississippi Rem Sens Lab
P.O. Drawer FR
Mississippi State, MS 39762
(601) 325-3760

John Kineman
University of Colorado
1185D Bear Mountain Dr.
Boulder, CO 80303
(303) 497-6900

Ron Koczor
ITT Aerospace/Optical Div.
3700 E Pontiac St, Box 3700
Fort Wayne, IN 46801
(219) 423-90636

Gerald Kreitner
Kansas State University
Dept. of Agronomy
Manhatten, KS 66506
(913) 532-7248

Lynne Painter Lanning
Wyoming Highway Dept.
P.O. Box 1708
Cheyenne, WY 82002-9019
(307) 777-4144

Robert Lees
Bechtel Civil & Minerals
P.O. Box 3965
San Francisco, CA 94123
(415) 768-6266

Thomas Lillesand
Univ. of Wisconsin/Madison
1225 West Dayton
Madison, WI 53706
(608) 263-3251

Bruce Jones
Resource Consultants, Inc.
P.O. Box Q
Fort Collins, CO 80522
(303) 482-8471

Bruce Keating
Bureau of Land Mangement
154 B6, Meridian Re
Cheyenne, WY 82009

Rich Kern
Nebraska Natural Res Com
P.O. Box 94876
Lincoln, NE 68509-4876
(402) 471-2081

Chris Kiikgaard
Oregon State University
5215 SW Blueberry
Corvallis, OR 97333

William Knorr
ITT Aerospace/Optical Div.
3700 E. Pontiac St. Box 3700
Fort Wayne, IN 46801
(219) 423-9636

Gregory Koeln
Ducks Unlimited, Inc.
1 Waterfowl Way/Gilmer Road
Long Grove, IL 60047-0216
(312) 438-4300

Glenn Landis
USGS/EROS Data Center
Mundt Federal Bldg.
Sioux Falls, SD 57198
(605) 594-6540

Donald Lauer
USGS/EROS Data Center
Mundt Federal Building
Sioux Falls, SD 57198
(605) 594-6540

Verne Lehmberg
Lee College
Baytown, TX 77520
(713) 427-5611

Thomas Logan
Jet Propulsion Laboratory
4800 Oak Grove Dr.
Pasadena, CA 91109
(818) 354-4032

Ray Lougeay
State University College
Dept. of Geography
Geneseo, NY 14454
(716) 245-5238

James Love
Santa Barbara Res. Ctr. EOSAT
7351 Eldorado Court
McLean, VA 22102
(805) 968-3511

Kim Lowell
University of Missouri
1-31 Agriculture
Columbia, MO 65211
(314) 882-4775

Fabian Lozano
Purdue University
Forestry & Nat Res Dept
West Lafayette, IN 47906
(317) 494-6305

David Lusch
Michigan State University
302 Berkey Hall
East Lansing, MI 48824-1111
(517) 353-7195

Carl Maass
USDA Forest Service
HC Box 1
Avery, ID 83802
(208) 245-4517

Jorge E. Macias
Sub. Forrestal Sarh.
Lomas De Sotelo #816
EDIF. E. DPTO 18
Mexico City, Mexico 11200
518-5156

Kabelo Gilbert Mafura
University of Arizona
#209 Nugent Blvd.
Tucson, AZ 85721
(602) 884-5196

Ann Maclean
University of Wisconsin
1245 Met. & Space Sci. Bldg.
Madison, WI 53714
(608) 262-0795

Lisa Jo Mann
University of CA/SB
4214 Encore Dr.
Santa Barbara, CA 93110
(805) 961-3845

Chris Markel
PA State University
Box 11218 Calder Square
State College, PA 16805
(814) 865-1132

Edward Martinko
University of Kansas
Space Tech Centr/W. Campus
Lawrence, KS 66045
(913) 864-4775

James Maslanik
University of Colorado
Nat. Snow & Ice Data Cntr
Boulder, CO 80309
(303) 492-5171

Robert Mathieu
Slippery Rock University
Dept. of Geography
Slippery Rock, PA 16101
(412) 794-7169

Keith McCaffery
WI Dept. of Natural Res.
729 Bruns Dr.
Rhinelander, WI 54501
(715) 369-3193

Ron McChesrey
University of Colorado
2227 Canyon Blvd. #461
Boulder, CO 80302
(303) 443-9250

Dudley G. McConnell
NASA
600 Independence Ave. SW
Washington, DC 20546
(202) 453-1420

Larrie McConnell
Petro-Canada
12th fl.,E.Tower, Box 2844
Calgary, Alb., Canada T2P 3E3
(403) 296-7441

Lew McCreery
USDA Forest Service
P.O. Box 640
Durham, NH 03824
(603) 868-5719

Theodore McDowell
California State University
5500 University Parking
San Bernardino, CA 92407
(714) 887-7287

Mark McGuire
Colorado State University
Dept. Forest & Wood Sci.
Fort Collins, CO 80523
(303) 491-5420

Randy McKinley
Tech. Gov. Services, Inc.
Denver Fed Ctr, Bldg. 50
Denver, CO 80225
(303) 236-6492

Peter McLaughlin
University of Arizona
3644 E. Fairmount Ave.
Tucson, AZ 85716

Delbert Mead
USDA Forest Service
201 E 9th Ave, Suite 303
Anchorage, AK 99501
(907) 261-2587

Romain Mees
USDA Forest Service
4955 Canyon Crest Dr.
Riverside, CA 92507
(714) 351-6558

Mel Mehl
USDA Forest Service
11177 W 8th Ave, Box 25127
Lakewood, CO 80225
(303) 236-3210

Douglas Meisner
University of Minnesota
1530 N Cleveland Ave 110 GRN
St. Paul, MN 55108
(612) 373-1412

James Merchant
University of Kansas
2291 Irving Hill Rd. Sp TCH
Lawrence, KS 66045
(913) 864-4775

Roger Mertens
Bureau of Land Management
802 Montclair Dr.
Cheyenne, WY 82009
(307) 772-2078

Kevin G. Meyer
USDI BLM
4700 E 72nd Ave.
Anchorage, AK 99507
(907) 267-1320

Louis Mihalyi
California State Univ/Chico
Department of Geography
Chico, CA 95929
(916) 895-5780

Lee D. Miller
University of Nebraska
113 Nebraska Hall
Lincoln, NE 68588-0517
(402) 472-3471

Wayne Miller
USGS/EROS Data Center
Mundt Federal Bldg.
Sioux Falls, SD 57198
(605) 594-6540

David Mills
Bureau of Land Management
P.O. Drawer AB
Price, UT 84501
(801) 637-4584

Adam Mixztal
US Fish & Wildlife Service
11029 Del Rio Dr.
Fairfax, VA 22030
(202) 343-4944

Katherine Mitchell
New Mexico State University
Box 614
Mesilla, NM 88046

Frederick Mokgwa
New Mexico State University
P.O. Box 3567
Las Cruces, NM 88003

Anthony Morse
ID Dept. of Water Resources
Statehouse
Boise, ID 83720
(208) 334-4456

Bruce W. Morse
Autometric, Inc.
2629 Redwing Rd. Suite 340
Fort Collins, CO 80526
(303) 226-3282

Richard Mroczynski
Earth Observ. Satellite Co.
2606 Talbot Court
Waldorf, MD 20601
(703) 558-4259

Paul Mueller
Purdue University
Dept. Forstry. & Nat. Res.
West Lafayette, IN 47907

R. Jay Murray
E.R.S.A.L.
Oregon State University
Corvallis, OR 97331-6703

Peter Murtha
University of B.C.
270-2357 Main Mall
Vancouver, BC V6T 1W5
(604) 226-6452

Douglas Myers
USDA Forest Service
324-25th Street
Ogden, UT 84401
(801) 785-3563

Richard J. Myhre
USDA Forest Service
1960 Sandalwood Lane
Fort Collins, CO 80526
(303) 224-1785

Deborah Negri
EG&G Energy Measurmnts, Inc.
Box 1912, MS H-02
Las Vegas, NV 89125
(702) 295-0656

Charles Nelson
GMIS/ESL TRW-BMD
Box 1310, Bldg 527, Rm 710
San Bernardino, CA 92402
(714) 382-2342

J. David Nichols
Jet Propulsion Laboratory
4800 Oak Grove Dr, MX 168/514
Pasadena, CA 91109
(818) 354-4994

Andrew Norton
Tech. Govmt. Services, Inc.
Bldg 50, D0442T, Fed. Centr.
Lakewood, CO 80225
(303) 236-6492

Maurice Nyquist
USDI National Park Service
P.O. Box 25287
Denver, CO 80225
(303) 236-8774

Phil Ogle
WY Dept. of Envir. Quality
122 W 25th St.
Cheyenne, WY 82002
(307) 777-6145

Uyato Omthile
University of Wyoming
Range Management Dept.
Laramie, WY 82071

Driss Oukassou
University of Idaho
College of Forestry
Moscow, ID 83843

Charles Palmer
TX Nat. Res. Info. System
P.O. Box 13087, Capitol Sta.
Austin, TX 78711

Angel Parker
Univ. of KS, Dept. Geography
1905 S. York, #103
Denver, CO 80210
(303) 236-0976

Ivan (Ike) Parker
USDA Forest Service
304 San Mateo Dr.
Vallejo, CA 94590
(415) 556-9407

H. Dennison Parker
Bureau of Land Management
Bldg 50, Denver Fed Ctr
Denver, CO 80225-0047
(303) 236-5841

Wynn Pecora
4572 Indian Terracerrace, NW
Washington, D.C. 20007

John Peters
USDA Forest Service
309 Woodland Ave.
Media, PA 19063
(215) 461-3021

Wilton Peterson
Bureau of Land Management
18th & C St.
Washington, DC 20240
(202) 653-9210

Lawrence Pettinger
USGS/EROS Data Center
Mundt Federal Bldg.
Sioux Falls, SD 57198
(605) 594-6540

Christopher Phelps
Agency for International Dev.
Training & Dev Div, Rm 489
Washington, DC 20523
(703) 235-9056

C.E. Poulton
121 Muir Ave.
Santa Clara, CA 95051

Joao Side Queivoz
Utah State University
USU Box 1230
Logan, UT 84322
(801) 750-2582

John W. Quimby
Dept Environ Res & Forestry
RD1, Box 120D
Hershey, PA 17033
(717) 787-5469

Klaus Radike
GEO Safety, Inc.
1462 Lachman Lane
Pacific Palisades, CA 90272
(213) 459-3426

Gerald Raisanen
Colorado State University
1713 Morningside Drive
Fort Collins, CO 80525

H.K. Ramapriyan
NASA GSFC
Code 636
Greenbelt, MD 20771
(301) 344-9496

Frank Ravet
NASA Johnson Space Center
PD2
Houston, TX 77058
(713) 483-6426

Juan Jose Reyes
SARH-Mexican Forest Service
Patricio Sanz 1747 B-504
Mexico City, Mexico 03100
(5) 512-4055

James Richardson
Canadian Forestry Service
351 Boulevard St. Joseph
Hull, Quebec, Canada K1A 1G5
(819) 997-1682

George Rosenfield
US Geological Survey
421 National Center
Reston, VA 22092
(703) 860-6345

Ralph R. Root
National Park Service/GIS
P.O. Box 25287
Denver, CO 80225
(303) 236-7936

William R. Rush
Bureau of Land Management
3948 Development Ave.
Boise, ID 83705
(208) 334-1582

Dr. Walter Russell
Earth Resource Data Corp.
10460 Mastin
Overland Park, KS 66212
(913) 492-0873

Bill Russon
University of Waterloo
Environmental Studies #1
Waterloo, Ont. Canada N2L 3G1

Steven A. Sader
NASA/Earth Resource Lab
114 E. 2nd St.
Long Beach, MS 39560
(601) 688-3830

Charlene Sailer
Univ. of CA/Santa Barbara
6606A Pasado
Goleta, CA 93117
(805) 961-3845

Antonio Santocono
Regione Veneto-Dipartimento Foreste (Italy)
Via Torino 110
Mestre (Venezia) Italy 30200
39-41-5310556

Gary Schmoeger
Technicolor Government Inc.
2042 Custer
Fort Collins, CO 80525
(303) 226-6974

Lucy Salazar
USDA Forest Service
4955 Canyon Crest Dr.
Riverside, CA 92507
(714) 351-6550

Hans Schreier
Univ. of B.C.
2357 Main Mall, Box 248
Vancouver, BC Canada V6T 2A2
(604) 228-4898

Wendy Scratch
US Geological Survey
Box 25046, MS 504
Denver, CO 80225
(303) 236-5829

Mark Shasby
Tech Govt Serv, EROS Field
4230 University Dr.
Anchorage, AK 99508
(907) 271-4065

Sharon Shin
University of Colorado/NPS
P.O. Box 25287
Denver, CO 80225
(303) 236-7938

Paul Snook
Colorado State University
2518 W. Laurel
Fort Collins, CO 80526
(303) 224-1878

Carter A. Steers
SCS-USDA
1100 Old Dominion
Mansfield, TX 76063
(817) 334-5231

George Stephens
NOAA/NES Data & Info.
NOAA/NESDIS, E/SP 22 Stop H
Washington, D.C. 20233
(301) 763-8142 (FTS)

Susan Stitt
USDI/National Park Service
P.O. Box 25287
Denver, CO 80225
(303) 236-7943

Thomas Stone
Ecosystems Str.
Marine Biology Lab.
Wood Hole, MA 02543
(617) 540-2514

Pamela Schnurr
Colorado Div of Wildlife
711 Independent Ave.
Grand Junction, CO 81505
(303) 248-7175

Donald L. Schrupp
6060 N. Broadway
Denver, CO 80216
(303) 297-1192

Eddie L. Schwertz, Jr.
Intergraph Corp.
13027 Camelot Dr., SE
Huntsville, AL 35803
(205) 772-2144

Yosio E. Shimabukuro
Colorado State University
500 W. Prospect #28-J
Fort Collins, CO 80526
(303) 491-8122

Philip Smith
Consultant
1930 N.W. 143rd, #70
Portland, OR 97229
(503) 230-5759

Henry F. Soehngen
Polytechnic Inst. NY
333 Jay St.
Brooklyn, NY 11201
(212) 653-2322

Janine Stenback
University of California
1089 Sycamore Dr.
Millbrae, CA 94030
(415) 694-5593

David A. Stockbridge
USDA SCS
P.O. Box 6567
Ft. Worth, TX 76028
(817) 295-3662

Harlan Stoin
University of Arkansas
Agronomy Dept Plant Sci. Bldg
Fayetteville, AR 72701
(501) 575-5721

S.L. Stover
Dept. of Geography
Kansas State University
Manhattan, KS 66506
(913) 532-6727

Stephen Talbot
US Fish & Wildlife Service
1011 East Tudor Road
Anchorage, AK 99503
(907) 786-3381

Charles Tenei
University of Arizona
203 Nugent Bldg.
Tucson, AZ 85721
(212) 427-8100

Randall Thomas
Univ. of CA/Berkeley
260 Space Sciences Lab
Berkeley, CA 94720
(415) 642-4270

Gene Thorley
US Geological Survey
2415 Myrtle Lane
Reston, VA 22091
(703) 860-7471

Franklin Titlow
USDI National Park Service
P.O. Box 25287
Denver, CO 80225
(303) 236-7937

John Tonelli
International Paper Co.
P.O. Box 797
Tuxedo Park, NY 10987
(914) 351-2101

Dean Treadwell
Consultant
P.O. Box 93
Moundsville, WV 26041
(304) 845-3146

Carl C. Trettin
Michigan Tech. University
Rt. 2, Box 736
L'Anse, MI 49946
(906) 524-6181

Paul T. Tueller
University of NV/Reno
400 Tarn Way
Reno, NV 89503
(702) 784-4053

Jose Urena
Jet Propulsion Lab.
4800 Oak Gr. Dr. MS 241-210
Pasadena, CA 91109
(818) 354-0122

Gary Waggoner
USDI National Park Service
P.O. Box 25287
Denver, CO 80225
(303) 236-8798

Ralph Warbington
USDA Forest Service
630 Sansome St.
San Francisco, CA 94111
(415) 556-2654

T. Victor Ward
Moniteq Ltd.
630 Rivermede Rd.
Concord, Ont. Canada L4K 2H7
(416) 669-5334

Richard Waring
Oregon State University
College of Forestry
Corvallis, OR 97331
(503) 754-2244

Howard Warriner
US Dept. of Commerce
EROS Data Center
Sioux Falls, SD 57198
(605) 594-2261

Allen Watkins
USGS/EROS Data Center
Mundt Federal Bldg.
Sioux Falls, SD 57198
(605) 594-6540

Keitlyn Watson
Pacific Gas & Elect. Co.
3400 Crow Canyon Road
San Ramon, CA 94520
(415) 820-2000

Bill Watters
Dept of Interior/BIA
Northern Cheyenne Agency
Lame Deer, MT 59043
(406) 477-6211

John Weaver
USDA Forest Service
Box 7669
Missoula, MT 59807

Wayne Weaver
USDA SCS
1407 Gibbon St.
Laramie, WY 82070
(307) 766-3955

K.C. Wehde
Tech. Govt. Serv/EROS Data
Mundt Federal Bldg.
Sioux Falls, SD 57198
(605) 594-6114

Lee Werth
Bureau of Land Management
Bld 50, DFC, Box 25047
Denver, CO 80225-0047
(303) 236-0824

Carol Wettstein
USDA·SCS
P.O. Box 775
Gainesville, FL 32602-0775
(904) 377-1092

David Wherry
Washington State University
Computing Service Center
Pullman, WA 99164-1220

Jeff Whiting
Saskatchewan Res. Council
30 Campus Drive
Saskatoon, Canada S7N 0X1
(306) 664-5423

Diane Wickland
NASA
NASA Code EEL
Washington, DC 20546
(202) 453-1720

Janet Wiley
Wyoming Water Research Cen
Box 3067, University Sta.
Laramie, WY 82071
(307) 766-2143

Carroll B. Williams
USDA Forest Service
1960 Addison St.
Berkeley, CA 94701
(415) 486-3442

David Williams
Lockheed Eng & Mgt Service
P.O. Box 15027
Las Vegas, NV 89114
(702) 798-3161

Julie Willis
BLM/Anchorage
702 C St, Box 13
Anchorage, AK 99513
(907) 271-5063

Jeff Weinstein
Pub. Lands, State of Wyoming
122 W. 25th St.
Cheyenne, WY 82002-0600
(307) 777-6525

Carol Wessman
University of Wisconsin
925 Emerald St.
Madison, WI 53715
(608) 263-4578

William Wheeler
WI Dept. of Natural Res.
822 Rice St.
Horicon, WI 53032
(414) 485-4434

William White
USDA FS, FPM/MAG
3825 E. Mulberry
Fort Collins, CO 80524
(303) 224-1777

Roy Whitmore
University of Vermont
3 Sebring Road
South Burlington, VT 05401
(802) 656-2620

Peter E. Wikoff
Colorado State University
316 Wood St.
Fort Collins, CO 80521
(303) 224-2823

Kent Williams
ASCS/Aerial Photo Field Off
1398 E. Downington
Salt Lake City, UT 84105
(801) 524-5856

C.P. Williams
EOSAT
1901 N. Moore St.
Arlington, VA 22209
(703) 558-4259

Samuel Williamson
US Fish & Wildlife Service
2627 Redwing Rd. Creekside I
Fort Collins, CO 80526
(303) 226-9362

William Wilson
Jet Propulsion Lab.
4800 Oak Grove
Pasadena, CA 91109
(818) 354-5699

Daniel Winner
USDA Forest Service
1705 Celina Ct.
Rio Rancho, NM 87124
(505) 842-3352

Gene Wirsig
Potlatch Corp.
P.O. Box 1016
Lewiston, ID 83501
(208) 799-1216

Gene Wirsig
Alberta Energy & Public Res
Bag 3014, 530-8 St. S.
Lethbridge, Alb,
Canada TlJ 2J
(403) 381-5486

Edgar Work, Jr.
BLM, Bldg. 50, DFC
Box 25047
Denver, CO 80225-0047
(303) 236-6377

Mon Yee
USDA
511 NW Broadway #514
Portland, OR 97209
(503) 221-2851

Doris Zemlicka
Denver Wildlife Research Ctr.
Bldg. 16 DFC, Box 25266
Denver, CO 80225-0266
(303) 236-6377

Paul J. Zinke
University of California
145 Mulford Hall
Berkeley, CA 94720
(415) 642-6472

Rex Zobell
Bureau of Land Management
2613 Chestnut Dr.
Cheyenne, WY 82001
(307) 772-2095